Foundations of Engineering Mechanics

B. Grigori Muravskii, Mechanics of Non-Homogeneous and Anisotropic Foundations

Springer
Berlin
Heidelberg
New York
Barcelona
Hong Kong
London
Milan
Paris
Singapore
Tokyo

B. Grigori Muravskii

Mechanics of Non-Homogeneous and Anisotropic Foundations

Translated by Boris Krasovitski

With 149 Figures

Springer

Series Editors:
Vladimir I. Babitsky, DSc
Louborough University
Department of Mechanical Engineering
LE11 3 TU Loughborough Leicestershire
United Kingdom

J. Wittenburg
Universität Karlsruhe (TH)
Institut für Technische Mechanik
Kaiserstr. 12
76128 Karlsruhe
Germany

Author:
Grigori B. Muravskii
Geotechnical Department
Faculty of Civil Engineering Technicon
32000 Haifa
Israel

Translator:
Boris Krasovitski
POB7843
36811 Nesher
Israel

ISBN 978-3-642-53602-1

Library of Congress Cataloging-in-Publication Data

Muravskii, Grigori B:
Mechanics of non-homogeneous and anisotropic foundations / B. Grigori Muravskii. Translated by Boris Krasovitski. - Berlin; Heidelberg; New York; Barcelona; Hong Kong; London; Milan; Paris; Singapore; Tokyo: Springer, 2001
(Foundations of engineering mechanics)
ISBN 978-3-642-53602-1 ISBN 978-3-540-44573-9 (eBook)
DOI 10.1007/978-3-540-44573-9

Springer-Verlag is a company in the BertelsmannSpringer publishing group
http://www.springer.de

Softcover reprint of the hardcover 1st edition 2001

Typesetting: Camera ready by author
Cover-Design: de'blik, Berlin
Printed on acid free paper SPIN: 10791920 62/3020/kk - 5 4 3 2 1 0

Preface

Although realistic soil and rock foundations reveal noticeable deviations in their properties from homogeneity and isotropy, the model of the homogeneous isotropic elastic half-space is widely used when studying static and dynamic interactions between a deformable foundation and structures. This is explained by significant mathematical difficulties inherent in problems concerning mechanics of anisotropic and heterogeneous elastic bodies. Solving the basic static and dynamic problems for heterogeneous and anisotropic half-spaces, such as different contact problems and problems of constructing Green's functions, has become possible in the last few decades due to the development of computer engineering techniques and numerical methods.

This book contains the results of investigations in the area of statics and dynamics of heterogeneous and anisotropic foundations, carried out by the author in the last five years while working in the Faculty of Civil Engineering at Technion – Israel Institute of Technology. The book is directed at engineers and scientists in the areas of soil mechanics, soil-structures interaction, seismology and geophysics.

Some characteristic features of the book are:

i) Constructing (Chap.1) solutions in a general form for the heterogeneous (in the depth direction) transversely isotropic elastic half-space subjected to different loadings, harmonic in time. Characteristics of the given half-space have an influence on functions (of depth z and parameter k of Hankel's transforms), which are determined from a system of ordinary differential equations.

ii) New dynamic solutions relating to the homogeneous transversely isotropic elastic half-space subjected to the action of vertical and horizontal forces applied to the half-space surface or below the surface. Solutions are presented relating to basic contact problems for this kind of half-space.

iii) Dynamic and static solutions for the linearly heterogeneous half-space, which, as applied to static problems, was intensively studied in works by R. E. Gibson and his coworkers. Solutions are presented for a half-space with exponentially varying stiffness. In these cases it is possible to find analytical solutions for the above-mentioned ordinary differential equations; thus the required amplitudes of vibrations are expressed in the form of integrals containing the determined functions.

iv) Development of numerical–analytical methods for the considered problems, which consist of a combinations of integral representations of the sought functions and numerical treatment of ordinary differential equations for the corresponding Fourier–Bessel transformations. The piecewise constant approximation for varying coefficients of these differential equations results in the well-known method of thin layers. Employing the general representations developed in Chap.1 and using simple solutions in the intervals (elements) with constant properties allows the construction of effective solutions. Another appropriate treatment is the use of Runge–Kutta method with additional application of Gogunov's method to eliminate computational difficulties connected with large values of the transformation parameter k.

The book presents, graphically, a large number of results of computations, which give a clear picture of the behavior of the mechanical systems considered. These results can serve for estimating accuracy of different simplified methods, e.g. combining the division of a half-space into thin layers with setting a form of displacements distribution within the layers.

I express my gratitude to my colleagues – Professors of Technion R. Baker and S. Frydman for useful discussions of geotechnical problems connected with the material of the book.

Haifa, April 2001. G. B. Muravskii

Contents

Introduction

The need to consider the heterogeneity and anisotropy of materials when solving various static and dynamic problems is common in engineering practice, especially when considering the properties of soil foundations. Numerous experimental results prove that the in-situ properties of soil and rock foundations differ from those of isotropic homogeneous media [25, 32, 48, 65, 88, 92, 96, 119].

The influence of the heterogeneity and anisotropy of soils on their strains and stresses has been a subject of concern, primarily in static problems [30, 44, 51, 126]. At the same time, the first reports appear, which deal with the dynamics of the heterogeneous isotropic elastic half-space (the medium is supposed to be heterogeneous in the vertical direction, i.e. in the direction normal to the boundary of the half-space). A comprehensive review of these reports is presented in [29]; in most cases, the authors deal with the free vibrations of a half-space, especially with the influence of different types of heterogeneity on characteristics of the Rayleigh and Love waves. A commonly accepted approach that enables heterogeneity to be incorporated into dynamic problems has been developed in [50, 111]. This approach is based on considering the elastic foundation as a set of homogeneous layers ([49, 62, 68] and others). For each layer, the solutions are relatively simple, while the complete solution of the problem is derived by employing contact conditions between the layers. Possible complications of this method are due to some terms in the solution for separate layers that grow exponentially, which can lead to an unacceptable loss of accuracy in the constructed total solution. However, these complications may be prevented following some special techniques [4, 9, 21, 27, 41, 62, 70, 83, 131]. The method using by stiffness matrices was shown to be effective in preventing the above-mentioned computational difficulties [11–13, 57]. The author of [11–13] presents a basic technique for solving dynamic problems in transversely isotropic layered half-spaces, considering initial stresses in the material.

Following a commonly accepted approach to the problems of the mechanics of layered foundations, the equations of motion (in Fourier's or Hankel's transformation space) are treated in the vertical direction using a finite-element technique ([58, 59, 60, 66, 72, 73, 85, 102, 108, 122–124] and others). These methods have been reported to be suitable for determining Rayleigh and Love wave parameters, when applied to the problems with free vibrations, and in some illustrative examples for vibrations under a given loading. However, the question concerning the range of where the finite-element method could be applied in the problems that require integrating by the transformation parameter (wave number) is still an open question. As wave numbers increase, the approximation of the displacements field by linear or parabolic functions deteriorates; this fact, for example, results in a singularity of Green's functions at those points where the functions are finite for the corresponding exact solution.

Compared with the problems for layered foundations, there is a significantly smaller number of reports concerning continuous heterogeneity. This is due to mathematical complications that appear in this case. Differential equations that contain the parameter of Fourier's or Hankel's transformation for functions of z-argument (that varies in the direction of the foundation depth) have explicit analytical solutions only for limited types of heterogeneity, even in the isotropic cases, but especially in dynamic problems. Intensive studies of the elastic foundations with the shear modulus, which depends linearly on the depth of foundation, commenced with the static problems solved by Gibson [39], and continued in the following publications ([8, 14–16, 20, 22–24, 55, 56]). Dynamic problems have been solved for a linearly heterogeneous half-space with Poisson's ratio $\nu = 0.5$ [6, 7, 78, 79, 106, 112], and for all acceptable values of ν [82, 114]. Note that in [106, 112, 114], the authors consider free vibrations of a half-space. Free shear vibrations are a subject of consideration in [112, 113]. Static problems have been solved for power-law behavior of shear modulus versus depth of half-space ($G = G_1 z^{\alpha}$), beginning with the zero value at the surface of the half-space [93, 98–100], [51, 55, 63, 64, 76]. In the latter group of references, the following assumption has been employed: the value of α and Poisson's ratio ν are related as $\alpha\nu = 1-2\nu$. This relationship provides a radial distribution of stresses in the half-space and enables a solution to be constructed for an axially symmetric case even with a non-zero shear modulus at the half-space surface [22, 56]. Some alternative relationships between ν and α that lead to relatively simple solutions were suggested in [89, 90]. In dynamic problems, the power-law function for shear modulus versus depth enables analytical solutions to be built for corresponding the Fourier's or Hankel's transformations only for $\alpha = 1$, or for $\alpha = 2$, $\nu = 1/4$ [46, 47] (for shear vibrations, analytical solutions exist, in addition, for $\alpha = 0.5$ [113] and for some other values). For $\alpha = 2$, $\nu = 1/4$, a series of static and dynamic problems have been investigated in the 2-D framework [31, 33]. An axially symmetric dynamic contact problem has also been solved [47]. Note that, in reality, the shear modulus increases with depth slower than by the quadratic law. Nevertheless, these results are important, as they provide a valuable estimate of the qualitative and quantitative effects that occur due to heterogeneity. In addition, they lead to the possibility of modeling an arbitrary heterogeneous foundation as a layered foundation with a parabolic approximation of the shear modulus within each layer. This option enables the discontinuity in material properties to be avoided at the boundaries of separate layers, which appears as a result of a rougher approximation with homogeneous layers [31].

An exponential relationship is an alternative type of function for shear modulus versus depth, which makes it possible to derive efficient analytical solutions. In dynamic problems, an exponential relationship has been chosen to describe the variation of shear modulus with depth in [127], where the author considers Love-type waves in the half-space; for the problem of anti-plane vibrations of a semi-plane with shear modulus, exponentially decreasing with depth [38]; and for the problem of vibrations of the half-space, subjected to a concentrated force (vertical and horizontal) at the surface of the half-space [80]. A relationship for the shear modulus and for the density of the medium that contains a decreasing exponent, and, therefore, provides limited values of these parameters with increasing depth, was suggested in [94, 95] (in one particular case, density can be constant). For the type of heterogeneity considered, analytical solutions may be derived in the form of a power series; solutions of numerous problems are presented in reports [81, 116–121].

A specific case of separation of the equations of motion with $\alpha = 2$, $\nu = 1/4$ [46, 47] is not unique. Other cases have been the subject of investigations elsewhere [3, 52, 54], although, in order to provide the possibility of separation of equations, a special form of relationship between the density and elastic parameters of the medium was required. In some cases, these relationships were found to be unrealistic.

The purpose of this book is to present a series of solutions (previously unpublished mostly) for the static and dynamic problems of anisotropic and heterogeneous elastic half-spaces. The investigation is limited to the case of transversal isotropy, and to the case where the characteristics of the half-space only in the direction of its depth vary. Special emphasis is laid on the studies of steady-state harmonic vibrations of half-spaces under given loads, acting at the surface of the half-space, or at a certain depth. In addition, a study of harmonic vibrations of a circular stiff disk, which is in contact with the surface of the half-space, is presented. When the solutions of the problems of harmonic vibrations are available, it is possible to construct appropriate solutions for arbitrary time-dependent loads.

The book contains five chapters. In Chap. 1, general solutions are presented for the problem of vibrations in the transversely isotropic half-space subjected to specific loads. Much attention has been paid to the case of concentrated vertical and horizontal forces, and thus the solutions are available for any arbitrary distribution of loads. Solutions which are derived in integral form contain functions that satisfy ordinary differential equations. The problem leads to the construction of the fundamental solutions of these differential equations, which contain the parameter k of Hankel's transformation of the required functions. In addition in Chap. 1, we present formulas, based on the principle of superposition, that allow calculation of linearly deformed foundations, subjected to loads which are distributed over the circular and rectangular domains. These formulas contain Green's functions (without further specification of their form), and they are presented as single integrals, instead of original double ones.

In the following three chapters, we consider various problems dealing with the homogeneous transversely isotropic half-space (Chap. 2), an isotropic linearly heterogeneous half-space (Chap. 3), and a transversely isotropic half-space varying exponentially with depth shear modulus (Chap. 4). In the latter case, two options are considered: first, the half-space with an infinite increase of shear modulus with depth, and, second, one with a finite limit for the shear modulus. Chapters. 2–4 contain the results of a vast number of computations needed for construction of exact solutions of the above-mentioned key differential equations. The efficient choice of contour of integration and the appropriate computational methods provide a precise determination of the amplitude of vibrations of points of the half-space. The solutions presented could serve as a standard for estimating the accuracy of numerical–analytical solutions, for which the differential equations (for Fourier's or Hankel's transformation) are solved numerically. One of the widely used numerical methods employs a multilayer half-space with piecewise constant characteristics within each layer as an approximation of the given continuously heterogeneous half-space.

In Chap. 5, numerical–analytical methods of constructing solutions for static and dynamic problems in a heterogeneous half-space are presented. Two approaches are employed in order to construct the solutions of the differential equations: first, a piecewise constant approximation of the coefficients of the equations, and, second, the Runge–Kutta method. Solutions of the differential equations for high values of k (the parameter in Hankel's transformation) should decrease rapidly, as the distance from the horizontal plane subjected to the load increases. Therefore, the bounds of the domain of interest are moved (as k-parameter grows) towards this plane. This method results in the saving of the computational resources. It enables us to construct the required solutions for the force applied to the surface of the half-space, without any additional measures. In those cases where the force is applied to an inner point of the half-space, an orthogonalization method due to Godunov [42] is used, in addition to the Runge-Kutta method. Similarly, when using the piecewise constant method of approximation, we apply a transformation of solutions to eliminate its increasing parts. The solutions obtained are verified by comparing them with those of corresponding problems developed for the linearly heterogeneous half-space in Chap. 3. Note that the determination of the integration contour when employing inverse Fourier's and Hankel's transformations, calculation of the appropriate integrals, and the algorithms of solutions of various contact problems are discussed in detail in Chap. 2, where these items are introduced for the first time.

Chapter 1. General Solutions of Harmonic Vibrations in Heterogeneous Transversely Isotropic Half-Space

1.1 Basic Relationships of Theory of Elasticity for Transversely Isotropic Body

Generally, Hooke's law for an anisotropic elastic body contains 36 coefficients A_{ij} [65]:

$$\begin{aligned}
\sigma_x &= A_{11}\varepsilon_x + A_{12}\varepsilon_y + A_{13}\varepsilon_z + A_{14}\gamma_{yz} + A_{15}\gamma_{xz} + A_{16}\gamma_{xy}, \\
\sigma_y &= A_{21}\varepsilon_x + A_{22}\varepsilon_y + A_{23}\varepsilon_z + A_{24}\gamma_{yz} + A_{25}\gamma_{xz} + A_{26}\gamma_{xy}, \\
&\cdots\cdots\cdots\cdots\cdots\cdots\cdots\cdots\cdots\cdots \\
\tau_{xy} &= A_{61}\varepsilon_x + A_{62}\varepsilon_y + A_{63}\varepsilon_z + A_{64}\gamma_{yz} + A_{65}\gamma_{xz} + A_{66}\gamma_{xy},
\end{aligned} \qquad (1.1)$$

where the components of the stress tensor $\sigma_x, \dots, \tau_{xy}$ are presented as linear homogeneous functions of strains $\varepsilon_x, \dots, \gamma_{xy}$. Assuming the existence of an elastic potential, we reduce the number of coefficients to 21 ($A_{ij} = A_{ji}$). Further reduction of their number is possible if the elastic properties are symmetrical. In the following, we consider a case with transverse isotropy, when the material is isotropic in the "horizontal" planes (in *X*, *Y*-axes), while its elastic properties in the "vertical" planes are not related to those in the horizontal ones. In this case, the strain properties of the material are stated with 5 constants [65]. Using cylindrical coordinates r, ϑ, z (Fig. 1.1), where coordinates r, ϑ vary in the plane of isotropy, we present relationships (1.1) for the transversely isotropic material as follows:

$$\boldsymbol{\sigma} = \mathbf{A}\,\boldsymbol{\varepsilon}, \qquad (1.2)$$

where the stress vector, $\boldsymbol{\sigma}$, the strain vector, $\boldsymbol{\varepsilon}$, and a matrix of elastic coefficients, $\mathbf{A}$, are given as

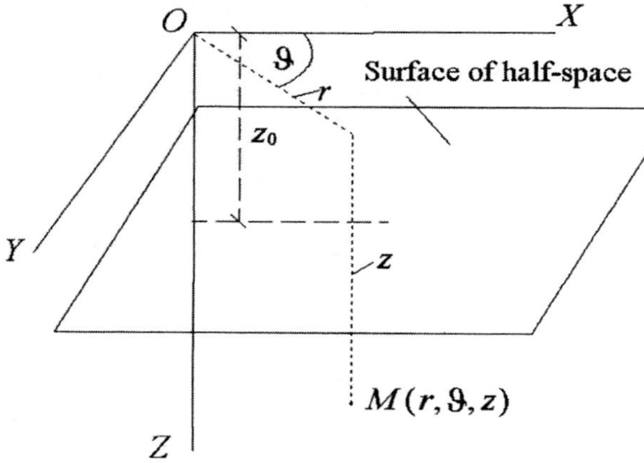

Fig. 1.1. Half-space referred to cylindrical coordinates

$$\boldsymbol{\sigma} = [\sigma_r, \sigma_\vartheta, \sigma_z, \tau_{rz}, \tau_{r\vartheta}, \tau_{\vartheta z}]^{\mathrm{T}}, \tag{1.3a}$$

$$\boldsymbol{\varepsilon} = [\varepsilon_r, \varepsilon_\vartheta, \varepsilon_z, \gamma_{rz}, \gamma_{r\vartheta}, \gamma_{\vartheta z}]^{\mathrm{T}}, \tag{1.3b}$$

$$\mathbf{A} = \begin{bmatrix} A_{rr} & A_{r\vartheta} & A_{rz} & & & \\ A_{r\vartheta} & A_{rr} & A_{rz} & & \mathbf{0} & \\ A_{rz} & A_{rz} & A_{zz} & & & \\ & & & G_{rz} & & \\ & \mathbf{0} & & & G_{r\vartheta} & \\ & & & & & G_{rz} \end{bmatrix}, \tag{1.3c}$$

and

$$G_{r\vartheta} = \frac{1}{2}(A_{rr} - A_{r\vartheta}). \tag{1.4}$$

Here, $G_{r\vartheta}$ and G_{rz} are shear modules for the plane of isotropy and for any plane normal to the plane of isotropy, respectively. In the specific case with isotropic material, only 2 of 5 constants remain independent,

$$G_{rz} = G_{r\vartheta} = G; \; A_{rz} = A_{r\vartheta} = \lambda; \; A_{rr} = A_{zz} = \lambda + 2G, \tag{1.5}$$

where G is the shear modulus of the material, and λ is Lamé's constant.

A relationship inverse to (1.2) may be presented as

$$\boldsymbol{\varepsilon} = \mathbf{a}\boldsymbol{\sigma}, \tag{1.6}$$

where

$$\mathbf{a}=\begin{bmatrix} a_{rr} & a_{r\vartheta} & a_{rz} & & & \\ a_{r\vartheta} & a_{rr} & a_{rz} & & \mathbf{0} & \\ a_{rz} & a_{rz} & a_{zz} & & & \\ & & & G_{rz}^{-1} & & \\ & \mathbf{0} & & & G_{r\vartheta}^{-1} & \\ & & & & & G_{rz}^{-1} \end{bmatrix}, \tag{1.7}$$

$$G_{r\vartheta}^{-1}=2(a_{rr}-a_{r\vartheta}). \tag{1.8}$$

The elements of matrix **a** may be derived by using engineering parameters [65]; E, E' are Young's modulus for the plane of isotropy and for any direction normal to this plane, ν is Poisson's ratio, which characterizes a transverse constriction in the plane of isotropy (r,ϑ) for the case of stretching in this plane, and ν' is Poisson's ratio for the case of stretching in the z-direction:

$$a_{rr}=\frac{1}{E},\quad a_{r\vartheta}=-\frac{\nu}{E},\quad a_{zz}=\frac{1}{E'},\quad a_{rz}=-\frac{\nu'}{E'},\quad G_{r\vartheta}=\frac{E}{2(1+\nu)}. \tag{1.9}$$

The elastic coefficients A_{ij}, which constitute matrix **A** (1.3c), may be represented in terms of a_{ij} in (1.7):

$$A_{rr}=\frac{a_{rr}a_{zz}-a_{rz}^2}{(a_{rr}-a_{r\vartheta})m}, \tag{1.10a}$$

$$A_{r\vartheta}=-\frac{a_{r\vartheta}a_{zz}-a_{rz}^2}{(a_{rr}-a_{r\vartheta})m}, \tag{1.10b}$$

$$A_{rz}=-\frac{a_{rz}}{m}, \tag{1.10c}$$

$$A_{zz}=\frac{a_{rr}+a_{r\vartheta}}{m}, \tag{1.10d}$$

where

$$m=(a_{rr}+a_{r\vartheta})a_{zz}-2a_{rz}^2. \tag{1.11}$$

A positively defined elastic potential leads to certain requirements for the coefficients of matrices **A** and **a**. Thus, the following conditions follow from a positive definition of the matrices **A** and **a**:

$$A_{rr}>0,\quad A_{zz}>0,\quad A_{rr}^2>A_{r\vartheta}^2,\quad A_{rr}A_{zz}>A_{rz}^2, \tag{1.12a}$$

$$a_{rr}>0,\quad a_{zz}>0,\quad a_{rr}^2>a_{r\vartheta}^2,\quad a_{rr}a_{zz}>a_{rz}^2. \tag{1.12b}$$

Employing relationships (1.10), we obtain, in addition,

$$m > 0,\ a_{rr} - a_{r\vartheta} > 0,\ a_{rr} + a_{r\vartheta} > 0 . \tag{1.13}$$

In what follows, we need relationships between the strains (1.3b) and the displacements U_r, U_z, U_ϑ along the coordinate lines of the cylindrical system of coordinates. In matrix form, these relations may be presented as follows:

$$\boldsymbol{\varepsilon} = \mathbf{D}\ \mathbf{u} , \tag{1.14}$$

where

$$\mathbf{u} = [u_r, u_z, u_\vartheta]^{\mathrm{T}} , \tag{1.15}$$

$$\mathbf{D} = \begin{bmatrix} \dfrac{\partial}{\partial r} & 0 & 0 \\ \dfrac{1}{r} & 0 & \dfrac{1}{r}\dfrac{\partial}{\partial \vartheta} \\ 0 & \dfrac{\partial}{\partial z} & 0 \\ \dfrac{\partial}{\partial z} & \dfrac{\partial}{\partial r} & 0 \\ \dfrac{1}{r}\dfrac{\partial}{\partial \vartheta} & 0 & -\dfrac{1}{r}+\dfrac{\partial}{\partial r} \\ 0 & \dfrac{1}{r}\dfrac{\partial}{\partial \vartheta} & \dfrac{\partial}{\partial z} \end{bmatrix} . \tag{1.16}$$

1.2 Particular Solutions for Transversely Isotropic Heterogeneous Elastic Medium

Consider a half-space ($z > z_0$), or a layer ($z_0 < z < z_0 + H$) with boundary planes that are infinite in directions normal to the Z-axis (Fig. 1.1). A time dependence for all considered values is assumed to have an exponential form $\exp(\mathrm{i}\,\omega\, t)$, where ω is a circular frequency of vibrations, and $\mathrm{i} = \sqrt{-1}$. As a rule, the multiplier $\exp(\mathrm{i}\,\omega\, t)$ is reduced in most of the equations considered below; we shall deal with the amplitudes of the studied parameters. We also assume that the planes, which are parallel to the coordinate planes $(x,\ y)$, or $(r,\ \vartheta)$, are planes of isotropy. Properties of material, i.e. elastic coefficients (elements of matrix $\mathbf{A}$) and density ρ, may vary in the direction of the Z-axis (the vertical axis). Application of Hankel's transformation to the functions in cylindrical coordinates with respect to the variable r, and Fourier series with respect to ϑ, are efficient measures that enable us to solve problems of vibrations of half-spaces and layers under the conditions described above.

For the case of an isotropic and homogeneous medium, application of the vector and scalar potentials (Lamé - type solutions) has become common practice in solving dynamic problems. This approach leads to the wave equations that, as a rule, are interconnected through boundary conditions of the problem. For an anisotropic homogeneous body, such a separation of equations requires application of some additional constraints on the elastic parameters of the material (e.g. Carrier's condition [19] that was used [61] to solve the problem of vibrations of a stiff disk on a transversely isotropic half-space). For the isotropic heterogeneous half-space, transformation of the problem into separated equations for the potentials is possible for some specific types of heterogeneity (e.g. for a quadratic variation of shear modulus $G(z)$ and Poisson's ratio $\nu = 0.25$, potentials were employed to solve a plane problem in [31, 33, 46], and for an axisymmetric problem in [47]). Various aspects concerning the separation of equations of motion have been discussed in detail in [3, 52, 54]. Note that in numerous cases, a successful solution of dynamic problems for an elastic medium is possible even by using non-separated equations. As shown in [124], Sezawa's representation [103] for displacements, which was used for an isotropic homogeneous medium, may be generalized for the case of transverse isotropy and heterogeneity along the Z-axis (depth direction). Note that Sezawa's solutions were employed in numerous works to solve dynamic problems for an elastic isotropic homogeneous half-space and for a set of layers (e.g. [17, 57, 58, 128]).

For the transversely isotropic medium which is heterogeneous with depth, the following representation is used for the amplitudes of displacements, U_r, U_z, U_ϑ, along the coordinate lines of the cylindrical system of coordinates [124]:

$$U_r = -\left[q(z,k)\frac{\mathrm{d}\,\mathrm{J}_n(\eta)}{\mathrm{d}\eta} + p(z,k)\frac{n}{\eta}\mathrm{J}_n(\eta) \right]\mathbf{\Gamma}_1 , \tag{1.17a}$$

$$U_z = w(z,k)\,\mathrm{J}_n(\eta)\mathbf{\Gamma}_1 , \tag{1.17b}$$

$$U_\vartheta = \left[p(z,k)\frac{\mathrm{d}\,\mathrm{J}_n(\eta)}{\mathrm{d}\eta} + q(z,k)\frac{n}{\eta}\mathrm{J}_n(\eta) \right]\mathbf{\Gamma}_2 , \tag{1.17c}$$

where

$$\eta = kr , \tag{1.18a}$$

$$\mathbf{\Gamma}_1 = \begin{bmatrix} \cos(n\vartheta) \\ \sin(n\vartheta) \end{bmatrix} , \tag{1.18b}$$

$$\mathbf{\Gamma}_2 = \begin{bmatrix} \sin(n\vartheta) \\ -\cos(n\vartheta) \end{bmatrix} . \tag{1.18c}$$

($n = 0,1,2,\ldots$).

Representations (1.17) contain Bessel's functions of the order n, and functions q, w, p that depend on coordinate z and parameter k; apparently, two options of the trigonometric dependence on the angle ϑ are possible. For a more general dependence on ϑ, the problem's solution is presented in the form of a Fourier series, i.e. n-summation of expressions (1.17) is performed. Note that numerous important solutions may be derived at specific values of n. For example, at $n = 0$, we obtain an axisymmetric problem. When choosing the upper line in Γ_1 and Γ_2, function p is excluded from the solution, while the choice of a lower line leads to the problem of torsional vibrations. In the last case, the only non-zero function is U_ϑ, expressed in terms of the function p. At $n = 1$, solutions may be obtained, first, for the case with a horizontal load distributed uniformly over a circular domain on the surface of a half-space or a layer, and, second, for the case of a vertical momentum load, which tends to rotate a circular domain about its horizontal Y-axis.

While the form of dependence on the coordinates r and ϑ is known a priori (it corresponds to Sezawa's representation), a Z-dependence should be found based on the following requirement: functions (1.17) and the corresponding stresses obtained from equations (1.2) and (1.14) should satisfy the equations of motion for the elastic medium (assuming the absence of volume forces):

$$\begin{aligned}
\rho\frac{\partial^2 u_r}{\partial t^2} &= \frac{\partial\sigma_{rr}}{\partial r}+\frac{1}{r}\frac{\partial\sigma_{r\vartheta}}{\partial\vartheta}+\frac{\partial\sigma_{rz}}{\partial z}+\frac{\sigma_{rr}-\sigma_{\vartheta\vartheta}}{r},\\
\rho\frac{\partial^2 u_\vartheta}{\partial t^2} &= \frac{\partial\sigma_{r\vartheta}}{\partial r}+\frac{1}{r}\frac{\partial\sigma_{\vartheta\vartheta}}{\partial\vartheta}+\frac{\partial\sigma_{\vartheta z}}{\partial z}+\frac{2}{r}\sigma_{r\vartheta},\\
\rho\frac{\partial^2 u_z}{\partial t^2} &= \frac{\partial\sigma_{rz}}{\partial r}+\frac{1}{r}\frac{\partial\sigma_{\vartheta z}}{\partial\vartheta}+\frac{\partial\sigma_{zz}}{\partial z}+\frac{\sigma_{rz}}{r}.
\end{aligned} \tag{1.19}$$

At this stage, functions (1.17) are considered as particular solutions containing the parameter k. In the following, our intention is to integrate these solutions with respect to k from zero to infinity, taking into account possible points of singularities. As shown below, using the properties of Hankel's transformation enables us to satisfy the given conditions specified for the planes z = const. Note that for a homogeneous isotropic medium functions $q(z,k)$, $w(z,k)$, $p(z,k)$ are expressed through the exponential functions containing radicals:

$$\alpha_1=\left(k^2-\frac{\omega^2}{C_s^2}\right)^{1/2},\quad \alpha_2=\left(k^2-\frac{\omega^2}{C_p^2}\right)^{1/2}, \tag{1.20}$$

where

$$C_s=\left(\frac{G}{\rho}\right)^{1/2},\quad C_p=\left(\frac{\lambda+2G}{\rho}\right)^{1/2}. \tag{1.21}$$

Here, C_p and C_s correspond to the velocities of propagation of compression and shear waves, respectively.

Having found the strains corresponding to the displacements (1.17) by using relationship (1.14), we now construct the stress tensor following (1.2) for the particular solution (1.17). When calculating derivatives, we keep in mind the following relationships:

$$\frac{\mathrm{d}\mathbf{\Gamma}_1}{\mathrm{d}\vartheta} = -n\mathbf{\Gamma}_2\,, \tag{1.22a}$$

$$\frac{\mathrm{d}\mathbf{\Gamma}_2}{\mathrm{d}\vartheta} = n\mathbf{\Gamma}_1\,, \tag{1.22b}$$

$$\frac{\mathrm{d}^2\,\mathrm{J}_n(\eta)}{\mathrm{d}\eta^2} = -\frac{1}{\eta}\frac{\mathrm{d}\,\mathrm{J}_n(\eta)}{\mathrm{d}\eta} + \left(\frac{n^2}{\eta^2} - 1\right)\mathrm{J}_n(\eta)\,. \tag{1.22c}$$

Expression (1.22c) corresponds to Bessel's equation [1]. The component amplitudes of the stress tensor are presented in the following form (instead of σ_r, σ_ϑ, σ_z, τ_{rz}, $\tau_{r\vartheta}$, $\tau_{\vartheta z}$, we use for the particular solution considered S_r, S_ϑ, S_z, T_{rz}, $T_{r\vartheta}$, $T_{\vartheta z}$, respectively):

$$\begin{aligned}
S_r &= \left\{\mathrm{J}_n(\eta)\left[\frac{kn(A_{rr} - A_{r\vartheta})(p - nq)}{\eta^2} + A_{rr}kq + A_{rz}\frac{\mathrm{d}w}{\mathrm{d}z}\right]\right.\\
&\quad \left. + \frac{\mathrm{dJ}_n(\eta)}{\mathrm{d}\eta}\frac{k(A_{rr} - A_{r\vartheta})(q - np)}{\eta}\right\}\mathbf{\Gamma}_1\,,\\
S_\vartheta &= \left\{\mathrm{J}_n(\eta)\left[-\frac{kn(A_{rr} - A_{r\vartheta})(p - nq)}{\eta^2} + A_{r\vartheta}kq + A_{rz}\frac{\mathrm{d}w}{\mathrm{d}z}\right]\right.\\
&\quad \left. - \frac{\mathrm{dJ}_n(\eta)}{\mathrm{d}\eta}\frac{k(A_{rr} - A_{r\vartheta})(q - np)}{\eta}\right\}\mathbf{\Gamma}_1\,,\\
S_z &= \mathrm{J}_n(\eta)\left[A_{rz}kq + A_{zz}\frac{\mathrm{d}w}{\mathrm{d}z}\right]\mathbf{\Gamma}_1\,,\\
T_{rz} &= G_{rz}\left[-\mathrm{J}_n(\eta)\frac{n}{\eta}\frac{\mathrm{d}p}{\mathrm{d}z} + \frac{\mathrm{dJ}_n(\eta)}{\mathrm{d}\eta}\left(kw - \frac{\mathrm{d}q}{\mathrm{d}z}\right)\right]\mathbf{\Gamma}_1\,,\\
T_{r\vartheta} &= -G_{r\vartheta}k\left[\mathrm{J}_n(\eta)\left(p + \frac{2n}{\eta^2}(q - np)\right) + \frac{\mathrm{dJ}_n(\eta)}{\mathrm{d}\eta}\frac{2}{\eta}(p - nq)\right]\mathbf{\Gamma}_2\,,\\
T_{\vartheta z} &= G_{rz}\left[\mathrm{J}_n(\eta)\left(\frac{n}{\eta}\left(\frac{\mathrm{d}q}{\mathrm{d}z} - kw\right)\right) + \frac{\mathrm{dJ}_n(\eta)}{\mathrm{d}\eta}\frac{\mathrm{d}p}{\mathrm{d}z}\right]\mathbf{\Gamma}_2\,.
\end{aligned} \tag{1.23}$$

In order to obtain equations which are satisfied by the previously introduced functions $q(z,k)$, $w(z,k)$, $p(z,k)$, we apply the equations of motion (1.19); for the steady-state harmonic vibrations, which are the subject of our study, these relationships are transformed into the equations for amplitudes. Substituting solutions (1.17) and the corresponding amplitudes of stresses (1.23) into these equations, we apply the formulas (1.22) again. Note that in the first two equations, all terms have the common multiplier Γ_1, and in the third one, Γ_2. In each equation, all terms are transformed to the form which contains $J_n(\eta)$, or $dJ_n(\eta)/d\eta$. The requirement of a vanishing sum of terms containing $J_n(\eta)$ in the first equation (1.19) leads to

$$G_{rz}\frac{d^2 p}{d z^2}+\frac{dG_{rz}}{dz}\frac{dp}{dz}+(\rho\omega^2-k^2G_{r\vartheta})p=0\,. \tag{1.24}$$

In turn, consideration of the terms containing $dJ_n(\eta)/d\eta$ yields

$$G_{rz}\frac{d^2 q}{d z^2}+\frac{dG_{rz}}{dz}\frac{dq}{dz}+(\rho\omega^2-k^2A_{rr})q-k(A_{rz}+G_{rz})\frac{dw}{dz}-k\frac{dG_{rz}}{dz}w=0\,. \tag{1.25}$$

Studying the second equation (1.19), we can see that the terms with the derivative of Bessel's function cancel, and equating the sum of the terms with $J_n(\eta)$ to zero, we obtain

$$A_{zz}\frac{d^2 w}{d z^2}+\frac{dA_{zz}}{dz}\frac{dw}{dz}+(\rho\omega^2-k^2G_{rz})w+k(A_{rz}+G_{rz})\frac{dq}{dz}+k\frac{dA_{rz}}{dz}q=0\,. \tag{1.26}$$

The last equation of (1.19) leads to the previously obtained equations (1.24) and (1.25). Thus, displacements (1.17) satisfy all the requirements that define the behavior of a linearly elastic transversely isotropic heterogeneous medium, under the following condition: functions $q(z,k)$, $w(z,k)$, $p(z,k)$ satisfy the system of equations (1.24), (1.25), (1.26). These equations were developed in [124]. Note that the first equation does not depend on the other two, which are interconnected. The equations do not contain parameter n (the number of harmonics with respect to the angle coordinate ϑ).

Next, we transform equations (1.24)–(1.26) by using the following dimensionless variables:

$$\tilde{z}=\frac{z}{z_r},\quad \tilde{k}=kz_r,\quad \tilde{p}=\frac{p}{z_r},\quad \tilde{q}=\frac{q}{z_r},\quad \tilde{w}=\frac{w}{z_r},\quad \tilde{\rho}=\frac{\rho}{\rho(z_0)},\quad \tilde{G}=\frac{G_{rz}}{G_{rz}(z_0)},$$

$$\tilde{G}_{r\vartheta}=\frac{G_{r\vartheta}}{G_{rz}(z_0)},\quad \tilde{A}_{zz}=\frac{A_{zz}}{G_{rz}(z_0)},\quad \tilde{A}_{rr}=\frac{A_{rr}}{G_{rz}(z_0)},\quad \tilde{A}_{rz}=\frac{A_{rz}}{G_{rz}(z_0)}, \tag{1.27}$$

where a reference length, z_r, has been introduced; all elastic coefficients and density are related to the shear modulus and density in the plane $z = z_0$, respectively. Let us take into account the dissipative properties of the medium. In order to incorporate the dissipation of energy in the material, we apply, keeping in mind processes harmonic in time, complex elastic coefficients. In its simplest form, the friction within the material may be accounted for by considering the value $G_{rz}(z_0)$ as the complex one:

$$G_{rz}(z_0) = G_{rz0}(1 + \mathrm{i}\varepsilon)\,, \tag{1.28}$$

where a small, constant, positive parameter ε enables us to account for the dissipation of energy in the material of the half-space, and G_{rz0} is the shear modulus of the material at $z = z_0$, for the case of an ideally elastic material. Assume that the dimensionless coefficients of elasticity denoted with "~" in (1.27) are real. Hence, the frequency-independent mechanism of energy dissipation is taken to be identical for various types of deformations. Taking into account relationships (1.27) and (1.28), we rewrite equations (1.24)–(1.26) as follows:

$$\widetilde{G}\frac{\mathrm{d}^2\widetilde{p}}{\mathrm{d}\widetilde{z}^2} + \frac{\mathrm{d}\widetilde{G}}{\mathrm{d}\widetilde{z}}\frac{\mathrm{d}\widetilde{p}}{\mathrm{d}\widetilde{z}} + (\widetilde{\rho}\beta^2\theta^2 - \widetilde{k}^2\widetilde{G}_{r\vartheta})\widetilde{p} = 0\,, \tag{1.29}$$

$$\widetilde{G}\frac{\mathrm{d}^2\widetilde{q}}{\mathrm{d}\widetilde{z}^2} + \frac{\mathrm{d}\widetilde{G}}{\mathrm{d}\widetilde{z}}\frac{\mathrm{d}\widetilde{q}}{\mathrm{d}\widetilde{z}} + (\widetilde{\rho}\beta^2\theta^2 - \widetilde{k}^2\widetilde{A}_{rr})\widetilde{q} - \widetilde{k}(\widetilde{A}_{rz} + \widetilde{G})\frac{\mathrm{d}\widetilde{w}}{\mathrm{d}\widetilde{z}} - \widetilde{k}\frac{\mathrm{d}\widetilde{G}}{\mathrm{d}\widetilde{z}}\widetilde{w} = 0\,, \tag{1.30}$$

$$\widetilde{A}_{zz}\frac{\mathrm{d}^2\widetilde{w}}{\mathrm{d}\widetilde{z}^2} + \frac{\mathrm{d}\widetilde{A}_{zz}}{\mathrm{d}\widetilde{z}}\frac{\mathrm{d}\widetilde{w}}{\mathrm{d}\widetilde{z}} + (\widetilde{\rho}\beta^2\theta^2 - \widetilde{k}^2\widetilde{G})\widetilde{w} + \widetilde{k}(\widetilde{A}_{rz} + \widetilde{G})\frac{\mathrm{d}\widetilde{q}}{\mathrm{d}\widetilde{z}} + \widetilde{k}\frac{\mathrm{d}\widetilde{A}_{rz}}{\mathrm{d}\widetilde{z}}\widetilde{q} = 0\,, \tag{1.31}$$

where

$$\theta = \omega z_r\sqrt{\frac{\rho(z_0)}{G_{rz0}}}\,, \tag{1.32a}$$

$$\beta = \sqrt{\frac{1}{1 + \mathrm{i}\varepsilon}}\,. \tag{1.32b}$$

For parameter β, a radical with a positive real part has been chosen. A frequency parameter, θ, may reflect the influence of heterogeneity via value z_r, if the function that characterizes heterogeneity contains z_r as a parameter.

1.3 Statement of Boundary Conditions for Planes z = const

Integration of expressions (1.17) for the amplitudes of vibrations and (1.23) for the stress amplitudes with respect to parameter k from zero to infinity yields integral representations which satisfy all the equations of the theory of elasticity in the domains, where no volume forces exist. If stresses or displacements are specified for some planes z = const, they result in the generation of the corresponding boundary conditions for functions $q(z,k), w(z,k), p(z,k)$. Consider stresses $S_z, T_{rz}, T_{\vartheta z}$ which occur in the planes of z = const. Following equations (1.23) we have:

$$S_z = \hat{S}_z \mathbf{\Gamma}_1, \tag{1.33a}$$

$$T_{rz} = \hat{T}_{rz} \mathbf{\Gamma}_1, \tag{1.33b}$$

$$T_{\vartheta z} = \hat{T}_{\vartheta z} \mathbf{\Gamma}_2, \tag{1.33c}$$

$$\hat{S}_z = \mathrm{J}_n(\eta) \left[A_{rz} kq + A_{zz} \frac{\mathrm{d}w}{\mathrm{d}z} \right], \tag{1.33d}$$

$$\hat{T}_{rz} = G_{rz} \left[-\mathrm{J}_n(\eta) \frac{n}{\eta} \frac{\mathrm{d}p}{\mathrm{d}z} + \frac{\mathrm{dJ}_n(\eta)}{\mathrm{d}\eta} \left(kw - \frac{\mathrm{d}q}{\mathrm{d}z} \right) \right], \tag{1.33e}$$

$$\hat{T}_{\vartheta z} = G_{rz} \left[\mathrm{J}_n(\eta) \frac{n}{\eta} \left(\frac{\mathrm{d}q}{\mathrm{d}z} - kw \right) + \frac{\mathrm{dJ}_n(\eta)}{\mathrm{d}\eta} \frac{\mathrm{d}p}{\mathrm{d}z} \right]. \tag{1.33f}$$

In order to use the properties of Hankel's transformation for studying the above-mentioned integrals with respect to parameter k, we have to obtain the combinations of stresses which contain Bessel's function only as a multiplier (value $\hat{S}_z$ already has the sought form). The last two equations (1.33) yield

$$\hat{T}_{rz} + \hat{T}_{\vartheta z} = G_{rz} \left(-kw + \frac{\mathrm{d}q}{\mathrm{d}z} - \frac{\mathrm{d}p}{\mathrm{d}z} \right) \mathrm{J}_{n+1}(\eta), \tag{1.34a}$$

$$\hat{T}_{rz} - \hat{T}_{\vartheta z} = G_{rz} \left(kw - \frac{\mathrm{d}q}{\mathrm{d}z} - \frac{\mathrm{d}p}{\mathrm{d}z} \right) \mathrm{J}_{n-1}(\eta), \tag{1.34b}$$

where the following recurrent relationships for Bessel's functions [1] have been accounted for:

$$\mathrm{J}_{n+1}(\eta) = \mathrm{J}_n(\eta) \frac{n}{\eta} - \frac{\mathrm{dJ}_n(\eta)}{\mathrm{d}\eta}, \tag{1.35a}$$

$$\mathrm{J}_{n-1}(\eta) = \mathrm{J}_n(\eta) \frac{n}{\eta} + \frac{\mathrm{dJ}_n(\eta)}{\mathrm{d}\eta}. \tag{1.35b}$$

Suppose that for some value $z = z_\sigma$ the amplitudes of stresses have the following form:

$$\sigma_{zz} = \hat{\sigma}_{zz}(r) \mathbf{\Gamma}_1, \tag{1.36a}$$

$$\tau_{rz} = \hat{\tau}_{rz}(r)\mathbf{\Gamma}_1, \tag{1.36b}$$

$$\tau_{\vartheta z} = \hat{\tau}_{\vartheta z}(r)\mathbf{\Gamma}_2 . \tag{1.36c}$$

The integral representations of the solution of the problem satisfy condition (1.36), if at $z = z_\sigma$

$$\int_0^\infty \hat{S}_z \,\mathrm{d}k = \hat{\sigma}_z , \tag{1.37a}$$

$$\int_0^\infty (\hat{T}_{rz} + \hat{T}_{\vartheta z})\,\mathrm{d}k = \hat{\tau}_{rz} + \hat{\tau}_{\vartheta z} , \tag{1.37b}$$

$$\int_0^\infty (\hat{T}_{rz} - \hat{T}_{\vartheta z})\,\mathrm{d}k = \hat{\tau}_{rz} - \hat{\tau}_{\vartheta z} \tag{1.37c}$$

or

$$\int_0^\infty \left(A_{rz}kq + A_{zz}\frac{\mathrm{d}w}{\mathrm{d}z} \right) \mathrm{J}_n(kr)\,\mathrm{d}k = \hat{\sigma}_z(r) , \tag{1.38a}$$

$$G_{rz}\int_0^\infty \left(-kw + \frac{\mathrm{d}q}{\mathrm{d}z} - \frac{\mathrm{d}p}{\mathrm{d}z} \right) \mathrm{J}_{n+1}(kr)\,\mathrm{d}k = \hat{\tau}_{rz}(r) + \hat{\tau}_{\vartheta z}(r) , \tag{1.38b}$$

$$G_{rz}\int_0^\infty \left(kw - \frac{\mathrm{d}q}{\mathrm{d}z} - \frac{\mathrm{d}p}{\mathrm{d}z} \right) \mathrm{J}_{n-1}(kr)\,\mathrm{d}k = \hat{\tau}_{rz}(r) - \hat{\tau}_{\vartheta z}(r). \tag{1.38c}$$

Using the formulas relating to the direct and inverse Hankel's transformation,

$$\int_0^\infty rf(r)\,\mathrm{J}_s(kr)\,\mathrm{d}r = F(k) , \tag{1.39a}$$

$$\int_0^\infty kF(k)\,\mathrm{J}_s(kr)\,\mathrm{d}k = f(r) , \tag{1.39b}$$

we obtain the conditions for functions $q(z,k)$, $w(z,k)$, $p(z,k)$, which must hold at $z = z_\sigma$:

$$A_{rz}(z_\sigma)kq + A_{zz}(z_\sigma)\frac{\mathrm{d}w}{\mathrm{d}z} = k\int_0^\infty r\hat{\sigma}_z(r)J_n(kr)\,\mathrm{d}r , \tag{1.40}$$

$$-kw + \frac{\mathrm{d}q}{\mathrm{d}z} - \frac{\mathrm{d}p}{\mathrm{d}z} = \frac{k}{G_{rz}(z_\sigma)}\int_0^\infty r[\hat{\tau}_{rz}(r) + \hat{\tau}_{\vartheta z}(r)]\mathrm{J}_{n+1}(kr)\,\mathrm{d}r , \tag{1.41a}$$

$$kw - \frac{\mathrm{d}q}{\mathrm{d}z} - \frac{\mathrm{d}p}{\mathrm{d}z} = \frac{k}{G_{rz}(z_\sigma)} \int_0^\infty r[\hat{\tau}_{rz}(r) - \hat{\tau}_{\vartheta z}(r)] \mathrm{J}_{n-1}(kr) \mathrm{d}r . \tag{1.41b}$$

Equations (1.41) give

$$\frac{\mathrm{d}p}{\mathrm{d}z} = -\frac{k}{2G_{rz}(z_\sigma)} \left[\int_0^\infty r[\hat{\tau}_{rz}(r) + \hat{\tau}_{\vartheta z}(r)] \mathrm{J}_{n+1}(kr) \mathrm{d}r + \int_0^\infty r[\hat{\tau}_{rz}(r) - \hat{\tau}_{\vartheta z}(r)] \mathrm{J}_{n-1}(kr) \mathrm{d}r \right], \tag{1.42a}$$

$$\frac{\mathrm{d}q}{\mathrm{d}z} - kw = \frac{k}{2G_{rz}(z_\sigma)} \left[\int_0^\infty r[\hat{\tau}_{rz}(r) + \hat{\tau}_{\vartheta z}(r)] \mathrm{J}_{n+1}(kr) \mathrm{d}r - \int_0^\infty r[\hat{\tau}_{rz}(r) - \hat{\tau}_{\vartheta z}(r)] \mathrm{J}_{n-1}(kr) \mathrm{d}r \right]. \tag{1.42b}$$

These equations, in addition to equation (1.40), may serve as boundary conditions for the functions $q(z,k), w(z,k), p(z,k)$ when stresses are specified on the plane $z = z_\sigma$.

In order to determine the boundary conditions for these functions, in the case with known amplitudes of displacements in the plane $z = z_u$, we develop from equations (1.17):

$$U_r = \hat{U}_r \mathbf{\Gamma}_1 \tag{1.43a}$$

$$U_z = \hat{U}_z \mathbf{\Gamma}_1 , \tag{1.43b}$$

$$U_\vartheta = \hat{U}_\vartheta \mathbf{\Gamma}_2 , \tag{1.43c}$$

$$\hat{U}_r + \hat{U}_\vartheta = (q - p) \mathrm{J}_{n+1}(\eta) , \tag{1.43d}$$

$$\hat{U}_r - \hat{U}_\vartheta = -(q + p) \mathrm{J}_{n-1}(\eta) , \tag{1.43e}$$

$$\hat{U}_z = w \mathrm{J}_n(\eta) . \tag{1.43f}$$

Let the amplitudes of displacements in the plane $z = z_u$ have the form

$$u_r = \hat{u}_r(r) \mathbf{\Gamma}_1 , \tag{1.44a}$$

$$u_z = \hat{u}_z(r) \mathbf{\Gamma}_1 , \tag{1.44b}$$

$$u_\vartheta = \hat{u}_\vartheta(r) \mathbf{\Gamma}_2 . \tag{1.44c}$$

Analogously to equations (1.37),

$$\int_0^\infty \hat{U}_z \, \mathrm{d}k = \hat{u}_z \,, \tag{1.45a}$$

$$\int_0^\infty (\hat{U}_r + \hat{U}_\vartheta) \mathrm{d}k = \hat{u}_r + \hat{u}_\vartheta \,, \tag{1.45b}$$

$$\int_0^\infty (\hat{U}_r - \hat{U}_\vartheta) \mathrm{d}k = \hat{u}_r - \hat{u}_\vartheta \tag{1.45c}$$

or

$$\int_0^\infty w \, \mathrm{J}_n(kr) \mathrm{d}k = \hat{u}_z(r), \tag{1.46a}$$

$$\int_0^\infty (q-p) \mathrm{J}_{n+1}(kr) \mathrm{d}k = \hat{u}_r(r) + \hat{u}_\vartheta(r), \tag{1.46b}$$

$$-\int_0^\infty (q+p) \mathrm{J}_{n-1}(kr) \mathrm{d}k = \hat{u}_r(r) - \hat{u}_\vartheta(r). \tag{1.46c}$$

Further employment of the direct and inverse Hankel's transformations (1.39) results in the following conditions for the functions $q(z,k)$, $w(z,k)$, $p(z,k)$ at $z = z_u$:

$$q - p = k \int_0^\infty r[\hat{u}_r(r) + \hat{u}_\vartheta(r)] \mathrm{J}_{n+1}(kr) \mathrm{d}r \,, \tag{1.47a}$$

$$q + p = -k \int_0^\infty r[\hat{u}_r(r) - \hat{u}_\vartheta(r)] \mathrm{J}_{n-1}(kr) \mathrm{d}r \,, \tag{1.47b}$$

$$w = k \int_0^\infty r\hat{u}_z(r) \mathrm{J}_n(kr) \mathrm{d}r \,. \tag{1.47c}$$

From the first two equations (1.47) we obtain:

$$q = \frac{k}{2}\left\{ \int_0^\infty r[\hat{u}_r(r) + \hat{u}_\vartheta(r)] \mathrm{J}_{n+1}(kr) \mathrm{d}r - \int_0^\infty r[\hat{u}_r(r) - \hat{u}_\vartheta(r)] \mathrm{J}_{n-1}(kr) \mathrm{d}r \right\}, \tag{1.48a}$$

$$p = -\frac{k}{2}\left\{ \int_0^\infty r[\hat{u}_r(r) + \hat{u}_\vartheta(r)] \mathrm{J}_{n+1}(kr) \mathrm{d}r + \int_0^\infty r[\hat{u}_r(r) - \hat{u}_\vartheta(r)] \mathrm{J}_{n-1}(kr) \mathrm{d}r \right\}. \tag{1.48a}$$

Boundary conditions (1.42a) and (1.48b) for the function p do not contain functions q and w. Accounting for the form of equations (1.24), (1.25), and (1.26), we come to the conclusion that the problems of determining the function p, and the functions q and w, are completely independent.

1.4 General Solution for Harmonic Vibrations of Half-Space Subjected to Surface Loads

In this section we consider some cases of vertical and horizontal loads applied over the circular domain of radius R on the surface of the half-space. In particular, we shall obtain the solutions for the case of concentrated forces, which enable us to construct solutions for various forms of distribution of applied loads.

1.4.1 Axisymmetric Vertical Load Applied over Circular Domain

Consider an axisymmetric vertical load harmonic in time applied to the surface of a half-space. An example of a load is shown in Fig. 1.2a: the force $P_0 \exp(\mathrm{i}\,\omega t)$ is uniformly distributed over the circle of radius R at the surface of the half-space.

Assuming $n = 0$ (the upper line in Γ_j), we present the amplitudes of vibrations as a result of integrating expressions (1.17):

$$u_r = \int_0^\infty q(z,k)\,\mathrm{J}_1(kr)\,\mathrm{d}k \ , \tag{1.49a}$$

$$u_z = \int_0^\infty w(z,k)\,\mathrm{J}_0(kr)\,\mathrm{d}k \,, \tag{1.49b}$$

$$u_\vartheta = 0 \,. \tag{1.49c}$$

At the surface of the half-space ($z = z_0$), tangential stresses vanish, and $\hat{\sigma}_z = -P_0/(\pi R^2)$ for $r < R$ (for the example presented in Fig. 1.2a) and $\hat{\sigma}_z = 0$ for $r>R$. The boundary conditions at $z = z_0$, in accordance with equation (1.40) at $z_\sigma = z_0$, and with equation (1.42b), have the following form:

$$\widetilde{A}_{rz}(z_0)kq + \widetilde{A}_{zz}(z_0)\frac{\mathrm{d}w}{\mathrm{d}z} = -\frac{kP_0}{\pi R^2 G_{rz}(z_0)}\int_0^R r\,\mathrm{J}_0(kr)\,\mathrm{d}r = -\frac{P_0}{\pi R G_{rz}(z_0)}\mathrm{J}_1(kR)\,, \tag{1.50a}$$

$$\frac{\mathrm{d}q}{\mathrm{d}z} - kw = 0 \ . \tag{1.50b}$$

Next, we consider equations (1.25) and (1.26) (or (1.30), (1.31)), which should be satisfied by the functions $q(z,k)$ and $w(z,k)$. This system of equations has four fundamental solutions, two of which (q_1, w_1 and q_1, w_1) satisfy the condition of absence of sources at infinity ($z \to \infty$). These solutions are not necessarily decreasing, but, for sufficiently high values of parameter k, solutions corresponding to the condition of absence of sources at infinity decrease as

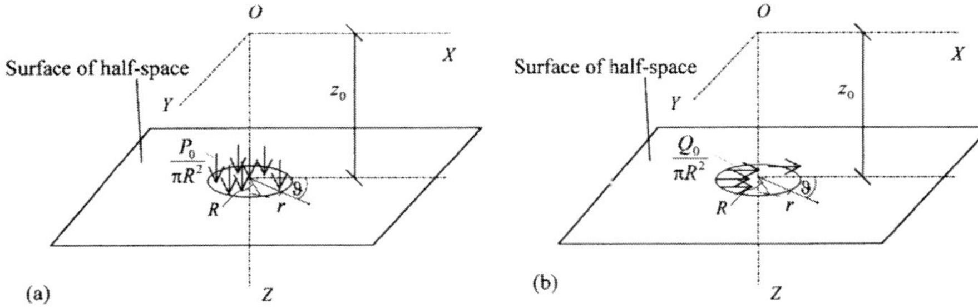

Fig. 1.2a,b. Harmonic vertical (a) and horizontal (b) forces applied to circular area on half-space surface

$z \to \infty$. For example, the sought solutions for the case of an isotropic homogeneous half-space with elastic properties, determined by (1.5), take the form:

$$q_1 = \alpha_1 \exp(-\alpha_1 z), \tag{1.51a}$$

$$w_1 = k \exp(-\alpha_1 z), \tag{1.51b}$$

$$q_2 = k \exp(-\alpha_2 z), \tag{1.51c}$$

$$w_2 = \alpha_2 \exp(-\alpha_2 z), \tag{1.51d}$$

where α_1 and α_2 are calculated from (1.20). Introducing two arbitrary coefficients A_1 and A_2 we present functions $q(z,k)$ and $w(z,k)$ as a linear combination of the above-mentioned fundamental solutions:

$$q(z,k) = A_1(k) q_1(z,k) + A_2(k) q_2(z,k), \tag{1.52a}$$

$$w(z,k) = A_1(k) w_1(z,k) + A_2(k) w_2(z,k). \tag{1.52b}$$

Substitution of these expressions into the boundary conditions (1.50) yields the following system of equations for the coefficients A_1 and A_2:

$$\begin{aligned} c_{11} A_1 + c_{12} A_2 &= d_1, \\ c_{21} A_1 + c_{22} A_2 &= d_2, \end{aligned} \tag{1.53}$$

where

$$c_{1j} = \widetilde{A}_{rz} k q_j + \widetilde{A}_{zz} \frac{\mathrm{d} w_j}{\mathrm{d} z} = \widetilde{A}_{rz} \widetilde{k} \widetilde{q}_j + \widetilde{A}_{zz} \frac{\mathrm{d} \widetilde{w}_j}{\mathrm{d} \widetilde{z}}, \tag{1.54a}$$

$$c_{2j} = \frac{\mathrm{d}q_j}{\mathrm{d}z} - kw_j = \frac{\mathrm{d}\tilde{q}_j}{\mathrm{d}\tilde{z}} - \tilde{k}\tilde{w}_j, \tag{1.54b}$$

$$d_1 = -\frac{P_0}{\pi R G_{rz}(z_0)} \mathrm{J}_1(kR), \tag{1.54c}$$

$$d_2 = 0. \tag{1.54d}$$

Here, the fundamental solutions $q_j(z,k)$, $w_j(z,k)$ $(j = 1,2)$ and the elastic parameters are taken for the surface of the half-space ($z = z_0$). A complete solution of the problem is obtained from (1.49), (1.52), (1.53) in the form

$$u_r = -\frac{P_0}{\pi R G_{rz}(z_0)} \int_0^{\infty} \frac{c_{22}q_1(z,k) - c_{21}q_2(z,k)}{D} \mathrm{J}_1(kR)\,\mathrm{J}_1(kr)\,\mathrm{d}k, \tag{1.55a}$$

$$u_z = -\frac{P_0}{\pi R G_{rz}(z_0)} \int_0^{\infty} \frac{c_{22}w_1(z,k) - c_{21}w_2(z,k)}{D} \mathrm{J}_1(kR)\,\mathrm{J}_0(kr)\,\mathrm{d}k, \tag{1.55b}$$

where

$$D = c_{11}c_{22} - c_{12}c_{21}. \tag{1.56}$$

The solution corresponding to the concentrated vertical force is constructed from (1.55), after replacing $\mathrm{J}_1(kR)/R$ with $k/2$:

$$u_r = -\frac{P_0}{2\pi G_{rz}(z_0)} \int_0^{\infty} \frac{c_{22}q_1(z,k) - c_{21}q_2(z,k)}{D} k\,\mathrm{J}_1(kr)\,\mathrm{d}k, \tag{1.57a}$$

$$u_z = -\frac{P_0}{2\pi G_{rz}(z_0)} \int_0^{\infty} \frac{c_{22}w_1(z,k) - c_{21}w_2(z,k)}{D} k\,\mathrm{J}_0(kr)\,\mathrm{d}k. \tag{1.57b}$$

A contour of integration in the expressions (1.55), (1.57) is not specified. For static problems and for some cases with dynamic problems dealing with the heterogeneous half-space vibrating at low frequencies, integration along the real axis (with respect to the real values of parameter k) is possible. As a rule, in dynamic problems, a value of D (the determinant of the system of equations (1.53)) may become zero at some positive values of k when damping is absent. This fact leads to the appearance of simple poles for the integrands in the expressions (1.55), (1.57). For these values of k, a non-zero solution of (1.53) exists, when the right side is equal to zero, and, respectively, a non-zero solution of the form (1.52) exists too. From a physical point of view, this means that free harmonic vibrations of the half-space (without any external loads) are possible. For the case of the homogeneous isotropic half-space, the single pole, which corresponds to the Rayleigh waves, exists on the real axis. On the contrary, for the heterogeneous half-space, numerous poles, or even an infinite number of poles, may exist. Complications arising from the singularities in expressions (1.55),

(1.57) may be overcome either by accounting for the dissipative properties of the medium (this results in the shifting of the poles from the real axis), or by application of an appropriate contour of integration. An interesting fact in this connection is the following: for the case of harmonic vibrations of the isotropic half-space with a shear modulus, which grows exponentially in the Z-direction, coupled zero points of the value D may occur at certain frequencies of vibrations. These frequencies correspond to the resonance: when damping is neglected, they result in infinite amplitudes of vibrations. Additional questions connected with the possibility of a zero value of D at real values of parameter k are discussed below in more detail for specific forms of half-spaces.

Integration of expression (1.23) with respect to parameter k by taking (1.52) and (1.53) into account yields the amplitudes of stresses corresponding to the considered load. For example, expressions for the stresses σ_z and τ_{rz} take the form

$$\sigma_z = -\frac{P_0}{\pi R}\int_0^\infty \left\{ c_{22}\left[k\tilde{A}_{rz} q_1(z,k) + \tilde{A}_{zz}\frac{\mathrm{d}w_1(z,k)}{\mathrm{d}z}\right] - c_{21}\left[k\tilde{A}_{rz} q_2(z,k) + \tilde{A}_{zz}\frac{\mathrm{d}w_2(z,k)}{\mathrm{d}z}\right]\right\}\frac{\mathrm{J}_1(kR)\,\mathrm{J}_0(kr)}{D}\,\mathrm{d}k\,, \tag{1.58a}$$

$$\tau_{rz} = -\frac{P_0\tilde{G}}{\pi R}\int_0^\infty \left\{ c_{22}\left[\frac{\mathrm{d}q_1(z,k)}{\mathrm{d}z} - kw_1(z,k)\right] - c_{21}\left[\frac{\mathrm{d}q_2(z,k)}{\mathrm{d}z} - kw_2(z,k)\right]\right\}\frac{\mathrm{J}_1(kR)\,\mathrm{J}_1(kr)}{D}\,\mathrm{d}k\,. \tag{1.58b}$$

A similar technique is used in order to construct solutions for different types of load. For example, let the force $P_0 \exp(\mathrm{i}\omega t)$ be distributed over the circle of radius R, according to the static problem for a circular stamp with the smooth base pressed against an isotropic homogeneous half-space with force P_0. Then, at $z = z_0$ and $r < R$ we have

$$\sigma_z = -\frac{P_0}{2\pi R(R^2 - r^2)^{1/2}}\,. \tag{1.59}$$

According to equation (1.40), equation (1.50a) becomes

$$\tilde{A}_{rz} kq + \tilde{A}_{zz}\frac{\mathrm{d}w}{\mathrm{d}z} = -\frac{P_0}{2\pi R G_{rz}(z_0)}\sin(kR)\,. \tag{1.60}$$

As a result, in solution (1.55), (1.58), $\mathrm{J}_1(kR)$ should be replaced with $\sin(kR)/2$.

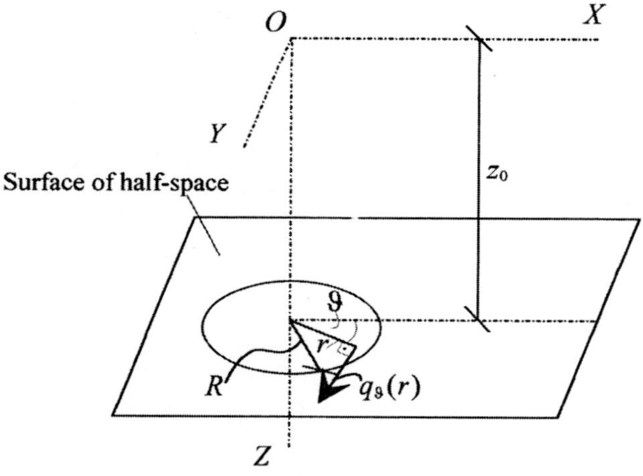

Fig. 1.3. Torque applied to circular area on half-space surface

1.4.2 Tangential Axisymmetric Load Applied to Circular Domain

An example for this type of load is shown in Fig. 1.3: a tangential load $q_\vartheta(r)\exp(\mathrm{i}\,\omega t)$ is applied to the circle of radius R causing a torque with respect to the Z-axis with amplitude

$$M_z = 2\pi \int_0^R q_\vartheta(r) r^2 \,\mathrm{d}r\,. \tag{1.61}$$

Consider the following case:

$$q_\vartheta = q_0 r\,. \tag{1.62}$$

Here, we have

$$M_z = \frac{\pi q_0 R^4}{2}\,, \tag{1.63a}$$

$$q_0 = \frac{2M_z}{\pi R^4}\,. \tag{1.63b}$$

In solution (1.17), we assume $n = 0$ (a lower line in $\mathbf{\Gamma}_1$):

$$U_\vartheta = -p(z,k)\frac{\mathrm{dJ}_0(\eta)}{\mathrm{d}\eta} = p(z,k)\,\mathrm{J}_1(\eta)\,, \tag{1.64a}$$

$$U_r = 0\,, \tag{1.64b}$$

$$U_z = 0. \tag{1.64c}$$

Amplitudes of vibrations of the points in the half-space in a tangential direction are presented in the form of the following integral:

$$u_\vartheta = \int_0^\infty p(z,k)\,\mathrm{J}_1(kr)\,\mathrm{d}k. \tag{1.65}$$

The function $p(z, k)$ satisfies equation (1.24) and the following boundary condition, which follows from equation (1.42a) at $z_\sigma = z_0$:

$$\frac{\mathrm{d}p}{\mathrm{d}z} = -\frac{k}{G_{rz}(z_0)}\int_0^\infty r\hat{\tau}_{\vartheta z}(r)\,\mathrm{J}_1(kr)\,\mathrm{d}r = -\frac{k}{G_{rz}(z_0)}\int_0^R r q_\vartheta(r)\,\mathrm{J}_1(kr)\,\mathrm{d}r. \tag{1.66}$$

Here, we take into account that at $z = z_0$

$$\hat{\tau}_{rz}(r) = 0, \tag{1.67a}$$

$$\hat{\tau}_{\vartheta z}(r) = -\tau_{\vartheta z}(r) = \begin{cases} q_\vartheta(r) & (r < R) \\ 0 & (r > R). \end{cases} \tag{1.67b}$$

Equation (1.24) has two fundamental solutions, one of which, $p_1(z,k)$, satisfies the condition of absence of sources at infinity. For example, in the case of an homogeneous isotropic half-space, this solution has the form $\exp(-\alpha_1 z)$ where α_1 is found from (1.20). Using the solution $p_1(z,k)$ we present the sought function, $p(z,k)$, in the following form:

$$p(z,k) = C_1 p_1(z,k). \tag{1.68}$$

Constant C_1 is determined from the boundary condition (1.66). For loads of the type (1.62), we obtain, by using the properties of Bessel's functions [1], the following expressions:

$$C_1\frac{\mathrm{d}p_1(z_0,k)}{\mathrm{d}z} = -\frac{2M_z k}{G_{rz}(z_0)\pi R^4}\int_0^R r^2\,\mathrm{J}_1(kr)\,\mathrm{d}r = -\frac{2M_z}{G_{rz}(z_0)\pi R^2}\mathrm{J}_2(kR). \tag{1.69}$$

Following (1.65), the amplitudes of displacements become

$$u_\vartheta = -\frac{2M_z}{G_{rz}(z_0)\pi R^2}\int_0^\infty p_1(z,k)\left[\frac{\mathrm{d}p_1(z_0,k)}{\mathrm{d}z}\right]^{-1}\mathrm{J}_2(kR)\,\mathrm{J}_1(kr)\,\mathrm{d}k. \tag{1.70}$$

For the homogeneous isotropic half-space, accounting for the above-mentioned expression for $p_1(z,k)$ yields

$$u_\vartheta = \frac{2M_z}{G_{rz}(z_0)\pi R^2}\int_0^\infty \frac{\exp[-\alpha_1(z-z_0)]}{\alpha_1} \mathrm{J}_2(kR)\mathrm{J}_1(kr)\,\mathrm{d}k\,. \tag{1.71}$$

For the case of a concentrated force, $\mathrm{J}_2(kR)/R^2$ should be replaced with $k^2/8$. This substitution results in diverging integrals in the case with $z = z_0$, i.e. for the surface of the half-space (solution (1.71), which is related to the homogeneous isotropic half-space, where the multiplier of Bessel's functions is equivalent to k^{-1} when $k \to \infty$, may serve as a vivid illustration of this fact). However, the possibility of obtaining the solution in the case with $z = z_0$ exists. Its construction can be performed in the following order: first, we find the form of the multiplier of Bessel's functions in (1.70) at high values of *k*; the result obtained is subtracted from the multiplier of Bessel's functions, and added to them; these expressions yield a tabulated integral and the second integral, convergence of which is sufficient for switching to the limit as $R \to 0$. Thus, using (1.71) at $z = z_0$ yields

$$u_\vartheta = \frac{2M_z}{G_{rz}(z_0)\pi R^2}\left\{\int_0^\infty \left[\frac{1}{\alpha_1} - \frac{1}{k}\right] \mathrm{J}_2(kR)\mathrm{J}_1(kr)\,\mathrm{d}k + \int_0^\infty \frac{\mathrm{J}_2(kR)\mathrm{J}_1(kr)}{k}\,\mathrm{d}k\right\}. \tag{1.72}$$

For $r > R$, the second integral equals [1]

$$\int_0^\infty \frac{\mathrm{J}_2(kR)\mathrm{J}_1(kr)}{k}\,\mathrm{d}k = \frac{R^2}{8r^2}\,{}_2F_1\!\left(\frac{3}{2},\frac{1}{2};3;\frac{R^2}{r^2}\right). \tag{1.73}$$

The hypergeometric function entering (1.73) reaches unity as $R \to 0$. Amplitudes of vibrations of the surface of the homogeneous isotropic half-space, subjected to the concentrated torque, may be presented as

$$u_\vartheta = \frac{M_z}{4G_{rz}(z_0)\pi}\left\{\int_0^\infty \left[\frac{1}{\alpha_1} - \frac{1}{k}\right] k^2\,\mathrm{J}_1(kr)\,\mathrm{d}k + \frac{1}{r^2}\right\}. \tag{1.74}$$

Accounting for the expression for α_1, we see that a static value of displacements, u_ϑ, is derived from (1.74) when the integral-containing term is neglected.

1.4.3 Horizontal Load Distributed Uniformly over Circular Domain

Consider a horizontal force $Q_0 \exp(\mathrm{i}\,\omega t)$ applied to the surface of the half-space and distributed uniformly over a circular area of radius *R* (Fig. 1.2b). This type of load corresponds to $n = 1$ in $\boldsymbol{\Gamma}_j$ in equations (1.17), (1.23), where the upper line is selected. Amplitudes of vibrations of the half-space are expressed as the integrated particular solutions (1.17). Using the expression for the derivative of Bessel's functions [1]

$$\frac{\mathrm{d}J_n(\eta)}{\mathrm{d}\eta} = J_{n-1}(\eta) - \frac{n}{\eta}J_n(\eta), \tag{1.75}$$

we obtain

$$u_r = \hat{u}_r \cos\vartheta,\ u_\vartheta = \hat{u}_\vartheta \sin\vartheta,\ u_z = \hat{u}_z \cos\vartheta, \tag{1.76a}$$

$$\hat{u}_r = \int_0^\infty \left[[q(z,k) - p(z,k)] \frac{J_1(kr)}{kr} - q(z,k) J_0(kr) \right] \mathrm{d}k, \tag{1.76b}$$

$$\hat{u}_\vartheta = \int_0^\infty \left[[q(z,k) - p(z,k)] \frac{J_1(kr)}{kr} + p(z,k) J_0(kr) \right] \mathrm{d}k, \tag{1.76c}$$

$$\hat{u}_z = \int_0^\infty w(z,k) J_1(kr) \,\mathrm{d}k. \tag{1.76d}$$

At the surface of the half-space ($z = z_0$) values of $\tau_{rz} = \hat{\tau}_{rz}(r)\mathbf{\Gamma}_1$, $\tau_{\vartheta z} = \hat{\tau}_{\vartheta z}(r)\mathbf{\Gamma}_2$ in equations (1.36) are equal to zero at $r > R$; at $r < R$ they become

$$\hat{\tau}_{rz}(r) = -\frac{Q_0}{\pi R^2}, \tag{1.77a}$$

$$\hat{\tau}_{\vartheta z}(r) = \frac{Q_0}{\pi R^2}. \tag{1.77b}$$

The value of $\hat{\sigma}_{zz}(r)$ is zero at the surface. The boundary conditions (1.40), (1.42) for the functions $q(z,k), w(z,k), p(z,k)$ take the form

$$\frac{\mathrm{d}p}{\mathrm{d}z} = \frac{Q_0 k}{G_{rz}(z_0)\pi R^2} \int_0^\infty r\, J_0(kr)\,\mathrm{d}r = \frac{Q_0}{G_{rz}(z_0)\pi R} J_1(kR), \tag{1.78}$$

$$\widetilde{A}_{rz} kq + \widetilde{A}_{zz} \frac{\mathrm{d}w}{\mathrm{d}z} = 0, \tag{1.79a}$$

$$\frac{\mathrm{d}q}{\mathrm{d}z} - kw = \frac{kQ_0}{G_{rz}(z_0)\pi R^2} \int_0^R r\, J_0(kr)\,\mathrm{d}r = \frac{Q_0}{G_{rz}(z_0)\pi R} J_1(kR). \tag{1.79b}$$

Furthermore, we employ representation of the form (1.52) for $q(z,k), w(z,k)$ resulting in the system of equations (1.53) for the coefficients A_1 and A_2 with the modified right sides:

$$d_1 = 0, \tag{1.80a}$$

$$d_2 = \frac{Q_0}{G_{rz}(z_0)\pi R} \mathrm{J}_1(kR) . \tag{1.80b}$$

Function $p(z,k)$ is presented in the form (1.68). By using equation (1.78), we obtain the following result for the constant C_1 :

$$C_1 = d_2 / F , \tag{1.81}$$

where

$$F = \frac{\mathrm{d}\, p_1(z_0,k)}{\mathrm{d}\, z} = \frac{\mathrm{d}\,\tilde{p}_1(\tilde{z}_0,\tilde{k})}{\mathrm{d}\tilde{z}} . \tag{1.82}$$

Dimensionless variables (in terms of (1.27)) are denoted with "~". Following determination of coefficients A_1 and A_2 from the system of equations (1.53) with right sides (1.80), we present expressions for the amplitudes of vibrations entering equations (1.76) in the form

$$\hat{u}_r = \frac{Q_0}{G_{rz}(z_0)\pi R} \int_0^\infty \mathrm{J}_1(kR) H_r(z,r,k)\,\mathrm{d}\,k , \tag{1.83a}$$

$$H_r(z,r,k) = \left[\frac{c_{11}q_2(z,k) - c_{12}q_1(z,k)}{D} - \frac{p_1(z,k)}{F}\right] \frac{\mathrm{J}_1(kr)}{kr}$$

$$- \frac{c_{11}q_2(z,k) - c_{12}q_1(z,k)}{D} \mathrm{J}_0(kr) , \tag{1.83b}$$

$$\hat{u}_\vartheta = \frac{Q_0}{G_{rz}(z_0)\pi R} \int_0^\infty \mathrm{J}_1(kR) H_\vartheta(z,r,k)\,\mathrm{d}\,k , \tag{1.84a}$$

$$H_\vartheta(z,r,k) = \left[\frac{c_{11}q_2(z,k) - c_{12}q_1(z,k)}{D} - \frac{p_1(z,k)}{F}\right] \frac{\mathrm{J}_1(kr)}{kr}$$

$$+ \frac{p_1(z,k)}{F} \mathrm{J}_0(kr) , \tag{1.84b}$$

$$\hat{u}_z = \frac{Q_0}{G_{rz}(z_0)\pi R} \int_0^\infty \mathrm{J}_1(kR) H_z(z,r,k)\,\mathrm{d}\,k , \tag{1.85a}$$

$$H_z(z,r,k) = \frac{c_{11}w_2(z,k) - c_{12}w_1(z,k)}{D} \mathrm{J}_1(kr) . \tag{1.85b}$$

A solution corresponding to the concentrated force is derived from these expressions after replacement of $\mathrm{J}_1(kR)/R$ with $k/2$:

$$\hat{u}_r = \frac{Q_0}{2G_{rz}(z_0)\pi} \int_0^\infty k H_r(z,r,k)\,\mathrm{d}\,k , \tag{1.86}$$

$$\hat{u}_\vartheta = \frac{Q_0}{2G_{rz}(z_0)\pi}\int_0^\infty kH_\vartheta(z,r,k)\,\mathrm{d}k\,, \tag{1.87}$$

$$\hat{u}_z = \frac{Q_0}{2G_{rz}(z_0)\pi}\int_0^\infty kH_z(z,r,k)\,\mathrm{d}k\,. \tag{1.88}$$

In a similar way, the construction of solutions is possible for other cases with loads acting at the surface of half-spaces. Note that having obtained the solutions related to the concentrated forces (the Green's functions), we may construct solutions for any loads by using the principle of superposition. Integrals appearing due to the application of this principle are much less complicated than those presented in (1.55), (1.57), (1.58), (1.70), (1.83)–(1.88). The Green's functions may be computed with a sufficiently small step of varying argument r, or of the parameter proportional to r. Values of these functions at the intermediate points that are required for integration, in accordance with the principle of superposition, may be determined through interpolation.

1.4.4 Loads Distributed Uniformly along Infinite Line on Surface of Half-Space (2-D Case)

In the following, solutions corresponding to the concentrated forces are applied for the construction of solutions for the case with loads distributed uniformly along an infinite line. These solutions could be derived by using Fourier's transformation for the appropriate 2-D problems, resulting in equations for the functions $q(z,k), w(z,k), p(z,k)$, which are similar to those obtained above. However, application of the principle of superposition is a more efficient method that, in addition, enables us to clarify the connection between the 3-D and 2-D solutions. For the case of a vertical load p_0 applied to an infinite line at the surface of a half-space (Fig. 1.4a), an integration of the above-obtained solutions (1.57) is performed by using the following tabulated integral [43]:

$$\int_0^\infty \frac{y^{2\mu+1}\,\mathrm{J}_\nu(k\sqrt{x^2+y^2})}{\sqrt{(x^2+y^2)^\nu}}\mathrm{d}y = \frac{2^\mu\,\Gamma(\mu+1)}{k^{\mu+1}x^{\nu-\mu-1}}\mathrm{J}_{\nu-\mu-1}(kx)\,. \tag{1.89}$$

From (1.89), we obtain the required integrals:

$$\int_0^\infty \mathrm{J}_0(k\sqrt{x^2+y^2})\,\mathrm{d}y = \frac{\cos(kx)}{k}, \tag{1.90a}$$

$$\int_0^\infty \mathrm{J}_1(k\sqrt{x^2+y^2})\cos(\gamma)\,\mathrm{d}y = \int_0^\infty \frac{x\,\mathrm{J}_1(k\sqrt{x^2+y^2})}{\sqrt{x^2+y^2}}\mathrm{d}y = \frac{\sin(kx)}{k}\,. \tag{1.90b}$$

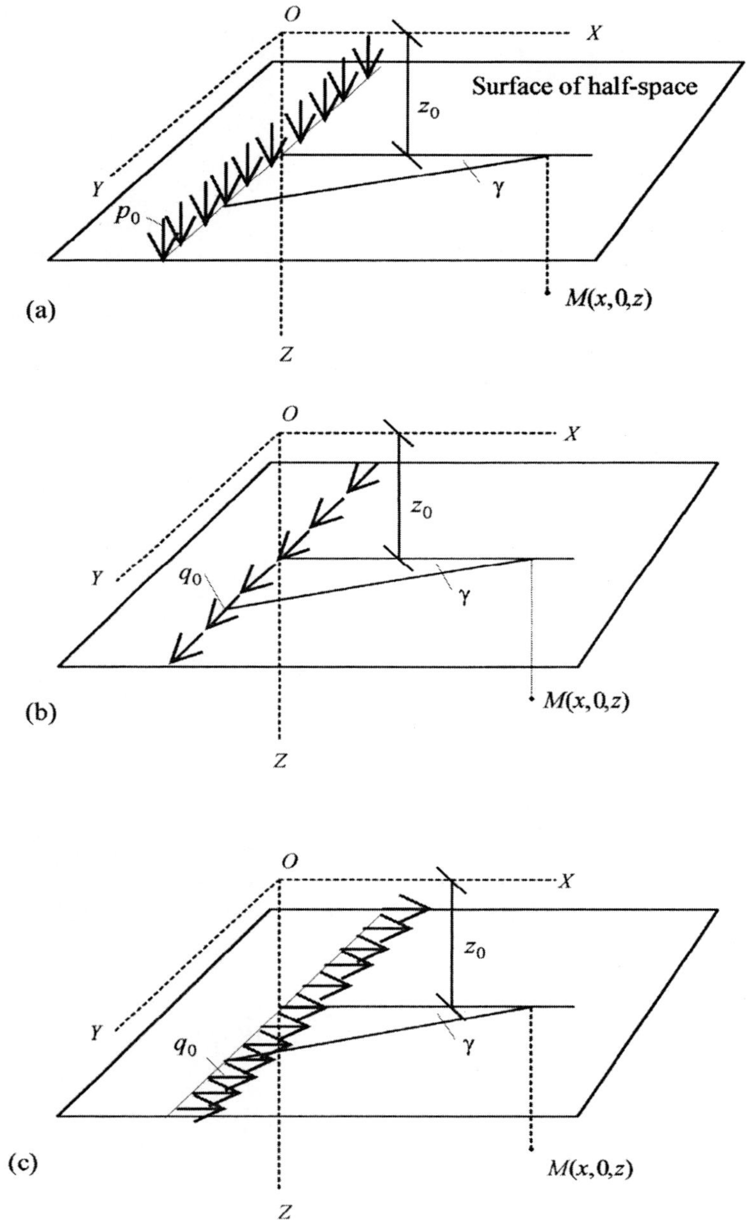

Fig. 1.4a,b,c. Loads distributed uniformly along infinite line parallel to Y-axis

For the point $M(x, 0, z)$, the amplitudes of vibrations $u_x(x,z)$ and $u_z(x,z)$ along the X-axis and Z -axis are obtained by integrating, with respect to y, the expressions $u_r[(x^2+y^2)^{1/2}]\cos(\gamma)$ and $u_z[(x^2+y^2)^{1/2}]$ in equations (1.57), respectively (P_0 is replaced by $p_0\,\mathrm{d}y$). Using equations (1.90) yields

$$u_x(x,z)=-\frac{p_0}{\pi G_{rz}(z_0)}\int_0^\infty \sin(kx)\frac{c_{22}q_1(z,k)-c_{21}q_2(z,k)}{D}\mathrm{d}k\,, \tag{1.91a}$$

$$u_z(x,z)=-\frac{p_0}{\pi G_{rz}(z_0)}\int_0^\infty \cos(kx)\frac{c_{22}w_1(z,k)-c_{21}w_2(z,k)}{D}\mathrm{d}k\,. \tag{1.91b}$$

In order to calculate the stresses σ_z for the 2-D problem, we obtain from equation (1.58a), keeping in mind the analogy with $u_z(x,z)$,

$$\sigma_z=-\frac{p_0}{\pi}\int_0^\infty\left\{k\tilde{A}_{rz}[c_{22}q_1(z,k)-c_{21}q_2(z,k)]+\tilde{A}_{zz}\left[c_{22}\frac{\mathrm{d}w_1(z,k)}{\mathrm{d}z}\right.\right.$$

$$\left.\left.-c_{21}\frac{\mathrm{d}w_2(z,k)}{\mathrm{d}z}\right]\right\}\frac{\cos(kx)}{D}\mathrm{d}k\,. \tag{1.92}$$

For the case of a constant horizontal load q_0 acting along the Y-axis (Fig. 1.4b), displacements occur along the Y-axis only (anti-plane vibrations). The amplitudes of vibrations $u_y(x,z)$ at the point $M(x, 0, z)$ are built following the expressions for $\hat{u}_r$ and $\hat{u}_\vartheta$, from equations (1.76), which are determined by using equations (1.86), (1.87).

According to the principle of superposition, we obtain

$$u_y(x,z)=2\int_0^\infty[-\hat{u}_\vartheta(\sqrt{x^2+y^2},z)\cos^2(\gamma)+\hat{u}_r(\sqrt{x^2+y^2},z)\sin^2(\gamma)]\mathrm{d}y\,. \tag{1.93}$$

The difference between $\hat{u}_r$ and $\hat{u}_\vartheta$, given in (1.93) and in (1.86)–(1.87), is as follows: in the latter expression, Q_0 is replaced with q_0, and coordinate r is replaced by $\sqrt{x^2+y^2}$. In accordance with (1.93), terms that contain solutions q_j are combined yielding the multiplier

$$\sin^2(\gamma)\mathrm{J}_2(k\sqrt{x^2+y^2})-\frac{\mathrm{J}_1(k\sqrt{x^2+y^2})}{k\sqrt{x^2+y^2}}\,. \tag{1.94}$$

Integration of the last expression with respect to y by using (1.89) leads to a zero result. Collecting terms that contain function p_1 and integrating with respect to y, while keeping in mind (1.89), yields

$$u_y(x,z)=\frac{q_0}{G_{rz}(z_0)\pi}\int_0^\infty \frac{kp_1(z,k)}{F}\left\{\int_0^\infty\left[\cos^2(\gamma)\mathrm{J}_2(k\sqrt{x^2+y^2})\right.\right.$$

$$\left.\left.-\frac{\mathrm{J}_1(k\sqrt{x^2+y^2})}{k\sqrt{x^2+y^2}}\mathrm{d}\,y\right]\mathrm{d}\,y\right\}\mathrm{d}\,k=-\frac{q_0}{G_{rz}(z_0)\pi}\int_0^\infty \frac{p_1(z,k)}{F}\cos(kx)\,\mathrm{d}\,k\,. \tag{1.95}$$

Thus, the considered shear vibrations are determined only by the function $p_1(z,k)$ that satisfies equation (1.24).

Next, we consider a constant horizontal load q_0 (Fig. 1.4c), which is directed parallel to the *X*-axis and applied in a straight line parallel to the *Y*-axis (in-plane vibrations). In analogy with equation (1.93), we obtain expressions for the amplitudes of vibrations along the axes *X* and *Z* of the points that belong to the plane (*X*, *Z*):

$$u_x(x,z)=2\int_0^\infty[-\hat{u}_\vartheta(\sqrt{x^2+y^2},z)\sin^2(\gamma)+\hat{u}_r(\sqrt{x^2+y^2},z)\cos^2(\gamma)]\,\mathrm{d}\,y\,, \tag{1.96}$$

$$u_z(x,z)=2\int_0^\infty \hat{u}_z(\sqrt{x^2+y^2},z)\cos(\gamma)\,\mathrm{d}\,y\,. \tag{1.97}$$

The note following equation (1.93) remains valid for the values $\hat{u}_r$, $\hat{u}_z$ and $\hat{u}_\vartheta$. The transformation of the expression for u_x is performed by using integral (1.89), similarly to the value u_y in equation (1.93). Now, terms containing the function p_1 vanish (they gain multiplier (1.94)), while integration with respect to y of terms containing q_1 and q_2 leads to the following result (similarly to (1.95)):

$$u_x(x,z)=\frac{q_0}{G_{rz}(z_0)\pi}\int_0^\infty \frac{c_{12}q_1(z,k)-c_{11}q_2(z,k)}{D}\cos(kx)\,\mathrm{d}\,k\,. \tag{1.98}$$

Amplitudes u_z from (1.97) are transformed according to (1.88), (1.90):

$$u_z(x,z)=-\frac{q_0}{G_{rz}(z_0)\pi}\int_0^\infty \frac{c_{12}w_1(z,k)-c_{11}w_2(z,k)}{D}\sin(kx)\,\mathrm{d}\,k\,. \tag{1.99}$$

A relation between solutions of 2-D and 3-D problems has been discussed in the literature for the case of a vertical load (e.g. in [39]); in this case, the structures of solutions are rather close to each other (compare solutions (1.57) and (1.91)). In the case with horizontal loads, separate parts of the 3-D solution correspond to 2-D solutions, while Bessel's functions are replaced with trigonometric functions, as for the case with vertical loads. Fundamental solutions $q_j(z,k), w_j(z,k), p_1(z,k)$ that satisfy the system of equations (1.24)–(1.26) remain valid for both 2-D and 3-D problems.

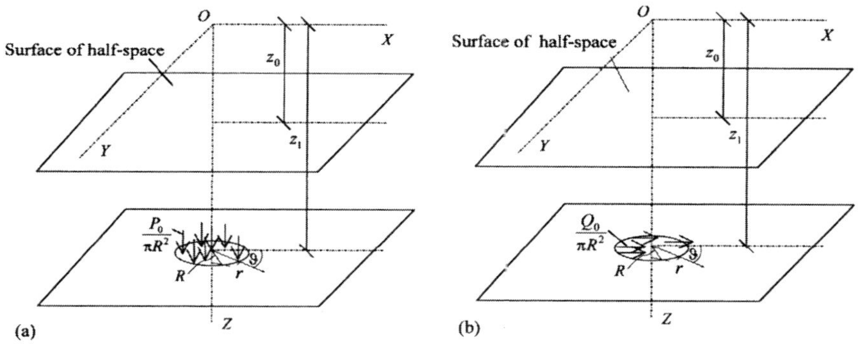

Fig.1.5a,b. Harmonic vertical (a) and horizontal (b) forces applied to circular area below half-space surface

1.5 General Solution for Harmonic Vibrations of Half-Space Subjected to Loads Applied below its Surface

In the following, we consider harmonic forces applied to a heterogeneous transversely isotropic half-space at some depth below the surface. First, the force is assumed to be distributed uniformly over a circle of radius R in a horizontal plane, and second, we consider the concentrated forces.

1.5.1 Action of Vertical Force

Let the vertical force, $P_0 \exp(\mathrm{i}\omega t)$, be applied at depth H within a heterogeneous transversely isotropic half-space (Fig. 1.5a). As for the case of a load applied at the surface of a half-space, we take $n = 0$, and the upper line in $\mathbf{\Gamma}_j$. For the parts of the half-space above and below the plane $z = z_1 = z_0 + H$, representations (1.49) may be applied, and functions $q(z,k), w(z,k)$ must be continuous when crossing the plane where the load is applied. Furthermore, equation (1.42b), which links the tangential stresses in some plane and functions $q(z,k), w(z,k)$, shows that, in addition, the derivative $\mathrm{d}q/\mathrm{d}z$ must be continuous. Application of condition (1.40) for the planes $z = z_1 + 0$ and $z = z_1 - 0$ (infinitely close to the plane $z = z_1$, both above and below it), and taking into account the type of acting load, lead to the conclusion that the discontinuity in derivative $\mathrm{d}w/\mathrm{d}z$ satisfies the equation

$$\frac{\mathrm{d}\,w(z_1+0,k)}{\mathrm{d}\,z} - \frac{\mathrm{d}\,w(z_1-0,k)}{\mathrm{d}\,z} = -b_w\,, \tag{1.100}$$

where

$$b_w = \frac{P_0}{\pi R A_{zz}(z_1)} \mathrm{J}_1(kR) . \tag{1.101}$$

At the surface of a half-space ($z = z_0$) where stresses are absent, the boundary condition takes the following form:

$$\widetilde{A}_{rz} kq + \widetilde{A}_{zz} \frac{\mathrm{d}w}{\mathrm{d}z} = 0 , \tag{1.102a}$$

$$\frac{\mathrm{d}q}{\mathrm{d}z} - kw = 0 . \tag{1.102b}$$

In addition to the previously mentioned conditions, the condition of absence of sources at infinity must hold, similarly to the above-considered case with loads applied to the surface of a half-space. We construct a particular solution $q_a(z,k), w_a(z,k)$ that satisfies all the required conditions, excluding conditions (1.102). For the part of a half-space $z > z_1$ (domain II), we assume this solution to be equal to zero. Employing the conditions of continuity on passing through the plane $z = z_1$, we obtain the following initial conditions for the rest of the half-space ($z < z_1$) (domain I):

$$q_a(z_1,k) = 0 , \tag{1.103a}$$

$$w_a(z_1,k) = 0 , \tag{1.103b}$$

$$\frac{\mathrm{d}q_a(z_1,k)}{\mathrm{d}z} = 0 , \tag{1.103c}$$

$$\frac{\mathrm{d}w_a(z_1,k)}{\mathrm{d}z} = b_w . \tag{1.103d}$$

Employing 4 fundamental solutions $q_j(z,k)$, $w_j(z,k)$ $(j = 1,...,4)$ of the system of differential equations (1.25), (1.26), we express the particular solution in domain I in the form of the following linear combination:

$$\begin{Bmatrix} q_a \\ w_a \\ \dfrac{\mathrm{d}q_a}{\mathrm{d}z} \\ \dfrac{\mathrm{d}w_a}{\mathrm{d}z} \end{Bmatrix} = b_w \sum_{j=1}^{4} B_j \begin{Bmatrix} q_j \\ w_j \\ \dfrac{\mathrm{d}q_j}{\mathrm{d}z} \\ \dfrac{\mathrm{d}w_j}{\mathrm{d}z} \end{Bmatrix} , \tag{1.104}$$

where the coefficients $B_j = B_j(k)$ satisfy the system of equations corresponding to conditions (1.103):

$$\sum_{j=1}^{4} B_j \begin{Bmatrix} q_j(z_1,k) \\ w_j(z_1,k) \\ \dfrac{\mathrm{d}\, q_j(z_1,k)}{\mathrm{d}\, z} \\ \dfrac{\mathrm{d}\, w_j(z_1,k)}{\mathrm{d}\, z} \end{Bmatrix} = \begin{Bmatrix} 0 \\ 0 \\ 0 \\ 1 \end{Bmatrix}. \tag{1.105}$$

In the following, we assume, as previously, that the first two fundamental solutions satisfy the condition of absence of sources at infinity.

Introducing two additional coefficients $A_1(k)$, $A_2(k)$, we present functions $q(z,k)$, $w(z,k)$ and their derivatives in the following form:

$$\begin{Bmatrix} q \\ w \\ \dfrac{\mathrm{d}\, q}{\mathrm{d}\, z} \\ \dfrac{\mathrm{d}\, w}{\mathrm{d}\, z} \end{Bmatrix} = \sum_{j=1}^{2} A_j \begin{Bmatrix} q_j \\ w_j \\ \dfrac{\mathrm{d}\, q_j}{\mathrm{d}\, z} \\ \dfrac{\mathrm{d}\, w_j}{\mathrm{d}\, z} \end{Bmatrix} + \begin{Bmatrix} q_a \\ w_a \\ \dfrac{\mathrm{d}\, q_a}{\mathrm{d}\, z} \\ \dfrac{\mathrm{d}\, w_a}{\mathrm{d}\, z} \end{Bmatrix}, \tag{1.106}$$

where the last term vanishes for $z > z_1$. Coefficients A_j are found from the boundary condition (1.102) for the surface of the half-space. As a result, we obtain a system of equations which differs from (1.53) in its right sides only:

$$d_1 = -\left[\widetilde{A}_{rz}(z_0) k q_a(z_0,k) + \widetilde{A}_{zz}(z_0) \frac{\mathrm{d}\, w_a(z_0,k)}{\mathrm{d}\, z} \right], \tag{1.107a}$$

$$d_2 = -\left[\frac{\mathrm{d}\, q_a(z_0,k)}{\mathrm{d}\, z} - k w_a(z_0,k) \right]. \tag{1.107b}$$

As seen from equations (1.103), values (1.107) have the following limits when $z_1 \to z_0$:

$$d_1 \to -b_w \widetilde{A}_{zz}(z_0) = -\frac{P_0}{\pi R G_{rz}(z_0)} \mathrm{J}_1(kR), \tag{1.108a}$$

$$d_2 \to 0. \tag{1.108b}$$

Here, the right sides are equal to those of system (1.53) which determines coefficients A_j for the case of a vertical force applied to the surface of a half-space.

Amplitudes of vibrations of the points within the half-space may be expressed as the following integrals:

$$u_r = \int_0^{\infty} [A_1(k)q_1(z,k) + A_2(k)q_2(z,k) + q_a(z,k)]\,\mathrm{J}_1(kr)\,\mathrm{d}\,k$$

$$= \frac{P_0}{A_{zz}(z_1)\pi R} \int_0^{\infty} \mathrm{J}_1(kR)\,\mathrm{J}_1(kr)[A_1^*(k)q_1(z,k) + A_2^*(k)q_2(z,k) + q_a^*(z,k)]\,\mathrm{d}\,k\,, \tag{1.109a}$$

$$u_z = \int_0^{\infty} [A_1(k)w_1(z,k) + A_2(k)w_2(z,k) + w_a(z,k)]\,\mathrm{J}_0(kr)\,\mathrm{d}\,k =$$

$$= \frac{P_0}{A_{zz}(z_1)\pi R} \int_0^{\infty} \mathrm{J}_1(kR)\,\mathrm{J}_0(kr)[A_1^*(k)w_1(z,k) + A_2^*(k)w_2(z,k) + w_a^*(z,k)]\,\mathrm{d}\,k, \tag{1.109b}$$

where the asterisk means that the particular solution is now determined by using (1.104), but without the multiplier before the sum. In addition, this solution is used to determine the right sides in (1.107), and the corresponding solution of system (1.53) is denoted as $A_j^*(k)$. Here, it becomes possible to take the multiplier b_w out of the parentheses.

1.5.2 Action of Horizontal Force

Next, we consider a horizontal force $Q_0 \exp(\mathrm{i}\,\omega t)$ applied at depth H within a heterogeneous transversely isotropic half-space (Fig. 1.5b). For the parts of the half-space below and above the plane $z = z_1 = z_0 + H$, we apply representations (1.76); functions $q(z,k), w(z,k), p(z,k)$ are required to be continuous on passing through the plane where the load is applied. Determination of $q(z,k), w(z,k)$ is almost the same as for the case with the vertical force. Equation (1.40), which links normal stresses in some plane and functions $q(z,k), w(z,k)$, shows that, in addition, derivative $\mathrm{d}\,w/\mathrm{d}\,z$ must be continuous. By using equation (1.42b) (at n = 1) for the planes $z = z_1 + 0$ and $z = z_1 - 0$, and taking into account the type of acting load, we find that the jump in derivative $\mathrm{d}\,q/\mathrm{d}\,z$ will be

$$\frac{\mathrm{d}\,q(z_1+0,k)}{\mathrm{d}\,z} - \frac{\mathrm{d}\,q(z_1-0,k)}{\mathrm{d}\,z} = b_q\,, \tag{1.110}$$

where the analogy with equation (1.79b) yields

$$b_q = \frac{Q_0}{G_{rz}(z_1)\pi R}\,\mathrm{J}_1(kR)\,. \tag{1.111}$$

As for the case of the vertical force, we construct the particular solution which is equal to zero in domain II, and has the following form in domain I (see (1.104)):

$$\begin{Bmatrix} q_a \\ w_a \\ \dfrac{\mathrm{d} q_a}{\mathrm{d} z} \\ \dfrac{\mathrm{d} w_a}{\mathrm{d} z} \end{Bmatrix} = b_q \sum_{j=1}^{4} B_j \begin{Bmatrix} q_j \\ w_j \\ \dfrac{\mathrm{d} q_j}{\mathrm{d} z} \\ \dfrac{\mathrm{d} w_j}{\mathrm{d} z} \end{Bmatrix}, \tag{1.112}$$

where the coefficients $B_j = B_j(k)$ satisfy the following system of equations (similar to (1.105)):

$$\sum_{j=1}^{4} B_j \begin{Bmatrix} q_j \\ w_j \\ \dfrac{\mathrm{d} q_j}{\mathrm{d} z} \\ \dfrac{\mathrm{d} w_j}{\mathrm{d} z} \end{Bmatrix} = \begin{Bmatrix} 0 \\ 0 \\ -1 \\ 0 \end{Bmatrix}. \tag{1.113}$$

Equation (1.106), representing the functions q, w and their derivatives, remains unchanged, as well as the system of equations (1.53) with right sides (1.107).

Next, we seek the part of solution expressed through the function p. In accordance with equation (1.42a) and with the type of applied load, we derive a relationship, which controls the discontinuity in the derivative of the function p, on passing through the plane $z = z_1$:

$$\frac{\mathrm{d}\, p(z_1+0)}{\mathrm{d} z} - \frac{\mathrm{d}\, p(z_1-0)}{\mathrm{d} z} = b_p, \tag{1.114}$$

where analogously to equation (1.78)

$$b_p = \frac{Q_0}{\pi R G_{rz}(z_1)} \mathrm{J}_1(kR) = b_q. \tag{1.115}$$

At $z = z_0$, where no stresses exist, the derivative of the function $p(z,k)$ with respect to z must be equal to zero, and, in addition, a condition of absence of sources as $z \to \infty$ should hold. Our intention is to construct a particular solution p_a, which is equal to zero in domain II ($z > z_1$) and satisfies condition (1.114). In the lower point of the domain $z < z_1$, function p_a is equal to zero, and its derivative with respect to z equals $-b_p$. Employing two fundamental solutions of equation (1.24), we present the solution p_a for domain I in the following form:

$$p_a = b_p(C_1 p_1 + C_2 p_2). \tag{1.116}$$

Coefficients $C_j = C_j(k)$ satisfy the following system of equations expressing the continuity of the function p_a and the behavior of its derivative, as it passes through the plane $z = z_1$:

$$p_1(z_1,k)C_1 + p_2(z_1,k)C_2 = 0, \tag{1.117a}$$

$$\frac{\mathrm{d}\,p_1(z_1,k)}{\mathrm{d}\,z}C_1 + \frac{\mathrm{d}\,p_2(z_1,k)}{\mathrm{d}\,z}C_2 = -1. \tag{1.117b}$$

Let the fundamental solution p_1 satisfy the condition of absence of sources at infinity. Then, an expression for the function p which satisfies all the required criteria is taken in the following form:

$$p = C(k)p_1 + p_a, \tag{1.118}$$

where an additional coefficient $C(k)$ is introduced. This coefficient may be calculated from the condition of absence of stresses at the surface of a half-space:

$$C(k) = -\frac{\mathrm{d}\,p_a(z_0,k)}{\mathrm{d}\,z}\bigg/\frac{\mathrm{d}\,p_1(z_0,k)}{\mathrm{d}\,z}. \tag{1.119}$$

Finally, we obtain expressions for the amplitudes of vibrations of the half-space subjected to the load shown in Fig. 1.5b:

$$u_r = \hat{u}_r \cos\vartheta,\ u_\vartheta = \hat{u}_\vartheta \sin\vartheta,\ u_z = \hat{u}_z \cos\vartheta, \tag{1.120a}$$

$$\hat{u}_r = \frac{Q_0}{\pi R G_{rz}(z_1)}\int_0^\infty\left[[q^*(z,k) - p^*(z,k)]\frac{\mathrm{J}_1(kr)}{kr} - q^*(z,k)\mathrm{J}_0(kr)\right]\mathrm{J}_1(kR)\mathrm{d}\,k, \tag{1.120b}$$

$$\hat{u}_\vartheta = \frac{Q_0}{\pi R G_{rz}(z_1)}\int_0^\infty\left[[q^*(z,k) - p^*(z,k)]\frac{\mathrm{J}_1(kr)}{kr} + p^*(z,k)\mathrm{J}_0(kr)\right]\mathrm{J}_1(kR)\mathrm{d}\,k, \tag{1.120c}$$

$$\hat{u}_z = \frac{Q_0}{\pi R G_{rz}(z_1)}\int_0^\infty w^*(z,k)\mathrm{J}_1(kr)\mathrm{J}_1(kR)\mathrm{d}\,k, \tag{1.120d}$$

where

$$q^*(z,k) = A_1^*(k)q_1(z,k) + A_2^*(k)q_2(z,k) + q_a^*(z,k), \tag{1.121a}$$

$$w^*(z,k) = A_1^*(k)w_1(z,k) + A_2^*(k)w_2(z,k) + w_a^*(z,k), \tag{1.121b}$$

$$p^*(z,k) = C^*(k)p_1(z,k) + p_a^*(z,k). \tag{1.121c}$$

Here, the asterisk means that the particular solutions (denoted with subscript a) are determined following (1.112) and (1.116), but without multipliers b_q and b_p, respectively (this leads, in particular, to the conclusion that the derivative of the function p_a^* with respect to z at $z = z_1 - 0$ is equal to -1). These transformed particular solutions are used in the right sides (1.107) of the system of equations (1.53) in order to calculate coefficients $A_j^*(k)$ and in equation (1.119) for the calculation of $C^*(k)$.

As shown previously, a transition to the case of concentrated force in solutions (1.109), (1.120) is performed by replacing $\mathrm{J}_1(kR)/R$ with $k/2$.

1.6 Application of Functions Related to Dilatation and Rotation of Displacement Field

Calculation of dilatation e corresponding to particular solution (1.17) yields

$$e = \left[kq(z,k) + \frac{\mathrm{d}\, w(z,k)}{\mathrm{d}\, z} \right] \mathrm{J}_n(\eta)\boldsymbol{\Gamma}_1 . \tag{1.122}$$

Considering components of those parts of the field's rotor, which are generated due to functions q and w, we note that the z-component is equal to zero, while the components lying in horizontal planes are proportional to the value

$$kw(z,k) + \frac{\mathrm{d}\, q(z,k)}{\mathrm{d}\, z} . \tag{1.123}$$

In some cases, in order to solve vibration problems in the heterogeneous half-space, it is reasonable to introduce functions related to (1.122) and (1.123) as follows:

$$\overline{e} = A_{zz}(z) \left[kq(z,k) + \frac{\mathrm{d}\, w(z,k)}{\mathrm{d}\, z} \right], \tag{1.124}$$

$$\overline{\chi} = G_{rz}(z) \left[kw(z,k) + \frac{\mathrm{d}\, q(z,k)}{\mathrm{d}\, z} \right]. \tag{1.125}$$

Use of the function $\overline{e}$ is especially convenient in the case of an incompressible medium, when coefficient A_{zz} tends to infinity (as for an isotropic medium with Poisson's ratio equal to 0.5), and dilatation tends to zero; function $\overline{e}$ remains limited. By using functions (1.124), (1.125), the system of two second-order equations (1.25) and (1.26) may be transformed to the following system of first-order equations:

$$\frac{\mathrm{d}\,\overline{\chi}}{\mathrm{d}\, z} = \frac{k\overline{e}}{A_{zz}}(2G_{rz} + A_{rz}) + 2k\frac{\mathrm{d}\, G_{rz}}{\mathrm{d}\, z} w + k^2 q(A_{rr} - A_{rz} - 2G_{rz}) - \rho\omega^2 q , \tag{1.126}$$

$$\frac{\mathrm{d}\,\bar{e}}{\mathrm{d}\,z} = k\bar{\chi}\frac{A_{zz} - G_{rz} - A_{rz}}{G_{rz}} + k\left[\frac{\mathrm{d}\,A_{zz}}{\mathrm{d}\,z} - \frac{\mathrm{d}\,A_{rz}}{\mathrm{d}\,z}\right]q + k^2 w(A_{rz} - A_{zz} + 2G_{rz}) - \rho\omega^2 w\,, \tag{1.127}$$

$$\frac{\mathrm{d}\,q}{\mathrm{d}\,z} = \frac{\bar{\chi}}{G_{rz}} - kw\,, \tag{1.128}$$

$$\frac{\mathrm{d}\,w}{\mathrm{d}\,z} = \frac{\bar{e}}{A_{zz}} - kq\,. \tag{1.129}$$

For the isotropic case, this system of equations is simplified following relationships (1.5):

$$\frac{\mathrm{d}\,\bar{\chi}}{\mathrm{d}\,z} = k\bar{e} + 2k\frac{\mathrm{d}\,G}{\mathrm{d}\,z}w - \rho\omega^2 q\,, \tag{1.130}$$

$$\frac{\mathrm{d}\,\bar{e}}{\mathrm{d}\,z} = k\bar{\chi} + 2k\frac{\mathrm{d}\,G}{\mathrm{d}\,z}q - \rho\omega^2 w\,, \tag{1.131}$$

$$\frac{\mathrm{d}\,q}{\mathrm{d}\,z} = \frac{\bar{\chi}}{G} - kw\,, \tag{1.132}$$

$$\frac{\mathrm{d}\,w}{\mathrm{d}\,z} = \frac{\tau^2\bar{e}}{G} - kq\,, \tag{1.133}$$

where

$$\tau^2 = \frac{G}{\lambda + 2G} = \frac{1 - 2\nu}{2(1-\nu)} = \frac{C_s^2}{C_p^2}\,. \tag{1.134}$$

Here, ν is Poisson's ratio, C_s and C_p are calculated from (1.21).
In addition to the dimensionless variables given in (1.27), we introduce

$$\tilde{\chi} = \frac{\bar{\chi}}{G_{rz}(z_0)}\,, \tag{1.135a}$$

$$e = \frac{\bar{e}}{G_{rz}(z_0)}\,. \tag{1.135b}$$

Equations (1.126)–(1.129) may be presented in the following form:

$$\frac{\mathrm{d}\,\tilde{\chi}}{\mathrm{d}\,\tilde{z}} = \frac{\tilde{k}\tilde{e}}{\tilde{A}_{zz}}(2\tilde{G} + \tilde{A}_{rz}) + 2\tilde{k}\frac{\mathrm{d}\,\tilde{G}}{\mathrm{d}\,\tilde{z}}\tilde{w} + \tilde{k}^2\tilde{q}(\tilde{A}_{rr} - \tilde{A}_{rz} - 2\tilde{G}) - \tilde{\rho}\beta^2\theta^2\tilde{q}\,, \tag{1.136}$$

$$\frac{\mathrm{d}\,\tilde{e}}{\mathrm{d}\,\tilde{z}} = \frac{\tilde{k}\tilde{\chi}}{\tilde{G}}(\tilde{A}_{zz} - \tilde{G} - \tilde{A}_{rz}) + \tilde{k}\left[\frac{\mathrm{d}\,\tilde{A}_{zz}}{\mathrm{d}\,\tilde{z}} - \frac{\mathrm{d}\,\tilde{A}_{rz}}{\mathrm{d}\,\tilde{z}}\right]\tilde{q} + \tilde{k}^2\tilde{w}(\tilde{A}_{rz} - \tilde{A}_{zz} + 2\tilde{G}) - \tilde{\rho}\beta^2\theta^2\tilde{w}, \tag{1.137}$$

$$\frac{d\tilde{q}}{d\tilde{z}} = \frac{\tilde{\chi}}{\tilde{G}(\tilde{z})} - \tilde{k}\tilde{w}, \tag{1.138}$$

$$\frac{d\tilde{w}}{d\tilde{z}} = \frac{\tilde{e}}{\tilde{A}_{zz}} - \tilde{k}\tilde{q}, \tag{1.139}$$

where the value of β related to the energy dissipation is determined by (1.32b). The equations for dimensionless values for the isotropic case become

$$\frac{d\tilde{\chi}}{d\tilde{z}} = \tilde{k}\tilde{e} + 2\tilde{k}\frac{d\tilde{G}}{d\tilde{z}}\tilde{w} - \tilde{\rho}\beta^2\theta^2\tilde{q}, \tag{1.140}$$

$$\frac{d\tilde{e}}{d\tilde{z}} = \tilde{k}\tilde{\chi} + 2\tilde{k}\frac{d\tilde{G}}{d\tilde{z}}\tilde{q} - \tilde{\rho}\beta^2\theta^2\tilde{w}, \tag{1.141}$$

$$\frac{d\tilde{q}}{d\tilde{z}} = \frac{\tilde{\chi}}{\tilde{G}(\tilde{z})} - \tilde{k}\tilde{w}, \tag{1.142}$$

$$\frac{d\tilde{w}}{d\tilde{z}} = \frac{\tau^2\tilde{e}}{\tilde{G}(\tilde{z})} - \tilde{k}\tilde{q}. \tag{1.143}$$

Conditions (1.40), (1.42b) that relate stresses in horizontal planes through functions q, w, are transformed in terms of functions $\tilde{\chi}$, $\tilde{e}$. Then, instead of (1.40), (1.42b) we have:

$$\tilde{e} - (\tilde{A}_{zz}(z_\sigma) - \tilde{A}_{rz}(z_\sigma))\tilde{k}\tilde{q} = \frac{k}{G_{rz}(z_0)}\int_0^\infty r\hat{\sigma}_z(r)\mathrm{J}_n(kr)\,\mathrm{d}r, \tag{1.144}$$

$$\tilde{\chi} - 2\tilde{k}\tilde{G}(\tilde{z}_\sigma)\tilde{w} = \frac{k}{2G_{rz}(z_0)}\left[\int_0^\infty r[\hat{\tau}_{rz}(r) + \hat{\tau}_{\vartheta z}(r)]\mathrm{J}_{n+1}(kr)\,\mathrm{d}r - \int_0^\infty r[\hat{\tau}_{rz}(r) - \hat{\tau}_{\vartheta z}(r)]\mathrm{J}_{n-1}(kr)\,\mathrm{d}r\right]. \tag{1.145}$$

Coefficients c_{ij} determined by (1.54) and entering solutions (1.55), (1.57), (1.58), (1.83)–(1.88) become (note that at the surface of a half-space, $\tilde{G}(z_0) = 1$)

$$c_{1j} = \tilde{e}_j(\tilde{z}_0, \tilde{k}) - [\tilde{A}_{zz}(z_0) - \tilde{A}_{rz}(z_0)]\tilde{k}\tilde{q}_j(\tilde{z}_0, \tilde{k}), \tag{1.146a}$$

$$c_{2j} = \tilde{\chi}_j(\tilde{z}_0, \tilde{k}) - 2\tilde{k}\tilde{w}_j(\tilde{z}_0, \tilde{k}). \tag{1.146b}$$

For the isotropic case

$$c_{1j} = \tilde{e}_j(\tilde{z}_0, \tilde{k}) - 2\tilde{k}\tilde{q}_j(\tilde{z}_0, \tilde{k}), \tag{1.147a}$$

$$c_{2j} = \widetilde{\chi}_j(\widetilde{z}_0, \widetilde{k}) - 2\widetilde{k}\widetilde{w}_j(\widetilde{z}_0, \widetilde{k}) . \tag{1.147b}$$

Thus, solutions of the vibration problems for the half-space considered in section 1.4 may be rewritten by using new expressions for coefficients c_{ij}.

1.6.1 Vertical Force Applied within Half-Space

Next, we consider the application of functions $\widetilde{\chi}$, $\widetilde{e}$ in the formulation of problems dealing with vibrations of a half-space subjected to loads applied within the half-space. In analogy with equation (1.100), we write the following equation that determines the discontinuity of function $\widetilde{e}$ when passing through the plane $z = z_1$ of acting the vertical load:

$$\widetilde{e}(z_1 + 0, k) - \widetilde{e}(z_1 - 0, k) = -b_e , \tag{1.148}$$

where

$$b_e = \frac{P_0}{\pi R G_{rz}(z_0)} \mathrm{J}_1(kR) . \tag{1.149}$$

Functions $\widetilde{\chi}, \widetilde{q}, \widetilde{w}$ must be continuous. The system of equations (1.102), which expresses the absence of stresses at the surface of a half-space, may be presented as

$$\widetilde{e} - (\widetilde{A}_{zz} - \widetilde{A}_{rz})\widetilde{k}\widetilde{q} = 0 , \tag{1.150}$$

$$\widetilde{\chi} - 2\widetilde{k}\widetilde{w} = 0 . \tag{1.151}$$

Similarly to the description presented in section 1.5.1, we construct a particular solution $\widetilde{\chi}_a, \widetilde{e}_a, \widetilde{q}_a, \widetilde{w}_a$ of the system of equations (1.136)–(1.139) or (1.140)–(1.143), which equals zero at $z > z_1$ and satisfies the condition of passing through the plane $z = z_1$. For $z < z_1$, this particular solution is constructed by taking into account the following initial data at $z = z_1$:

$$\begin{aligned} &\widetilde{\chi}_a(\widetilde{z}_1, \widetilde{k}) = 0, \\ &\widetilde{e}_a(\widetilde{z}_1, \widetilde{k}) = b_e , \\ &\widetilde{q}_a(\widetilde{z}_1, \widetilde{k}) = 0, \\ &\widetilde{w}_a(\widetilde{z}_1, \widetilde{k}) = 0 . \end{aligned} \tag{1.152}$$

Employing fundamental solutions $\widetilde{\chi}_j, \widetilde{e}_j, \widetilde{q}_j, \widetilde{w}_j$ (j = 1,...,4) we present a particular solution in the form similar to equation (1.104):

$$\begin{Bmatrix} \tilde{\chi}_a \\ \tilde{e}_a \\ \tilde{q}_a \\ \tilde{w}_a \end{Bmatrix} = b_e \sum_{j=1}^{4} B_j \begin{Bmatrix} \tilde{\chi}_j \\ \tilde{e}_j \\ \tilde{q}_j \\ \tilde{w}_j \end{Bmatrix}, \tag{1.153}$$

where the coefficients B_j satisfy the following system of equations:

$$\sum_{j=1}^{4} B_j \begin{Bmatrix} \tilde{\chi}_j(\tilde{z}_1) \\ \tilde{e}_j(\tilde{z}_1) \\ \tilde{q}_j(\tilde{z}_1) \\ \tilde{w}_j(\tilde{z}_1) \end{Bmatrix} = \begin{Bmatrix} 0 \\ 1 \\ 0 \\ 0 \end{Bmatrix}. \tag{1.154}$$

A complete solution becomes

$$\begin{Bmatrix} \tilde{\chi} \\ \tilde{e} \\ \tilde{q} \\ \tilde{w} \end{Bmatrix} = \sum_{j=1}^{2} A_j \begin{Bmatrix} \tilde{\chi}_j \\ \tilde{e}_j \\ \tilde{q}_j \\ \tilde{w}_j \end{Bmatrix} + \begin{Bmatrix} \tilde{\chi}_a \\ \tilde{e}_a \\ \tilde{q}_a \\ \tilde{w}_a \end{Bmatrix}. \tag{1.155}$$

Here, the coefficients A_j must satisfy the system of equations of form (1.53) with coefficients c_{ij} according to (1.146), (1.147), and with the following right sides:

$$d_1 = -[\tilde{e}_a(\tilde{z}_0, \tilde{k}) - (\tilde{A}_{zz} - \tilde{A}_{rz})\tilde{k}\tilde{q}_a(\tilde{z}_0, \tilde{k})], \tag{1.156a}$$

$$d_2 = -[\tilde{\chi}_a(\tilde{z}_0, \tilde{k}) - 2\tilde{k}\tilde{w}_a(\tilde{z}_0, \tilde{k})]. \tag{1.156b}$$

For the isotropic case, $\tilde{A}_{zz} - \tilde{A}_{rz} = 2$.

Following equation (1.153), a modified particular solution (denoted with an asterisk) may be considered. This solution is derived by replacing a multiplier b_e with unity, in order to take this multiplier out of the parentheses in the representation of the complete solution (1.155). The modified particular solution has a component $\tilde{e}_a^* = 1$ at $z = z_1 - 0$, while the rest of the components are equal to zero. This solution is also used for calculation of values (1.156), which contribute to the determination of coefficients A_{ij} (also denoted further with an asterisk) from the system of equations (1.53). As a result, a solution of the problem for the case of vertical force acting within a half-space may be presented in the following form:

$$u_r = \frac{P_0}{G_{rz}(z_0)\pi R} \int_0^\infty \mathrm{J}_1(\tilde{k}\tilde{R})\,\mathrm{J}_1(\tilde{k}\tilde{r})[A_1^*(\tilde{k})\tilde{q}_1(\tilde{z}, \tilde{k})$$

$$+ A_2^*(\tilde{k})\tilde{q}_2(\tilde{z}, \tilde{k}) + \tilde{q}_a^*(\tilde{z}, \tilde{k})]\,\mathrm{d}\tilde{k}, \tag{1.157a}$$

$$u_z = \frac{P_0}{G_{rz}(z_0)\pi R}\int_0^{\infty} \mathrm{J}_1(\widetilde{k}\widetilde{R})\,\mathrm{J}_0(\widetilde{k}\widetilde{r})[A_1^{*}(\widetilde{k})\widetilde{w}_1(\widetilde{z},\widetilde{k})$$

$$+ A_2^{*}(k)\widetilde{w}_2(\widetilde{z},\widetilde{k}) + \widetilde{w}_a^{*}(\widetilde{z},\widetilde{k})]\,\mathrm{d}\widetilde{k}, \tag{1.157b}$$

where

$$\widetilde{r} = \frac{r}{z_r},\ \widetilde{z} = \frac{z}{z_r},\ \widetilde{R} = \frac{R}{z_r}\ . \tag{1.158}$$

When radius R tends to zero, the formulas yield a result corresponding to the concentrated force.

1.6.2 Horizontal Force Applied within Half-Space

Consider a renewed formulation (compared with the one given in section 1.5.2) of the problem of vibrations of a half-space subjected to the action of a horizontal force applied within the half-space. Now, functions $\widetilde{e},\widetilde{q},\widetilde{w}$ are continuous within the half-space, while function $\widetilde{\chi}$ has a discontinuity when passing through the plane of application of the load. Taking into account equation (1.145), we have similarly to (1.110),

$$\widetilde{\chi}(z_1+0,k) - \widetilde{\chi}(z_1-0,k) = b_\chi\,, \tag{1.159}$$

where

$$b_\chi = \frac{Q_0\widetilde{G}(z_1)}{\pi R G_{rz}(z_1)}\mathrm{J}_1(kR) = \frac{Q_0}{\pi R G_{rz}(z_0)}\mathrm{J}_1(kR)\,. \tag{1.160}$$

As previously, we construct a particular solution, which equals zero at $z > z_1$, and at $z < z_1$ has the following form:

$$\begin{Bmatrix} \widetilde{\chi}_a \\ \widetilde{e}_a \\ \widetilde{q}_a \\ \widetilde{w}_a \end{Bmatrix} = b_\chi \sum_{j=1}^{4} B_j \begin{Bmatrix} \widetilde{\chi}_j \\ \widetilde{e}_j \\ \widetilde{q}_j \\ \widetilde{w}_j \end{Bmatrix}. \tag{1.161}$$

In accordance with equation (1.159), coefficients B_j satisfy the following system of equations:

$$\sum_{j=1}^{4} B_j \begin{Bmatrix} \widetilde{\chi}_j(\widetilde{z}_1) \\ \widetilde{e}_j(\widetilde{z}_1) \\ \widetilde{q}_j(\widetilde{z}_1) \\ \widetilde{w}_j(\widetilde{z}_1) \end{Bmatrix} = \begin{Bmatrix} -1 \\ 0 \\ 0 \\ 0 \end{Bmatrix}. \tag{1.162}$$

Representation (1.155) and the right sides (1.156) of the system of equations, which serve for the calculation of coefficients A_j, remain valid.

Function $p(z,k)$, required for the construction of a solution of the problem dealing with a horizontal force, has been given in section 1.5.2. In the following, we apply a dimensionless function $\widetilde{p} = p / z_r$, similarly to the values $\widetilde{q}$ and $\widetilde{w}$.

The form of solution (1.120), (1.121) remains almost unchanged. Dimensionless variables denoted by "~" appear, and the particular solution is now based on application of the functions $\widetilde{\chi}$ and $\widetilde{e}$. As in previous sections, we use a modification of the particular solution, where b_χ is replaced in (1.161) with unity, and the common multiplier b_χ is taken out of all terms in the sum. This multiplier differs from value b_p by (1.115), which results in an additional coefficient for the function $\widetilde{p}$. The solution of the problem may be written as follows:

$$u_r = \hat{u}_r \cos\vartheta,\; u_\vartheta = \hat{u}_\vartheta \sin\vartheta,\; u_z = \hat{u}_z \cos\vartheta, \tag{1.163}$$

$$\hat{u}_r = \frac{Q_0}{\pi R G_{rz}(\widetilde{z}_0)} \int_0^\infty \left\{ \left[\widetilde{q}(\widetilde{z},\widetilde{k}) - \frac{\widetilde{p}(\widetilde{z},\widetilde{k})}{\widetilde{G}(\widetilde{z}_1)} \right] \frac{\mathrm{J}_1(\widetilde{k}\widetilde{r})}{\widetilde{k}\widetilde{r}} - \widetilde{q}(\widetilde{z},\widetilde{k}) \mathrm{J}_0(\widetilde{k}\widetilde{r}) \right\} \mathrm{J}_1(\widetilde{k}\widetilde{R}) \mathrm{d}\widetilde{k}, \tag{1.164a}$$

$$\hat{u}_\vartheta = \frac{Q_0}{\pi R G_{rz}(\widetilde{z}_0)} \int_0^\infty \left\{ \left[\widetilde{q}(\widetilde{z},\widetilde{k}) - \frac{\widetilde{p}(\widetilde{z},\widetilde{k})}{\widetilde{G}(\widetilde{z}_1)} \right] \frac{\mathrm{J}_1(\widetilde{k}\widetilde{r})}{\widetilde{k}\widetilde{r}} + \frac{\widetilde{p}(\widetilde{z},\widetilde{k})}{\widetilde{G}(\widetilde{z}_1)} \mathrm{J}_0(\widetilde{k}\widetilde{r}) \right\} \mathrm{J}_1(\widetilde{k}\widetilde{R}) \mathrm{d}\widetilde{k}, \tag{1.164b}$$

$$\hat{u}_z = \frac{Q_0}{\pi R G_{rz}(\widetilde{z}_0)} \int_0^\infty \widetilde{w}(\widetilde{z},\widetilde{k}) \mathrm{J}_1(\widetilde{k}\widetilde{r}) \mathrm{J}_1(\widetilde{k}\widetilde{R}) \mathrm{d}\widetilde{k}, \tag{1.164c}$$

where

$$\widetilde{q}(\widetilde{z},\widetilde{k}) = A_1^*(\widetilde{k})\widetilde{q}_1(\widetilde{z},\widetilde{k}) + A_2^*(\widetilde{k})\widetilde{q}_2(\widetilde{z},\widetilde{k}) + \widetilde{q}_a^*(\widetilde{z},\widetilde{k}), \tag{1.165a}$$

$$\widetilde{w}(\widetilde{z},\widetilde{k}) = A_1^*(k)\widetilde{w}_1(\widetilde{z},\widetilde{k}) + A_2^*(k)\widetilde{w}_2(\widetilde{z},\widetilde{k}) + \widetilde{w}_a^*(\widetilde{z},\widetilde{k}), \tag{1.165b}$$

$$\widetilde{p}(\widetilde{z},\widetilde{k}) = C^*(\widetilde{k})\widetilde{p}_1(\widetilde{z},\widetilde{k}) + \widetilde{p}_a^*(\widetilde{z},\widetilde{k}). \tag{1.165c}$$

Formulation of the problems, as given in the present section, enables us, in particular, to construct analytical solutions of the dynamic problems for a heterogeneous isotropic half-space with a shear modulus varying linearly with depth.

1.7 Application of Superposition Principle to Loadings of Rectangular and Circular Domains

Isotropy of the studied half-space in horizontal planes enables us to derive simple formulas that express amplitudes of vibrations for various types of load applied to rectangular or circular domains on the surface of a half-space, or in a horizontal plane at some depth. These formulas are based on the integration of Green's functions, i.e. solutions related to the concentrated forces, and a significant part of the transformations may be performed independently of the form of Green's functions. Only the isotropy and homogeneity of the material are required in the horizontal planes. As a result, 2-D integrals become 1-D. The results obtained for various types of distributed load are significant, primarily due to a perspective of their further use for solving contact problems based on the following technique. A domain with an unknown distribution of contact stresses is presented as a set of small rectangular or circular elements with a piecewise uniform distribution of stresses. Unknown intensities of these contact stresses within the elements' boundaries are calculated from the following requirement: displacements of the mean points of the elements should equal the displacements specified in the conditions given for the contact problem. For the case of contact problems with a circular contact domain, a reasonable choice is to use ring-shaped elements with a specific distribution of stresses within the elements. In this case, besides uniform distributions, other types of distributions should be applied, depending on the form of displacements of the circular contact domain as a rigid disk. Below, different distributions are considered for circular domains, which enable us to construct solutions for ring-shaped domains.

For 2-D problems, striped elements, which are placed symmetrically with respect to the plane (Y, Z), are a reasonable choice; the corresponding results are derived from consideration of the loading of an entire strip subjected to various types of loading.

1.7.1 Vertical Load Distributed Uniformly over Rectangular Domain

Let a uniformly distributed vertical load $p_0 \exp(\mathrm{i}\,\omega t)$ be applied to a rectangular domain that belongs to a horizontal plane on the surface of a heterogeneous transversely isotropic half-space, or at some depth H. We introduce Green's functions $w_{vv}(z,r), w_{vh}(z,r)$ for the vertical force with unit amplitude applied to the plane of the specified rectangle (Fig. 1.6a). These functions express the amplitudes of vibrations in the vertical and horizontal directions, respectively, at the point with coordinate z and at a distance r from the line of acting force. For the load applied to the surface of a half-space, Green's functions are given by equations (1.57) (u_z and u_r), while for the load applied at some depth, they are expressed through equations (1.109) at $R = 0$ ($P_0 = 1$). For the considered half-space, which is isotropic and homogeneous in horizontal planes, the contribution of each element of the load depends, besides the coordinate z, on the

(a)

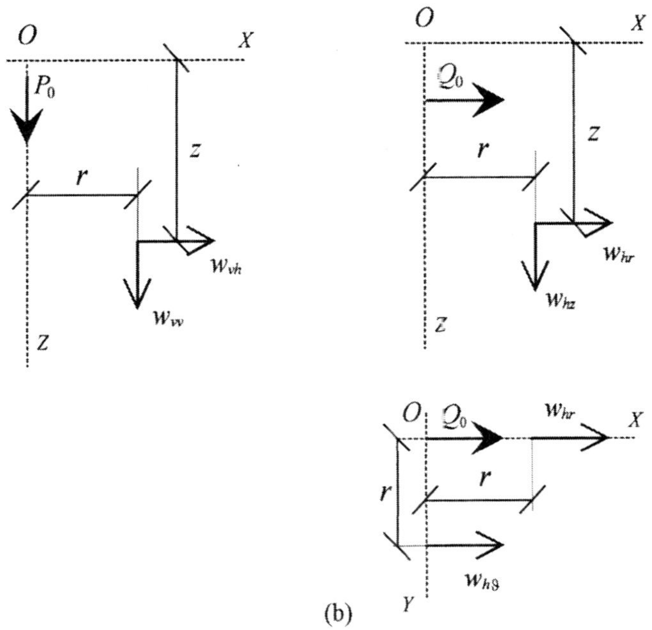

(b)

Fig. 1.6a,b. Green's functions expressing amplitudes of vibrations of half-space points under the action of vertical (a) and horizontal (b) forces of unit amplitude

distance between the vertical lines passing through the element and the studied point. Here, it is reasonable to apply a widely used method of superposition of rectangles [91] based on employing results for a point *A* that belongs to the same vertical line as a corner point of the rectangle (Fig. 1.7). In the following, for the sake of brevity we drop the coordinate *z* in Green's functions and in notations for parameters related to the considered point. Amplitudes of vibrations at the point *A* along axes (*X*, *Y*, *Z*) are presented in the form of integrals over a rectangular domain (*D*), where a load is applied:

$$\delta_x(\mathrm{A}) = p_0 \iint\limits_{(D)} w_{vh}(r)\sin(\psi) r \,\mathrm{d}r\,\mathrm{d}\psi \,, \tag{1.166a}$$

$$\delta_y(\mathrm{A}) = p_0 \iint\limits_{(D)} w_{vh}(r)\cos(\psi) r \,\mathrm{d}r\,\mathrm{d}\psi \,, \tag{1.166b}$$

$$\delta_z(\mathrm{A}) = p_0 \iint\limits_{(D)} w_{vv}(r) r \,\mathrm{d}r\,\mathrm{d}\psi \,. \tag{1.166c}$$

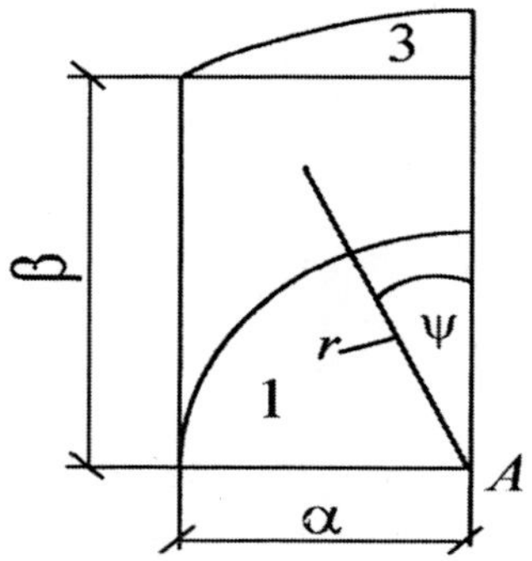

Fig. 1.7. Determination of displacements of half-space point belonging to same vertical as corner point of rectangular area in which load is applied

Next, we introduce an integral which yields, as a particular case, integrals entering relationships (1.166):

$$K_{ij}(\alpha,\beta;l,m) = \iint\limits_{(D)} w_{ij}(r)\sin^l(\psi)\cos^m(\psi) r\,\mathrm{d}r\mathrm{d}\psi\,, \tag{1.167}$$

where $i = v, j = h$ or v; l and m equal unity or zero; here, the dependence on the dimensions of a rectangle, α and β, is specified. When considering the action of a horizontal load in the following, additional values of l and m will be used. In the double integral (1.167), we integrate with respect to variable ψ. Following Fig. 1.7,

$$K_{ij}(\alpha,\beta;l,m) = \int_0^\alpha w_{ij}(r) r\Phi_{lm}(\pi/2)\mathrm{d}r + \int_\alpha^g w_{ij}(r) r\Phi_{lm}[\arcsin(\alpha/r)]\mathrm{d}r$$
$$- \int_\beta^g w_{ij}(r) r\Phi_{lm}[\arccos(\beta/r)]\mathrm{d}r\,, \tag{1.168}$$

where

$$g = (\alpha^2 + \beta^2)^{1/2}, \tag{1.169a}$$

$$\Phi_{lm}(\psi) = \int_0^\psi \sin^l(u)\cos^m(u)du\,. \tag{1.169b}$$

The first integral in expression (1.168) corresponds to integration over domain 1, and the third one over domain 3, which becomes excessive while evaluating the second integral. Expression (1.168) has been derived for the case $\alpha < \beta$, but it is valid for any relationship between these parameters. The amplitudes in equations (1.166) may be represented in the following form:

$$\delta_x(A) = p_0 K_{vh}(\alpha,\beta;1,0)\,, \tag{1.170a}$$

$$K_{vh}(\alpha,\beta;1,0) = \int_0^g w_{vh}(r) r\,\mathrm{d}r - \int_\alpha^g w_{vh}(r)\sqrt{(r^2-\alpha^2)}\,\mathrm{d}r - \int_\beta^g w_{vh}(r)(r-\beta)\,\mathrm{d}r, \tag{1.170b}$$

$$\delta_y(A) = p_0 K_{vh}(\alpha,\beta;0,1)\,, \tag{1.171a}$$

$$K_{vh}(\alpha,\beta;0,1) = \int_0^g w_{vh}(r) r\,\mathrm{d}r - \int_\alpha^g w_{vh}(r)(r-\alpha)\,\mathrm{d}r - \int_\beta^g w_{vh}(r)\sqrt{r^2-\beta^2}\,\mathrm{d}r, \tag{1.171b}$$

$$\delta_z(A) = p_0 K_{vv}(\alpha,\beta;0,0)\,, \tag{1.172a}$$

$$K_{vv}(\alpha,\beta;0,0) = \frac{\pi}{2}\int_0^\alpha w_{vv}(r) r\,\mathrm{d}r + \int_\alpha^g w_{vv}(r) r \arcsin\frac{\alpha}{r}\,\mathrm{d}r$$

$$- \int_\beta^g w_{vv}(r) r \arccos\frac{\beta}{r}\,\mathrm{d}r, \tag{1.172b}$$

Note that according to equations (1.170), (1.171), the following relationship takes place (it is also evident from symmetry):

$$K_{vh}(\alpha,\beta;1,0) = K_{vh}(\beta,\alpha;0,1)\,. \tag{1.173}$$

In order to perform numerical integration (e.g. by using Simpson's formula) following equations (1.170), (1.171), (1.172), one should account for the existence of points where derivatives become infinite (points α and β for radicals and for inverse trigonometric functions). In the vicinity of these points, a smaller integration step should be used.

As an illustrative example, we consider displacements of points that belong to the surface of a homogeneous isotropic half-space under the action of a load applied statically to the surface. For this case, we have [91]

$$w_{vh} = -\frac{1-2\nu}{4G\pi r}\,, \tag{1.174}$$

$$w_{vv} = \frac{1-\nu}{2G\pi r}\,, \tag{1.175}$$

where G and ν are the shear modulus and Poisson's ratio for the half-space material, respectively. Substitution of these expressions into relationships (1.170) – (1.172) and integration yield

$$K_{vh}(\alpha,\beta;1,0) = -\frac{1-2\nu}{4G\pi}\left[\alpha \arccos\frac{\alpha}{g} + \beta \ln\frac{g}{\beta}\right], \tag{1.176}$$

$$K_{vh}(\alpha,\beta;0,1) = -\frac{1-2\nu}{4G\pi}\left[\beta \arccos\frac{\beta}{g} + \alpha \ln\frac{g}{\alpha}\right], \tag{1.177}$$

$$K_{vv}(\alpha,\beta;0,0) = \frac{1-\nu}{2G\pi}\left[\alpha \ln\frac{g+\beta}{\alpha} + \beta \ln\frac{g+\alpha}{\beta}\right]. \tag{1.178}$$

Next, we turn to the derivation of expressions for the amplitudes of vibrations at a point with arbitrary coordinates (x, y) when a load is applied over a rectangle with sizes $2a$ and $2b$. We introduce four additional rectangles in the plane of application of the loaad, so that the considered point and one of the corners of these rectangles belong to the same vertical (Fig. 1.8). Each rectangle contains a diagonal, which passes through one of the corners of the rectangle, to which the given load is applied (these corners are labeled by indexes n and s, as shown in Fig. 1.8), and through the projection of a point (x, y, z) onto the plane of application of the load. Using indexes n and s to identify the rectangles is convenient for determining the values at the point (x, y, z). We shall sum the contributions of rectangles, taking into account the above-obtained results and the symmetry. The following values are introduced:

$$\xi_n = a + (-1)^n x, \quad \eta_s = b + (-1)^s y \quad (n, s = 1, 2). \tag{1.179}$$

The side lengths of additional rectangles along axes X and Y are equal to the absolute values of these parameters, respectively. The signs of parameters ξ_n, η_s, as well as the values of the indexes n and s that define the position of the rectangles (Fig. 1.8), influence the signs of terms contributed by the rectangles. A simple check shows that expressions for the amplitudes of vibrations for the points with arbitrary coordinates x, y along Cartesian axes may be written in the following form:

$$\delta_x(x,y) = p_0 \sum_{n,s=1}^{2} (-1)^n \operatorname{sign}(\eta_s) K_{vh}(|\xi_n|, |\eta_s|; 1, 0), \tag{1.180}$$

$$\delta_y(x,y) = p_0 \sum_{n,s=1}^{2} (-1)^s \operatorname{sign}(\xi_n) K_{vh}(|\xi_n|, |\eta_s|; 0, 1), \tag{1.181}$$

$$\delta_z(x,y) = p_0 \sum_{n,s=1}^{2} \operatorname{sign}(\xi_n \eta_s) K_{vv}(|\xi_n|, |\eta_s|; 0, 0). \tag{1.182}$$

The sign function is equal to 1, –1, 0 at positive, negative and zero values of its argument, respectively. Taking the symmetry for account, we have

$$\delta_y(x,y) = \tilde{\delta}_x(y,x) , \tag{1.183}$$

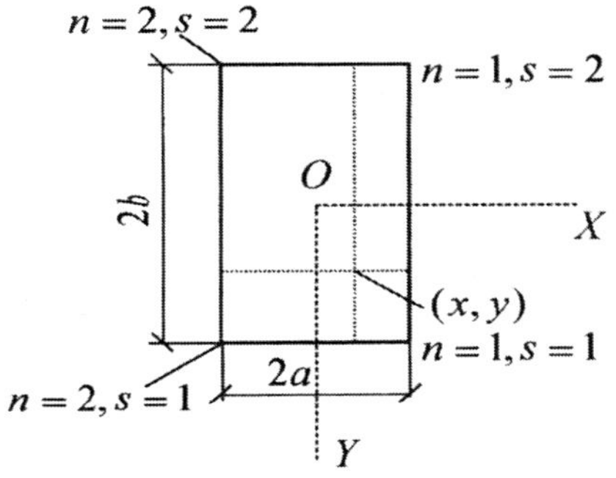

Fig. 1.8. Determination of displacements at arbitrary point of half-space

where function $\tilde{\delta}_x(y,x)$ is found from (1.180), where a is replaced by b and vice versa.

In order to build a result, related to this illustrative example (i.e. static displacements of points that belong to the surface of a homogeneous isotropic half-space), one should substitute expressions (1.176)–(1.178) into formulas (1.180)–(1.182):

$$\delta_x(x,y) = -\frac{p_0(1-2\nu)}{4G\pi}\sum_{n,s=1}^{2}(-1)^n \operatorname{sign}(\eta_s)\left[|\xi_n| \arccos\frac{|\xi_n|}{g_{ns}} + |\eta_s| \ln\frac{g_{ns}}{|\eta_s|}\right], \tag{1.184}$$

$$\delta_y(x,y) = -\frac{p_0(1-2\nu)}{4G\pi}\sum_{n,s=1}^{2}(-1)^s \operatorname{sign}(\xi_n)\left[|\eta_s| \arccos\frac{|\eta_s|}{g_{ns}} + |\xi_n| \ln\frac{g_{ns}}{|\xi_n|}\right], \tag{1.185}$$

$$\delta_z(x,y) = \frac{p_0(1-\nu)}{2G\pi}\sum_{n,s=1}^{2}\operatorname{sign}(\xi_n\eta_s)\left[|\xi_n| \ln\frac{g_{ns}+|\eta_s|}{|\xi_n|} + |\eta_s| \ln\frac{g_{ns}+|\xi_n|}{|\eta_s|}\right], \tag{1.186}$$

where

$$g_{ns} = \sqrt{\xi_n^2 + \eta_s^2}\,. \tag{1.187}$$

Equation (1.186) for $\delta_z(x,y)$ corresponds to the well-known Schleicher's formula [101].

1.7.2 Horizontal Load Distributed Uniformly over Rectangular Domain

Consider a load of type $q_0 \exp(\mathrm{i}\,\omega t)$ acting parallel to the X-axis and applied to the rectangular domain that belongs to the horizontal plane. We introduce three Green's functions $w_{hr}(z,r), w_{h\vartheta}(z,r), w_{hz}(z,r)$, as shown in Fig. 1.6b, for the horizontal force of amplitude equal to 1, applied in the plane where the given load acts. The first two functions express the amplitudes of vibrations in the direction of the applied force for the points that belong to the vertical plane containing the force ($w_{hr}(z,r)$), and for the points that belong to the plane passing through the point of application of the force and normal to its direction ($w_{h\vartheta}(z,r)$). Function $w_{hz}(z,r)$ corresponds to the amplitudes of vibrations in the direction of the Z-axis for the points that belong to the vertical plane that contains the force. These functions are associated with the displacements considered above when constructing solutions for the half-space, subjected to the action of the horizontal force, in their general form. For the case with a force applied to the surface of the half-space, values $w_{hr}(z,r), w_{h\vartheta}(z,r), w_{hz}(z,r)$ equal $\hat{u}_r, -\hat{u}_\vartheta, \hat{u}_z$, respectively, following (1.83)–(1.85) with $Q_0 = 1$, while for the case with a force applied within the half-space, we use (1.120) with $Q_0 = 1$ and $R = 0$. First, we construct, as previously, the solution for point A that corresponds to the corner of a rectangle (Fig. 1.7). Each elementary force is decomposed into components along the straight line connecting the given element with the corner point, and along the normal to this line. Application of the principle of superposition leads to the following result (for the sake of convenience, index z is dropped):

$$\delta_x(A) = q_0 \iint\limits_{(D)} [w_{hr}(r)\sin^2(\psi) + w_{h\vartheta}(r)\cos^2(\psi)]r\,\mathrm{d}r\,\mathrm{d}\psi\,, \tag{1.188}$$

$$\delta_y(A) = q_0 \iint\limits_{(D)} w_{-}(r)\sin(\psi)\cos(\psi)r\,\mathrm{d}r\,\mathrm{d}\psi\,, \tag{1.189}$$

$$\delta_z(A) = q_0 \iint\limits_{(D)} w_{hz}(h)\sin(\psi)r\,\mathrm{d}r\,\mathrm{d}\psi\,, \tag{1.190}$$

where

$$w_{-}(r) = w_{hr}(r) - w_{h\vartheta}(r)\,. \tag{1.191}$$

Further, we use notations (1.167) and results (1.168), (1.169). The amplitudes of vibrations $\delta_x(\mathrm{A})$ and $\delta_y(\mathrm{A})$ may be presented in the following form:

$$\delta_x(A) = q_0[K_{hr}(\alpha,\beta;2,0) + K_{h\vartheta}(\alpha,\beta;0,2)]\,, \tag{1.192a}$$

$$K_{hr}(\alpha,\beta;2,0)=\frac{\pi}{4}\int_0^{\alpha} w_{hr}(r)r\,\mathrm{d}r+\frac{1}{4}\int_{\alpha}^{g} w_{hr}(r)r[2\varphi_\alpha-\sin(2\varphi_\alpha)]\,\mathrm{d}r$$

$$-\frac{1}{4}\int_{\beta}^{g} w_{hr}(r)r[2\varphi_\beta-\sin(2\varphi_\beta)]\,\mathrm{d}r\,, \tag{1.192b}$$

$$K_{h\vartheta}(\alpha,\beta;0,2)=\frac{\pi}{4}\int_0^{\alpha} w_{h\vartheta}(r)r\,\mathrm{d}r+\frac{1}{4}\int_{\alpha}^{g} w_{h\vartheta}(r)r[2\varphi_\alpha+\sin(2\varphi_\alpha)]\,\mathrm{d}r$$

$$-\frac{1}{4}\int_{\beta}^{g} w_{h\vartheta}(r)r[2\varphi_\beta+\sin(2\varphi_\beta)]\,\mathrm{d}r\,, \tag{1.192c}$$

where

$$\varphi_\alpha=\arcsin\frac{\alpha}{r}\,, \tag{1.193a}$$

$$\varphi_\beta=\arccos\frac{\beta}{r}\,. \tag{1.193b}$$

$$\delta_y(A)=q_0K_-(\alpha,\beta;1,1)\,, \tag{1.194}$$

$$K_-(\alpha,\beta;1,1)=\frac{1}{2}\int_0^{\alpha} w_-(r)r\,\mathrm{d}r+\frac{1}{2}\int_{\alpha}^{g} w_-(r)r\frac{\alpha^2}{r^2}\mathrm{d}r-\frac{1}{2}\int_{\beta}^{g} w_-(r)r\left(1-\frac{\beta^2}{r^2}\right)\mathrm{d}r\,. \tag{1.194a}$$

Value $\delta_z(A)$ is similar to $\delta_x(A)$ by (1.170a):

$$\delta_z(\mathrm{A})=p_0K_{hz}(\alpha,\beta;1,0)\,, \tag{1.195}$$

$$K_{hz}(\alpha,\beta;1,0)=\int_0^{g} w_{hz}(r)r\,\mathrm{d}r-\int_{\alpha}^{g} w_{hz}(r)\sqrt{(r^2-\alpha^2)}\,\mathrm{d}r-\int_{\beta}^{g} w_{hz}(r)(r-\beta)\,\mathrm{d}r\,. \tag{1.195a}$$

For the illustrative example – a case with a load applied statically to the surface of a homogeneous isotropic half-space – Green's functions for the half-space surface are equal to [91]

$$w_{hr}=\frac{1}{2G\pi r}\,, \tag{1.196}$$

$$w_{h\vartheta}=\frac{1-\nu}{2G\pi r}\,, \tag{1.197}$$

$$w_{hz} = \frac{1-2\nu}{4G\pi r}. \tag{1.198}$$

Value $K_{hz}(\alpha,\beta;1,0) = -K_{vh}(\alpha,\beta;1,0)$ (see (1.176)), whereas values $K_{hr}(\alpha,\beta;2,0)$, $K_{h\vartheta}(\alpha,\beta;0,2)$, and $K_{-}(\alpha,\beta;1,1)$ have the form

$$K_{hr}(\alpha,\beta;2,0) = \frac{1}{2G\pi}\beta \ln\frac{g+\alpha}{\beta}, \tag{1.199}$$

$$K_{h\vartheta}(\alpha,\beta;2,0) = \frac{1-\nu}{2G\pi}\alpha \ln\frac{g+\beta}{\alpha}, \tag{1.200}$$

$$K_{-}(\alpha,\beta;2,0) = \frac{\nu}{2G\pi}[\alpha+\beta-\sqrt{\alpha+\beta}]. \tag{1.201}$$

In order to determine the amplitudes of vibrations at the points of the half-space that have arbitrary coordinates x, y, we employ an approach similar to that used for the case of a vertical load (Fig. 1.8). Taking the properties of symmetry into account, and analogously with equations (1.180)–(1.182), we obtain

$$\delta_x(x,y) = q_0 \sum_{n,s=1}^{2} \operatorname{sign}(\xi_n\eta_s)[K_{hr}(|\xi_n|,|\eta_s|;2,0) + K_{h\vartheta}(|\xi_n|,|\eta_s|;0,2)], \tag{1.202}$$

$$\delta_y(x,y) = q_0 \sum_{n,s=1}^{2} (-1)^{n+s} K_{-}(|\xi_n|,|\eta_s|;1,1), \tag{1.203}$$

$$\delta_z(x,y) = q_0 \sum_{n,s=1}^{2} (-1)^n \operatorname{sign}(\eta_s) K_{hz}(|\xi_n|,|\eta_s|;1,0). \tag{1.204}$$

For the surface of the homogeneous isotropic half-space, static displacements $\delta_z(x,y)$ are similar to $\delta_x(x,y)$ by (1.180), while displacements $\delta_x(x,y)$ and $\delta_y(x,y)$ have the following form:

$$\delta_x(x,y) = \frac{q_0}{2G\pi} \sum_{n,s=1}^{2} \operatorname{sign}(\xi_n\eta_s)\left[|\eta_s| \ln\frac{g_{ns}+|\xi_n|}{|\eta_s|} + (1-\nu)|\xi_n| \ln\frac{g_{ns}+|\eta_s|}{|\xi_n|}\right], \tag{1.205}$$

$$\delta_y(x,y) = \frac{q_0\nu}{2G\pi} \sum_{n,s=1}^{2} (-1)^{n+s}[|\xi_n|+|\eta_s|-g_{ns}]. \tag{1.206}$$

For a horizontal load directed parallel to the Y-axis, the corresponding results are obtained from equations (1.202)–(1.204) by a simple transformation of coordinates (suitable coordinate axes obtained by rotation of axes (X, Y) by 90° are introduced). In addition, the dimensions of a rectangle, a and b, should be

reciprocally interchanged. Designating the vibration amplitudes in this case by an additional superscript y, we can write

$$\delta_x^y(x,y) = -\tilde{\delta}_y(y,-x) = \tilde{\delta}_y(y,x)\,, \tag{1.207}$$

$$\delta_y^y(x,y) = \tilde{\delta}_x(y,-x) = \tilde{\delta}_x(y,x)\,, \tag{1.208}$$

$$\delta_z^y(x,y) = \tilde{\delta}_z(y,-x) = \tilde{\delta}_z(y,x)\ . \tag{1.209}$$

Values on the right sides of these relationships are determined by formulas (1.202)–(1.204) with reciprocal interchanging of dimensions a and b.

1.7.3 Vertical Load Distributed Uniformly over Circular Domain

In analogy with the case of a load applied to the rectangular domain, we consider a circular domain subjected to a uniformly distributed vertical load $p_0 \exp(\mathrm{i}\,\omega t)$ (Fig. 1.9). Point M is a projection of the point, where the amplitudes of vibrations have to be determined, onto the plane of application of the load. Again, we drop index z in the notations of Green's functions and of the considered amplitudes of vibrations. Following the principle of superposition, double integrals with respect to variables γ and s are transformed after integration by γ to 1-D integrals:

$$\delta_x(r) = 2p_0 \int\limits_{r-R}^{r+R}\int\limits_{0}^{\varphi} w_{vh}(s)\cos(\gamma) s\,\mathrm{d}\gamma\,\mathrm{d}s = 2p_0 \int\limits_{r-R}^{r+R} w_{vh}(s)\sin(\varphi) s\,\mathrm{d}s$$

$$= 2p_0 R^2 \int\limits_{\eta-1}^{\eta+1} w_{vh}(\lambda R)\sin(\varphi)\lambda\,\mathrm{d}\lambda\ , \tag{1.210}$$

$$\delta_z(r) = 2p_0 \int\limits_{r-R}^{r+R}\int\limits_{0}^{\varphi} w_{vv}(s) s\,\mathrm{d}\gamma\,\mathrm{d}s = 2p_0 \int\limits_{r-R}^{r+R} w_{vv}(s)\varphi s\,\mathrm{d}s$$

$$= 2p_0 R^2 \int\limits_{\eta-1}^{\eta+1} w_{vv}(\lambda R)\varphi\lambda\,\mathrm{d}\lambda\,, \tag{1.211}$$

where

$$\lambda = \frac{s}{R}\,, \tag{1.212a}$$

$$\eta = \frac{r}{R}\,, \tag{1.212b}$$

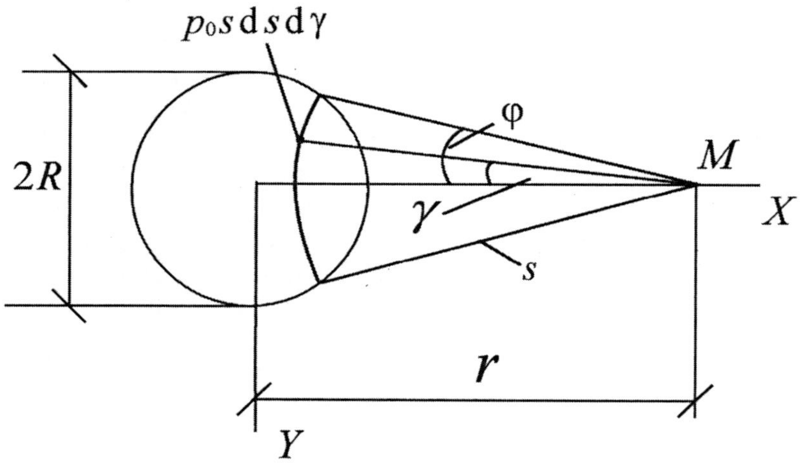

Fig. 1.9. Uniformly distributed vertical load applied to circular area

$$\varphi = \arccos\frac{s^2 + r^2 - R^2}{2sr} = \arccos\frac{\lambda^2 + \eta^2 - 1}{2\lambda\eta}\,. \tag{1.212c}$$

These formulas correspond to the case of $r \geq R$. For $r < R$, two integrals should be considered: the first one, between the limits $0 < s < R - r$ and with the angle $\varphi = \pi$ (this integral becomes zero for the case of determination of horizontal displacements δ_x), and the second one having a form similar to the integrals in (1.210), (1.211), but with the lower limit, with respect to variable *s*, equal $R - r$. Formulas which are valid for any *r* may be written as follows:

$$\delta_x(r) = 2p_0R^2 \int\limits_{|\eta-1|}^{\eta+1} w_{vh}(\lambda R)\sin(\varphi)\lambda\,\mathrm{d}\lambda\,, \tag{1.213a}$$

$$\delta_z(r) = 2p_0R^2 \int\limits_{|\eta-1|}^{\eta+1} w_{vv}(\lambda R)\varphi\lambda\,\mathrm{d}\lambda + \begin{cases} 2p_0\pi R^2 \int\limits_0^{1-\eta} w_{vv}(\lambda R)\lambda\,\mathrm{d}\lambda & (r < R) \\ 0 & (r \geq R) \end{cases} \tag{1.213b}$$

Note that derivatives of the functions under the integral signs become infinite at the ends of the interval of integration, due to function φ determined by (1.212c).

1.7.4 Horizontal Load Distributed Uniformly over Circular Domain

A horizontal load of amplitude q_0 is applied to the circular domain that belongs to a horizontal plane at the surface, or within a transversely isotropic heterogeneous half-space (Fig. 1.10). In order to find the displacements at a point M with cylindrical coordinates (r, ϑ, z), we must determine the amplitudes of vibrations at two characteristic points, M_1, with Cartesian coordinates (r, 0, z), and M_2, with coordinates (0, r, z). Projections of these points onto the plane of application of the load belong to the planes (X, Z), and (Y, Z), respectively. For the point M_1, one only has to determine the amplitudes along axes X (i.e. δ_{1x}) and Z (i.e. δ_{1z}), while for the point M_2, only the amplitudes along the X-axis (i.e. δ_{2x}) are non-zero. Consider a projection of the point M onto the plane where the given load is applied, and decompose the load at each point along two directions: first, parallel to the straight line passing through the center of the circle of loading and through the projection of the point M onto the plane of loading, and, second, along the direction which is normal to the first one. It is evident that

$$\delta_r(M) = \delta_{1x}\cos(\vartheta)\,, \tag{1.214a}$$

$$\delta_\vartheta(M) = -\delta_{2x}\sin(\vartheta)\,, \tag{1.214b}$$

$$\delta_z(M) = \delta_{1z}\cos(\vartheta)\,. \tag{1.214c}$$

Next, we determine the displacements at the points M_1 and M_2. In order to calculate value δ_{1x}, we decompose each element of the load as shown in Fig. 1.10, and apply the Green's functions which were introduced in the previous consideration of loadings for the rectangular domain:

$$\begin{aligned}\delta_{1x} &= 2q_0\int_{r-R}^{r+R}\int_0^{\varphi}[w_{hr}(s)\cos^2(\gamma)+w_{h\vartheta}(s)\sin^2(\gamma)]s\,\mathrm{d}\gamma\,\mathrm{d}s\\ &= q_0\int_{r-R}^{r+R}\left[w_{hr}(s)\left(\varphi+\frac{1}{2}\sin 2\varphi\right)+w_{h\vartheta}(h)\left(\varphi-\frac{1}{2}\sin 2\varphi\right)\right]s\,\mathrm{d}s\\ &= q_0R^2\int_{\eta-1}^{\eta+1}\left[w_{hr}(\lambda R)\left(\varphi+\frac{1}{2}\sin 2\varphi\right)+w_{h\vartheta}(\lambda R)\left(\varphi-\frac{1}{2}\sin 2\varphi\right)\right]\lambda\,\mathrm{d}\lambda\,.\end{aligned} \tag{1.215}$$

This result corresponds to the case of $r \geq R$. If $r < R$, the lower limit of the integrals in equation (1.215) should be replaced by $R-r$ or $(1-\eta)$, and an integral similar to those in (1.215), but between the limits of zero to $R-r$ or $1-\eta$ and with $\varphi = \pi$, should be added. A general expression for δ_{1x} may be

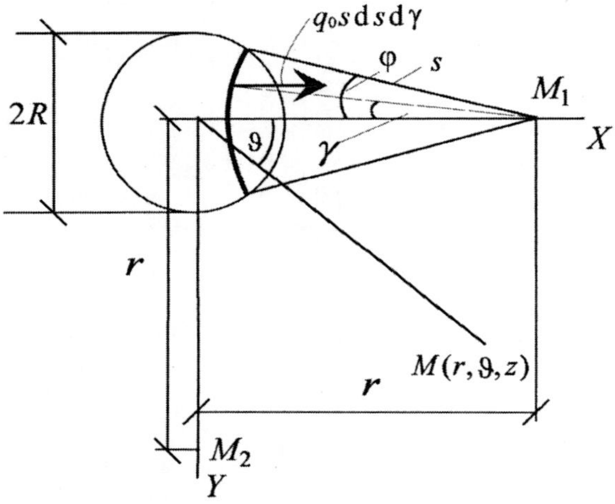

Fig. 1.10. Uniformly distributed horizontal load applied to circular area

written as follows:

$$\delta_{1x}(r) = q_0 R^2 \int\limits_{|\eta-1|}^{\eta+1} \left[w_{hr}(\lambda R)\left(\varphi + \frac{1}{2}\sin 2\varphi \right) + w_{h\vartheta}(\lambda R)\left(\varphi - \frac{1}{2}\sin 2\varphi \right) \right] \lambda\,\mathrm{d}\lambda$$

$$+ \begin{cases} q_0 \pi R^2 \int\limits_0^{1-\eta} [w_{hr}(\lambda R) + w_{h\vartheta}(\lambda R)] \lambda\,\mathrm{d}\lambda & (r < R) \\ 0 & (r \geq R). \end{cases} \tag{1.216}$$

Similarly, amplitudes δ_{2x} are determined; in the expression for δ_{1x}, values w_{hr} and $w_{h\vartheta}$ interchange:

$$\delta_{2x} = q_0 R^2 \int\limits_{|\eta-1|}^{\eta+1} w_{hr}(\lambda R)\left[\left(\varphi - \frac{1}{2}\sin 2\varphi \right) + w_{h\vartheta}(\lambda R)\left(\varphi + \frac{1}{2}\sin 2\varphi \right)\right] \lambda\,\mathrm{d}\lambda$$

$$+ \begin{cases} q_0 \pi R^2 \int\limits_0^{1-\eta} [w_{hr}(\lambda R) + w_{h\vartheta}(\lambda R)] \lambda\,\mathrm{d}\lambda & (r < R) \\ 0 & (r \geq R). \end{cases} \tag{1.217}$$

For the amplitudes δ_{1z}, the corresponding result is similar to δ_x from (1.210):

$$\delta_{1z} = 2q_0 R^2 \int_{|\eta-1|}^{\eta+1} w_{hz}(\lambda R)\sin(\varphi)\lambda\,\mathrm{d}\lambda\ . \tag{1.218}$$

1.7.5 Axisymmetric Radial Load Applied to Circular Domain

We consider a horizontal radial load of amplitude $q_0\rho$ (Fig. 1.11) applied to the circular domain (ρ is the distance between the point of loading and the circle center). This type of loading may be required when accounting for the welded contact between an elastic medium and a stiff disk, when a vertical translational displacement of the disk occurs. Employing the previously used approach, we decompose the load elements as shown in Fig. 1.11. The amplitudes at point M_1 along the X-axis are calculated as follows:

$$\begin{aligned}\delta_{1x} &= 2q_0 \int_{r-R}^{r+R}\int_{0}^{\varphi} [w_{hr}(s)\rho\cos(\psi)\cos(\gamma) + w_{h\vartheta}(s)\rho\sin(\psi)\sin(\gamma)]s\,\mathrm{d}\gamma\,\mathrm{d}s \\
&= 2q_0 \int_{r-R}^{r+R}\int_{0}^{\varphi} [w_{hr}(s)(r\cos(\gamma)-s)\cos(\gamma) + w_{h\vartheta}(s)r\sin^2(\gamma)]s\,\mathrm{d}\gamma\,\mathrm{d}s \\
&= q_0 R^3 \int_{\eta-1}^{\eta+1} \left\{ w_{hr}(\lambda R)\left[-2\lambda\sin(\varphi) + \varphi\eta + \frac{\eta}{2}\sin(2\varphi)\right]\right. \\
&\quad \left. + \eta w_{h\vartheta}(\lambda R)\left[\varphi - \frac{\sin(2\varphi)}{2}\right]\right\}\lambda\,\mathrm{d}\lambda\,.\end{aligned} \tag{1.219}$$

This result is valid for $r \geq R$. For the general case

$$\begin{aligned}\delta_{1x} &= q_0 R^3 \int_{|\eta-1|}^{\eta+1} \left\{ w_{hr}(\lambda R)\left[-2\lambda\sin(\varphi) + \varphi\eta + \frac{\eta}{2}\sin(2\varphi)\right]\right. \\
&\quad \left. + \eta w_{h\vartheta}(\lambda R)\left[\varphi - \frac{\sin(2\varphi)}{2}\right]\right\}\lambda\,\mathrm{d}\lambda\ + \\
&\quad + \begin{cases} q_0 \pi R^3 \eta \displaystyle\int_0^{1-\eta} [w_{hr}(\lambda R) + w_{h\vartheta}(\lambda R)]\lambda\,\mathrm{d}\lambda & (r < R) \\ 0 & (r \geq R). \end{cases}\end{aligned} \tag{1.220}$$

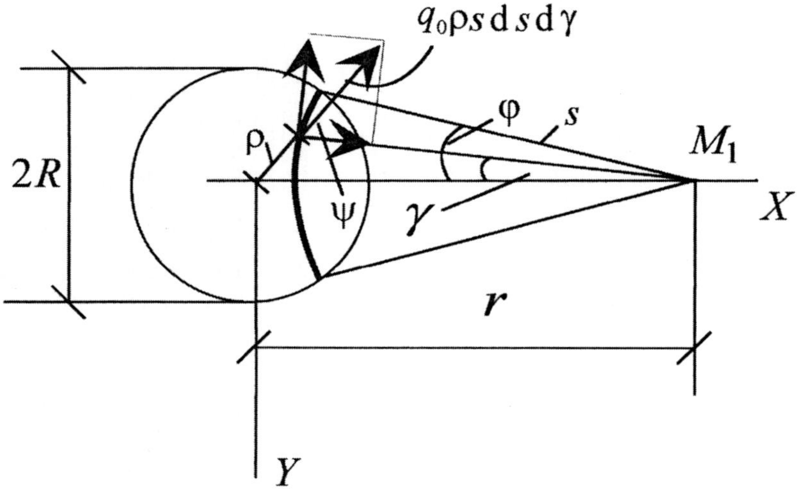

Fig. 1.11. Axisymmetric radial load proportional to distance from axis of symmetry

Consider, for example, the case with Green's functions w_{hr} having the form of *C/r*, where *C* is a constant. Such a relationship occurs for surface displacements, when the surface of a homogeneous isotropic or transversely isotropic half-space is subjected to the static action of a force (for this case, function $w_{h\vartheta}$ takes a similar form). The function in square brackets which serves as a multiplier for w_{hr} under the integral in equation (1.219) has the following anti-derivative function:

$$\eta\lambda\varphi - \frac{1}{2}\eta\lambda\sin(2\varphi)\,. \tag{1.221}$$

As a result, for this particular case, contributions of value w_{hr} both to displacements δ_{1x} in (1.219) and to the first integral in expression (1.220) vanish. An expression for the amplitudes of vibrations at point M_1 in the vertical direction may be written as follows ($r \geq R$):

$$\delta_{1z} = 2q_0 \int\limits_{r-R}^{r+R}\int\limits_{0}^{\varphi} w_{hz}(s)\rho\cos(\psi)s\,\mathrm{d}\gamma\,\mathrm{d}s = 2q_0 \int\limits_{r-R}^{r+R}\int\limits_{0}^{\varphi} w_{hz}(s)(r\cos(\gamma)-s)s\,\mathrm{d}\gamma\,\mathrm{d}s$$

$$= 2q_0R^3 \int_{\eta-1}^{\eta+1} \{w_{hz}(\lambda R)[\eta\sin(\varphi) - \varphi\lambda]\}\lambda\,\mathrm{d}\lambda\,. \tag{1.222}$$

For the general case:

$$\delta_{1z} = 2q_0R^3 \int_{|\eta-1|}^{\eta+1} \{w_{hz}(\lambda R)[\eta\sin(\varphi) - \varphi\lambda]\}\lambda\,\mathrm{d}\lambda$$
$$+\begin{cases} -2\pi q_0R^3 \displaystyle\int_0^{1-\eta} w_{hz}(\lambda R)\lambda^2\,\mathrm{d}\lambda & (r < R) \\ 0 & (r \ge R). \end{cases} \tag{1.223}$$

1.7.6 Antisymmetric Vertical Load Applied to Circular Domain

Consider a vertical load $p_0\rho\cos(\vartheta)\exp(\mathrm{i}\,\omega t)$ (Fig. 1.12) applied to the circular domain. This form of load may be needed to determine the rotational stiffness of a circular stamp. In order to calculate the displacements at various points of the half-space, one should determine the vertical (δ_{1z}) and horizontal (δ_{1x}) amplitudes of vibrations at point M_1, and, in addition, the horizontal amplitudes δ_{2x} at point M_2. For an arbitrary point M with cylindrical coordinates (r, ϑ, z), we apply equations (1.214). Consider the value of δ_{1x} $(r \ge R)$:

$$\delta_{1x} = 2p_0 \int_{r-R}^{r+R}\int_0^{\varphi} w_{vh}(s)\rho\cos(\vartheta)\cos(\gamma)s\,\mathrm{d}\gamma\,\mathrm{d}s$$

$$= 2p_0 \int_{r-R}^{r+R}\int_0^{\varphi} w_{vh}(s)[r - s\cos(\gamma)]\cos(\gamma)\,s\,\mathrm{d}\gamma\,\mathrm{d}s$$

$$= p_0R^3 \int_{\eta-1}^{\eta+1} w_{vh}(\lambda R)\left[2\eta\sin(\varphi) - \varphi\lambda - \frac{\lambda}{2}\sin(2\varphi)\right]\lambda\,\mathrm{d}\lambda\,. \tag{1.224}$$

For arbitrary values of r we have

$$\delta_{1x} = p_0R^3 \int_{|\eta-1|}^{\eta+1} w_{vh}(\lambda R)\left[2\eta\sin(\varphi) - \varphi\lambda - \frac{\lambda}{2}\sin(2\varphi)\right]\lambda\,\mathrm{d}\lambda$$

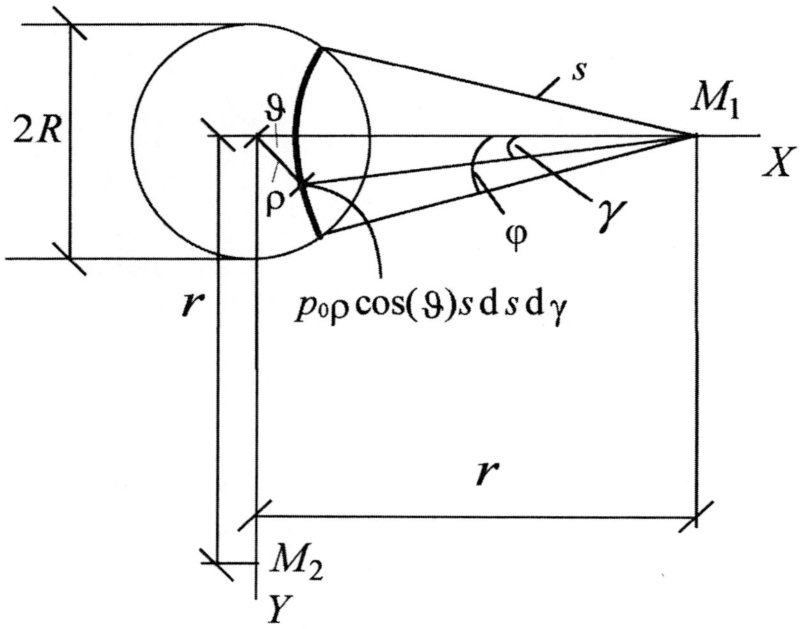

Fig. 1.12. Antisymmetric vertical load proportional to distance from plane (*Y*, *Z*)

$$+\begin{cases} -\pi p_0 R^3 \displaystyle\int_0^{1-\eta} w_{vh}(\lambda R)\lambda^2\, \mathrm{d}\lambda & (r < R) \\ 0 & (r \geq R). \end{cases} \tag{1.225}$$

For $r \geq R$, amplitudes δ_{1z} take the following form:

$$\delta_{1z} = 2p_0 \int_{r-R}^{r+R} \int_0^{\varphi} w_{vv}(s)\rho\cos(\vartheta)\, s\, \mathrm{d}\gamma\, \mathrm{d}s = 2p_0 \int_{r-R}^{r+R} \int_0^{\varphi} w_{vv}(s)[r - s\cos(\gamma)]\, s\, \mathrm{d}\gamma\, \mathrm{d}s$$

$$= 2p_0 R^3 \int_{\eta-1}^{\eta+1} w_{vv}(\lambda R)[\eta\varphi - \lambda\sin(\varphi)]\lambda\, \mathrm{d}\lambda\,, \tag{1.226}$$

while, for arbitrary values of *r*

$$\delta_{1z} = 2p_0R^3 \int_{|\eta-1|}^{\eta+1} w_{vv}(\lambda R)[\eta\varphi - \lambda\sin(\varphi)]\lambda\,\mathrm{d}\lambda$$

$$+\begin{cases} 2\pi p_0R^3\eta \displaystyle\int_0^{1-\eta} w_{vv}(\lambda R)\lambda\,\mathrm{d}\lambda & (r < R) \\ 0 & (r \geq R). \end{cases} \tag{1.227}$$

Consider the horizontal amplitudes of vibrations at point M_2. Analogously to relationship (1.224), we obtain for $r \geq R$

$$\delta_{2x} = -2p_0 \int_{r-R}^{r+R}\int_0^{\varphi} w_{vh}(s)\rho\cos(\vartheta)\sin(\gamma)s\,\mathrm{d}\gamma\,\mathrm{d}s$$

$$= -2p_0 \int_{r-R}^{r+R}\int_0^{\varphi} w_{vh}(s)s\sin^2(\gamma)s\,\mathrm{d}\gamma\,\mathrm{d}s$$

$$= -p_0R^3 \int_{\eta-1}^{\eta+1} w_{vh}(\lambda R)\left[\varphi - \frac{1}{2}\sin(2\varphi)\right]\lambda^2\,\mathrm{d}\lambda. \tag{1.228}$$

The general formula takes the form

$$\delta_{2x} = -p_0R^3 \int_{|\eta-1|}^{\eta+1} w_{vh}(\lambda R)\left[\varphi - \frac{1}{2}\sin(2\varphi)\right]\lambda^2\,\mathrm{d}\lambda$$

$$+\begin{cases} -\pi p_0R^3 \displaystyle\int_0^{1-\eta} w_{vh}(\lambda R)\lambda^2\,\mathrm{d}\lambda & (r < R) \\ 0 & (r \geq R). \end{cases} \tag{1.229}$$

1.7.7 Horizontal Load Acting in Tangential Direction

Consider a horizontal load of the form $q_0\rho\exp(\mathrm{i}\omega t)$, acting in a tangential direction (Fig. 1.13). This form of loading may be used to determine the torsional stiffness of a circular stamp. In this case, one should only calculate the horizontal amplitudes δ_{1y} at point M_1. Derivation of the formula for the displacements is analogous to (1.219). For $r \geq R$,

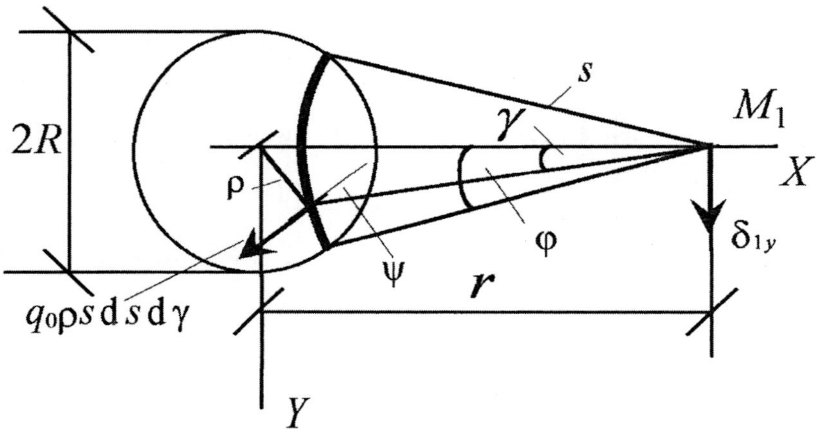

Fig. 1.13. Axisymmetric tangential load proportional to distance from axis of symmetry

$$\delta_{1y} = 2q_0 \int_{r-R}^{r+R} \int_0^{\varphi} [w_{hr}(s)\rho\cos(\psi)\sin(\gamma) + w_{h\vartheta}(s)\rho\sin(\psi)\cos(\gamma)]\,s\,\mathrm{d}\gamma\,\mathrm{d}s$$

$$= 2q_0 \int_{r-R}^{r+R} \int_0^{\varphi} [w_{hr}(s) r\sin^2(\gamma) + w_{h\vartheta}(s)(r\cos(\gamma) - s)\cos(\gamma)]\,s\,\mathrm{d}\gamma\,\mathrm{d}s$$

$$= q_0 R^3 \int_{\eta-1}^{\eta+1} \left\{ w_{h\vartheta}(\lambda R)\left[-2\lambda\sin(\varphi) + \varphi\eta + \frac{\eta}{2}\sin(2\varphi)\right]\right.$$

$$\left. + \eta w_{hr}(\lambda R)\left[\varphi - \frac{\sin(2\varphi)}{2}\right]\right\}\lambda\,\mathrm{d}\lambda\,. \tag{1.230}$$

For generalization, one should replace the lower limit in the last integral in relationship (1.230) with $|\eta - 1|$, and add the term equal to the last summand in equation (1.220).

1.7.8 Self-Balanced Horizontal Load

Consider a horizontal load having the radial component $q_r = q_0\rho^2\cos(\vartheta)$ and the tangential component $q_\vartheta = q_0\rho^2\sin(\vartheta)$ (Fig. 1.14). This form of load is self-balanced: the resultant force and the moment corresponding to this load vanish. Using a horizontal uniformly distributed load applied to the elementary ring-shaped domains (this type of load is considered for a circular area in section 1.7.4) is not sufficient for solving a contact problem for a circular stamp. In order to provide specified horizontal displacements over the entire contact domain, additional loadings are required. A considered self-balanced load is suitable for this purpose. As shown in the previous subsections, in order to find a solution for an arbitrary point, one has to calculate the amplitudes of vibrations δ_{1z}, δ_{1x} and δ_{2x} at points M_1 and M_2, with further application of relationships (1.214). Consider amplitudes δ_{1x} ($r \geq R$):

$$\delta_{1x} = 2q_0 \int_{r-R}^{r+R}\int_0^{\varphi} \{\rho^2 w_{hr}(s)[\cos(\vartheta)\cos(\psi) - \sin(\vartheta)\sin(\psi)]\cos(\gamma)$$

$$+ \rho^2 w_{h\vartheta}(s)[\cos(\vartheta)\sin(\psi) + \sin(\vartheta)\cos(\psi)]\sin(\gamma)\} s\,\mathrm{d}\gamma\,\mathrm{d}s$$

$$= 2q_0 \int_{r-R}^{r+R}\int_0^{\varphi} \{w_{hr}(s)[(r^2 + s^2)\cos(\gamma) - 2rs]\cos(\gamma)$$

$$+ w_{h\vartheta}(s)(r^2 - s^2)\sin^2(\gamma)\} s\,\mathrm{d}\gamma\,\mathrm{d}s$$

$$= q_0 R^4 \int_{|\eta-1|}^{\eta+1} \left\{ w_{hr}(\lambda R)\left[(\lambda^2 + \eta^2)\left(\varphi + \frac{1}{2}\sin(2\varphi)\right) - 4\lambda\eta\sin(\varphi)\right]\right.$$

$$\left. + w_{h\vartheta}(\lambda R)(\eta^2 - \lambda^2)\left(\varphi - \frac{1}{2}\sin(2\varphi)\right)\right\} \lambda\,\mathrm{d}\lambda. \tag{1.231}$$

When $r < R$, the following value β_1 should be added to the expression obtained:

$$\beta_1 = \pi q_0 R^4 \int_0^{1-\eta} [w_{hr}(\lambda R)(\eta^2 + \lambda^2) + w_{h\vartheta}(\lambda R)(\eta^2 - \lambda^2)]\lambda\,\mathrm{d}\lambda\ . \tag{1.232}$$

The amplitudes δ_{1z} may be written as follows ($r \geq R$):

$$\delta_{1z} = 2q_0 \int_{r-R}^{r+R}\int_0^{\varphi} \rho^2 w_{hz}(s)[\cos(\vartheta)\cos(\psi) - \sin(\vartheta)\sin(\psi)] s\,\mathrm{d}\gamma\,\mathrm{d}s$$

$$= 2q_0 \int_{r-R}^{r+R}\int_0^{\varphi} w_{hz}(s)[(r^2 + s^2)\cos(\gamma) - 2rs] s\,\mathrm{d}\gamma\,\mathrm{d}s$$

$$= 2q_0 R^4 \int_{|\eta-1|}^{\eta+1} w_{hz}(\lambda R)[(\lambda^2 + \eta^2)\sin(\varphi) - 2\lambda\eta o]\lambda\,\mathrm{d}\lambda\ . \tag{1.233}$$

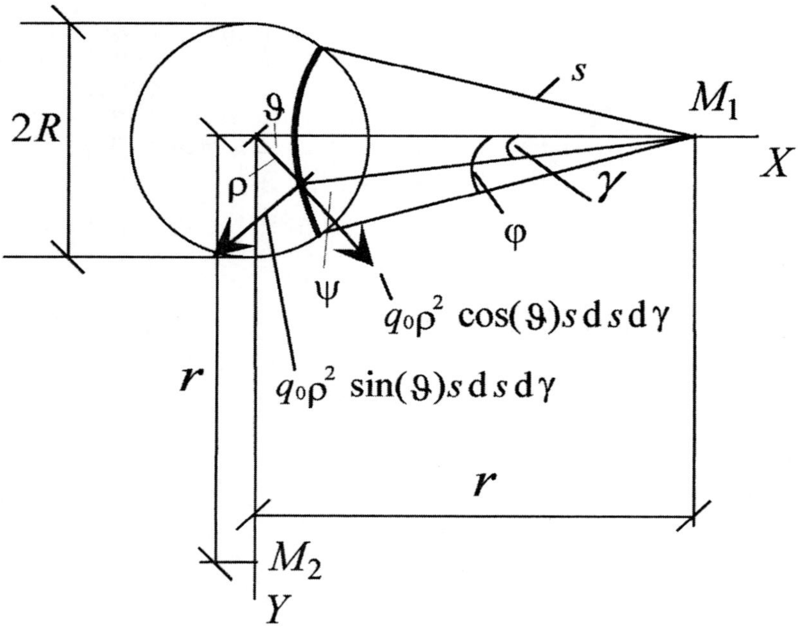

Fig. 1.14. Self-balanced horizontal load having radial and tangential components

The value of β_2, which should be added to the latter integral for $r < R$, has the following form:

$$\beta_2 = -4\pi q_0 R^4 \eta \int_0^{1-\eta} w_{hz}(\lambda R)\lambda^2 \, \mathrm{d}\lambda . \tag{1.234}$$

The value of δ_{2x} may be calculated by using transformations analogous to those employed for determining the amplitudes δ_{1x}. Finally, the result takes the form

$$\delta_{2x} = q_0 R^4 \int_{|\eta-1|}^{\eta+1} \left\{ -w_{h\vartheta}(\lambda R)\left[(\lambda^2+\eta^2)\left(\varphi + \frac{1}{2}\sin(2\varphi)\right) - 4\lambda\eta\sin(\varphi)\right] \right.$$
$$\left. - w_{hr}(\lambda R)(\eta^2-\lambda^2)\left(\varphi - \frac{1}{2}\sin(2\varphi)\right)\right\} \lambda \, \mathrm{d}\lambda$$

$$+\begin{cases} -\pi q_0 R^4 \int\limits_0^{1-\eta} [w_{h\vartheta}(\lambda R)(\lambda^2+\eta^2)+w_{hr}(\lambda R)(\eta^2-\lambda^2)]\lambda\,\mathrm{d}\lambda & (r<R) \\ 0 & (r\geq R). \end{cases} \tag{1.235}$$

Thus, in order to obtain an expression for δ_{2x} from the expression for δ_{1x}, one should interchange the positions of values w_{hr} and $w_{h\vartheta}$, and reverse the sign for all summands.

1.7.9 Loading of Infinite Strip

In this subsection, we seek solutions for the amplitudes of vibrations that occur due to loading of an infinite strip with edges parallel to the *Y*-axis, with the following types of loads: uniformly distributed vertical and horizontal (parallel to the *X*-axis), horizontal, acting parallel to the *X*-axis and varying in proportion to the distance to the central line of the strip (analogous to the load considered in section 1.7.5), vertical antisymmetric load varying in proportion to the distance to the central line of the strip (analogous to the load considered in section 1.7.6), and a uniformly distributed horizontal load parallel to the *Y*-axis. Here, Green's functions correspond to the loads of unit intensity distributed along the infinite line, parallel to the *Y*-axis: $w_{vx}(z,x)$ and $w_{vz}(z,x)$ represent the amplitudes of vibrations along axes *X* and *Z*, respectively, under the action of the vertical load; $w_{hx}(z,x)$ and $w_{hz}(z,x)$ represent the same amplitudes of vibrations that occur due to the action of the horizontal load, parallel to the *X*-axis, and $w_{hy}(z,x)$ the amplitudes of vibrations along the *Y*-axis under the action of a load parallel to this axis. For the amplitudes of vibrations that occur in the plane of application of loads z = const, a relationship $w_{hz}=-w_{vx}$ holds (this fact stems from the principle of reciprocity). For the case with loads applied to the surface of a half-space, the Green's functions might be determined by using the formulas given in section 1.4.4. In the following, we drop the index z in the notation for the Green's functions and for values related to the considered point.

Considering a uniformly distributed loading of an infinite strip of width $2R$ with a vertical load p_0, we employ formulas analogous to (1.213). Here, we denote the x-coordinate of a point in the half-space, where the displacements are determined, by r, similar to the polar coordinate in the 3-D case described above:

$$\delta_x(r) = p_0 R \int\limits_{|\eta-1|}^{\eta+1} w_{vx}(\lambda R)\,\mathrm{d}\lambda\,, \tag{1.236}$$

$$\delta_z(r) = p_0 R \int\limits_{|\eta-1|}^{\eta+1} w_{vz}(\lambda R)\,\mathrm{d}\lambda + \begin{cases} 2p_0 R \int\limits_0^{1-\eta} w_{vz}(\lambda R)\,\mathrm{d}\lambda & (r < R) \\ 0 & (r \geq R). \end{cases} \tag{1.237}$$

Here and in the following representation, we consider positive values of r and $\eta = r/R$ only; in the Green's functions, the coordinate x is replaced with λR. Results for negative values of r may be obtained from symmetry considerations.

For the case with a horizontal uniformly distributed load q_0, parallel to the X-axis, the corresponding relationships have a similar structure:

$$\delta_x(r) = q_0 R \int\limits_{|\eta-1|}^{\eta+1} w_{hx}(\lambda R)\,\mathrm{d}\lambda + \begin{cases} 2q_0 R \int\limits_0^{1-\eta} w_{hx}(\lambda R)\,\mathrm{d}\lambda & (r < R) \\ 0 & (r \geq R), \end{cases} \tag{1.238}$$

$$\delta_z(r) = q_0 R \int\limits_{|\eta-1|}^{\eta+1} w_{hz}(\lambda R)\,\mathrm{d}\lambda\,. \tag{1.239}$$

For the case with a uniformly distributed load q_0 parallel to the Y-axis, only displacements δ_y along the Y-axis are non-zero; the latter may be written in a form similar to relationship (1.238), where w_{hx} is replaced by w_{hy} :

$$\delta_y(r) = q_0 R \int\limits_{|\eta-1|}^{\eta+1} w_{hy}(\lambda R)\,\mathrm{d}\lambda + \begin{cases} 2q_0 R \int\limits_0^{1-\eta} w_{hy}(\lambda R)\,\mathrm{d}\lambda & (r < R) \\ 0 & (r \geq R). \end{cases} \tag{1.240}$$

Next, we consider loading of the strip with a symmetric load $q_0\rho$, parallel to the X-axis and varying in proportion to the distance ρ between the point at which the load is applied and the central line of the loaded strip. This form of load is analogous to the axisymmetric load, described in section 1.7.5. Application of the principle of superposition yields

$$\delta_x(r) = q_0 R^2 \int\limits_{|\eta-1|}^{\eta+1} (\eta - \lambda) w_{hx}(\lambda R)\,\mathrm{d}\lambda + \begin{cases} 2q_0 R^2 \eta \int\limits_0^{1-\eta} w_{hx}(\lambda R)\,\mathrm{d}\lambda & (r < R) \\ 0 & (r \geq R), \end{cases} \tag{1.241}$$

$$\delta_z(r) = q_0 R^2 \int\limits_{|\eta-1|}^{\eta+1} (\eta - \lambda) w_{hz}(\lambda R)\, d\lambda - \begin{cases} 2q_0 R^2 \int\limits_0^{1-\eta} \lambda w_{hz}(\lambda R)\, d\lambda & (r < R) \\ 0 & (r \geq R). \end{cases} \quad (1.242)$$

For the case with antisymmetric vertical load $p_0\rho$, varying in proportion to the distance $|\rho|$ between the point of application of the load and the axis of symmetry of the loaded strip, the corresponding formulas have a similar structure:

$$\delta_x(r) = p_0 R^2 \int\limits_{|\eta-1|}^{\eta+1} (\eta - \lambda) w_{vx}(\lambda R)\, d\lambda - \begin{cases} 2p_0 R^2 \int\limits_0^{1-\eta} \lambda w_{vx}(\lambda R)\, d\lambda & (r < R) \\ 0 & (r \geq R), \end{cases} \quad (1.243)$$

$$\delta_z(r) = p_0 R^2 \int\limits_{|\eta-1|}^{\eta+1} (\eta - \lambda) w_{vz}(\lambda R)\, d\lambda + \begin{cases} 2p_0 R^2 \eta \int\limits_0^{1-\eta} w_{vz}(\lambda R)\, d\lambda & (r < R) \\ 0 & (r \geq R). \end{cases} \quad (1.244)$$

The developed formulas are employed further in order to construct solutions of dynamic contact problems for a stamp having the form of the infinite strip.

Chapter 2. Static and Dynamic Problems for Homogeneous Transversely Isotropic Half-Space

Forced vibrations occurring in an anisotropic half-space have been the object of study in a small number of reports [61, 102, 124]. In most cases, free vibrations or solutions in the wave number space have been considered [11–13, 29, 59, 60]. A possibility of employing potentials to separate the equations of motion for a medium has been studied in [19]. In this report, the author presents an additional condition for the elastic constants, which enables separated wave equations for potentials to be found. Kirkner [61] considered vibrations of a circular stiff disk on a transversely isotropic half-space that satisfies the condition stated in [19]. However, construction of a solution for problems dealing with forced vibrations in a homogeneous transversely isotropic half-space is relatively simple (no additional constraints on elastic coefficients are needed) when using the equations presented in Chap. 1, or via formulations that employ stiffness matrices for a half-space, or for layers which form a layered half-space [11–13, 59, 60, 102, 124]. Solutions of static problems for the homogeneous transversely isotropic half-space are well represented in the literature, starting with the work of Michell [74]. Numerous solutions related to a half-space subjected to loads applied to its surface are presented in [32, 36, 65, 130]. Static Green's functions for the case of forces applied within the half-space have been constructed in [86, 87, 107]. In the present chapter, we consider some problems of the dynamics and statics for a homogeneous transversely isotropic half-space: namely, the action of vertical and horizontal forces applied to the surface of a half-space and in its depth; and vibrations of a stiff disk on a half-space.

2.1 Vibrations of Half-Space Subjected to Vertical Force Applied to Half-Space Surface

The corresponding solution for the case of a concentrated force is given in its general form by equations (1.57). Rewriting these equations in their dimensionless form, by using the dimensionless parameters given in (1.27) and assuming $r \neq 0$, yields

$$u_r = \frac{P_0(1-\nu')}{2\pi G_{rz0} r} S_{vh}, \tag{2.1a}$$

$$S_{vh} = -\frac{\beta^2 \tilde{r}}{1-\nu'} \int_0^\infty \frac{c_{22}\tilde{q}_1(\tilde{z},\tilde{k}) - c_{21}\tilde{q}_2(\tilde{z},\tilde{k})}{D} \tilde{k}\, \mathrm{J}_1(\tilde{k}\tilde{r})\,\mathrm{d}\tilde{k}\,, \tag{2.1b}$$

$$u_z = \frac{P_0(1-\nu')}{2\pi G_{rz0} r} S_{vv}\,, \tag{2.1c}$$

$$S_{vv} = -\frac{\beta^2 \tilde{r}}{1-\nu'} \int_0^\infty \frac{c_{22}\tilde{w}_1(\tilde{z},\tilde{k}) - c_{21}\tilde{w}_2(\tilde{z},\tilde{k})}{D} \tilde{k}\, \mathrm{J}_0(\tilde{k}\tilde{r})\,\mathrm{d}\tilde{k}\,, \tag{2.1d}$$

where

$$\tilde{r} = \frac{r}{z_r}\,. \tag{2.2}$$

Here, value D is determined from (1.56). The dimensionless values S_{vh} and S_{vv} in expressions (2.1a) and (2.1c) for the amplitudes of vibrations are multiplied by the vertical static displacements of points on the surface of an isotropic half-space with shear modulus G_{rz0} and Poisson's ratio ν' (see (1.9)). For the case of u_r, the corresponding horizontal static displacements could be employed; however, this is less convenient since these displacements become are zero at $\nu' = 0.5$. In order to calculate the amplitudes of vibrations at the points that belong to the Z-axis, i.e. at $r = 0$, one should employ a modified representation of the solution by replacing coordinate r with z in its static part with the corresponding modification of coefficients (this procedure concerns amplitudes u_v only; the horizontal amplitudes of vibrations are zero on the Z-axis). Realization of solution (2.1) is possible when two linearly independent solutions of the system of equations (1.25), (1.26), or (1.30), (1.31), satisfying the condition of absence of sources at infinity, are constructed. In addition, one could employ a system of first-order equations (1.126)–(1.129), or (1.136)–(1.139). Using equations (1.30), (1.31), and the fact that in the considered case of homogeneity all characteristics of the half-space are constant and $\tilde{G} = 1$, $\tilde{\rho} = 1$, we obtain

$$\frac{\mathrm{d}^2 \tilde{q}}{\mathrm{d}\tilde{z}^2} + (\beta^2\theta^2 - \tilde{k}^2 \tilde{A}_{rr})\tilde{q} - \tilde{k}(\tilde{A}_{rz} + 1)\frac{\mathrm{d}\tilde{w}}{\mathrm{d}\tilde{z}} = 0\,, \tag{2.3}$$

$$\tilde{A}_{zz} \frac{\mathrm{d}^2 \tilde{w}}{\mathrm{d}\tilde{z}^2} + (\beta^2\theta^2 - \tilde{k}^2)\tilde{w} + \tilde{k}(\tilde{A}_{rz} + 1)\frac{\mathrm{d}\tilde{q}}{\mathrm{d}\tilde{z}} = 0\,. \tag{2.4}$$

Taking an origin of coordinates at the surface of the half-space, we have $z_0 = 0$. Next, we search for a solution of the system of equations (2.3), (2.4) in the following form:

$$\tilde{q} = C_q \exp(-\eta\tilde{z})\,, \tag{2.5a}$$

$$\tilde{w} = C_w \exp(-\eta\tilde{z})\,. \tag{2.5b}$$

As a result, we have the following system of equations with respect to coefficients C_q, C_w:

$$(\eta^2 + \beta^2\theta^2 - \tilde{k}^2 \tilde{A}_{rr})C_q + \tilde{k}(\tilde{A}_{rz} + 1)\eta C_w = 0,$$
$$-\tilde{k}(\tilde{A}_{rz} + 1)\eta C_q + (\tilde{A}_{zz}\eta^2 + \beta^2\theta^2 - \tilde{k}^2)C_w = 0. \qquad (2.6)$$

Since we seek for a non-zero solution having the form (2.5), the determinant of the system of equations (2.6) should be equated to zero:

$$\tilde{A}_{zz}\eta^4 + B\eta^2 + C = 0, \qquad (2.7)$$

where

$$B = \tilde{k}^2 \tilde{B} + \beta^2\theta^2(1 + \tilde{A}_{zz}), \qquad (2.8a)$$

$$C = (\tilde{k}^2 - \beta^2\theta^2)(\tilde{A}_{rr}\tilde{k}^2 - \beta^2\theta^2), \qquad (2.8b)$$

$$\tilde{B} = 2\tilde{A}_{rz} + \tilde{A}_{rz}^2 - \tilde{A}_{zz}\tilde{A}_{rr}. \qquad (2.8c)$$

Consider the following roots of equation (2.7):

$$\eta_1 = \sqrt{\frac{-B + \sqrt{B^2 - 4\tilde{A}_{zz}C}}{2\tilde{A}_{zz}}}, \qquad (2.9a)$$

$$\eta_2 = \sqrt{\frac{-B - \sqrt{B^2 - 4\tilde{A}_{zz}C}}{2\tilde{A}_{zz}}}. \qquad (2.9b)$$

where the principal values of radicals are taken. It can be proved that these roots have a positive real part at large values of $\tilde{k}$. Indeed, for the case

$$\tilde{B} < 0 \qquad (2.10)$$

coefficient B becomes negative at large $\tilde{k}$, so that the statement concerning the roots η_1 and η_2 is obviously true (the latter radicals contain numbers with a positive real part). If $\tilde{B} \geq 0$, a consideration of the expression under the inner radical in equations (2.9) shows that the multiplier before $\tilde{k}^4$ takes the form

$$\tilde{B}^2 - 4\tilde{A}_{rr}\tilde{A}_{zz}. \qquad (2.11)$$

Since $\tilde{B} \geq 0$ and

$$\tilde{A}_{rz}^2 < \tilde{A}_{zz}\tilde{A}_{rr} \qquad (2.12)$$

(this stems from relationships (1.12)), the value in (2.11) must be negative. Indeed,

$$(2\widetilde{A}_{rz} + \widetilde{A}_{rz}^2 - \widetilde{A}_{zz}\widetilde{A}_{rr})^2 - 4\widetilde{A}_{rr}\widetilde{A}_{zz} < 4\widetilde{A}_{rz}^2 - 4\widetilde{A}_{rr}\widetilde{A}_{zz} < 0 . \tag{2.13}$$

Thus, for $\widetilde{B} \geq 0$ and for large values of $\widetilde{k}$, the expression under the inner radical in equations (2.9) becomes negative, which results in complex conjugate roots η_1 and η_2 with positive real parts. The other roots, $\eta_3 = -\eta_1$ and $\eta_4 = -\eta_2$, have a negative real part at large values of parameter $\widetilde{k}$.

We determine two needed solutions in the form (2.5) by taking $\eta = \eta_j$ (j=1,2); coefficient C_q is assumed to be equal to 1 for both solutions, while coefficient C_w is calculated for roots η_j from the first of equations (2.6):

$$C_{wj} = -\frac{\eta_j^2 + \beta^2\theta^2 - \widetilde{k}^2\widetilde{A}_{rr}}{\widetilde{k}(\widetilde{A}_{rz} + 1)\eta_j} . \tag{2.14}$$

For the isotropic case (1.5), relationships (2.9) yield

$$\eta_1 = \sqrt{\widetilde{k}^2 - \beta^2\theta^2} , \tag{2.15a}$$

$$\eta_2 = \sqrt{\widetilde{k}^2 - \beta^2\theta^2 / \widetilde{A}_{zz}} , \tag{2.15b}$$

which correspond to values (1.20). Coefficients C_{wj} for the isotropic case take the form

$$C_{w1} = \frac{\widetilde{k}}{\eta_1} , \tag{2.16a}$$

$$C_{w2} = \frac{\eta_2}{\widetilde{k}} . \tag{2.16b}$$

As a result, we obtain two solutions corresponding to those predented in equations (1.51).

For dynamic problems ($\omega \neq 0$), we employ the freedom to select the reference length z_r, and set the condition: $\theta = 1$. Thus, in accordance with (1.32a), the reference length varies with the vibration frequency:

$$z_r = \frac{C_{rz}}{\omega} , \tag{2.17}$$

where

$$C_{rz} = \sqrt{\frac{G_{rz0}}{\rho}} . \tag{2.18}$$

The value of z_r equals the distance which the shear wave corresponding to the shear modulus G_{rz0} and density ρ of the considered medium passes by the time ω^{-1}, equal to the vibration period divided by 2π. As a result, the dimensionless parameters $\widetilde{r}$ and $\widetilde{z}$ take the form

$$\widetilde{r} = \frac{\omega r}{C_{rz}}, \tag{2.19a}$$

$$\widetilde{z} = \frac{\omega z}{C_{rz}}. \tag{2.19b}$$

Thus, the sought results are obtained by formulas (2.1) taking $\theta = 1$ and using dimensionless distances calculated from (2.19). Values of c_{ij} and D are determined by employing (1.54), (1.56) and taking into account the fundamental solutions constructed above.

2.1.1 Static Action of Vertical Force

In order to solve a static problem, one should set the following values of parameters, namely: $\theta = 0, \beta = 1$. For this case, integrals entering the expressions for S_{vv} and S_{vh} may be represented in a finite form. Indeed, the dependence of values appearing in the integrals of variable $\widetilde{k}$ is significantly simplified:

$$\eta_1 = \widetilde{k}\widetilde{\eta}_1, \tag{2.20a}$$

$$\eta_2 = \widetilde{k}\widetilde{\eta}_2, \tag{2.20b}$$

$$C_{wj} = \widetilde{C}_{wj} = \frac{\widetilde{A}_{rr} - \widetilde{\eta}_j^2}{(\widetilde{A}_{rz} + 1)\widetilde{\eta}_j}, \tag{2.20c}$$

$$c_{ij} = \widetilde{k}\widetilde{c}_{ij}, \tag{2.20d}$$

$$D = \widetilde{k}^2 \widetilde{D}, \tag{2.20e}$$

$$\widetilde{q}_j(\widetilde{z},\widetilde{k}) = \exp(-\widetilde{k}\widetilde{\eta}_j\widetilde{z}), \tag{2.20f}$$

$$\widetilde{w}_j(\widetilde{z},\widetilde{k}) = \widetilde{C}_{wj}\exp(-\widetilde{k}\widetilde{\eta}_j\widetilde{z}), \tag{2.20g}$$

where the following constants are introduced:

$$\widetilde{\eta}_1 = \sqrt{\frac{-\widetilde{B} + \sqrt{\widetilde{B}^2 - 4\widetilde{A}_{zz}\widetilde{A}_{rr}}}{2\widetilde{A}_{zz}}}, \tag{2.21a}$$

$$\widetilde{\eta}_2 = \sqrt{\frac{-\widetilde{B} - \sqrt{\widetilde{B}^2 - 4\widetilde{A}_{zz}\widetilde{A}_{rr}}}{2\widetilde{A}_{zz}}}\,, \tag{2.21b}$$

$$\widetilde{c}_{1j} = \widetilde{A}_{rz} - \widetilde{A}_{zz}\widetilde{C}_{wj}\widetilde{\eta}_j\,, \tag{2.21c}$$

$$\widetilde{c}_{2j} = -\widetilde{\eta}_j - \widetilde{C}_{wj}\,, \tag{2.21d}$$

$$\widetilde{D} = \widetilde{c}_{11}\widetilde{c}_{22} - \widetilde{c}_{12}\widetilde{c}_{21}\,. \tag{2.21e}$$

For the case with a negative value under the inner radical, pairs of values $\widetilde{\eta}_1$ and $\widetilde{\eta}_2$, $\widetilde{c}_{11}$ and $\widetilde{c}_{12}$, $\widetilde{C}_{w1}$ and $\widetilde{C}_{w2}$ become complex conjugates. Static parameters S_{vv} and S_{vh} take the following form:

$$S_{vh} = -\frac{\widetilde{c}_{22}\widetilde{r}}{\widetilde{D}(1-\nu')}\int_0^\infty \exp(-\widetilde{z}\widetilde{\eta}_1\widetilde{k})\,\mathrm{J}_1(\widetilde{k}\widetilde{r})\,\mathrm{d}\widetilde{k} + \frac{\widetilde{c}_{21}\widetilde{r}}{\widetilde{D}(1-\nu')}\int_0^\infty \exp(-\widetilde{z}\widetilde{\eta}_2\widetilde{k})\,\mathrm{J}_1(\widetilde{k}\widetilde{r})\,\mathrm{d}\widetilde{k}, \tag{2.22a}$$

$$S_{vv} = -\frac{\widetilde{c}_{22}\widetilde{C}_{w1}\widetilde{r}}{\widetilde{D}(1-\nu')}\int_0^\infty \exp(-\widetilde{z}\widetilde{\eta}_1\widetilde{k})\,\mathrm{J}_0(\widetilde{k}\widetilde{r})\,\mathrm{d}\widetilde{k} + \frac{\widetilde{c}_{21}\widetilde{C}_{w2}\widetilde{r}}{\widetilde{D}(1-\nu')}\int_0^\infty \exp(-\widetilde{z}\widetilde{\eta}_2\widetilde{k})\,\mathrm{J}_0(\widetilde{k}\widetilde{r})\,\mathrm{d}\widetilde{k}. \tag{2.22b}$$

Employing tabulated integrals [43], we obtain

$$S_{vh} = \frac{1}{\widetilde{D}(1-\nu')}\left[-\widetilde{c}_{22}\left(1 - \frac{z\widetilde{\eta}_1}{\sqrt{r^2 + z^2\widetilde{\eta}_1^2}}\right) + \widetilde{c}_{21}\left(1 - \frac{z\widetilde{\eta}_2}{\sqrt{r^2 + z^2\widetilde{\eta}_2^2}}\right)\right], \tag{2.23a}$$

$$S_{vv} = \frac{r}{\widetilde{D}(1-\nu')}\left(-\frac{\widetilde{c}_{22}\widetilde{C}_{w1}}{\sqrt{r^2 + z^2\widetilde{\eta}_1^2}} + \frac{\widetilde{c}_{21}\widetilde{C}_{w2}}{\sqrt{r^2 + z^2\widetilde{\eta}_2^2}}\right). \tag{2.23b}$$

These results are in agreement with the solution derived by Michell [10, 74].

For the surface of the half-space ($z = 0$)

$$S_{vh} = \frac{\widetilde{c}_{21} - \widetilde{c}_{22}}{\widetilde{D}(1-\nu')}, \tag{2.24a}$$

$$S_{vv} = \frac{\widetilde{c}_{21}\widetilde{C}_{w2} - \widetilde{c}_{22}\widetilde{C}_{w1}}{\widetilde{D}(1-\nu')}\,. \tag{2.24b}$$

For the points that belong to the acting line of force (on the Z-axis), the displacements u_r vanish, while the displacements u_z may be written as follows:

$$u_z = \frac{P_0(3-2\nu')}{4\pi G_{rz0} z} S_{vz} , \tag{2.25a}$$

$$S_{vz} = \frac{2}{(3-2\nu')\widetilde{D}} \left(\frac{\widetilde{c}_{21}\widetilde{C}_{w2}}{\widetilde{\eta}_2} - \frac{\widetilde{c}_{22}\widetilde{C}_{w1}}{\widetilde{\eta}_1} \right), \tag{2.25b}$$

where the static displacements of an isotropic half-space are separated out in (2.25a) (the multiplier to S_{vz}) [91].

Note that a specific case is possible with $\widetilde{\eta}_1$ equal to $\widetilde{\eta}_2$ at $\widetilde{B}^2 - 4\widetilde{A}_{zz}\widetilde{A}_{rr} = 0$ (e.g. for an isotropic half-space $\widetilde{\eta}_1 = \widetilde{\eta}_2 = 1$). In this case, value $\widetilde{D}$ vanishes and expressions (2.23), (2.24) and (2.25) become indeterminate. Instead of constructing the appropriate limits, one may use the above-obtained formulas by setting a very small deviation of elastic constants (or one of them) from given values that result in the equality of roots (this approach requires sufficient precision of calculation).

Consider the following representation for the elastic constants (see equations (1.9)):

$$E' = \xi' G_{rz}, E = \xi E' . \tag{2.26}$$

Note that a material should be isotropic when $\nu = \nu'$, $\xi' = 2(1+\nu')$ and $\xi = 1$. Conditions $m > 0$ and $a_{rr}a_{zz} > a_{rz}^2$, which were mentioned in (1.12) and (1.13), yield the following constraints, respectively:

$$\xi < \frac{1-\nu}{2\nu'^2} , \tag{2.27a}$$

$$\xi < \frac{1}{\nu'^2} . \tag{2.27b}$$

Specifying ν', ν, ξ' and ξ, one may determine the ratios of modules E' and E to shear modulus G_{rz} and calculate by using formulas (1.10) the normalized parameters $\widetilde{A}_{rr}$, $\widetilde{A}_{zz}$, $\widetilde{A}_{rz}$ required for constructing the solution of the problem.

The spatial behavior of static displacements on the surface of a half-space and on the vertical line of the force acting is similar for isotropic and transversely isotropic half-spaces (the form of dependence on coordinates r and z is the same); the only difference is due to the corresponding coefficients. In Fig. 2.1, we present coefficients S_{vv}, S_{vh}, S_{vz} versus parameter ξ at $\nu = \nu'$, $\xi' = 2(1+\nu')$ (the degree of anisotropy is only determined by deviation of parameter ξ from unity) and for Poisson's ratios $\nu' = \nu = 0, 1/3, 1/2$. According to relationsips (2.27), in the first

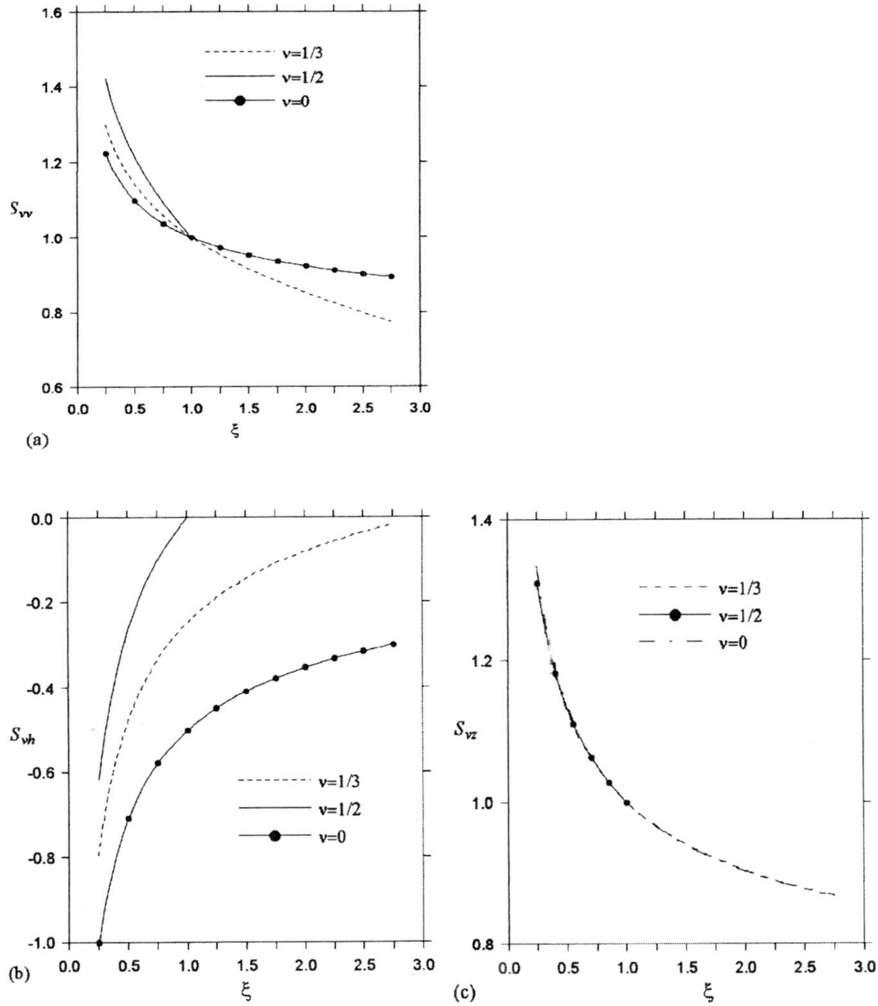

Fig. 2.1a,b,c. Normalized static displacements of points of transversely isotropic half-space subjected to vertical force

case parameter ξ has no upper bound, while in the second and third cases the bounds are 3 and 1, respectively. At $\xi = 1$, the values of S_{vv}, S_{vz} are equal to unity, and $S_{vh} = -0.5, -0.25, 0$ at $\nu' = 0, 1/3, 1/2$, respectively, being in agreement with the results for the isotropic case. As expected, increasing medium stiffness in the horizontal directions results in a reduction of displacements. The influence of anisotropy is more noticeable in the case of horizontal displacements.

The value of S_{vz} is practically independent of Poisson's ratio within the domain of definition of S_{vz} .

2.1.2 Free Vibrations of Half-Space

Next, we consider the possibility of the occurrence of undamped free vibrations of a half-space. In order to answer this question, one should study the value of D that is a determinant of the system of equations (1.53) for coefficients A_j. Zeros of this value that correspond to the occurrence of free vibrations (Rayleigh waves) result in the poles of integrands in solution (2.1) of the problem. The roots $\tilde{k}_R$ of the equation $D = 0$ at $\varepsilon = 0$ are calculated for the case of relationships (2.26) at $\nu = \nu'$, $\xi' = 2(1+\nu')$, similar to the calculations performed in the previous example. Parameter $\tilde{k}_R$ may be called a dimensionless wave number for Rayleigh waves (corresponding to dimensionless distances $\tilde{r}$ in (2.19a)). In Fig. 2.2, parameter $\tilde{k}_R$ is presented versus ξ for three different values $\nu = 0,\ 1/3,\ 1/2$. At $\xi = 1$ (the isotropic case), we obtain known results: $\tilde{k}_R$ =1.1441, 1.0724, 1.0468 for $\nu = 0,\ 1/3,\ 1/2$, respectively. Note that in the study of free vibrations, in order to construct a solution in the form of propagating waves, one should consider the dependence of solutions on coordinate r in the form of Hankel's functions $\mathrm{H}_0^{(2)}(\tilde{k}_R\tilde{r})$ or $\mathrm{H}_1^{(2)}(\tilde{k}_R\tilde{r})$, instead of in the form $\mathrm{J}_0(\tilde{k}_R\tilde{r})$ or $\mathrm{J}_1(\tilde{k}_R\tilde{r})$ (as in the integrands (2.1)); in 2-D problems, the function $\exp(-\mathrm{i}\,\tilde{k}_R\tilde{x})$ corresponding to solution (1.91) is used. Following (2.19), the spatial wave number with respect to variable r (or x) may be written as

$$k_R = \frac{\tilde{k}_R\omega}{C_{rz}} \tag{2.28}$$

with the corresponding velocity of Rayleigh waves:

$$V_R = \frac{\omega}{k_R} = \frac{C_{rz}}{\tilde{k}_R}. \tag{2.29}$$

The results presented in Fig. 2.2 indicate that a reduction in material stiffness in the horizontal direction is followed by a decreasing velocity of Rayleigh waves, as compared with the velocity of shear waves C_{rz}. For all values of ξ, we have $\tilde{k}_R >$ 1; at $\nu = 0$, the value of $\tilde{k}_R$ tends to unity, as ξ increases.

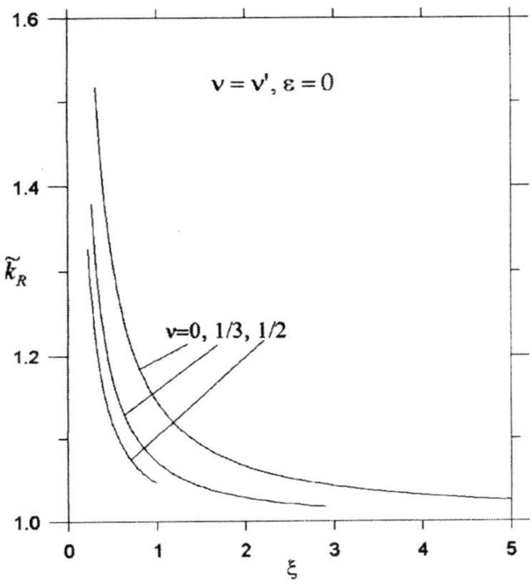

Fig. 2.2. Dimensionless wave numbers $\widetilde{k}_R$ versus anisotropy parameter ξ

2.1.3 Forced Vibrations of Half-Space

In order to calculate the amplitudes of vibrations by using formulas (2.1), one should perform the integration by applying a corresponding path of integration because of singularities in the integrands. If one takes into account the dissipative properties of the medium by specifying some value $\varepsilon > 0$ in equations (1.28) and (1.32b), then a zero of D moves down from the real axis in the complex plane $\widetilde{k}$. If $\widetilde{k}_R$ is a root of the equation $D = 0$ at $\varepsilon = 0$ ($\beta = 1$), then the complex value $\beta\,\widetilde{k}_R$ (that corresponds to a point in the complex plane $\widetilde{k}$, which is located below the real axis) makes D vanish at $\varepsilon > 0$. Indeed, in going from $\varepsilon = 0, \widetilde{k} = \widetilde{k}_R$ to $\varepsilon > 0$, $\widetilde{k} = \beta\widetilde{k}_R$, values B and C gain multipliers β^2 and β^4, respectively, and, following equation (2.7), new roots η_j are obtained from the previous ones by multiplying the latter by β. Furthermore, following (2.14), coefficients C_{wj} remain unchanged. Coefficients c_{ij}, determined from (1.54), gain the multiplier β, while a new value of D in the given transition becomes equal to its previous value, multiplied by β^2, i.e. zero. Taking into account the location of the roots of D in the complex plane at $\varepsilon > 0$, application of the path of integration shown in Fig. 2.3

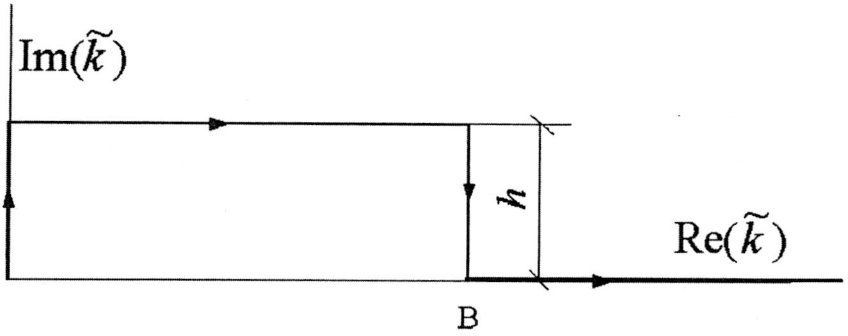

Fig. 2.3. Integration contour for dynamic problems

may serve as a reasonable approach to integrating expressions of the form (2.1). Here, h is a small positive value (h = 0.03–0.05), and a point B is located to the right from possible poles of the integrand at $\varepsilon = 0$ (in the considered case, a single pole $\widetilde{k}_R$ exists). In computations, an abscissa of point B was taken to be equal to 20. The convergence of the integrals at $z > 0$ is provided by exponentially decreasing multipliers entering solutions q_1, w_1 and q_2, w_2 (see (2.5)). For the points on the surface of the half-space, the integrals in (2.1) converge only due to decreasing oscillations of Bessel's functions. For the case with $z = 0$, the following procedure may be recommended. If the value $\widetilde{k} = \widetilde{k}_{\mathrm{B}}$ does not provide sufficient precision in the calculation of Bessel's functions by using the principal term in their asymptotic representation [1]

$$\mathrm{J}_n(\widetilde{k}\widetilde{r}) \approx \left(\frac{2}{\pi\widetilde{k}\widetilde{r}}\right)^{1/2} \cos\left(\widetilde{k}\widetilde{r} - \frac{n\pi}{2} - \frac{\pi}{4}\right), \tag{2.30}$$

we introduce a second point, B_1, with abscissa $\widetilde{k}_{\mathrm{B}_1} > \widetilde{k}_{\mathrm{B}}$. In our calculations, we assume $\widetilde{k}_{\mathrm{B}_1} = 90/\widetilde{r}$. If the latter value exceeds $\widetilde{k}_{\mathrm{B}}$, then one should add the integral between the limits $\widetilde{k}_{\mathrm{B}}$ and $\widetilde{k}_{\mathrm{B}_1}$ to the integral taken along that part of the integration path located to the left of point B. Finally, integrals between the limits $\widetilde{k}_m = \max(\widetilde{k}_{\mathrm{B}}, \widetilde{k}_{\mathrm{B}_1})$ and ∞ are determined approximately with the help of the following formula, derived by integration by parts and using representation (2.30):

$$\int_{\widetilde{k}_m}^{\infty} \Phi(\widetilde{k}) \left(\frac{2}{\pi\widetilde{k}\widetilde{r}}\right)^{1/2} \cos\left(\widetilde{k}\widetilde{r} - \frac{n\pi}{2} - \frac{\pi}{4}\right) \mathrm{d}\widetilde{k}$$

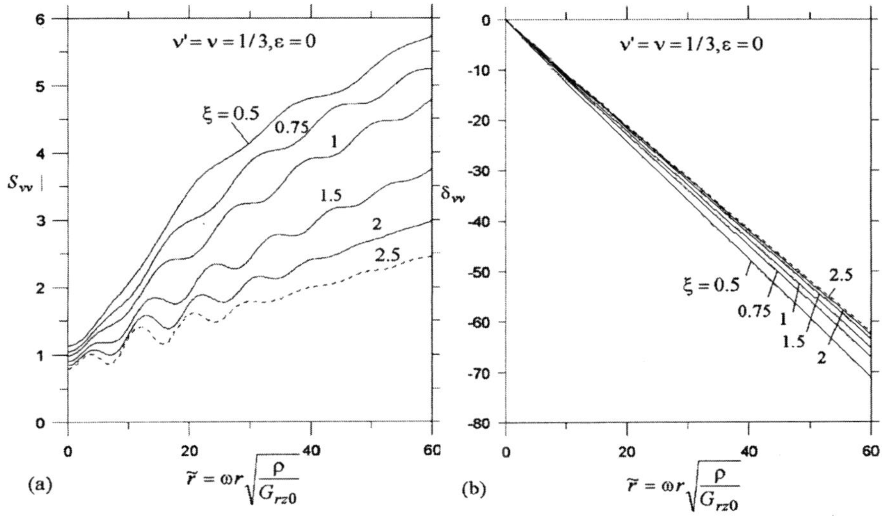

Fig. 2.4a,b. Absolute values (a) and phases (b) of complex normalized vertical amplitudes of vibrations under action of vertical force

$$\approx -\Phi(\tilde{k}_m)\left(\frac{2}{\pi\tilde{k}_m\tilde{r}}\right)^{1/2}\frac{\sin(\tilde{k}_m\tilde{r} - n\pi/2 - \pi/4)}{\tilde{r}}. \tag{2.31}$$

Here, function $\Phi(\tilde{k})$ corresponds to the multipliers before Bessel's functions in the integrals in (2.1); its behavior at high values of $\tilde{k}$ results in an expansion of negative powers of $\tilde{k}$. Note that continuing the procedure of integration by parts leads to an expansion of integral (2.31) in negative powers of $\tilde{k}_m\tilde{r} \geq 90$. It is best to increase the value of $\tilde{k}_m$ so that the magnitude of the sine in relationship (2.31) becomes unity. As a result, the next term appearing in (2.31), due to further integration by parts, vanishes. Computations of the integrals using Simpson's formula, taking into account the considerations above, provide a sufficiently high accuracy (4–5 significant digits); high values of variable $\tilde{r}$ can be reached.

In Fig. 2.4, magnitudes $|S_{vv}|$ and phases δ_{vv} are shown for complex normalized amplitudes of vibrations

$$S_{vv} = |S_{vv}|\exp(\mathrm{i}\,\delta_{vv}) \tag{2.32}$$

of the points on the surface of a transversely isotropic homogeneous half-space at $\xi' = 2(1+\nu')$, $\nu' = \nu = 1/3$, $\varepsilon = 0$, and for a number of values of parameter ξ. An increase in values $|S_{vv}|$ with increasing values of $\tilde{r}$ determined by (2.19a) is explained by considering the influence of the pole $\tilde{k} = \tilde{k}_R$, which results in a contribution to the value S_{vv} proportional to $(\tilde{r})^{1/2}$ (at high values of $\tilde{r}$). The results show that decreasing stiffness in a horizontal plane causes a significant increase in the dynamic effect related to Rayleigh waves. The values of normalized amplitudes at $\tilde{r} = 0$ correspond to the static action of the force. For the isotropic case ($\xi = 1$), the corresponding static value of $|S_{vv}|$ equals unity, decreasing (increasing) with increasing (decreasing) parameter ξ. These static values were calculated following (2.24). Next, an alternative technique is considered. We determine the limit (as $\tilde{k} \to \infty$) value of Φ_∞ of the multiplier before Bessel's function $J_0(\tilde{k}\tilde{r})$ under the integral (2.1d) at $\tilde{z} = 0$. Subtraction and addition of this limit value under the integral yield

$$S_{vv} = -\frac{\beta^2 \tilde{r}}{1-\nu'} \int_0^\infty [\Phi(\tilde{k}) - \Phi_\infty] J_0(\tilde{k}\tilde{r}) \mathrm{d}\tilde{k} - \frac{\beta^2 \tilde{r}}{1-\nu'} \int_0^\infty \Phi_\infty J_0(\tilde{k}\tilde{r}) \mathrm{d}\tilde{k} . \qquad (2.33)$$

Consideration of the behavior of Bessel's function at high values of its argument shows that the first summand in (2.33) tends to zero as $\tilde{r} \to \infty$, so that the static value of S_{vv} is obtained from the following relationship:

$$S_{vv}^{st} = -\frac{\beta^2 \tilde{r}}{1-\nu'} \Phi_\infty \int_0^\infty J_0(\tilde{k}\tilde{r}) \mathrm{d}\tilde{k} = -\frac{\beta^2}{1-\nu'} \Phi_\infty . \qquad (2.34)$$

The limit value Φ_∞ may be determined by calculating an integrand at extremely high values of $\tilde{k}$, e.g. for $\tilde{k} = 1000$. This approach leads (at $\beta = 1$) to results that agree with those obtained by the formula used above.

The behavior of phase δ_{vv}, shown in Fig. 2.4b, is in good agreement with the above values of $\tilde{k}_R$. As the value of $\tilde{r}$ increases, that part of the solution which appears due to the pole of the integrand in equations (2.1) (i.e. the part which contains the parameters of Rayleigh waves) should represent the complete solution in an increasingly accurate way. Therefore, at high values of $\tilde{r}$ the lines in Fig. 2.4b must tend to straight lines with a slope tending to $\tilde{k}_R$. Considering the secant curves in Fig. 2.4b for the interval $50 < \tilde{r} < 60$, we find the following slopes: 1.1709, 1.0998, 1.0720, 1.0426, 1.0317, 1.0246, respectively, for the values $\xi = 0.5, 0.75, 1, 1.5, 2, 2.5$. For the same values of ξ, the corresponding values of $\tilde{k}_R$ are equal to 1.1713, 1.1037, 1.0724, 1.0433, 1.0297, 1.0218, respectively.

The results of calculations of the normalized horizontal amplitudes S_{vh} by (2.1b) for the points on the surface of a half-space with the same values of parameters, as in the case with S_{vv}, are presented in Fig. 2.5. Here, a representation analogous to (2.32) is employed. In comparison with the results relating to vertical amplitudes, a dynamic effect is now clearly visible, and the

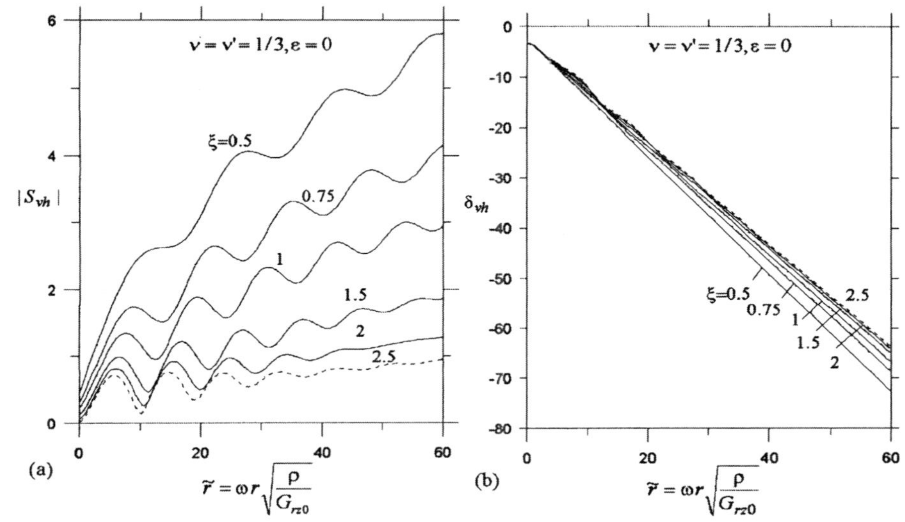

Fig. 2.5a,b. Absolute values (a) and phases (b) of complex normalized horizontal amplitudes of vibrations under action of vertical force

influence of parameter ξ on the results of calculations becomes more significant. The phase curves, corresponding to small and intermediate values of parameter $\tilde{r}$, differ from straight lines to a more considerable extent than in the case of amplitudes S_{vv}. Only at $\tilde{r}$ >30–40 do curves approach the straight lines; this corresponds to the prevailing influence of the Rayleigh waves at high values of $\tilde{r}$. Static values of amplitudes S_{vh} are negative (for the isotropic case, we obtain the known value, namely –0.25), so that the initial phase in Fig. 2.5b equals $-\pi$.

Next, we consider the influence of internal friction on the amplitudes of vibrations. Calculations indicate that for small values of parameter ε, phase angles δ_{vv} and δ_{vh} differ insignificantly from their values at $\varepsilon = 0$. However, a change occurring in the magnitudes of amplitudes is rather significant, as compared with the case of absence of energy dissipation. For $\varepsilon = 0.05$ (with corresponding damping ratio equal to $\varepsilon/2 = 0.025$), values of $|S_{vv}|$ and $|S_{vh}|$ are presented in Fig. 2.6, where values of the rest of parameters are kept the same as in previous calculations. A comparison with results, corresponding to the case $\varepsilon = 0$ (Fig. 2.4a and Fig. 2.5a), indicates that a radical change in behavior of the amplitudes of vibrations takes place. Instead of an increase of normalized amplitudes, which occurs due to the influence of Rayleigh waves, we observe a decrease, starting from values of parameter $\tilde{r}$ around 20. Also, one should account for an additional decreasing with r due to the multiplier in (2.1), which represents static displacements.

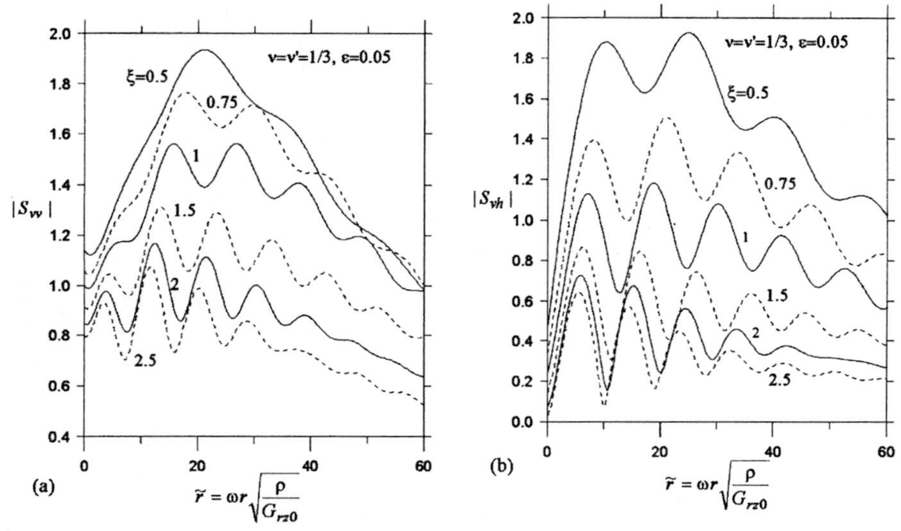

Fig. 2.6a,b. Absolute values of normalized vertical (a) and horizontal (b) amplitudes of vibrations of surface points of half-space with internal friction (ε =0.05) under action of vertical force

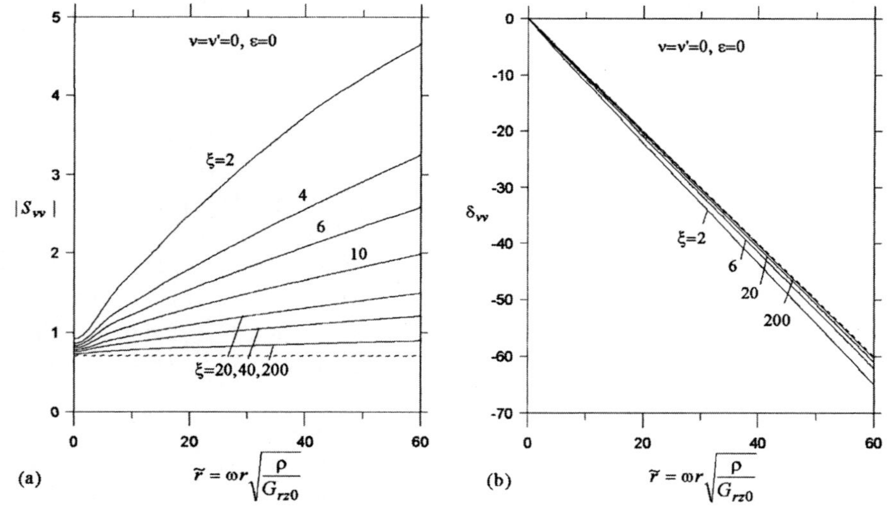

Fig. 2.7a,b. Absolute values (a) and phases (b) of complex normalized vertical amplitudes of vibrations at high values of half-space stiffness in horizontal direction

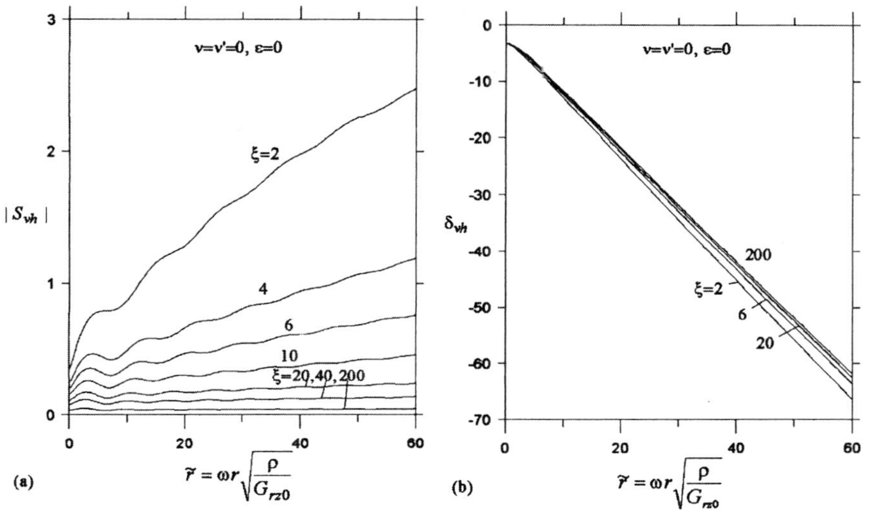

Fig. 2.8a,b. Absolute values (a) and phases (b) of complex normalized horizontal amplitudes of vibrations at high values of half-space stiffness in horizontal direction

The following sample calculations correspond to zero values of Poisson's ratios ν and ν'; dissipation parameter ε vanishes, too. For parameters ξ and ξ', one may take arbitrary positive numbers. The results of calculations are presented in Figs. 2.7–2.8 for a series of values of ξ and at $\xi' = 2(1+\nu') = 2$. With increasing ξ, horizontal displacements of the points in the half-space decrease and vanish when $\xi \to \infty$; normalized amplitudes increase slower with increasing $\widetilde{r}$, and as $\xi \to \infty$, the growth ceases (the influence of surface waves disappears). The limiting case of anisotropy represented by the dashed lines in Fig. 2.7 leads to a model of deformable foundations, which has relatively simple determining equations. This model is presented in section 2.2.

Next, we consider stresses σ_z, caused by a vertical force which is applied to the surface of a transversely isotropic half-space. For the case of a concentrated force, equation (1.58a) may be written in the following form:

$$\sigma_z = -\frac{3P_0}{2\pi z^2} S_\sigma , \tag{2.35a}$$

$$S_\sigma = \frac{\widetilde{z}^2}{3}\int_0^\infty \widetilde{k}\left\{ c_{22}\left[\widetilde{k}\widetilde{A}_{rz}\widetilde{q}_1(\widetilde{z},\widetilde{k}) + \widetilde{A}_{zz}\frac{\mathrm{d}\,\widetilde{w}_1(\widetilde{z},\widetilde{k})}{\mathrm{d}\,\widetilde{z}} \right] \right.$$
$$\left. - c_{21}\left[\widetilde{k}\widetilde{A}_{rz}\widetilde{q}_2(\widetilde{z},\widetilde{k}) + \widetilde{A}_{zz}\frac{\mathrm{d}\,\widetilde{w}_2(\widetilde{z},\widetilde{k})}{\mathrm{d}\,\widetilde{z}} \right]\right\} \frac{\mathrm{J}_0(\widetilde{k}\widetilde{r})}{D}\,\mathrm{d}\,\widetilde{k} . \tag{2.35b}$$

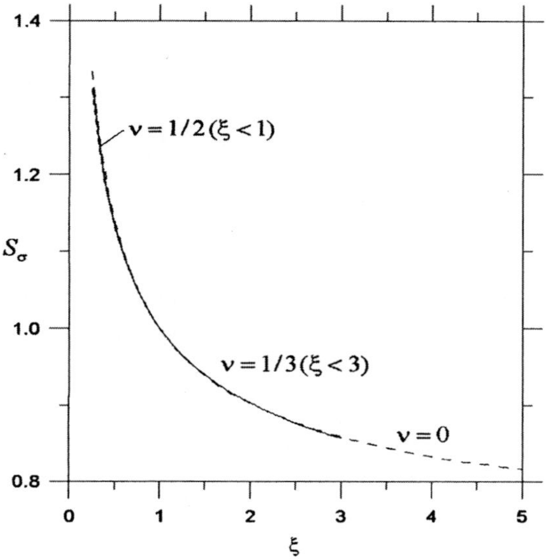

Fig. 2.9. Influence of anisotropy parameter ξ on normalized stresses σ_z at points of half-space, that belong to the line of action of vertical force applied to half-space surface

As previously noted, we assume $z_0 = 0$. The multiplier separated out in relationship (2.35a) represents static stresses that occur in points on the Z-axis in the isotropic homogeneous half-space subjected to the force P_0 [91]. Convergence of the integral in the expression for normalized stresses S_σ is provided at $\tilde{z} > 0$, by exponentially decreasing multipliers in fundamental solutions $\tilde{q}_j$, $\tilde{w}_j$ (see (2.5), (2.9)).

First, we consider stresses under the static action of force P_0. Applying relationships (2.20), (2.21), one may present parameter S_σ in the following form:

$$S_\sigma = \frac{1}{3\tilde{D}}\left[(\tilde{A}_{rz} - \tilde{A}_{zz}\tilde{C}_{w1}\tilde{\eta}_1)\tilde{c}_{22}\tilde{z}^2 \int_0^\infty \tilde{k}\exp(-\tilde{k}\tilde{\eta}_1\tilde{z})\,\mathrm{J}_0(\tilde{k}\tilde{r})\,\mathrm{d}\tilde{k}\right.$$

$$\left. -(\tilde{A}_{rz} - \tilde{A}_{zz}\tilde{C}_{w2}\tilde{\eta}_2)\tilde{c}_{21}\tilde{z}^2 \int_0^\infty \tilde{k}\exp(-\tilde{k}\tilde{\eta}_2\tilde{z})\,\mathrm{J}_0(\tilde{k}\tilde{r})\,\mathrm{d}\tilde{k}\right]. \tag{2.36}$$

Employing tabulated integrals and formulas (2.21c) for $\tilde{c}_{1j}$, we obtain the following expression for static normalized stresses S_σ:

$$S_\sigma = \frac{z^3}{3\tilde{D}}\left[\tilde{c}_{11}\tilde{c}_{22}\frac{\tilde{\eta}_1}{(\tilde{\eta}_1^2 z^2 + r^2)^{3/2}} - \tilde{c}_{12}\tilde{c}_{21}\frac{\tilde{\eta}_2}{(\tilde{\eta}_2^2 z^2 + r^2)^{3/2}}\right]. \tag{2.37}$$

On the line of action of the force ($r = 0$)

$$S_\sigma = \frac{1}{3\widetilde{D}}\left[\frac{\widetilde{c}_{11}\widetilde{c}_{22}}{\widetilde{\eta}_1^2} - \frac{\widetilde{c}_{12}\widetilde{c}_{21}}{\widetilde{\eta}_2^2}\right]. \tag{2.38}$$

Next, the results of calculations by using (2.38) with relationships between elastic coefficients given in equations (2.26) are presented for $\nu = \nu', \xi' = 2(1+\nu')$, $\nu = 0, 1/3, 0.5$ (Fig. 2.9). The corresponding values of S_σ are close to each other. Thus, in the considered case of anisotropy, Poisson's ratio has no significant influence on the relationship between stresses and the degree of anisotropy; recall that the permissible upper limit for parameter ξ depends on ν and ν', as follows from conditions (2.27). The influence of parameter ξ on stresses σ_z is similar to its influence on displacements: an increase of parameter ξ causes a decrease in the considered values.

Consider the behavior of dynamic stresses σ_z. In Fig. 2.10, the magnitudes of the normalized amplitudes of stresses, $|S_\sigma|$, on the line of action of the concentrated force ($\widetilde{r} = 0$) are presented for numerous values of parameter ξ at $\nu = \nu' = 1/3$, $\xi' = 2(1+\nu')$ (the same values were employed in Fig. 2.4); the dissipation parameter $\varepsilon = 0$ (Fig. 2.10a) and $\varepsilon = 0.05$ (Fig. 2.10b). The static values correspond to $\widetilde{z} \to 0$; at $\varepsilon = 0$, these values are equal to those calculated

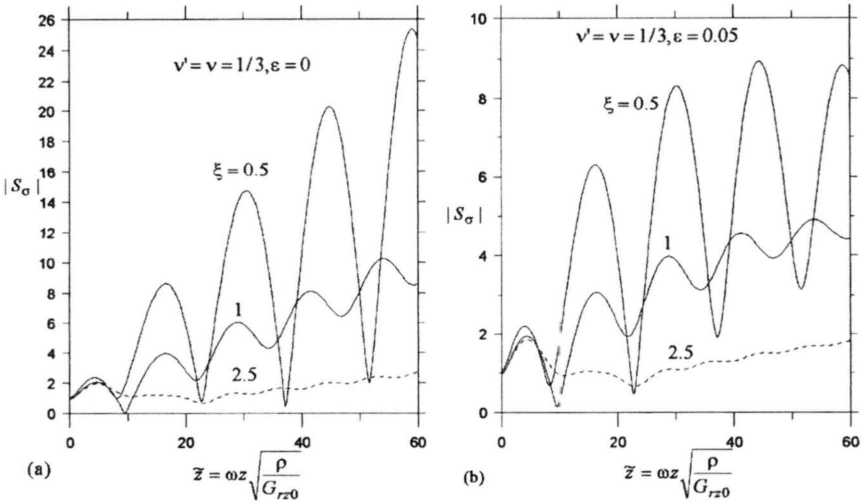

Fig. 2.10a,b. Absolute values of dynamic normalized stresses σ_z at the points of half-space, that belong to the line of action of vertical force applied to half-space surface at $\varepsilon = 0$ (a) and $\varepsilon = 0.05$ (b)

from (2.38). The influence of anisotropy on dynamic stresses is rather significant.

As a rule, an increase of parameter ξ leads to a reduction of dynamic stresses. Special emphasis should be lade on the difference in the extent of the influence of internal friction on the amplitudes of stresses and displacements. According to Fig. 2.6, at $\varepsilon = 0.05$, a noticeable decrease in the normalized amplitudes of vibrations occurs with an increase in dimensionless distance $\tilde{r}$ (at $\tilde{r} > 20$), though the normalized amplitudes of stresses continue to increase in the considered range of parameter $\tilde{z}$. Note that zones of increased and decreased amplitudes of stresses exist, and a gap between the minimal and maximal values of amplitudes becomes more significant with a decrease of parameter ξ. For the isotropic case (ξ = 1), the minimal value is in the vicinity of zero (for $\varepsilon = 0$ and $\tilde{z} = 9.4$, the value of $|S_\sigma|$ equals 0.0152).

Results presented correspond to the case of the concentrated force, so that they may be used for the forces distributed over a certain circular area of radius R on the surface of a half-space, if the ratio z/R is sufficiently large (in this case, static stresses comprise a small fraction of stresses on the half-space surface). In order to study stresses in the vicinity of the area of application of the load, we consider the action of a force distributed uniformly over a circular domain. Now, a dynamic coefficient should be associated with static stresses on the surface of the half-space. For $R \neq 0$, equation (1.58a) for points on the Z-axis takes the form

$$\sigma_z = -\frac{P_0}{\pi R^2} S_{\sigma R}, \tag{2.39a}$$

$$S_{\sigma R} = \frac{\tilde{z}R}{z} \int_0^\infty \left\{ c_{22} \left[\tilde{k}\tilde{A}_{rz}\tilde{q}_1(\tilde{z},\tilde{k}) + \tilde{A}_{zz} \frac{\mathrm{d}\tilde{w}_1(\tilde{z},\tilde{k})}{\mathrm{d}\tilde{z}} \right] \right.$$
$$\left. - c_{21} \left[\tilde{k}\tilde{A}_{rz}\tilde{q}_2(\tilde{z},\tilde{k}) + \tilde{A}_{zz} \frac{\mathrm{d}\tilde{w}_2(\tilde{z},\tilde{k})}{\mathrm{d}\tilde{z}} \right] \right\} \frac{\mathrm{J}_1(\tilde{k}\tilde{z}R/z)}{D} \mathrm{d}\tilde{k} . \tag{2.39b}$$

Compared with the case of the concentrated force, we note that the dynamic coefficient $S_{\sigma R}$ now depends on an additional parameter expressed by the ratio R/z.

The expression for static values of parameter $S_{\sigma R}$, which is built analogously to (2.36), (2.37), is

$$S_{\sigma R} = \frac{1}{\tilde{D}} \left[\tilde{c}_{11}\tilde{c}_{22} \left(1 - \frac{\tilde{\eta}_1}{\sqrt{\tilde{\eta}_1^2 + (R/z)^2}} \right) - \tilde{c}_{12}\tilde{c}_{21} \left(1 - \frac{\tilde{\eta}_2}{\sqrt{\tilde{\eta}_2^2 + (R/z)^2}} \right) \right]. \tag{2.40}$$

Employing formulas (2.39), (2.40) at $R \to 0$ yields the same expression for σ_z, as that obtained by using (2.35) with S_σ according to (2.38).

Furthermore, we present the results of calculations related to dynamic problems. The magnitudes of coefficient $S_{\sigma R}$ corresponding to ratios $R/z = 0.2$,

0.5, 1, $\nu = \nu' = 1/3$, $\xi' = 2(1+\nu')$, $\varepsilon = 0$, are presented in Fig. 2.11; three values were taken for parameter $\xi = 0.5$, 1, 2.5. At $R/z = 0.2$, the behavior of the curves is similar to that of the curves in Fig. 2.9a. If parameters $S_{\sigma R}$ were to be associated with static stresses in the isotropic half-space at depth z (this could be done by dividing the ordinates of curves in Fig. 2.11a by 0.05713, equal to the initial ordinate of the corresponding curve for the isotropic case), rather than with

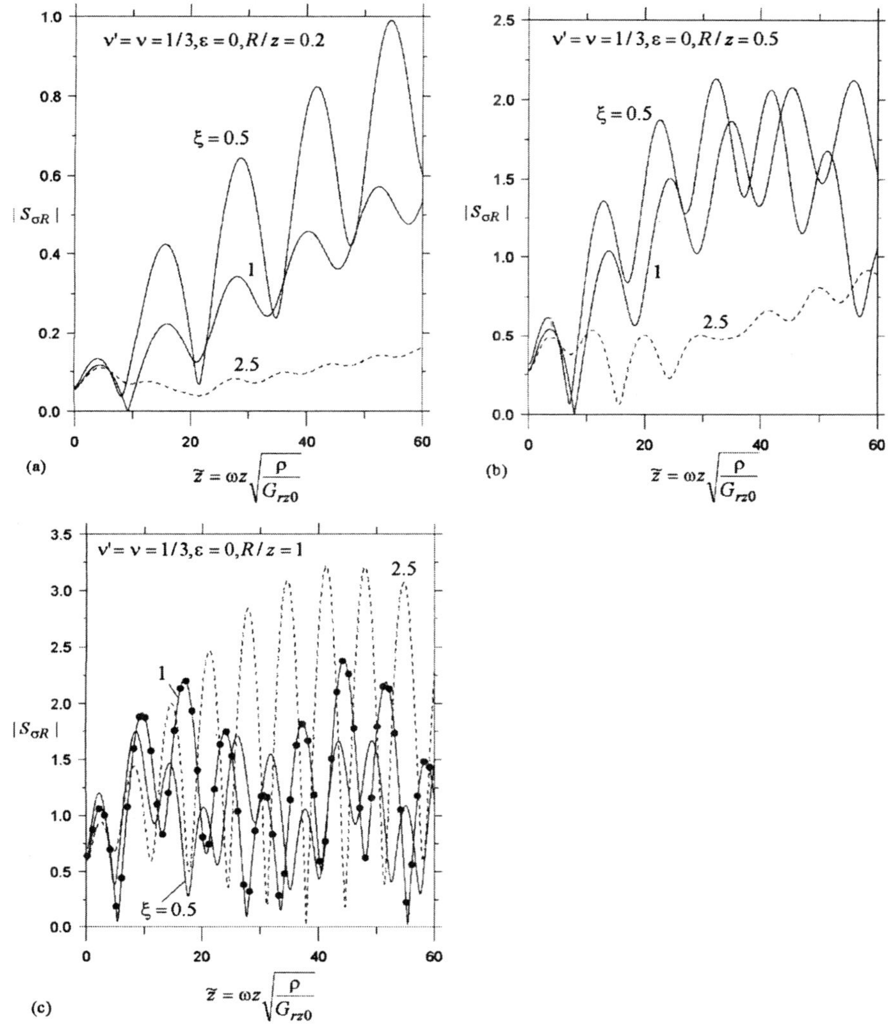

Fig. 2.11a,b,c. Dynamic normalized stresses σ_z at points on Z-axis for various sizes of circular area to which vertical load is applied

static stresses on the surface of the half-space, a proximity would take place for the corresponding ordinates of curves in Figs. 2.9a and 2.11a. Decreasing ratio R/z leads to more proximity with results corresponding to the action of the oncentrated force. A change in the influence of anisotropy (parameter ξ) on the behavior of stresses on approaching the surface of the half-space is shown in Fig. 2.11b and 2.11c. Whereas for small R/z the stiffness (i.e. parameter ξ), increasing in the horizontal direction, results in decreasing stresses, for $R/z = 1$, higher values of stresses correspond to higher values of ξ, as shown in Figs. 2.11c. Note that for a fixed ratio R/z, one should consider variations of parameter $\tilde{z}$ as a result of varying frequency of vibrations.

2.2 Model of Deformable Foundation as Limiting Case of Transversely Isotropic Half-Space

Consider equation (2.4), assuming that the summands containing function $\tilde{q}(\tilde{z},\tilde{k})$ may be neglected, i.e. horizontal displacements of the points of a half-space may be neglected:

$$\tilde{A}_{zz}\frac{d^2\tilde{w}}{d\tilde{z}^2}+(\beta^2\theta^2-\tilde{k}^2)\tilde{w}=0\,. \tag{2.41}$$

Boundary condition (1.50a) for the half-space surface, subjected to the action of force $P_0\exp(\mathrm{i}\omega t)$ distributed uniformly over a circular area of radius R takes the following form:

$$\tilde{A}_{zz}(z_0)\frac{d\tilde{w}}{d\tilde{z}}=-\frac{kP_0}{\pi R^2 G_{rz}(z_0)}\int_0^R r\,\mathrm{J}_0(kr)\,\mathrm{d}r=-\frac{P_0}{\pi R G_{rz}(z_0)}\mathrm{J}_1(kR)\,. \tag{2.42}$$

In the more general case of a heterogeneous half-space, we obtain the following equation, replacing (2.41) (according to (1.31)):

$$\tilde{A}_{zz}\frac{d^2\tilde{w}}{d\tilde{z}^2}+\frac{d\tilde{A}_{zz}}{d\tilde{z}}\frac{d\tilde{w}}{d\tilde{z}}+(\tilde{\rho}\beta^2\theta^2-\tilde{k}^2\tilde{G})\tilde{w}=0\,, \tag{2.43}$$

while boundary condition (2.42) is still valid. As shown in the discussion given in the previous section, increasing stiffness of elements of the half-space which belong to horizontal planes results in equations (2.41)–(2.43). On the contrary, these equations may be considered as a model of a deformable foundation, suitable for studying interactions between beams, plates, stamps, and the foundation in specific cases, when only vertical displacements are important. A similar model was proposed by Westergaard [126] for solving static problems. In [78] it was employed for studying dynamic problems under the name “wave model”.

In order to construct a solution of the problem, one may still apply equation (1.49b) to determine the vertical amplitudes of vibrations. Let $w_1(\tilde{z},\tilde{k})$ be one of the solutions of equations (2.43) or (2.41) (for a homogeneous half-space), satisfying the condition of absence of sources at $z \to \infty$. Assuming

$$\tilde{w} = C_w \tilde{w}_1(\tilde{z},\tilde{k}) \tag{2.44}$$

we calculate coefficient C_w according to boundary condition (2.42):

$$C_w = -\frac{P_0}{\pi R \tilde{A}_{zz}(z_0) G_{rz}(z_0)} \mathrm{J}_1(kR) \Big/ \frac{\mathrm{d}\tilde{w}_1(\tilde{z}_0,\tilde{k})}{\mathrm{d}\tilde{z}} . \tag{2.45}$$

A solution of the problem for the case of the concentrated force may be written as follows:

$$u_z = \frac{P_0}{2\pi G_{rz0} r} S_{vv} , \tag{2.46a}$$

$$S_{vv} = -\frac{\beta^2 \tilde{r}}{\tilde{A}_{zz}(z_0)} \int_0^\infty \tilde{k}\tilde{w}_1(\tilde{z},\tilde{k}) \mathrm{J}_0(\tilde{k}\tilde{r}) \Big/ \frac{\mathrm{d}\tilde{w}_1(\tilde{z}_0,\tilde{k})}{\mathrm{d}\tilde{z}} \mathrm{d}\tilde{k} . \tag{2.46b}$$

Now, we consider equation (2.41). Let $z_0 = 0$. A solution which satisfies the condition of absence of sources at infinity takes the form

$$\tilde{w}_1 = \exp\left(-\frac{\tilde{z}}{\sqrt{\tilde{A}_{zz}}} \sqrt{\tilde{k}^2 - \beta^2\theta^2} \right). \tag{2.47}$$

In the case of a homogeneous half-space, the normalized amplitude of vibrations, S_{vv}, may be expressed as

$$S_{vv} = \frac{\beta^2 \tilde{r}}{\sqrt{\tilde{A}_{zz}}} \int_0^\infty \tilde{k} \exp\left(-\frac{\tilde{z}}{\sqrt{\tilde{A}_{zz}}} \sqrt{\tilde{k}^2 - \beta^2\theta^2} \right) \frac{\mathrm{J}_0(\tilde{k}\tilde{r})}{\sqrt{\tilde{k}^2 - \beta^2\theta^2}} \mathrm{d}\tilde{k}, \tag{2.48}$$

where one may assume $\theta = 1$ and take variables (2.19) as $\tilde{r}$ and $\tilde{z}$. The integral in expression (2.48) may be considered as a particular case of the following tabulated integral [43]:

$$\int_0^\infty t^{\nu+1}(b^2 + t^2)^{-1/2} \exp[-a(b^2 + t^2)^{1/2}] \mathrm{J}_\nu(ut) \mathrm{d}t$$

$$= \left(\frac{2b}{\pi} \right)^{1/2} (ub)^\nu \alpha^{-\nu-1/2} K_{\nu+1/2}(b\alpha), \tag{2.49}$$

where

$$\alpha = (a^2 + u^2)^{1/2}, \ \mathrm{Re}\, a > 0, \ \mathrm{Re}\, b > 0, \ \mathrm{Re}\, \nu > -1 . \tag{2.50}$$

On the right side of (2.49), the modified Bessel's function has been employed. The integral in equation (2.48) is obtained from (2.49) at $\nu = 0$, $b = i\beta$, $u = \tilde{r}$, $a = \tilde{z}/\sqrt{\tilde{A}_{zz}}$:

$$S_{vv} = \frac{\beta^2 \tilde{r}}{\sqrt{\tilde{A}_{zz}\tilde{r}^2 + \tilde{z}^2}} \exp(-i\beta\sqrt{\tilde{r}^2 + \tilde{z}^2/\tilde{A}_{zz}}). \tag{2.51}$$

For the half-space surface at $\beta = 1$ ($\varepsilon = 0$), we obtain

$$S_{vv} = \frac{\exp(-i\tilde{r})}{\sqrt{\tilde{A}_{zz}}}. \tag{2.52}$$

This result corresponds to the wave having constant amplitude $|S_{vv}| = 1/\sqrt{\tilde{A}_{zz}}$ and phase $\delta_{vv} = -\tilde{r}$; its velocity is equal to C_{rz} by (2.18). For other directions, the values of velocity are different, e.g. the velocity of wave propagation along the Z-axis equals $C_{rz}(\tilde{A}_{zz})^{1/2}$.

Let us compare the results of calculations for high values of parameter ξ by using formula (2.52), and the expression for S_{vv} (2.1d) at $\tilde{z} = 0$, $\nu = \nu' = 0$, $\xi' = 2$ (multipliers before the normalized amplitudes of vibrations in expressions (2.46a) and (2.1c) are equal to each other). Following formulas (1.9)–(1.11), at $\nu' = 0$ we have $A_{zz} = E'$ and $\tilde{A}_{zz} = \xi'$. For the calculations presented in Fig. 2.7, parameter $\tilde{A}_{zz} = 2$. In Fig. 2.7, the dashed lines represent results corresponding to formula (2.41) ($|S_{vv}| = \sqrt{0.5}$, $\delta_{vv} = -\tilde{r}$); at high values of ξ, the curves are close to these dashed lines.

2.3 Vibrations of Half-Space Subjected to Action of Horizontal Force Applied to Half-Space Surface

Next, we present a solution corresponding to the action of a horizontal concentrated force $Q_0 \exp(i\omega t)$. As follows from expressions (1.86)–(1.88), on separating out corresponding static values:

$$\hat{u}_r = \frac{Q_0}{2G_{rz0}\pi r} S_{hr}, \tag{2.53a}$$

$$S_{hr} = \beta^2 \tilde{r} \int_0^\infty \tilde{k}\tilde{H}_r(\tilde{z}, \tilde{r}, \tilde{k})\, d\tilde{k}, \tag{2.53b}$$

$$\hat{u}_\vartheta = -\frac{Q_0(1-\nu')}{2G_{rz0}\pi r} S_{h\vartheta}, \tag{2.54a}$$

$$S_{h\vartheta} = -\frac{\beta^2 \tilde{r}}{1-\nu'} \int_0^{\infty} \tilde{k} \tilde{H}_\vartheta(\tilde{z},\tilde{r},\tilde{k}) \mathrm{d}\tilde{k} , \tag{2.54b}$$

$$\hat{u}_z = -\frac{Q_0(1-\nu')}{2G_{rz0}\pi r} S_{hz} , \tag{2.55a}$$

$$S_{hz} = -\frac{\beta^2 \tilde{r}}{1-\nu'} \int_0^{\infty} \tilde{k} \tilde{H}_z(\tilde{z},\tilde{r},\tilde{k}) \mathrm{d}\tilde{k} . \tag{2.55b}$$

Expressions $\tilde{H}_r$, $\tilde{H}_\vartheta$, $\tilde{H}_z$ are deduced from H_r, H_ϑ, H_z following (1.83)–(1.85) by replacing functions $q_j(z,k), w_j(z,k)$ with $\tilde{q}_j(\tilde{z},\tilde{k}), \tilde{w}_j(\tilde{z},\tilde{k})$ ($j = 1,2$), k with $\tilde{k}$, r with $\tilde{r}$, where variables marked with "~" are given in relationships (1.27) and (2.2). Function $p_1(z,k)$ is replaced by $\tilde{p}_1(\tilde{z},\tilde{k})$, while its derivative with respect to z is replaced by the derivative of $\tilde{p}_1(\tilde{z},\tilde{k})$ with respect to $\tilde{z}$.

Multipliers before S_{hr} and $S_{h\vartheta}$ represent static values $\hat{u}_r$ and $\hat{u}_\vartheta$ for the surface of an isotropic half-space having shear modulus G_{rz0} and Poisson's ratio ν'. The coefficient for $\hat{u}_z$ is taken equal to that for $\hat{u}_\vartheta$, in order to avoid complications occurring due to the vanishing of the static value of $\hat{u}_z$ in the case of incompressible isotropic medium. Note that at $\tilde{z} = 0$, values of $\hat{u}_z$ by (2.55) and $\hat{u}_r$ by (2.1) are identical when $Q_0 = -P_0$, due to Rayleigh's principle of reciprocity.

In section 2.1, solutions $\tilde{q}_j$, $\tilde{w}_j$ ($j = 1,2$) have already been constructed for the considered half-space. In order to determine function p_1, we employ equation (1.29) which, in the case of homogeneity, takes the following form:

$$\frac{\mathrm{d}^2 \tilde{p}}{\mathrm{d}\tilde{z}^2} + (\beta^2\theta^2 - \tilde{k}^2 \tilde{G}_{r\vartheta})\tilde{p} = 0 . \tag{2.56}$$

Function p_1 is expressed as follows:

$$\tilde{p}_1 = \exp(-\tilde{z}\eta_0) , \tag{2.57}$$

where

$$\eta_0 = \sqrt{\tilde{G}_{r\vartheta}\tilde{k}^2 - \beta^2\theta^2} . \tag{2.58}$$

Value F entering functions $\tilde{H}_r$, $\tilde{H}_\vartheta$ equals (at $z_0 = 0$)

$$F = \frac{\mathrm{d}\, p_1(z_0,k)}{\mathrm{d}z} = \frac{\mathrm{d}\tilde{p}_1(0,\tilde{k})}{\mathrm{d}\tilde{z}} = -\eta_0 . \tag{2.59}$$

As in the case of the vertical force, we assume for the dynamic case that $\theta = 1$, and variables (2.19) are applied as dimensionless coordinates.

2.3.1 Static Action of Horizontal Force

Let $\theta = 0$, $\beta = 1$; in addition to relationships (2.20), (2.21), we apply expressions related to function $\widetilde{p}_1$:

$$\widetilde{p}_1 = \exp(-\widetilde{z}\widetilde{k}\widetilde{\eta}_0), \tag{2.60a}$$

$$\widetilde{F} = -\widetilde{k}\widetilde{\eta}_0, \tag{2.60b}$$

where

$$\widetilde{\eta}_0 = \sqrt{\widetilde{G}_{r\vartheta}}\,. \tag{2.61}$$

The static normalized displacements may be written in the following form:

$$S_{hr} = \frac{1}{\widetilde{D}}\left\{\widetilde{c}_{11}\widetilde{r}\int_0^\infty e^{-\widetilde{k}\widetilde{\eta}_2\widetilde{z}}\left[\frac{J_1(\widetilde{k}\widetilde{r})}{\widetilde{k}\widetilde{r}} - J_0(\widetilde{k}\widetilde{r})\right]d\widetilde{k}\right.$$
$$\left. - \widetilde{c}_{12}\widetilde{r}\int_0^\infty e^{-\widetilde{k}\widetilde{\eta}_1\widetilde{z}}\left[\frac{J_1(\widetilde{k}\widetilde{r})}{\widetilde{k}\widetilde{r}} - J_0(\widetilde{k}\widetilde{r})\right]d\widetilde{k}\right\} + \frac{\widetilde{r}}{\widetilde{\eta}_0}\int_0^\infty e^{-\widetilde{k}\widetilde{\eta}_0\widetilde{z}}\frac{J_1(\widetilde{k}\widetilde{r})}{\widetilde{k}\widetilde{r}}d\widetilde{k}\,, \tag{2.62}$$

$$S_{h\vartheta} = -\frac{1}{1-\nu'}\left\{\frac{\widetilde{c}_{11}\widetilde{r}}{\widetilde{D}}\int_0^\infty e^{-\widetilde{k}\widetilde{\eta}_2\widetilde{z}}\frac{J_1(\widetilde{k}\widetilde{r})}{\widetilde{k}\widetilde{r}}d\widetilde{k} - \frac{\widetilde{c}_{12}\widetilde{r}}{\widetilde{D}}\int_0^\infty e^{-\widetilde{k}\widetilde{\eta}_1\widetilde{z}}\frac{J_1(\widetilde{k}\widetilde{r})}{\widetilde{k}\widetilde{r}}d\widetilde{k}\right.$$
$$\left. + \frac{\widetilde{r}}{\widetilde{\eta}_0}\int_0^\infty e^{-\widetilde{k}\widetilde{\eta}_0\widetilde{z}}\left[\frac{J_1(\widetilde{k}\widetilde{r})}{\widetilde{k}\widetilde{r}} - J_0(\widetilde{k}\widetilde{r})\right]d\widetilde{k}\right\}, \tag{2.63}$$

$$S_{hz} = -\frac{1}{\widetilde{D}(1-\nu')}\left[\widetilde{c}_{11}\widetilde{C}_{w2}\widetilde{r}\int_0^\infty e^{-\widetilde{k}\widetilde{\eta}_2\widetilde{z}}J_1(\widetilde{k}\widetilde{r})d\widetilde{k} - \widetilde{c}_{12}\widetilde{C}_{w1}\widetilde{r}\int_0^\infty e^{-\widetilde{k}\widetilde{\eta}_1\widetilde{z}}J_1(\widetilde{k}\widetilde{r})d\widetilde{k}\right]. \tag{2.64}$$

Integrals containing exponents and Bessel's functions may be expressed in their finite form [43], resulting in the following equations for S_{hr}, $S_{h\vartheta}$, S_{hz}:

$$S_{hr} = \frac{zr}{\widetilde{D}}\left[-\frac{\widetilde{c}_{11}\widetilde{\eta}_2}{F_1(r, z\widetilde{\eta}_2)} + \frac{\widetilde{c}_{12}\widetilde{\eta}_1}{F_1(r, z\widetilde{\eta}_1)}\right] + \frac{r}{\widetilde{\eta}_0 F_2(r, z\widetilde{\eta}_0)}, \tag{2.65}$$

$$S_{h\vartheta} = -\frac{r}{1-\nu'}\left[\frac{\widetilde{c}_{11}}{\widetilde{D}F_2(r, z\widetilde{\eta}_2)} - \frac{\widetilde{c}_{12}}{\widetilde{D}F_2(r, z\widetilde{\eta}_1)} - \frac{z}{F_1(r, z\widetilde{\eta}_0)}\right], \tag{2.66}$$

$$S_{hz} = -\frac{1}{\widetilde{D}(1-\nu')}\left[\widetilde{c}_{11}\widetilde{C}_{w2}\left(1-\frac{z\widetilde{\eta}_2}{F_3(r,z\widetilde{\eta}_2)}\right)-\widetilde{c}_{12}\widetilde{C}_{w1}\left(1-\frac{z\widetilde{\eta}_1}{F_3(r,z\widetilde{\eta}_1)}\right)\right], \tag{2.67}$$

where

$$F_1(r,s) = \sqrt{r^2+s^2}\,(s+\sqrt{r^2+s^2}), \tag{2.68}$$

$$F_2(r,s) = s+\sqrt{r^2+s^2}, \tag{2.69}$$

$$F_3(r,s) = \sqrt{r^2+s^2}. \tag{2.70}$$

On the surface of a half-space:

$$S_{hr} = \frac{1}{\widetilde{\eta}_0} = \frac{1}{\sqrt{\widetilde{G}_{r\vartheta}}}, \tag{2.71}$$

$$S_{h\vartheta} = \frac{\widetilde{c}_{12}-\widetilde{c}_{11}}{\widetilde{D}(1-\nu')}, \tag{2.72}$$

$$S_{hz} = \frac{\widetilde{c}_{12}\widetilde{C}_{w1}-\widetilde{c}_{11}\widetilde{C}_{w2}}{\widetilde{D}(1-\nu')}. \tag{2.73}$$

In accordance with the principle of reciprocity, the last expression may be reduced to the form identical to S_{vh} by (2.24a). At points that belong to the Z-axis, displacements $\hat{u}_z$ vanish, while displacements $\hat{u}_r$ and $\hat{u}_\vartheta$ are equal in magnitude but opposite in sign. Next, we give an expression for $\hat{u}_r$, separating out the static horizontal displacements on the Z-axis ($r = 0$):

$$\hat{u}_r = \frac{Q_0(3-2\nu')}{8G_{rz0}\pi z}S_{hrz}, \tag{2.74a}$$

$$S_{hrz} = \frac{2}{3-2\nu'}\left(\frac{\widetilde{c}_{12}}{\widetilde{\eta}_1\widetilde{D}}-\frac{\widetilde{c}_{11}}{\widetilde{\eta}_2\widetilde{D}}+\frac{1}{\widetilde{\eta}_0^2}\right). \tag{2.74b}$$

Note that the case of equal roots, $\widetilde{\eta}_1$ and $\widetilde{\eta}_2$, should be considered separately (see the corresponding comments in section 2.1.1).

Calculations were performed for the following relations between parameters: $\nu = \nu' = 1/3$, $\xi' = 2(1+\nu')$ (see (2.26)). The normalized shear modulus $\widetilde{G}_{r\vartheta}$ may be presented as

$$\widetilde{G}_{r\vartheta} = \frac{E}{2(1+\nu)G_{rz}} = \frac{\xi\xi'}{2(1+\nu)}. \tag{2.75}$$

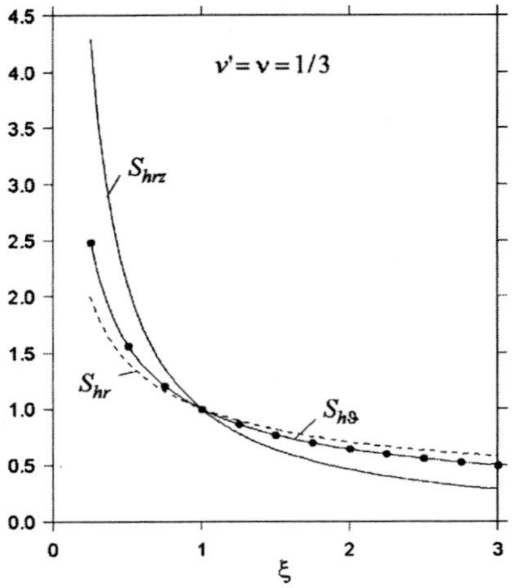

Fig. 2.12. Normalized static horizontal displacements of points of transversely isotropic half-space subjected to action of horizontal force

Taking into account the assumed relationships between elastic coefficients, we obtain $\widetilde{G}_{r\vartheta}=\xi$, $\widetilde{\eta}_0=\sqrt{\xi}$. In Fig. 2.12, values S_{hr} and $S_{h\vartheta}$ are for the half-space surface, and S_{hrz} is for points on the Z-axis. Note that the influence of parameter ξ is stronger for horizontal displacements, which occur due to a horizontal force, than for vertical ones caused by a vertical force (Fig. 2.1a).

2.3.2 Analysis of Amplitudes of Vibrations

In this subsection, we study amplitudes of vibrations in a half-space. Consider the quantity $S_{h\vartheta}$ determined in (2.54b). The part of this expression that contains function $\widetilde{p}_1$ and multiplier $\mathrm{J}_0(\widetilde{k}\widetilde{r})$ (we denote this part as $S'_{h\vartheta}$) may be expressed in a finite form by using tabulated integral (2.49):

$$S'_{h\vartheta}=-\frac{\beta^2\widetilde{r}}{1-\nu'}\int_0^{\infty}\frac{\widetilde{k}\widetilde{p}_1(\widetilde{z},\widetilde{k})}{F}\mathrm{J}_0(\widetilde{k}\widetilde{r})\mathrm{d}\widetilde{k}=\frac{\beta^2\widetilde{r}}{1-\nu'}\int_0^{\infty}\frac{\widetilde{k}\exp(-\widetilde{z}\sqrt{\widetilde{G}_{r\vartheta}}\sqrt{\widetilde{k}^2-\beta^2/\widetilde{G}_{r\vartheta}})}{\sqrt{\widetilde{G}_{r\vartheta}}\sqrt{\widetilde{k}^2-\beta^2/\widetilde{G}_{r\vartheta}}}\mathrm{J}_0(\widetilde{k}\widetilde{r})\mathrm{d}\widetilde{k}$$

$$=\frac{\beta^2\widetilde{r}}{(1-\nu')\widetilde{G}_{r\vartheta}\sqrt{\widetilde{z}^2+\widetilde{r}^2/\widetilde{G}_{r\vartheta}}}\exp(-\mathrm{i}\beta\sqrt{\widetilde{z}^2+\widetilde{r}^2/\widetilde{G}_{r\vartheta}})\ . \qquad (2.76)$$

Analyzing the structure of the integrand in the integral, which expresses $S_{h\vartheta}$ (see (2.54b) and (1.84)), one may conclude that with increasing $\widetilde{r}$, a contribution of

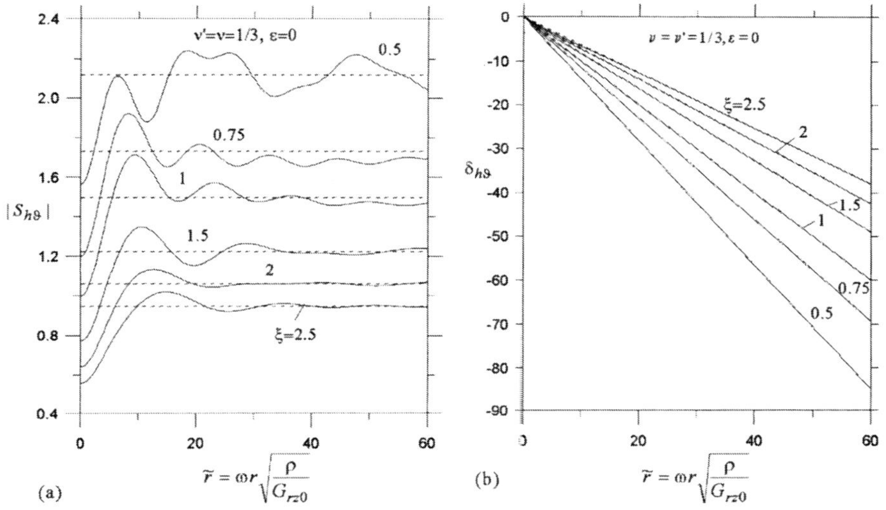

Fig. 2.13a,b. Absolute values (a) and phases (b) of complex normalized tangential amplitudes of vibrations $S_{h\vartheta}$ under action of horizontal force

the part of the integrand, which enters the integral in relationship (2.76), to the amplitudes of vibrations should prevail (the rest of the terms under the integral contain an additional multiplier $\tilde{r}^{-1}$). Consider the results of calculations. For the data used in the previous section, $\nu = \nu' = 1/3,\ \xi' = 2(1+\nu')$, the magnitudes and phase angles of normalized amplitudes of vibrations $S_{h\vartheta}$ on the half-space surface are as shown in Fig. 2.13 for several values of parameter ξ at $\varepsilon = 0$. The dashed lines in Fig. 2.13 represent the results of calculations when formula (2.76) is employed. At $\tilde{z} = 0,\ \beta = 1$, this relationship yields

$$| S'_{h\vartheta} | = \frac{1}{(1-\nu')\sqrt{\widetilde{G}_{r\vartheta}}} , \tag{2.77a}$$

$$\delta'_{h\vartheta} = -\frac{\tilde{r}}{\sqrt{\widetilde{G}_{r\vartheta}}} . \tag{2.77b}$$

The phase curves merge practically with the corresponding dashed lines. The convergence of quantity $S_{h\vartheta}$ to the approximate result by (2.77a) improves with increasing parameter ξ, i.e. stiffness in the horizontal direction.

In Fig. 2.14a, the magnitudes of complex amplitudes S_{hr} are presented for the same values of parameters as in the above case. Now, in contrast to the

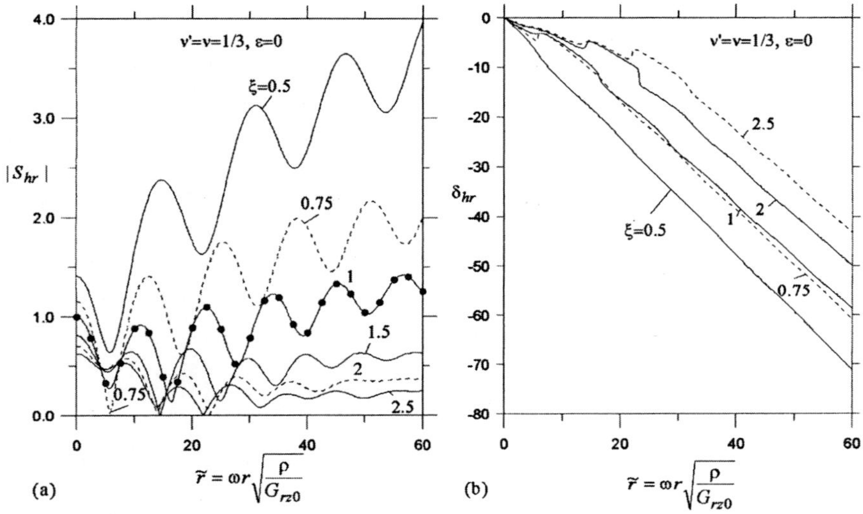

Fig. 2.14a,b. Absolute values (a) and phases (b) of complex normalized radial amplitudes of vibrations S_{hr} under action of horizontal force

case of $S_{h\vartheta}$, the influence of Rayleigh waves becomes significant, causing an increase in the normalized amplitudes of vibrations with increasing $\tilde{r}$. As in the case of vibrations which occur due to a vertical force, this influence becomes stronger with decreasing parameter ξ. The behavior of phase δ_{hr} is shown in Fig. 2.14b; note that intervals with abrupt changes of phase correspond to those values of parameter $\tilde{r}$ for which amplitude S_{hr} is close to zero (Fig. 2.14a).

2.4 Torsional Vibrations of Half-Space

Let a torque $M_z \exp(\mathrm{i}\,\omega t)$ uniformly distributed over a circular area of radius R be applied to the half-space surface $z = 0$, in accordance with relationship (1.62). Application of expression (1.70) using dimensionless parameters (1.27) and solution $\tilde{p}_1(\tilde{z},\tilde{k})$ from (2.57) yields

$$u_\vartheta = \frac{2M_z\beta^2}{G_{rz0}\pi R^2}\int_0^\infty \exp(-\eta_0\tilde{z})\frac{\mathrm{J}_2(\tilde{k}\tilde{R})\mathrm{J}_1(\tilde{k}\tilde{r})}{\eta_0}\mathrm{d}\tilde{k}$$

$$= \frac{2M_z\beta^2}{G_{rz0}\sqrt{\tilde{G}_{r\vartheta}}\,\pi R^2}\int_0^\infty \exp(-\eta_0\tilde{z}_1)\frac{\mathrm{J}_2(\tilde{k}\tilde{R})\mathrm{J}_1(\tilde{k}\tilde{r})}{\overline{\eta}_0}\mathrm{d}\tilde{k}\,, \qquad (2.78)$$

where

$$\overline{\eta}_0 = \sqrt{\widetilde{k}^2 - \beta^2\theta^2 / \widetilde{G}_{r\vartheta}}\,, \tag{2.79}$$

$$\widetilde{R} = \frac{R}{z_r}\,, \tag{2.80}$$

$$\widetilde{z}_1 = \widetilde{z}\sqrt{\widetilde{G}_{r\vartheta}}\,. \tag{2.81}$$

In the present case, we use reference length z_r in a different way than in (2.17), requiring

$$\theta = \sqrt{\widetilde{G}_{r\vartheta}}\,. \tag{2.82}$$

In this case, variable $\overline{\eta}_0$ does not contain any characteristics of the half-space, and the structure of the integral in expression (2.78) is identical for different values of elastic constants. Taking into account expression (1.32a) for θ, we obtain

$$z_r = \frac{C_{r\vartheta}}{\omega}\,, \tag{2.83}$$

where

$$C_{r\vartheta} = \sqrt{\frac{G_{r\vartheta 0}}{\rho}}\,. \tag{2.84}$$

Here, in analogy with notation G_{rz0} in relationship (1.28), $G_{r\vartheta 0}$ represents the shear modulus corresponding to shear deformations in a horizontal plane, while energy dissipation is neglected. The dimensionless distances $\widetilde{r}$, $\widetilde{R}$, $\widetilde{z}_1$ take the form

$$\widetilde{r} = \omega r\sqrt{\frac{\rho}{G_{r\vartheta 0}}}\,, \tag{2.85a}$$

$$\widetilde{R} = \omega R\sqrt{\frac{\rho}{G_{r\vartheta 0}}}\,, \tag{2.85b}$$

$$\widetilde{z}_1 = \omega z\sqrt{\frac{\rho}{G_{rz0}}}\,. \tag{2.85c}$$

The results obtained indicate that an intimate connection exists between the solutions of the given problem for isotropic and transversely isotropic half-spaces: the amplitudes of vibrations will be identical when dimensionless distances

$\tilde{r}$, $\tilde{R}$, $\tilde{z}_1$ are identical in these two cases, and the shear modulus G_0 of the isotropic half-space is taken as

$$G_{rz0}\sqrt{\tilde{G}_{r\vartheta}} = \sqrt{G_{rz0}G_{r\vartheta 0}} \tag{2.85d}$$

(when determining the coefficient before the integral in equation (2.78)). A similar correspondence holds in the case of action of a tangential load distributed uniformly along a circle on the surface of a half-space, and for the contact problem, which may be solved by employing an integral equation containing the solution for the load applied to the circle as a kernel. Consider, for example, the dynamic torsional stiffness for a disk of radius R on the isotropic half-space [67]. It contains shear modulus G_0 entering as a coefficient in the expression depending on parameter a_0:

$$a_0 = \omega R\sqrt{\frac{\rho}{G_0}}, \tag{2.86}$$

which is similar to parameter $\tilde{R}$ in (2.85b). In order to obtain the stiffness for the case of transverse isotropy (keeping the value of the dissipation parameter ε unchanged), one should replace (in the solution for the isotropic half-space) the above-mentioned multiplier (G_0) with $\sqrt{G_{rz0}G_{r\vartheta 0}}$ and replace a_0 with $\tilde{R}$ from (2.85b).

Going to the limit in expression (2.78) when $R \to 0$, we obtain a solution for the concentrated torque:

$$u_\vartheta = \frac{M_z}{4\sqrt{G_{rz0}G_{r\vartheta 0}}\,\pi r^2} S_\vartheta, \tag{2.87}$$

$$\begin{aligned} S_\vartheta &= \beta^2 \tilde{r}^2 \int_0^\infty \exp(-\eta_0 \tilde{z}_1)\frac{\tilde{k}^2\, \mathrm{J}_1(\tilde{k}\tilde{r})}{\eta_0}\mathrm{d}\tilde{k} \\ &= \beta^2 \tilde{r}^2 \int_0^\infty \frac{\tilde{k}^2 \exp(-\tilde{z}_1\sqrt{\tilde{k}^2-\beta^2})}{\sqrt{\tilde{k}^2-\beta^2}} \mathrm{J}_1(\tilde{k}\tilde{r})\,\mathrm{d}\tilde{k}\,. \end{aligned} \tag{2.88}$$

According to (1.74), the coefficient of S_ϑ in (2.87) equals the static displacements on the surface of an isotropic half-space with shear modulus $\sqrt{G_{rz0}G_{r\vartheta 0}}$. The integral in (2.88) may be calculated by using (2.49) at $\nu = 1$, $u = \tilde{r}$, $a = \tilde{z}_1$, $b = \mathrm{i}\beta$, so that

$$S_\vartheta = \frac{\beta^2\tilde{r}^2}{\tilde{r}^2+\tilde{z}_1^2}\exp\left(-\mathrm{i}\beta\sqrt{\tilde{r}^2+\tilde{z}_1^2}\right)\left[\mathrm{i}\tilde{r}\beta + \frac{\tilde{r}}{\sqrt{\tilde{r}^2+\tilde{z}_1^2}}\right]. \tag{2.89}$$

For the half-space surface we obtain

$$S_\vartheta = \beta^2 \exp(-i\beta\tilde{r})[i\tilde{r}\beta + 1] \ . \tag{2.90}$$

Note the increasing value of normalized amplitudes of vibrations as parameter $\tilde{r}$ increases. If no energy dissipation occurs ($\beta = 1$), the amplitude of vibrations u_ϑ at the fixed point of the half-space increases linearly with increasing frequency of vibrations.

2.5 Action of Force Applied in Infinite Space

2.5.1 Action of Vertical Force

We consider the action of force $P_0 \exp(i\omega t)$ applied in the plane $z = 0$ and directed along the vertical Z-axis; at first, the force is assumed to be uniformly distributed over the area of a circle of radius R. As for the case of a force applied to the surface of a half-space, we use a representation of the solution in the form (1.49). Dimensionless variables given in (1.27) are used. Taking into account the symmetry of the problem, one may rewrite functions $\tilde{q}(\tilde{z},\tilde{k})$ and $\tilde{w}(\tilde{z},\tilde{k})$ for $z \ge 0$ in the following form:

$$\tilde{q}(\tilde{z},\tilde{k}) = A(\tilde{k})[\tilde{q}_1(\tilde{z},\tilde{k}) - \tilde{q}_2(\tilde{z},\tilde{k})] , \tag{2.91a}$$

$$w(\tilde{z},\tilde{k}) = A(\tilde{k})[\tilde{w}_1(\tilde{z},\tilde{k}) - \tilde{w}_2(\tilde{z},\tilde{k})] , \tag{2.91b}$$

where coefficient $A(\tilde{k})$ is introduced; functions $\tilde{q}_j$, $\tilde{w}_j$ are determined by (2.5) with $\eta = \eta_1, \eta_2$; $C_q = 1$ and $C_w = C_{wj}$ by (2.14). We construct the solution for $z < 0$ by continuing functions $\tilde{w}$ and $\tilde{q}$ in the symmetric and antisymmetric ways, respectively. Since $\tilde{q} = 0$ at $\tilde{z} = 0$, function $\tilde{q}(\tilde{z},\tilde{k})$ and its first derivative are continuous over the entire space; function $\tilde{w}(\tilde{z},\tilde{k})$ is also continuous. Thus, the assumed form of the solution provides continuity of displacements when crossing the plane of application of the external load. Moreover, following equation (1.42b), the requirement of continuity of the value on the left side of this equation is satisfied (this requirement is related to the absence of horizontal external loads in the plane $\tilde{z} = 0$). Therefore, we have to consider only one requirement, related to condition (1.40), which determines discontinuity in the value of derivative, by using an analogy with equation (1.100). Taking into account the evenness of function $\tilde{w}$, the following boundary condition for solution (2.91) may be written at $\tilde{z} = 0$:

$$2\frac{\mathrm{d}\widetilde{w}}{\mathrm{d}\widetilde{z}} = -b_w = -\frac{P_0}{\pi R A_{zz}} \mathrm{J}_1(kR). \tag{2.92}$$

Hence, we find coefficient A in solution (2.91):

$$A = \frac{P_0 \, \mathrm{J}_1(kR)}{2\pi R A_{zz}(C_{w1}\eta_1 - C_{w2}\eta_2)}. \tag{2.93}$$

The amplitudes of vibrations for the domain $z \geq 0$ may be found from (1.49) by using (2.91) and (2.93). For the case of the concentrated force, we obtain

$$u_r = \frac{P_0}{4\pi G_{rz0} r} S_{vh}, \tag{2.94a}$$

$$S_{vh} = -\frac{\beta^2 \widetilde{r}(\widetilde{A}_{rz}+1)}{\widetilde{A}_{zz}} \int_0^\infty \frac{\exp(-\eta_1\widetilde{z}) - \exp(-\eta_2\widetilde{z})}{\eta_1^2 - \eta_2^2} \widetilde{k}^2 \, \mathrm{J}_1(\widetilde{k}\widetilde{r}) \mathrm{d}\widetilde{k}, \tag{2.94b}$$

$$u_z = \frac{P_0}{4\pi G_{rz0} r} S_{vv}, \tag{2.95a}$$

$$S_{vv} = -\frac{\beta^2 \widetilde{r}(\widetilde{A}_{rz}+1)}{\widetilde{A}_{zz}} \int_0^\infty \frac{C_{w1}\exp(-\eta_1\widetilde{z}) - C_{w2}\exp(-\eta_2\widetilde{z})}{\eta_1^2 - \eta_2^2} \widetilde{k}^2 \, \mathrm{J}_0(\widetilde{k}\widetilde{r}) \mathrm{d}\widetilde{k}. \tag{2.95b}$$

At $r = 0,\ z \neq 0$, the following transformation of these expressions is reasonable: one should divide by z, rather than by r, in the multipliers of S_{vh} and S_{vv}, and replace $\widetilde{r}$ with $\widetilde{z}$ before the integrals. Here, as in previous sections, we assume for dynamic problems that $\theta = 1$ and apply equations (2.19).

Particular Case: Isotropic Space

Consider the case of an isotropic space when roots η_j and coefficients C_{wj} have the form as given in (2.15), (2.16). Integrals (2.94b), (2.95b) are rewritten as follows:

$$S_{vh} = \widetilde{r} \int_0^\infty [\exp(-\widetilde{z}\sqrt{\widetilde{k}^2 - \beta^2}) - \exp(-\widetilde{z}\sqrt{\widetilde{k}^2 - \beta^2 / \widetilde{A}_{zz}})]\widetilde{k}^2 \, \mathrm{J}_1(\widetilde{k}\widetilde{r}) \mathrm{d}\widetilde{k}, \tag{2.96}$$

$$S_{vv} = \widetilde{r} \int_0^\infty \left[\frac{\widetilde{k}}{\sqrt{\widetilde{k}^2 - \beta^2}} \exp(-\widetilde{z}\sqrt{\widetilde{k}^2 - \beta^2}) - \frac{\sqrt{\widetilde{k}^2 - \beta^2 / \widetilde{A}_{zz}}}{\widetilde{k}} \exp(-\widetilde{z}\sqrt{\widetilde{k}^2 - \beta^2 / \widetilde{A}_{zz}}) \right] \widetilde{k}^2 \, \mathrm{J}_0(\widetilde{k}\widetilde{r}) \mathrm{d}\widetilde{k}. \tag{2.97}$$

The integrals in equations (2.96), (2.97) may be expressed in finite form by employing tabulated integral (2.49) ($u=\tilde{r}$, $a=\tilde{z}$, $b=\mathrm{i}\beta$ or $b=\mathrm{i}\beta/(\tilde{A}_{zz})^{1/2}$). In order to obtain the integrals in (2.96), one should apply (2.49) at $\nu=0$ and then differentiate both parts of this equations with respect to a and u. This results in

$$S_{vh}=\tilde{r}^2\tilde{z}\left[\left(\frac{1}{\alpha}\frac{\mathrm{d}}{\mathrm{d}\alpha}\right)^2\left(\frac{1}{\alpha}[\exp(-\mathrm{i}\beta\alpha)-\exp(-\mathrm{i}\beta\alpha/\sqrt{\tilde{A}_{zz}})]\right)\right], \tag{2.98}$$

where

$$\alpha=(\tilde{z}^2+\tilde{r}^2)^{1/2}. \tag{2.99}$$

One may rewrite the first summand in square brackets in integral (2.97) as follows:

$$\frac{\tilde{k}}{\sqrt{\tilde{k}^2-\beta^2}}\exp(-\tilde{z}\sqrt{\tilde{k}^2-\beta^2})=\left[\frac{\sqrt{\tilde{k}^2-\beta^2}}{\tilde{k}}+\frac{\beta^2}{\tilde{k}\sqrt{\tilde{k}^2-\beta^2}}\right]\exp(-\tilde{z}\sqrt{\tilde{k}^2-\beta^2}), \tag{2.100}$$

so that the problem of determination of S_{vv} reduces to the evaluation of three integrals: the first two integrals, obtained from (2.49) at $\nu=0$ by double differentiation with respect to a, are identical; the last integral is determined directly from (2.49). The expression for S_{vv} takes the form

$$S_{vv}=\tilde{r}\left\{\frac{1}{\alpha}\beta^2\exp(-\mathrm{i}\beta\alpha)\right.$$

$$\left.+\left[\frac{1}{\alpha}\frac{\mathrm{d}}{\mathrm{d}\alpha}+\tilde{z}^2\left(\frac{1}{\alpha}\frac{\mathrm{d}}{\mathrm{d}\alpha}\right)^2\right]\left(\frac{1}{\alpha}[\exp(-\mathrm{i}\beta\alpha)-\exp(-\mathrm{i}\beta\alpha/\sqrt{\tilde{A}_{zz}})]\right)\right\}. \tag{2.101}$$

The derived solution (2.98), (2.101) for an isotropic half-space agrees with the known solution by Stokes [29].

Static Action of Vertical Force

Consider the static solution for a transversely isotropic space subjected to a vertical force. Employing relationships (2.20), (2.21), we rewrite the expressions for normalized displacements in (2.94b), (2.95b) in the following form ($z\geq 0$):

$$S_{vh}=-\frac{\tilde{r}(\tilde{A}_{rz}+1)}{\tilde{A}_{zz}(\tilde{\eta}_1^2-\tilde{\eta}_2^2)}\int_0^\infty[\exp(-\tilde{k}\tilde{\eta}_1\tilde{z})-\exp(-\tilde{k}\tilde{\eta}_2\tilde{z})]\mathrm{J}_1(\tilde{k}\tilde{r})\,\mathrm{d}\tilde{k}, \tag{2.102}$$

$$S_{vv}=-\frac{\tilde{r}(\tilde{A}_{rz}+1)}{\tilde{A}_{zz}(\tilde{\eta}_1^2-\tilde{\eta}_2^2)}\int_0^\infty[\tilde{C}_{w1}\exp(-\tilde{k}\tilde{\eta}_1\tilde{z})-\tilde{C}_{w2}\exp(-\tilde{k}\tilde{\eta}_2\tilde{z})]\mathrm{J}_0(\tilde{k}\tilde{r})\,\mathrm{d}\tilde{k} \tag{2.103}$$

or using tabulated integrals:

$$S_{vh} = \frac{\widetilde{z}(\widetilde{A}_{rz}+1)}{\widetilde{A}_{zz}(\widetilde{\eta}_1^2-\widetilde{\eta}_2^2)}\left(\frac{\widetilde{\eta}_1}{\sqrt{\widetilde{r}^2+(\widetilde{\eta}_1\widetilde{z})^2}}-\frac{\widetilde{\eta}_2}{\sqrt{\widetilde{r}^2+(\widetilde{\eta}_2\widetilde{z})^2}}\right), \tag{2.104}$$

$$S_{vv} = -\frac{\widetilde{r}(\widetilde{A}_{rz}+1)}{\widetilde{A}_{zz}(\widetilde{\eta}_1^2-\widetilde{\eta}_2^2)}\left(\frac{\widetilde{C}_{w1}}{\sqrt{\widetilde{r}^2+(\widetilde{\eta}_1\widetilde{z})^2}}-\frac{\widetilde{C}_{w2}}{\sqrt{\widetilde{r}^2+(\widetilde{\eta}_2\widetilde{z})^2}}\right). \tag{2.105}$$

These expressions also hold for negative values of z. Clearly, normalized coordinates $\widetilde{z}$ and $\widetilde{r}$ may be replaced by coordinates z and r, respectively. The indeterminacy of expressions that represent the solution of the problem at $\widetilde{\eta}_1 = \widetilde{\eta}_2$ has been discussed in section 2.1.1. One may obtain the components of the stress tensor by differentiating the expressions for the displacements.

2.5.2 Action of Horizontal Force

Here, we consider the action of a horizontal force $Q_0 \exp(i\omega t)$ applied to the plane $z = 0$ within a transversely isotropic space. In the following, the representation of the solution given in equation (1.76) is used. Instead of (2.91), we search for a solution at $z > 0$ in the following form (function $\widetilde{p}(\widetilde{z},\widetilde{k})$ has been added):

$$\widetilde{q}(\widetilde{z},\widetilde{k}) = B(\widetilde{k})[C_{w2}\widetilde{q}_1(\widetilde{z},\widetilde{k}) - C_{w1}\widetilde{q}_2(\widetilde{z},\widetilde{k})], \tag{2.106a}$$

$$\widetilde{w}(\widetilde{z},\widetilde{k}) = B(\widetilde{k})[C_{w2}\widetilde{w}_1(\widetilde{z},\widetilde{k}) - C_{w1}\widetilde{w}_2(\widetilde{z},\widetilde{k})], \tag{2.106b}$$

$$\widetilde{p}(\widetilde{z},\widetilde{k}) = C(\widetilde{k})\widetilde{p}_1(\widetilde{z},\widetilde{k}), \tag{2.106c}$$

where the same fundamental solutions $\widetilde{q}_j$, $\widetilde{w}_j$ as in equations (2.91) are used, and function $\widetilde{p}_1$ has the form (2.57). The proposed form of solution for $\widetilde{q}$ and $\widetilde{w}$ is selected in order to satisfy the condition $\widetilde{w}(0,\widetilde{k}) = 0$, which stems from symmetry considerations. For $z < 0$, functions $\widetilde{q}(\widetilde{z},\widetilde{k})$ and $\widetilde{p}(\widetilde{z},\widetilde{k})$ are continued symmetrically, and function $\widetilde{w}(\widetilde{z},\widetilde{k})$ antisymmetrically. Condition $\widetilde{w}(0,\widetilde{k}) = 0$ results in the continuity of function $\widetilde{w}(\widetilde{z},\widetilde{k})$ and its derivative with respect to z over the entire space; functions $\widetilde{q}(\widetilde{z},\widetilde{k})$ and $\widetilde{p}(\widetilde{z},\widetilde{k})$ are also continuous. Thus, continuity of displacements, and, in accordance with equation (1.40), of stresses σ_z, is provided when passing the plane of application of the load. One may calculate coefficients $B(\widetilde{k})$ and $C(\widetilde{k})$ by providing the required jump in the value of derivatives with respect to z of functions $\widetilde{q}(\widetilde{z},\widetilde{k})$ and $\widetilde{p}(\widetilde{z},\widetilde{k})$ when passing the plane $z = 0$. Taking into account both the evenness of these functions and equations (1.110), (1.114), one may formulate the following boundary condition at $z=0$ for functions $\tilde{q}(\tilde{z},\tilde{k})$ and $\tilde{p}(\tilde{z},\tilde{k})$ in the case of the uniform distribution of the force over a circular area of radius R:

$$2\frac{\mathrm{d}\tilde{q}}{\mathrm{d}\tilde{z}}=b_q=\frac{Q_0}{\pi R G_{rz}}\mathrm{J}_1(kR)\,, \tag{2.107}$$

$$2\frac{\mathrm{d}\tilde{p}}{\mathrm{d}\tilde{z}}=b_p=\frac{Q_0}{\pi R G_{rz}}\mathrm{J}_1(kR)\,. \tag{2.108}$$

Hence we determine coefficients B and C in solution (2.106):

$$B=-\frac{Q_0\,\mathrm{J}_1(kR)}{2\pi R G_{rz}(C_{w2}\eta_1-C_{w1}\eta_2)}\,, \tag{2.109}$$

$$C=-\frac{Q_0\,\mathrm{J}_1(kR)}{2\pi R G_{rz}\sqrt{\tilde{G}_{r\vartheta}\tilde{k}^2-\beta^2\theta^2}}\,. \tag{2.110}$$

The amplitudes of vibrations at $z\geq 0$ are presented in the form of integrals (1.76) taking into account (2.106), (2.109), (2.110). In the case of a concentrated force, we obtain

$$u_r=\hat{u}_r\cos\vartheta,\ u_\vartheta=\hat{u}_\vartheta\sin\vartheta,\ u_z=\hat{u}_z\cos\vartheta\,, \tag{2.111}$$

$$\hat{u}_r=\frac{Q_0}{4\pi G_{rz0}r}S_{hr}\,, \tag{2.112a}$$

$$S_{hr}=\beta^2\tilde{r}\int_0^\infty\left[[\tilde{q}(\tilde{z},\tilde{k})-\tilde{p}(\tilde{z},\tilde{k})]\frac{\mathrm{J}_1(\tilde{k}\tilde{r})}{\tilde{k}\tilde{r}}-\tilde{q}(\tilde{z},\tilde{k})\mathrm{J}_0(\tilde{k}\tilde{r})\right]\mathrm{d}\tilde{k}\,, \tag{2.112b}$$

$$\hat{u}_\vartheta=-\frac{Q_0}{4\pi G_{rz0}r}S_{h\vartheta}\,, \tag{2.113a}$$

$$S_{h\vartheta}=-\beta^2\tilde{r}\int_0^\infty\left[[\tilde{q}(\tilde{z},\tilde{k})-\tilde{p}(\tilde{z},\tilde{k})]\frac{\mathrm{J}_1(\tilde{k}\tilde{r})}{\tilde{k}\tilde{r}}+\tilde{p}(\tilde{z},\tilde{k})\mathrm{J}_0(\tilde{k}\tilde{r})\right]\mathrm{d}\tilde{k}\,, \tag{2.113b}$$

$$\hat{u}_z=-\frac{Q_0}{4\pi G_{rz0}r}S_{hz}\,, \tag{2.114a}$$

$$S_{hz}=-\beta^2\tilde{r}\int_0^\infty\frac{C_{w1}C_{w2}}{C_{w2}\eta_1-C_{w1}\eta_2}[\exp(-\eta_1\tilde{z})-\exp(-\eta_2\tilde{z})]\tilde{k}\,\mathrm{J}_1(\tilde{k}\tilde{r})\,\mathrm{d}\tilde{k}\,, \tag{2.114b}$$

where

$$\overline{q}(\widetilde{z},\widetilde{k}) = -\frac{\widetilde{k}[C_{w2}\exp(-\eta_1\widetilde{z}) - C_{w1}\exp(-\eta_2\widetilde{z})]}{C_{w2}\eta_1 - C_{w1}\eta_2}, \tag{2.115a}$$

$$\overline{p}(\widetilde{z},\widetilde{k}) = -\frac{\widetilde{k}}{\sqrt{\widetilde{k}^2\widetilde{G}_{r\vartheta} - \beta^2\theta^2}}\exp[-\widetilde{z}\sqrt{\widetilde{k}^2\widetilde{G}_{r\vartheta} - \beta^2\theta}\ ]. \tag{2.115b}$$

As in previous sections, we take $\theta = 1$ when solving dynamic problems and employ dimensionless coordinates (2.19). Replacing r with z in the coefficients of the normalized amplitudes and replacing $\widetilde{r}$ with $\widetilde{z}$ in the multipliers before the integrals are reasonable when calculating displacements on the Z-axis ($r = 0$).

Static Action of Horizontal Force

Consider static displacements that occur due to the action of a horizontal force. Taking $\beta = 1, \theta = 0$, we apply relationships (2.20). Normalized displacements are expressed by using the same integrals as in equations (2.62)–(2.64):

$$S_{hr} = \frac{\widetilde{r}}{\widetilde{C}_{w1}\widetilde{\eta}_2 - \widetilde{C}_{w2}\widetilde{\eta}_1}\left\{\widetilde{C}_{w2}\int_0^\infty \exp(-\widetilde{k}\widetilde{\eta}_1\widetilde{z})\left[\frac{J_1(\widetilde{k}\widetilde{r})}{\widetilde{k}\widetilde{r}} - J_0(\widetilde{k}\widetilde{r})\right]d\widetilde{k}\right.$$
$$\left. - \widetilde{C}_{w1}\int_0^\infty \exp(-\widetilde{k}\widetilde{\eta}_2\widetilde{z})\left[\frac{J_1(\widetilde{k}\widetilde{r})}{\widetilde{k}\widetilde{r}} - J_0(\widetilde{k}\widetilde{r})\right]d\widetilde{k}\right\}$$
$$+\frac{\widetilde{r}}{\widetilde{\eta}_0}\int_0^\infty \exp(-\widetilde{k}\widetilde{\eta}_0\widetilde{z})\frac{J_1(\widetilde{k}\widetilde{r})}{\widetilde{k}\widetilde{r}}d\widetilde{k}, \tag{2.116}$$

$$S_{h\vartheta} = \frac{\widetilde{r}}{\widetilde{C}_{w2}\widetilde{\eta}_1 - \widetilde{C}_{w1}\widetilde{\eta}_2}\left[\widetilde{C}_{w2}\int_0^\infty \exp(-\widetilde{k}\widetilde{\eta}_1\widetilde{z})\frac{J_1(\widetilde{k}\widetilde{r})}{\widetilde{k}\widetilde{r}}d\widetilde{k}\right.$$
$$\left. - \widetilde{C}_{w1}\int_0^\infty \exp(-\widetilde{k}\widetilde{\eta}_2\widetilde{z})\frac{J_1(\widetilde{k}\widetilde{r})}{\widetilde{k}\widetilde{r}}d\widetilde{k}\right]$$
$$-\frac{\widetilde{r}}{\widetilde{\eta}_0}\int_0^\infty \exp(-\widetilde{k}\widetilde{\eta}_0\widetilde{z})\left[\frac{J_1(\widetilde{k}\widetilde{r})}{\widetilde{k}\widetilde{r}} - J_0(\widetilde{k}\widetilde{r})\right]d\widetilde{k}, \tag{2.117}$$

$$S_{hz} = \frac{\widetilde{r}\widetilde{C}_{w1}\widetilde{C}_{w2}}{\widetilde{C}_{w2}\widetilde{\eta}_1 - \widetilde{C}_{w1}\widetilde{\eta}_2}\left[\int_0^\infty \exp(-\widetilde{k}\widetilde{\eta}_1\widetilde{z})J_1(\widetilde{k}\widetilde{r})d\widetilde{k} - \int_0^\infty \exp(-\widetilde{k}\widetilde{\eta}_2\widetilde{z})J_1(\widetilde{k}\widetilde{r})d\widetilde{k}\right] \tag{2.118}$$

The subsequent integration yields

$$S_{hr} = \frac{zr}{\widetilde{C}_{w1}\widetilde{\eta}_2 - \widetilde{C}_{w2}\widetilde{\eta}_1}\left[-\frac{\widetilde{C}_{w2}\widetilde{\eta}_1}{F_1(r, z\widetilde{\eta}_1)} + \frac{\widetilde{C}_{w1}\widetilde{\eta}_2}{F_1(r, z\widetilde{\eta}_2)}\right] + \frac{r}{\widetilde{\eta}_0 F_2(r, z\widetilde{\eta}_0)}, \tag{2.119}$$

$$S_{h\vartheta} = \frac{r}{\widetilde{C}_{w2}\widetilde{\eta}_1 - \widetilde{C}_{w1}\widetilde{\eta}_2}\left[\frac{\widetilde{C}_{w2}}{F_2(r, z\widetilde{\eta}_1)} - \frac{\widetilde{C}_{w1}}{F_2(r, z\widetilde{\eta}_2)}\right] + \frac{zr}{F_1(r, z\widetilde{\eta}_0)}, \tag{2.120}$$

$$S_{hz} = \frac{\widetilde{C}_{w1}\widetilde{C}_{w2}z}{\widetilde{C}_{w2}\widetilde{\eta}_1 - \widetilde{C}_{w1}\widetilde{\eta}_2}\left[-\frac{\widetilde{\eta}_1}{F_3(r, z\widetilde{\eta}_1)} + \frac{\widetilde{\eta}_2}{F_3(r, z\widetilde{\eta}_2)}\right], \tag{2.121}$$

where the nomenclature given in (2.68)–(2.70) is used. Consider a possible case of coincidence of values $\widetilde{\eta}_1$ and $\widetilde{\eta}_2$ when calculated static displacements (2.119)–(2.121) and (2.104), (2.105) become indeterminacies. As in section 2.1.1, we recommend applying (instead of evaluating the indeterminities) a small variation of elastic parameters to avoid the equality of roots $\widetilde{\eta}_1$ and $\widetilde{\eta}_2$.

The solutions of dynamic and static problems concerning the action of a concentrated force on an infinite transversely isotropic space are significant for applications of the method of boundary elements in dynamic and static problems for a transversely isotropic body.

2.6 Vibrations of Transversely Isotropic Half-Space under Action of Force Applied within Half-Space

2.6.1 Action of Vertical Force

Let a vertical force $P_0 \exp(\mathrm{i}\,\omega t)$ be applied in the plane $z = z_1$ within a transversely isotropic half-space. Instead of constructing a solution for the homogeneous half-space in its general form (given in relationships (1.109)), one could simplify the construction of the solution by employing a solution for the infinite space, with the addition of corresponding corrections, to satisfy the conditions of vanishing stresses on the surface of the half-space. This technique has been employed by Mindlin [75] to solve the problem dealing with the static action of a force within an isotropic homogeneous elastic half-space. For the sake of generality, we apply the form of solution given in relationships (1.109). Parts of the solution, which correspond to the action of a force in an infinite space, are separated out in the process of building the solution, so that a reliable check of the performed transformations can be provided.

The application of the dimensionless parameters given in (1.27) yields, for the case of the concentrated force,

$$u_r = \frac{P_0}{2G_{rz0}\pi r} S_{vr}, \tag{2.122a}$$

$$S_{vr} = \frac{\tilde{r}\beta^2}{\tilde{A}_{zz}} \int_0^\infty \tilde{k}\, \mathrm{J}_1(\tilde{k}\tilde{r})[A_1^*(\tilde{k})\tilde{q}_1(\tilde{z},\tilde{k}) + A_2^*(\tilde{k})\tilde{q}_2(\tilde{z},\tilde{k}) + \tilde{q}_a^*(\tilde{z},\tilde{k})]\mathrm{d}\tilde{k}, \tag{2.122b}$$

$$u_z = \frac{P_0}{2G_{rz0}\pi r} S_{vv}, \tag{2.123a}$$

$$S_{vv} = \frac{\tilde{r}\beta^2}{\tilde{A}_{zz}} \int_0^\infty \tilde{k}\, \mathrm{J}_0(\tilde{k}\tilde{r})[A_1^*(\tilde{k})\tilde{w}_1(\tilde{z},\tilde{k}) + A_2^*(\tilde{k})\tilde{w}_2(\tilde{z},\tilde{k}) + w_a^*(\tilde{z},\tilde{k})]\mathrm{d}\tilde{k}. \tag{2.123b}$$

Here, in accordance with the technique presented in section 1.5.1, we employ the particular solution (denoted by index *a*), which vanishes at $z > z_1$ and may be expressed as a linear combination of fundamental solutions at $z < z_1$ $(\tilde{z} < \tilde{z}_1 = z_1 / z_r)$:

$$\tilde{q}_a^* = \sum_j^4 B_j \tilde{q}_j, \tag{2.124a}$$

$$\tilde{w}_a^* = \sum_j^4 B_j \tilde{w}_j. \tag{2.124b}$$

Following (2.5), the fundamental solutions have the form

$$\tilde{q}_j = C_{qj} \exp(-\eta_j \tilde{z}), \tag{2.125a}$$

$$\tilde{w}_j = C_{wj} \exp(-\eta_j \tilde{z}), \tag{2.125b}$$

where coefficients $C_{qj} = 1$, while coefficients C_{wj} are determined by (2.14), in accordance with the number of roots η_j of the characteristic equation (2.7). The first two roots, η_1 and η_2, are found according to (2.9); $\eta_3 = -\eta_1$, $\eta_4 = -\eta_2$. Next, coefficients B_j are determined by employing the system of equations (1.105) with dimensionless values $\tilde{q}_j, \tilde{w}_j, \tilde{z}_1, \tilde{k}$. The last transition has no impact on the form of the system of equations. In order to remove exponential multipliers in the coefficients of the system of equations, one could introduce modified coefficients B_j^*:

$$B_j^* = B_j \exp(-\eta_j \tilde{z}_1). \tag{2.126}$$

Following (1.105), we obtain

$$B_1^* + B_2^* + B_3^* + B_4^* = 0,$$

$$C_{w1}B_1^* + C_{w2}B_2^* + C_{w3}B_3^* + C_{w4}B_4^* = 0\,,$$
$$\eta_1 B_1^* + \eta_2 B_2^* + \eta_3 B_3^* + \eta_4 B_4^* = 0\,, \tag{2.127}$$
$$\eta_1 C_{w1}B_1^* + \eta_2 C_{w2}B_2^* + \eta_3 C_{w3}B_3^* + \eta_4 C_{w4}B_4^* = -1\,.$$

Employing (2.14) and the relationship between the roots η_j, we have

$$C_{w3} = -C_{w1}\,, \tag{2.128a}$$

$$C_{w4} = -C_{w2}\,. \tag{2.128b}$$

Keeping in mind relationships (2.128), we obtain the solution of system (2.127):

$$B_1^* = B_3^* = -B_2^* = -B_4^* = -\frac{1}{2(\eta_1 C_{w1} - \eta_2 C_{w2})} = \frac{\widetilde{k}(\widetilde{A}_{rz}+1)}{2(\eta_1^2 - \eta_2^2)}\,. \tag{2.129}$$

Coefficients A_1^*, A_2^* are determined from the system of equations (1.53) with the right sides equal to the following (according to relationships (1.107) at $z_0 = 0$) values:

$$d_1 = -\left[\widetilde{A}_{rz}\widetilde{k}\widetilde{q}_a^*(0,\widetilde{k}) + \widetilde{A}_{zz}\frac{\mathrm{d}\,\widetilde{w}_a^*(0,\widetilde{k})}{\mathrm{d}\,\widetilde{z}}\right] = -\sum_{j=1}^{4} c_{1j}B_j\,, \tag{2.130a}$$

$$d_2 = -\left[\frac{\mathrm{d}\,\widetilde{q}_a^*(0,\widetilde{k})}{\mathrm{d}\,\widetilde{z}} - \widetilde{k}\widetilde{w}_a^*(0,\widetilde{k})\right] = -\sum_{j=1}^{4} c_{2j}B_j\,. \tag{2.130b}$$

where the values of c_{ij} are calculated in accordance with equations (1.54). Note that the following relationships hold:

$$c_{13} = c_{11},\; c_{14} = c_{12},\; c_{23} = -c_{21},\; c_{24} = -c_{22}\,. \tag{2.131}$$

In order to clarify the behavior of the integrands in solution (2.122), (2.123), it is reasonable to group summands in brackets ($\widetilde{z} < \widetilde{z}_1$):

$$A_1^*(\widetilde{k})\widetilde{q}_1(\widetilde{z},\widetilde{k}) + A_2^*(\widetilde{k})\widetilde{q}_2(\widetilde{z},\widetilde{k}) + \widetilde{q}_a^*(\widetilde{z},\widetilde{k})$$
$$= \sum_{j=1}^{2} A_j^{\wedge}(\widetilde{k})\widetilde{q}_j(\widetilde{z},\widetilde{k}) + \sum_{j=3}^{4} B_j(\widetilde{k})\widetilde{q}_j(\widetilde{z},\widetilde{k}) \tag{2.132a}$$

$$A_1^*(\widetilde{k})\widetilde{w}_1(\widetilde{z},\widetilde{k}) + A_2^*(\widetilde{k})\widetilde{w}_2(\widetilde{z},\widetilde{k}) + w_a^*(\widetilde{z},\widetilde{k})$$
$$= \sum_{j=1}^{2} A_j^{\wedge}(\widetilde{k})\widetilde{w}_j(\widetilde{z},\widetilde{k}) + \sum_{j=3}^{4} B_j(\widetilde{k})\widetilde{w}_j(\widetilde{z},\widetilde{k}) \tag{2.132b}$$

where

$$A_j^{\wedge}(\tilde{k}) = A_j^{*}(\tilde{k}) + B_j(\tilde{k}). \tag{2.133}$$

Taking into account the form of the system of equations (1.53) with right sides (2.130), we obtain the following system of equations for the new coefficients $A_j^{\wedge}$:

$$c_{11}A_1^{\wedge} + c_{12}A_2^{\wedge} = -\sum_{j=3}^{4} c_{1j}B_j = -B_1^{*}[c_{11}\exp(-\eta_1\tilde{z}_1) - c_{12}\exp(-\eta_2\tilde{z}_1)],$$

$$c_{21}A_1^{\wedge} + c_{22}A_2^{\wedge} = -\sum_{j=3}^{4} c_{2j}B_j = B_1^{*}[c_{21}\exp(-\eta_1\tilde{z}_1) - c_{22}\exp(-\eta_2\tilde{z}_1)], \tag{2.134}$$

resulting in

$$A_1^{\wedge} = B_1^{*}[K_1\exp(-\eta_1\tilde{z}_1) + K_2\exp(-\eta_2\tilde{z}_1)],$$

$$A_2^{\wedge} = B_1^{*}[K_3\exp(-\eta_1\tilde{z}_1) + K_1\exp(-\eta_2\tilde{z}_1)] \tag{2.135}$$

with

$$K_1 = -\frac{c_{11}c_{22} + c_{12}c_{21}}{D}, \tag{2.136a}$$

$$K_2 = \frac{2c_{12}c_{22}}{D}, \tag{2.136b}$$

$$K_3 = \frac{2c_{11}c_{21}}{D}, \tag{2.136c}$$

$$D = c_{11}c_{22} - c_{12}c_{21}. \tag{2.136d}$$

Coefficients $A_j^{\wedge}$ decrease exponentially at high values of $\tilde{k}$ and $\tilde{z}_1 > 0$.

The normalized amplitudes of vibrations (2.122a), (2.123a) at $\tilde{z} \le \tilde{z}_1$ are presented as

$$S_{vr} = \frac{\tilde{r}\beta^2}{\tilde{A}_{zz}} \int_0^{\infty} \tilde{k}\, \mathrm{J}_1(\tilde{k}\tilde{r})\{A_1^{\wedge}(\tilde{k})\exp(-\eta_1\tilde{z}) + A_2^{\wedge}(\tilde{k})\exp(-\eta_2\tilde{z})$$
$$+ B_1^{*}[\exp(-\eta_1(\tilde{z}_1 - \tilde{z})) - \exp(-\eta_2(\tilde{z}_1 - \tilde{z}))]\}\, \mathrm{d}\tilde{k} \tag{2.137a}$$

$$S_{vv} = \frac{\tilde{r}\beta^2}{\tilde{A}_{zz}} \int_0^{\infty} \tilde{k}\, \mathrm{J}_0(\tilde{k}\tilde{r})\{A_1^{\wedge}(\tilde{k})C_{w1}\exp(-\eta_1\tilde{z}) + A_2^{\wedge}(\tilde{k})C_{w2}\exp(-\eta_2\tilde{z})$$
$$- B_1^{*}[C_{w1}\exp(-\eta_1(\tilde{z}_1 - \tilde{z})) - C_{w2}\exp(-\eta_2(\tilde{z}_1 - \tilde{z}))]\}\, \mathrm{d}\tilde{k}. \tag{2.137b}$$

For the part of the half-space $z > z_1$, we apply representation (2.122), (2.123) with $\tilde{q}_a = \tilde{w}_a = 0$. Using for relationship (2.133), one may rewrite the expressions for normalized amplitudes for $z > z_1$ as follows:

$$S_{vr} = \frac{\tilde{r}\beta^2}{\tilde{A}_{zz}} \int_0^\infty \tilde{k}\, \mathrm{J}_1(\tilde{k}\tilde{r})\{A_1^\wedge(\tilde{k})\exp(-\eta_1\tilde{z}) + A_2^\wedge(\tilde{k})\exp(-\eta_2\tilde{z})$$

$$- B_1^*[\exp(-\eta_1(\tilde{z}-\tilde{z}_1)) - \exp(-\eta_2(\tilde{z}-\tilde{z}_1))]\}\,\mathrm{d}\tilde{k} \tag{2.138a}$$

$$S_{vv} = \frac{\tilde{r}\beta^2}{\tilde{A}_{zz}} \int_0^\infty \tilde{k}\, \mathrm{J}_0(\tilde{k}\tilde{r})\{A_1^\wedge(\tilde{k})C_{w1}\exp(-\eta_1\tilde{z}) + A_2^\wedge(\tilde{k})C_{w2}\exp(-\eta_2\tilde{z})$$

$$- B_1^*[C_{w1}\exp(-\eta_1(\tilde{z}-\tilde{z}_1)) - C_{w2}\exp(-\eta_2(\tilde{z}-\tilde{z}_1))]\}\,\mathrm{d}\tilde{k}\ . \tag{2.138b}$$

The last summands in expressions (2.137) and (2.138) (containing multiplier B_1^*) represent the solution for infinite space given in equations (2.94), (2.95) for $z \geq 0$; summands containing coefficients $A_1^\wedge$, $A_2^\wedge$ express a correction, required in order to satisfy the condition of vanishing stresses in the plane $z = 0$. The results obtained indicate that at $\tilde{z} \neq \tilde{z}_1$ the integrands contain a multiplier which decreases exponentially with increasing $\tilde{k}$. At $\tilde{z} = \tilde{z}_1$, the multiplier of Bessel's function in expression S_{vv} tends to a finite limit as $\tilde{k} \to \infty$, and only decreasing and oscillation of Bessel's function $\mathrm{J}_0(\tilde{k}\tilde{r})$ provide a relatively slow convergence of the integrals. It is reasonable to take parameter θ equal to unity (in dynamic problems) and use values (2.19) as dimensionless coordinates. Considerations employed for the calculation of integrals, which express the amplitudes of vibrations of a half-space subjected to a force applied to its surface, still hold for the given case (see section 2.1). The structure of coefficients $A_1^\wedge$, $A_2^\wedge$ indicates that Rayleigh waves, related to the vanishing of determinant D, develop progressively less with increasing depth z_1 of the application point of the force.

Static Action of Vertical Force

Consider the static problem ($\theta = 0,\ \beta = 1$). By employing values (2.20), (2.21), one may express displacements of the points in a half-space in finite form, in the same way as for solving other static problems dealing with the transversely isotropic half-space. For $0 \leq \tilde{z} \leq \tilde{z}_1$:

$$S_{vr} = \frac{(\tilde{A}_{rz}+1)\tilde{r}}{2(\tilde{\eta}_1^2 - \tilde{\eta}_2^2)\tilde{A}_{zz}} \left[\tilde{K}_1 \int_0^\infty \mathrm{J}_1(\tilde{k}\tilde{r})\exp(-\tilde{k}\tilde{\eta}_1(\tilde{z}+\tilde{z}_1))\,\mathrm{d}\tilde{k} \right.$$

$$+\widetilde{K}_2\int_0^{\infty}\mathrm{J}_1(\widetilde{k}\widetilde{r})\exp(-\widetilde{k}(\widetilde{\eta}_1\widetilde{z}+\widetilde{\eta}_2\widetilde{z}_1))\,\mathrm{d}\widetilde{k}$$

$$+\widetilde{K}_3\int_0^{\infty}\mathrm{J}_1(\widetilde{k}\widetilde{r})\exp(-\widetilde{k}(\widetilde{\eta}_2\widetilde{z}+\widetilde{\eta}_1\widetilde{z}_1))\,\mathrm{d}\widetilde{k}+\widetilde{K}_1\int_0^{\infty}\mathrm{J}_1(\widetilde{k}\widetilde{r})\exp(-\widetilde{k}\widetilde{\eta}_2(\widetilde{z}+\widetilde{z}_1))\,\mathrm{d}\widetilde{k}$$

$$\left.+\int_0^{\infty}\mathrm{J}_1(\widetilde{k}\widetilde{r})\exp(-\widetilde{k}\widetilde{\eta}_1(\widetilde{z}_1-\widetilde{z}))\,\mathrm{d}\widetilde{k}-\int_0^{\infty}\mathrm{J}_1(\widetilde{k}\widetilde{r})\exp(-\widetilde{k}\widetilde{\eta}_2(\widetilde{z}_1-\widetilde{z}))\,\mathrm{d}\widetilde{k}\right], \tag{2.139a}$$

$$S_{vv}=\frac{(\widetilde{A}_{rz}+1)\widetilde{r}}{2(\widetilde{\eta}_1^2-\widetilde{\eta}_2^2)\widetilde{A}_{zz}}\left[\widetilde{K}_1\widetilde{C}_{w1}\int_0^{\infty}\mathrm{J}_0(\widetilde{k}\widetilde{r})\exp(-\widetilde{k}\widetilde{\eta}_1(\widetilde{z}+\widetilde{z}_1))\,\mathrm{d}\widetilde{k}\right.$$

$$+\widetilde{K}_2\widetilde{C}_{w1}\int_0^{\infty}\mathrm{J}_0(\widetilde{k}\widetilde{r})\exp(-\widetilde{k}(\widetilde{\eta}_1\widetilde{z}+\widetilde{\eta}_2\widetilde{z}_1))\,\mathrm{d}\widetilde{k}$$

$$+\widetilde{K}_3\widetilde{C}_{w2}\int_0^{\infty}\mathrm{J}_0(\widetilde{k}\widetilde{r})\exp(-\widetilde{k}(\widetilde{\eta}_2\widetilde{z}+\widetilde{\eta}_1\widetilde{z}_1))\,\mathrm{d}\widetilde{k}$$

$$+\widetilde{K}_1\widetilde{C}_{w2}\int_0^{\infty}\mathrm{J}_0(\widetilde{k}\widetilde{r})\exp(-\widetilde{k}\widetilde{\eta}_2(\widetilde{z}+\widetilde{z}_1))\,\mathrm{d}\widetilde{k}$$

$$-\widetilde{C}_{w1}\int_0^{\infty}\mathrm{J}_0(\widetilde{k}\widetilde{r})\exp(-\widetilde{k}\widetilde{\eta}_1(\widetilde{z}_1-\widetilde{z}))\,\mathrm{d}\widetilde{k}\,,$$

$$\left.+\widetilde{C}_{w2}\int_0^{\infty}\mathrm{J}_0(\widetilde{k}\widetilde{r})\exp(-\widetilde{k}\widetilde{\eta}_2(\widetilde{z}_1-\widetilde{z}))\,\mathrm{d}\widetilde{k}\right] \tag{2.139b}$$

where the following constants are introduced:

$$\widetilde{K}_1=-\frac{\widetilde{c}_{11}\widetilde{c}_{22}+\widetilde{c}_{12}\widetilde{c}_{21}}{\widetilde{D}}, \tag{2.140a}$$

$$\widetilde{K}_2=\frac{2\widetilde{c}_{12}\widetilde{c}_{22}}{\widetilde{D}}, \tag{2.140b}$$

$$\widetilde{K}_3=\frac{2\widetilde{c}_{11}\widetilde{c}_{21}}{\widetilde{D}}. \tag{2.140c}$$

The values of $\widetilde{c}_{ij}$, $\widetilde{D}$ are given in (2.21). Integrating yields [43]

$$S_{vr}=\frac{(\widetilde{A}_{rz}+1)}{2(\widetilde{\eta}_1^2-\widetilde{\eta}_2^2)\widetilde{A}_{zz}}\left\{\widetilde{K}_1\left[1-\frac{\widetilde{\eta}_1(z+z_1)}{F_3(r,(z+z_1)\widetilde{\eta}_1)}\right]+\widetilde{K}_2\left[1-\frac{\widetilde{\eta}_1 z+\widetilde{\eta}_2 z_1}{F_3(r,z\widetilde{\eta}_1+z_1\widetilde{\eta}_2)}\right]\right.$$

$$+\widetilde{K}_3\left[1-\frac{\widetilde{\eta}_2 z+\widetilde{\eta}_1 z_1}{F_3(r, z\widetilde{\eta}_2+z_1\widetilde{\eta}_1)}\right]+\widetilde{K}_1\left[1-\frac{\widetilde{\eta}_2(z+z_1)}{F_3(r,(z+z_1)\widetilde{\eta}_2)}\right]$$

$$\left.-\frac{\widetilde{\eta}_1(z_1-z)}{F_3(r,(z_1-z)\widetilde{\eta}_1)}+\frac{\widetilde{\eta}_2(z_1-z)}{F_3(r,(z_1-z)\widetilde{\eta}_2)}\right\}, \tag{2.141}$$

$$S_{vv}=\frac{(\widetilde{A}_{rz}+1)r}{2(\widetilde{\eta}_1^2-\widetilde{\eta}_2^2)\widetilde{A}_{zz}}\left[\frac{\widetilde{K}_1\widetilde{C}_{w1}}{F_3(r,(z_1+z)\widetilde{\eta}_1)}+\frac{\widetilde{K}_2\widetilde{C}_{w1}}{F_3(r,z\widetilde{\eta}_1+z_1\widetilde{\eta}_2)}+\frac{\widetilde{K}_3\widetilde{C}_{w2}}{F_3(r,z\widetilde{\eta}_2+z_1\widetilde{\eta}_1)}\right.$$

$$\left.+\frac{\widetilde{K}_1\widetilde{C}_{w2}}{F_3(r,(z_1+z)\widetilde{\eta}_2)}-\frac{\widetilde{C}_{w1}}{F_3(r,(z_1-z)\widetilde{\eta}_1)}+\frac{\widetilde{C}_{w2}}{F_3(r,(z_1-z)\widetilde{\eta}_2)}\right], \tag{2.142}$$

where symbol F_3 given in (2.70) is used. Expressions for S_{vr}, S_{vv} at $z > z_1$ have the same form (2.141), (2.142). In order to obtain the static solution for the isotropic case, when the roots $\widetilde{\eta}_1$ and $\widetilde{\eta}_2$ are equal, one should evaluate an indeterminacy of the form 0/0, resulting in a known Mindlin's solution, or perform calculations by using formulas for a transversely isotropic half-space with a small deviation of elastic properties from those of an isotropic body.

2.6.2 Action of Horizontal Force

Let a horizontal force $Q_0 \exp(\mathrm{i}\,\omega t)$ be applied in the plane $z = z_1$ located within a transversely isotropic half-space. We shall employ the solution presented in relationships (1.120), (1.121). A transition to the case of the concentrated force and using the dimensionless variables given in (1.27) yields

$$u_r=\hat{u}_r\cos\vartheta,\ u_\vartheta=\hat{u}_\vartheta\sin\vartheta,\ u_z=\hat{u}_z\cos\vartheta\,, \tag{2.143}$$

$$\hat{u}_r=\frac{Q_0}{2\pi G_{rz0} r}S_{hr}\,, \tag{2.144a}$$

$$S_{hr}=\beta^2\widetilde{r}\int_0^\infty\left[[\widetilde{q}^*(\widetilde{z},\widetilde{k})-\widetilde{p}^*(\widetilde{z},\widetilde{k})]\frac{\mathrm{J}_1(\widetilde{k}\widetilde{r})}{\widetilde{k}\widetilde{r}}-\widetilde{q}^*(\widetilde{z},\widetilde{k})\mathrm{J}_0(\widetilde{k}\widetilde{r})\right]\widetilde{k}\,\mathrm{d}\widetilde{k}, \tag{2.144b}$$

$$\hat{u}_\vartheta=-\frac{Q_0}{2\pi G_{rz0} r}S_{h\vartheta}\,, \tag{2.145a}$$

$$S_{h\vartheta}=-\beta^2\widetilde{r}\int_0^\infty\left[[\widetilde{q}^*(\widetilde{z},\widetilde{k})-\widetilde{p}^*(\widetilde{z},\widetilde{k})]\frac{\mathrm{J}_1(\widetilde{k}\widetilde{r})}{\widetilde{k}\widetilde{r}}+\widetilde{p}^*(\widetilde{z},\widetilde{k})\mathrm{J}_0(\widetilde{k}\widetilde{r})\right]\widetilde{k}\,\mathrm{d}\widetilde{k}, \tag{2.145b}$$

$$\hat{u}_z = -\frac{Q_0}{2\pi G_{rz0} r} S_{hz}, \tag{2.146a}$$

$$S_{hz} = -\beta^2 \widetilde{r} \int_0^\infty \widetilde{w}^*(\widetilde{z}, \widetilde{k}) \mathrm{J}_1(\widetilde{k}\widetilde{r}) \widetilde{k} \, \mathrm{d}\widetilde{k}, \tag{2.146b}$$

where

$$\widetilde{q}^*(\widetilde{z}, \widetilde{k}) = A_1^*(\widetilde{k}) \widetilde{q}_1(\widetilde{z}, \widetilde{k}) + A_2^*(\widetilde{k}) \widetilde{q}_2(\widetilde{z}, \widetilde{k}) + \widetilde{q}_a^*(\widetilde{z}, \widetilde{k}), \tag{2.147}$$

$$\widetilde{w}^*(\widetilde{z}, \widetilde{k}) = A_1^*(\widetilde{k}) \widetilde{w}_1(\widetilde{z}, \widetilde{k}) + A_2^*(\widetilde{k}) \widetilde{w}_2(\widetilde{z}, \widetilde{k}) + \widetilde{w}_a^*(\widetilde{z}, \widetilde{k}), \tag{2.148}$$

$$\widetilde{p}^*(\widetilde{z}, \widetilde{k}) = C^*(\widetilde{k}) \widetilde{p}_1(\widetilde{z}, \widetilde{k}) + \widetilde{p}_a^*(\widetilde{z}, \widetilde{k}). \tag{2.149}$$

For $\widetilde{z} > \widetilde{z}_1$, the summands with index a are dropped. Here, we keep the notations introduced when considering action of the vertical force for new values, and use representations (2.124)–(2.126) as before. In contrast to the system of equations (2.127), the following system of equations takes place, according to (1.113):

$$\begin{aligned}
&B_1^* + B_2^* + B_3^* + B_4^* = 0, \\
&C_{w1} B_1^* + C_{w2} B_2^* + C_{w3} B_3^* + C_{w4} B_4^* = 0, \\
&\eta_1 B_1^* + \eta_2 B_2^* + \eta_3 B_3^* + \eta_4 B_4^* = 1, \\
&\eta_1 C_{w1} B_1^* + \eta_2 C_{w2} B_2^* + \eta_3 C_{w3} B_3^* + \eta_4 C_{w4} B_4^* = 0.
\end{aligned} \tag{2.150}$$

By using the previously stated relationships between the roots of the characteristic equation and between coefficients C_{wj}, the derivation of the solution for system (2.150) is straightforward:

$$\begin{aligned}
&B_1^* = -B_3^* = \frac{C_{w2}}{2(\eta_1 C_{w2} - \eta_2 C_{w1})}, \\
&B_2^* = -B_4^* = -\frac{C_{w1}}{2(\eta_1 C_{w2} - \eta_2 C_{w1})} = -\frac{C_{w1}}{C_{w2}} B_1^*.
\end{aligned} \tag{2.151}$$

Transformation of (2.147) and (2.148) in accordance with (2.132), (2.133) and application of the system of equations analogous to (2.134) yields

$$\begin{aligned}
&c_{11} A_1^\wedge + c_{12} A_2^\wedge = -\sum_{j=3}^{4} c_{1j} B_j = B_1^* \left(c_{11} \exp(-\eta_1 \widetilde{z}_1) - \frac{C_{w1}}{C_{w2}} c_{12} \exp(-\eta_1 \widetilde{z}_1) \right), \\
&c_{21} A_1^\wedge + c_{22} A_2^\wedge = -\sum_{j=3}^{4} c_{2j} B_j = -B_1^* \left(c_{21} \exp(-\eta_1 \widetilde{z}_1) - \frac{C_{w1}}{C_{w2}} c_{22} \exp(-\eta_2 \widetilde{z}_1) \right).
\end{aligned} \tag{2.152}$$

Hence we obtain coefficients $A_j^{\wedge}$:

$$A_1^{\wedge} = -B_1^{*}\left[K_1 \exp(-\eta_1 \tilde{z}_1) + \frac{C_{w1}}{C_{w2}} K_2 \exp(-\eta_2 \tilde{z}_1)\right],$$

$$A_2^{\wedge} = -B_1^{*}\left[K_3 \exp(-\eta_1 \tilde{z}_1) + \frac{C_{w1}}{C_{w2}} K_1 \exp(-\eta_2 \tilde{z}_1)\right], \tag{2.153}$$

where the values of K_j from (2.136) are used. Now, we may rewrite functions $\tilde{q}^{*}(\tilde{z},\tilde{k})$ and $\tilde{w}^{*}(\tilde{z},\tilde{k})$ given in (2.147), (2.148) in a more convenient form for our calculations. For $z \le z_1$

$$\tilde{q}^{*}(\tilde{z},\tilde{k}) = A_1^{\wedge}(\tilde{k})\exp(-\eta_1 \tilde{z}) + A_2^{\wedge}(\tilde{k})\exp(-\eta_2 \tilde{z}) - B_1^{*}\left[\exp(-\eta_1(\tilde{z}_1 - \tilde{z})) - \frac{C_{w1}}{C_{w2}}\exp(-\eta_2(\tilde{z}_1 - \tilde{z}))\right], \tag{2.154a}$$

$$\tilde{w}^{*}(\tilde{z},\tilde{k}) = A_1^{\wedge}(\tilde{k})C_{w1}\exp(-\eta_1 \tilde{z}) + A_2^{\wedge}(\tilde{k})C_{w2}\exp(-\eta_2 \tilde{z}) + B_1^{*}C_{w1}[\exp(-\eta_1(\tilde{z}_1 - \tilde{z})) - \exp(-\eta_2(\tilde{z}_1 - \tilde{z}))]. \tag{2.154b}$$

For $z \ge z_1$, we apply (2.147), (2.148) where the last summand is dropped. Using (2.133), we obtain

$$\tilde{q}^{*}(\tilde{z},\tilde{k}) = A_1^{\wedge}(\tilde{k})\exp(-\eta_1 \tilde{z}) + A_2^{\wedge}(\tilde{k})\exp(-\eta_2 \tilde{z}) - B_1^{*}\left[\exp(-\eta_1(\tilde{z} - \tilde{z}_1)) - \frac{C_{w1}}{C_{w2}}\exp(-\eta_2(\tilde{z} - \tilde{z}_1))\right] \tag{2.155a}$$

$$\tilde{w}^{*}(\tilde{z},\tilde{k}) = A_1^{\wedge}(\tilde{k})C_{w1}\exp(-\eta_1 \tilde{z}) + A_2^{\wedge}(\tilde{k})C_{w2}\exp(-\eta_2 \tilde{z}) - B_1^{*}C_{w1}[\exp(-\eta_1(\tilde{z} - \tilde{z}_1)) - \exp(-\eta_2(\tilde{z} - \tilde{z}_1))]. \tag{2.155b}$$

Next, we consider function $\tilde{p}^{*}(\tilde{z},\tilde{k})$ determined in (2.149), (1.116)–(1.119). For $\tilde{z} \le \tilde{z}_1$:

$$\tilde{p}^{*}(\tilde{z},\tilde{k}) = C^{*}(\tilde{k})\tilde{p}_1(\tilde{z},\tilde{k}) + C_1\tilde{p}_1(\tilde{z},\tilde{k}) + C_2\tilde{p}_2(\tilde{z},\tilde{k}) = C^{\wedge}(\tilde{k})\tilde{p}_1(\tilde{z},\tilde{k}) + C_2\tilde{p}_2(\tilde{z},\tilde{k}), \tag{2.156}$$

where a notation analogous to (2.133) is used:

$$C^{\wedge}(\tilde{k}) = C^{*}(\tilde{k}) + C_1(\tilde{k}). \tag{2.157}$$

Functions $\tilde{p}_j(\tilde{z},\tilde{k})$ ($j = 1, 2$) represent fundamental solutions of equation (2.56):

$$\widetilde{p}_1(\widetilde{z},\widetilde{k}) = \exp(-\eta_0\widetilde{z})\,, \tag{2.158a}$$

$$\widetilde{p}_2(\widetilde{z},\widetilde{k}) = \exp(\eta_0\widetilde{z})\,, \tag{2.158b}$$

with the value of η_0 given in (2.58). Coefficients C_1, C_2 satisfy the system of equations (1.117), in which dimensionless variables $\widetilde{p}_1(\widetilde{z}_1,\widetilde{k})$, $\widetilde{p}_2(\widetilde{z}_1,\widetilde{k})$, $\widetilde{z}$ should be introduced. From a computational point of view, it is convenient to use the following variables analogous to (2.126):

$$C_1^* = C_1 \exp(-\eta_0\widetilde{z}_1)\,, \tag{2.159a}$$

$$C_2^* = C_2 \exp(\eta_0\widetilde{z}_1)\,. \tag{2.159b}$$

The system of equations for these coefficients follows from system (1.117):

$$\begin{aligned} &C_1^* + C_2^* = 0\,, \\ &-\eta_0 C_1^* + \eta_0 C_2^* = -1\,. \end{aligned} \tag{2.160}$$

Hence,

$$C_1^* = -C_2^* = \frac{1}{2\eta_0}\,. \tag{2.161}$$

According to (1.119) and (2.157), we have

$$C^{\wedge}(\widetilde{k}) = -C_2(\widetilde{k})\frac{\mathrm{d}\widetilde{p}_2(0,\widetilde{k})}{\mathrm{d}\widetilde{z}}\bigg/\frac{\mathrm{d}\widetilde{p}_1(0,\widetilde{k})}{\mathrm{d}\widetilde{z}} = C_2(\widetilde{k}) = -\frac{1}{2\eta_0}\exp(-\eta_0\widetilde{z}_1)\,. \tag{2.162}$$

Taking into account (2.156), (2.159), (2.162) gives the expression for $\widetilde{p}^*(\widetilde{z},\widetilde{k})$ in its final form. For $\widetilde{z} \le \widetilde{z}_1$:

$$\widetilde{p}^*(\widetilde{z},\widetilde{k}) = -\frac{1}{2\eta_0}[\exp(-\eta_0(\widetilde{z}+\widetilde{z}_1)) + \exp(-\eta_0(\widetilde{z}_1-\widetilde{z}))]\,. \tag{2.163}$$

For $\widetilde{z} \ge \widetilde{z}_1$, one should omit the second term in (2.149). Thus, using (2.157), we obtain

$$\widetilde{p}^*(\widetilde{z},\widetilde{k}) = -\frac{1}{2\eta_0}[\exp(-\eta_0(\widetilde{z}+\widetilde{z}_1)) + \exp(-\eta_0(\widetilde{z}-\widetilde{z}_1))]\,. \tag{2.164}$$

The substitution of the obtained expressions (2.154), (2.155), (2.163), (2.164) into (2.143)–(2.146) results in the form of solution suitable for computations. Note that the last terms in (2.164) and (2.155) (having coefficient B_1^*) correspond to the action of the force in an infinite space; the contribution of these terms to the amplitudes of vibrations leads to results which agree with those obtained from (2.111)–(2.115).

Static Action of Horizontal Force

In the following, we take in results of the previous formulas $\theta = 0, \beta = 1$ and employ the values given in relationships (2.20), (2.21), (2.61). The displacements of a half-space are reduced to integrals of the same form as those in relationships (2.62)–(2.64), but with other powers in the exponential functions. Using notations (2.68)–(2.70), we obtain the following expressions for the normalized displacements of points in a transversely isotropic half-space subjected to the action of a horizontal force within the half-space ($z \le z_1$):

$$S_{hr} = \frac{\widetilde{C}_{w2} r}{2(\widetilde{\eta}_1 \widetilde{C}_{w2} - \widetilde{\eta}_2 \widetilde{C}_{w1})} \left[\frac{\widetilde{K}_1 \widetilde{\eta}_1 (z+z_1)}{F_1(r, \widetilde{\eta}_1 (z+z_1))} + \frac{\widetilde{C}_{w1} \widetilde{K}_2 (z\widetilde{\eta}_1 + z_1 \widetilde{\eta}_2)}{\widetilde{C}_{w2} F_1(r, z\widetilde{\eta}_1 + z_1 \widetilde{\eta}_2)} \right.$$
$$+ \frac{\widetilde{K}_3 (z\widetilde{\eta}_2 + z_1 \widetilde{\eta}_1)}{F_1(r, z\widetilde{\eta}_2 + z_1 \widetilde{\eta}_1)} + \frac{\widetilde{C}_{w1} \widetilde{K}_1 \widetilde{\eta}_2 (z+z_1)}{\widetilde{C}_{w2} F_1(r, \widetilde{\eta}_2 (z+z_1))} + \frac{\widetilde{\eta}_1 (z_1 - z)}{F_1(r, \widetilde{\eta}_1 (z_1 - z))}$$
$$\left. - \frac{\widetilde{C}_{w1} \widetilde{\eta}_2 (z_1 - z)}{\widetilde{C}_{w2} F_1(r, \widetilde{\eta}_2 (z_1 - z))} \right] + \frac{r}{2\widetilde{\eta}_0 F_2(r, \widetilde{\eta}_0 (z+z_1))} + \frac{r}{2\widetilde{\eta}_0 F_2(r, \widetilde{\eta}_0 (z_1 - z))}, \tag{2.165}$$

$$S_{h\vartheta} = \frac{\widetilde{C}_{w2} r}{2(\widetilde{\eta}_1 \widetilde{C}_{w2} - \widetilde{\eta}_2 \widetilde{C}_{w1})} \left[\frac{\widetilde{K}_1}{F_2(r, \widetilde{\eta}_1 (z+z_1))} + \frac{\widetilde{C}_{w1} \widetilde{K}_2}{\widetilde{C}_{w2} F_2(r, z\widetilde{\eta}_1 + z_1 \widetilde{\eta}_2)} \right.$$
$$+ \frac{\widetilde{K}_3}{F_2(r, z\widetilde{\eta}_2 + z_1 \widetilde{\eta}_1)} + \frac{\widetilde{C}_{w1} \widetilde{K}_1}{\widetilde{C}_{w2} F_2(r, \widetilde{\eta}_2 (z+z_1))} + \frac{1}{F_2(r, \widetilde{\eta}_1 (z_1 - z))}$$
$$\left. - \frac{\widetilde{C}_{w1}}{\widetilde{C}_{w2} F_2(r, \widetilde{\eta}_2 (z_1 - z))} \right] + \frac{r(z+z_1)}{2\widetilde{\eta}_0 F_1(r, \widetilde{\eta}_0 (z+z_1))} + \frac{r(z_1 - z)}{2\widetilde{\eta}_0 F_1(r, \widetilde{\eta}_0 (z_1 - z))}, \tag{2.166}$$

$$S_{hz} = \frac{\widetilde{C}_{w2}}{2(\widetilde{\eta}_1 \widetilde{C}_{w2} - \widetilde{\eta}_2 \widetilde{C}_{w1})} \left\{ \widetilde{C}_{w1} \widetilde{K}_1 \left[1 - \frac{\widetilde{\eta}_1 (z+z_1)}{F_3(r, \widetilde{\eta}_1 (z+z_1))} \right] \right.$$
$$+ \frac{\widetilde{C}_{w1}^2 \widetilde{K}_2}{\widetilde{C}_{w2}} \left[1 - \frac{z\widetilde{\eta}_1 + z_1 \widetilde{\eta}_2}{F_3(r, z\widetilde{\eta}_1 + z_1 \widetilde{\eta}_2)} \right] + \widetilde{C}_{w2} \widetilde{K}_3 \left[1 - \frac{z\widetilde{\eta}_2 + z_1 \widetilde{\eta}_1}{F_3(r, z\widetilde{\eta}_2 + z_1 \widetilde{\eta}_1)} \right]$$
$$\left. + \widetilde{C}_{w1} \widetilde{K}_1 \left[1 - \frac{\widetilde{\eta}_2 (z+z_1)}{F_3(r, \widetilde{\eta}_2 (z+z_1))} \right] + \widetilde{C}_{w1} \left[\frac{\widetilde{\eta}_1 (z_1 - z)}{F_3(r, \widetilde{\eta}_1 (z_1 - z))} - \frac{\widetilde{\eta}_2 (z_1 - z)}{F_3(r, \widetilde{\eta}_2 (z_1 - z))} \right] \right\}. \tag{2.167}$$

In order to obtain expressions for the displacements in the case $z \geq z_1$, one should modify, in (2.165) and (2.166), only terms containing $z_1 - z$ by replacing (in accordance with (2.155) and (2.164)) $z_1 - z$ with $z - z_1$. With substitution $|z - z_1|$, formulas (2.165) and (2.166) hold for any relationship between z and z_1. Expression (2.167) for S_{hz} is still valid for $z \geq z_1$. A comparison of the obtained results with the solution for an infinite space presented in equations (2.119)–(2.121) shows complete agreement between the part of solution for the half-space, which contains $z - z_1$, and the solution for the infinite space (an "extra" 2 in the denominators of relationships (2.165)–(2.167) is explained by the fact that the coefficients of normalized displacements in the solutions for the half-space are half as great as those for the space).

2.7 Contact Problems for Transversely Isotropic Half-Space

2.7.1 Static Stiffnesses for Circular Disk on Transversely Isotropic Half-Space

Consideration of displacements of the surface of a transversely isotropic half-space, subjected to the static action of vertical and horizontal forces applied to the surface, leads to the following conclusion: the relationship between the displacements and distance r between the considered point on the surface and the force application point has a form similar to that for the case of an isotropic half-space. Hence, some formulas for stiffness taken from the theory of the isotropic half-space may be easily generalized for the case of transverse isotropy. This holds for the vertical and rocking (about the horizontal axis) stiffness for a circular disk having a relaxed contact with the half-space and for the torsional stiffness. For the isotropic case, the vertical stiffness takes the following form [129]:

$$K_{zi} = \frac{4GR}{1-\nu}, \tag{2.168}$$

where R is the radius of the disk, and G and ν are the shear modulus and Poisson's ratio of the isotropic elastic medium, respectively. In the case of the relaxed contact, the vertical stiffness is determined by Green's function, which expresses vertical displacements of the surface of the half-space resulting from the vertical unit force applied to the surface. In accordance with equation (2.1b), this Green's function has the following form:

$$w_{vv}(r) = \frac{1-\nu'}{2G_{rz0}\pi r} S_{vv}, \tag{2.169}$$

where for the static case constant S_{vv} is given in (2.24b). The Green's function for the isotropic half-space is expressed as

$$w_{vv}(r) = \frac{1-\nu}{2G\pi r}. \tag{2.170}$$

As follows from a comparison of expressions (2.169) and (2.170), the formulas for the stiffness K_z in the case of transverse isotropy may be obtained by multiplying the value in equation (2.168) by the ratio of Green's functions in (2.170) and (2.169). This results in

$$K_z = \frac{4G_{rz0}R}{(1-\nu')S_{vv}} = \frac{4G_{rz0}\widetilde{D}R}{\widetilde{c}_{21}\widetilde{C}_{w2} - \widetilde{c}_{22}\widetilde{C}_{w1}}. \tag{2.171}$$

Analogously, employing the expression for the rocking stiffness for an isotropic half-space [129]

$$K_{\varphi_y i} = \frac{8GR^3}{3(1-\nu)}, \tag{2.172}$$

we obtain the corresponding relationship for the transversely isotropic half-space in the case of relaxed contact (i.e. neglecting the influence of tangential stresses in the contact area on the vertical displacements of points):

$$K_{\varphi_y} = \frac{8G_{rz0}R^3}{3(1-\nu')S_{vv}} = \frac{8G_{rz0}\widetilde{D}R^3}{3(\widetilde{c}_{21}\widetilde{C}_{w2} - \widetilde{c}_{22}\widetilde{C}_{w1})}. \tag{2.173}$$

The torsional static stiffness for the isotropic half-space may be written as [129]

$$K_{\varphi_z i} = \frac{16GR^3}{3}. \tag{2.174a}$$

According to the presentation in section 2.4, one may calculate the torsional stiffness for the transversely isotropic half-space by replacing G with

$$G_{rz0}\sqrt{\widetilde{G}_{r\vartheta}} = \sqrt{G_{rz0}G_{r\vartheta 0}} \tag{2.174b}$$

in expression (2.174a):

$$K_{\varphi_z} = \frac{16\sqrt{G_{rz0}G_{r\vartheta 0}}\,R^3}{3}. \tag{2.175}$$

The torsional stiffness for the disk located on a transversely isotropic half-space is identical to that calculated for the case of an isotropic half-space with the shear modulus equal to the geometric mean of shear modulus values of the given transversely isotropic half-space.

The values of stiffness given in relationships (2.175), (2.173), (2.171) were found in [36] by using a method of dual integral equations. Expressions for K_z

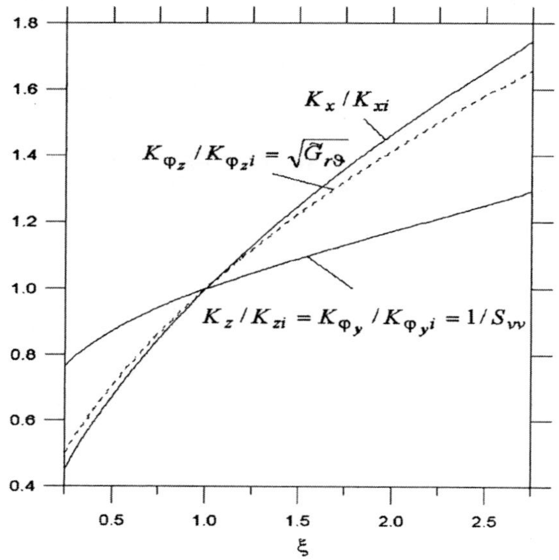

Fig. 2.15. Ratio of static stiffnesses of disk resting on transversely isotropic half-space to corresponding stiffnesses for isotropic half-space (relaxed contact)

and K_{φ_y} were derived in [36] in a different form to (2.171) and (2.173), with separate consideration for the cases of real and imaginary roots $\tilde{\eta}_j$.

Ratios $K_z / K_{zi} = K_{\varphi_y} / K_{\varphi_y i} = 1/S_{vv}$ and $K_{\varphi_z} / K_{\varphi_z i} = (\tilde{G}_{r\vartheta})^{1/2}$ at $\nu = \nu'$, $G = G_{rz0}$ are shown in Fig. 2.15 for the relationships between parameters of the anisotropic half-space given in equations (2.26) with $\nu' = 1/3$, $\xi' = 2(1+\nu')$, $\nu = \nu'$ (in this case $\tilde{G}_{r\vartheta} = \xi$).

In order to determine the stiffness for a disk resting on a half-space in other cases, we shall employ a numerical method based on dividing the contact area into elements (ring-shaped in the considered case) with a specified form of load distribution within each element [71, 128]. Unknown coefficients of loads acting on the elements are determined from the requirement of equality between displacements at the middle points of the elements and the specified displacements, according to the form of disk motion considered. Since contact stresses increase without bound when approaching the boundary of the contact area, it is appropriate to reduce the widths of the ring-shaped elements in the direction from the center of the contact area to its boundary. In our calculations, the width of the ring-shaped elements decreased according to a geometric progression with common ratio $q = 0.9$ for the ring number $N = 40$. The accuracy of solutions was estimated by using known solutions for the homogeneous isotropic half-space, as well as by comparison with results corresponding to various N (10, 20, 40, 80).

Vertical Stiffness in Case of Welded Contact between Disk and Half-Space

We introduce $2N$ unknowns p_j, q_{0j} $(j=1,...,N)$, where p_j represent normal pressures acting on the ring-shaped elements, which are kept constant in each element; $q_{0j}r$ is the radial horizontal load for the *j*th element proportional to the distance r from the disk center. The corresponding results for displacements are given in equations (1.213), (1.220), (1.223) for a circular loading area; for a ring-shaped area, the results are obtained by subtraction of the values for the corresponding circular areas. The unknowns are determined following the requirement of equality between vertical displacements at the element mean points and the vertical displacement of the disk, while horizontal displacements should vanish. The resultant vertical load corresponding to the unit vertical displacement of the disk represents the sought vertical stiffness.

Taking into account expressions (2.1), (2.53)–(2.55) for displacements of the transversely isotropic half-space subjected to the action of concentrated forces, we rewrite the expressions for displacements δ_x, δ_z from (1.213) and δ_{1x}, δ_{1z} from (1.220) and (1.223) in the following form:

$$\delta_x(r,R)=\frac{p_0R(1-\nu')}{2\pi G_{rz0}}\tilde{\delta}_x(\eta,R), \tag{2.176a}$$

$$\tilde{\delta}_x(\eta,R)=\int\limits_{|\eta-1|}^{\eta+1}2S_{vh}\sin(\varphi)\,d\lambda\ , \tag{2.176b}$$

$$\delta_z(r,R)=\frac{p_0R(1-\nu')}{2\pi G_{rz0}}\tilde{\delta}_z(\eta,R), \tag{2.177a}$$

$$\tilde{\delta}_z(\eta,R)=\int\limits_{|\eta-1|}^{\eta+1}2S_{vv}\varphi\,d\lambda+\begin{cases}\int\limits_{0}^{1-\eta}2S_{vv}\pi\,d\lambda & (\eta\le 1)\\ 0 & (\eta>1),\end{cases} \tag{2.177b}$$

$$\delta_{1x}(r,R)=\frac{q_0R^2}{2\pi G_{rz0}}\tilde{\delta}_{1x}(\eta,R), \tag{2.178a}$$

$$\tilde{\delta}_{1x}(\eta,R)=\int\limits_{|\eta-1|}^{\eta+1}\left\{S_{hr}\left[\varphi\eta+\frac{\eta}{2}\sin(2\varphi)-2\lambda\sin(\varphi)\right]-(1-\nu')S_{h\vartheta}\eta\left[\varphi-\frac{\sin(2\varphi)}{2}\right]\right\}d\lambda$$

$$+\begin{cases}\int\limits_0^{1-\eta} \pi\eta[S_{hr}+(1-\nu')S_{h\vartheta}]\,d\lambda & (\eta\le 1)\\ 0 & (\eta>1),\end{cases} \tag{2.178b}$$

$$\delta_{1z}(r,R)=\frac{q_0R^2}{2\pi G_{rz0}}\tilde{\delta}_{1z}(\eta,R), \tag{2.179a}$$

$$\tilde{\delta}_{1z}(\eta,R)=-\int\limits_{|\eta-1|}^{\eta+1}2S_{vh}(1-\nu')[\eta\sin(\varphi)-\varphi\lambda]\,d\lambda+\begin{cases}\int\limits_0^{1-\eta}2S_{vh}(1-\nu')\pi\lambda\,d\lambda & (\eta\le 1)\\ 0 & (\eta>1).\end{cases} \tag{2.179b}$$

The expression for $\delta_{1z}(r,R)$ is derived by taking into account the relationship $w_{hz}(r)=-w_{vh}(r)$, which holds due to the principle of reciprocity. In static problems, one should recall that values of $S_{vv}, S_{vh}, S_{hr}, S_{h\vartheta}$ are constant (see formulas (2.24), (2.71)–(2.73)) and integrate the complementary terms in the equations for displacements. The second argument, R, denoted by "~" in normalized displacements, should be dropped in the static case; it was written in order to preserve generality. In dynamic problems and in problems dealing with the heterogeneous half-space argument R enters functions $S_{vv}, S_{vh}, S_{hr}, S_{h\vartheta}$, which contain, in addition, integration variable λ. Recall that in the final formulas constructed in sections. 1.7.3–1.7.8 for the circular loading area, argument r in Green's functions is replaced with λR. Let the disk radius be equal to R. Denoting the radii of ring-shaped elements as R_j ($R_0=0,\ R_N=R$), the radii of the mean points of the elements as $r_i=(R_i+R_{i-1})/2\ (i=1,2,\ldots,N)$, we introduce the following relative variables:

$$\bar{r}_i=\frac{r_i}{R}, \tag{2.180a}$$

$$\bar{R}_j=\frac{R_j}{R}. \tag{2.180b}$$

The system of equations, which serves for the determination of unknowns p_j, q_{0j}, may be written in the following form:

$$\begin{gathered}\sum_{j=1}^{N}K_{ij}\tilde{p}_j+\sum_{j=1}^{N}L_{ij}\tilde{q}_{0j}=1,\\ \sum_{j=1}^{N}M_{ij}\tilde{p}_j+\sum_{j=1}^{N}N_{ij}\tilde{q}_{0j}=0\ (i=1,\ldots,N),\end{gathered} \tag{2.181}$$

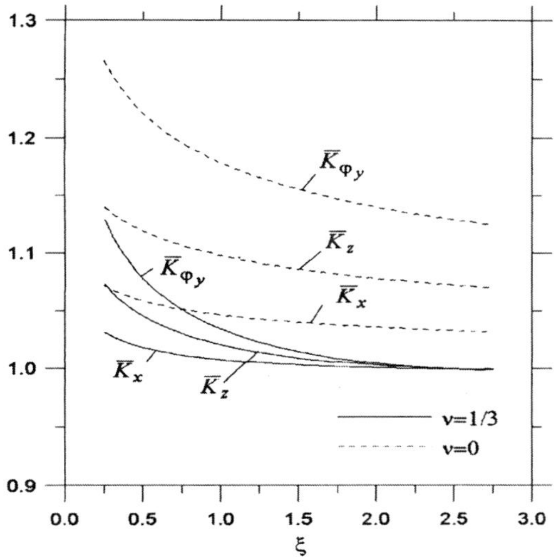

Fig. 2.16. Dependence of coefficients accounting for welded contact on degree of anisotropy

where

$$\widetilde{p}_j = \frac{p_j R(1-\nu')}{2\pi G_{rz0}}, \tag{2.182}$$

$$\widetilde{q}_{0j} = \frac{q_{0j} R^2}{2\pi G_{rz0}}, \tag{2.183}$$

$$K_{ij} = \overline{R}_j \widetilde{\delta}_z(\overline{r}_i / \overline{R}_j, R_j) - \overline{R}_{j-1} \widetilde{\delta}_z(\overline{r}_i / \overline{R}_{j-1}, R_{j-1}), \tag{2.184}$$

$$L_{ij} = \overline{R}_j^2 \widetilde{\delta}_{1z}(\overline{r}_i / \overline{R}_j, R_j) - \overline{R}_{j-1}^2 \widetilde{\delta}_{1z}(\overline{r}_i / \overline{R}_{j-1}, R_{j-1}), \tag{2.185}$$

$$M_{ij} = \overline{R}_j \widetilde{\delta}_x(\overline{r}_i / \overline{R}_j, R_j) - \overline{R}_{j-1} \widetilde{\delta}_x(\overline{r}_i / \overline{R}_{j-1}, R_{j-1}), \tag{2.186}$$

$$N_{ij} = \overline{R}_j^2 \widetilde{\delta}_{1x}(\overline{r}_i / \overline{R}_j, R_j) - \overline{R}_{j-1}^2 \widetilde{\delta}_{1x}(\overline{r}_i / \overline{R}_{j-1}, R_{j-1}). \tag{2.187}$$

In order to calculate the vertical stiffness K_z, one should determine the resultant vertical force corresponding to pressures p_j:

$$K_z = \pi \sum_{j=1}^{N} p_j (R_j^2 - R_{j-1}^2) = \frac{2\pi^2 R G_{rz0}}{1-\nu'} \sum_{j=1}^{N} \widetilde{p}_j (\overline{R}_j^2 - \overline{R}_{j-1}^2) = \overline{K}_z \frac{4 G_{rz0} R}{(1-\nu') S_{\nu\nu}}, \tag{2.188}$$

where

$$\overline{K}_z = \frac{\pi^2 S_{vv}}{2} \sum_{j=1}^{N} \widetilde{p}_j (\overline{R}_j^2 - \overline{R}_{j-1}^2). \tag{2.189}$$

The multiplier of $\overline{K}_z$ in equation (2.188) is equal to the stiffness corresponding to the relaxed contact between the disk and the half-space (see (2.171)), while value $\overline{K}_z$ represents a correction at the expense of adhesion between the disk and foundation. Normalized stiffness $\overline{K}_z$ is presented in Fig. 2.16 for the relationships between parameters given in equations (2.26) for two values of Poisson's ratio: namely, $\nu' = 0$ and $\nu' = 1/3$, $\xi' = 2(1+\nu')$, $\nu = \nu'$. As seen from Fig. 2.16, the influence of adhesion between the disk and foundation on the vertical stiffness decreases with increasing parameter ξ.

In the case of relaxed contact between the disk and foundation, we obtain the following system of equations instead of (2.181):

$$\sum_{j=1}^{N} K_{ij} \widetilde{p}_j = 1 \quad (i = 1,\dots,N). \tag{2.190}$$

In this case, the value of $\overline{K}_z$ is independent of the elastic properties of the material, and should equal unity, resulting, in accordance with (2.188), in the expression for stiffness presented in equation (2.171). Calculations performed with $q = 0.9$ and $N = 10, 20, 40$ yield the values of 0.9895, 0.9971, 0.99944, respectively. These results indicate that the method of ring-shaped elements, besides its simplicity, provides sufficient computational accuracy.

Horizontal Stiffness

Due to the translational horizontal displacement of a circular area, normal stresses occur on the surface of a half-space in the contact zone, in addition to tangential ones. The most complete statement of the present problem requires the introduction of three groups of unknown loads applied to the ring-shaped elements. For these we take uniformly distributed horizontal loads q_j, for which the corresponding displacements of the surface points are determined by using formulas (1.216)–(1.218) (for the case of a circular loading area); self-balanced horizontal loads, considered in section 1.7.8, which comprise radial components $q_{0j} r^2 \cos(\vartheta)$ and tangential components $q_{0j} r^2 \sin(\vartheta)$,displacements are given in formulas (1.231)–(1.235); and antisymmetric vertical loads $p_{0j} r \cos(\vartheta)$, the displacements are calculated by using formulas (1.225), (1.227), (1.229). Determination of $3N$ unknowns is carried out in such a way that the following conditions are satisfied: the displacements of mean points of the ring-shaped elements on the axes X and Y along the X-axis (i.e. displacements δ_{1x} and δ_{2x})

must be equal to the horizontal displacement of the disk, while the vertical displacements of the mean points of the elements on the X-axis (i.e. δ_{1z}) must vanish. The resultant horizontal force, which occurs due to loads q_j, is equal to the required horizontal stiffness (for unit displacement of the disk). According to relationships (1.214) (which hold, clearly, for the considered loadings), if one makes displacements δ_{1x} and δ_{2x} equal, then displacements along the X-axis will be equal at all points that belong to the contact area, while displacements along the Y-axis will vanish; at zero vertical displacements δ_{1z}, vertical displacements will vanish over the entire contact area. The Green's functions entering the formulas in section 1.7.8 may be determined by using expressions for the displacements given in equations (2.53)–(2.55), (2.1). Displacements of the surface of a transversely isotropic half-space, with a circular loading area, can be written analogously to formulas dealing with the vertical stiffness (see equations (2.176)–(2.179)). For the case of the horizontal uniformly distributed load:

$$\delta_{1x}(r,R)=\frac{q_0R}{2G_{rz0}\pi}\tilde{\delta}_{1x}^{(1)}(\eta,R)\,, \tag{2.191a}$$

$$\tilde{\delta}_{1x}^{(1)}(\eta,R)=\int\limits_{|\eta-1|}^{\eta+1}\{S_{hr}[\varphi+0.5\sin(2\varphi)]+S_{h\vartheta}(1-\nu')[\varphi-0.5\sin(2\varphi)]\}\mathrm{d}\lambda$$
$$+\begin{cases}\int\limits_{0}^{1-\eta}\pi[S_{hr}+S_{h\vartheta}(1-\nu')]\mathrm{d}\lambda & (\eta\le 1)\\ 0 & (\eta>1),\end{cases} \tag{2.191b}$$

$$\delta_{2x}(r,R)=\frac{q_0R}{2G_{rz0}\pi}\tilde{\delta}_{2x}^{(1)}(\eta,R)\,, \tag{2.192a}$$

$$\tilde{\delta}_{2x}^{(1)}(\eta,R)=\int\limits_{|\eta-1|}^{\eta+1}\{S_{hr}[\varphi-0.5\sin(2\varphi)]+S_{h\vartheta}(1-\nu')[\varphi+0.5\sin(2\varphi)]\}\mathrm{d}\lambda$$
$$+\begin{cases}\int\limits_{0}^{1-\eta}\pi[S_{hr}+S_{h\vartheta}(1-\nu')]\mathrm{d}\lambda & (\eta\le 1)\\ 0 & (\eta>1),\end{cases} \tag{2.192b}$$

$$\delta_{1z}(r,R)=\frac{q_0R}{2G_{rz0}\pi}\tilde{\delta}_{1z}^{(1)}(\eta,R)\,, \tag{2.193a}$$

$$\widetilde{\delta}_{1z}^{(1)}(\eta,R) = -\int_{|\eta-1|}^{\eta+1} 2(1-\nu')S_{vh}\sin(\varphi)\,\mathrm{d}\lambda\,. \tag{2.193b}$$

For the case of the horizontal self-balanced load:

$$\delta_{1x}(r,R) = \frac{q_0 R^3}{2G_{rz0}\pi}\widetilde{\delta}_{1x}^{(2)}(\eta,R)\,, \tag{2.194a}$$

$$\begin{aligned}\widetilde{\delta}_{1x}^{(2)}(\eta,R) = &\int_{|\eta-1|}^{\eta+1}\Big\{S_{hr}\big[(\lambda^2+\eta^2)(\varphi+0.5\sin(2\varphi))\\ &-4\lambda\eta\sin(\varphi)\big]+S_{h\vartheta}(1-\nu')(\eta^2-\lambda^2)(\varphi-0.5\sin(2\varphi))\Big\}\mathrm{d}\lambda\\ &+\begin{cases}\displaystyle\int_0^{1-\eta}\pi[S_{hr}(\lambda^2+\eta^2)+S_{h\vartheta}(1-\nu')(\eta^2-\lambda^2)]\,\mathrm{d}\lambda & (\eta\le 1)\\ 0 & (\eta>1),\end{cases}\end{aligned} \tag{2.194b}$$

$$\delta_{2x}(r,R) = \frac{q_0 R^3}{2G_{rz0}\pi}\widetilde{\delta}_{2x}^{(2)}(\eta,R)\,, \tag{2.195a}$$

$$\begin{aligned}\widetilde{\delta}_{2x}^{(2)}(\eta,R) = -&\int_{|\eta-1|}^{\eta+1}\Big\{S_{h\vartheta}(1-\nu')[(\lambda^2+\eta^2)(\varphi+0.5\sin(2\varphi))-4\lambda\eta\sin(\varphi)]\\ &+S_{hr}(\eta^2-\lambda^2)(\varphi-0.5\sin(2\varphi))\Big\}\mathrm{d}\lambda\\ &-\begin{cases}\displaystyle\int_0^{1-\eta}\pi[S_{h\vartheta}(1-\nu')(\lambda^2+\eta^2)+S_{hr}(\eta^2-\lambda^2)]\,\mathrm{d}\lambda & (\eta\le 1)\\ 0 & (\eta>1),\end{cases}\end{aligned} \tag{2.195b}$$

$$\delta_{1z}(r,R) = \frac{q_0 R^3}{2G_{rz0}\pi}\widetilde{\delta}_{1z}^{(2)}(\eta,R)\,, \tag{2.196a}$$

$$\widetilde{\delta}_{1z}^{(2)}(\eta,R) = -\int_{|\eta-1|}^{\eta+1} 2(1-\nu')S_{vh}[(\lambda^2+\eta^2)\sin(\varphi)$$

$$-2\lambda\eta\varphi]\,d\lambda + \begin{cases} \int\limits_0^{1-\eta} 4(1-\nu')S_{vh}\pi\eta\lambda\,d\lambda & (\eta \le 1) \\ 0 & (\eta > 1). \end{cases} \tag{2.196b}$$

For the case of the vertical antisymmetric load:

$$\delta_{1x}(r,R) = \frac{p_0(1-\nu')R^2}{2G_{rz0}\pi}\tilde{\delta}_{1x}^{(3)}(\eta,R)\ , \tag{2.197a}$$

$$\tilde{\delta}_{1x}^{(3)}(\eta,R) = \int\limits_{|\eta-1|}^{\eta+1} S_{vh}[2\eta\sin(\varphi) - \varphi\lambda - 0.5\lambda\sin(2\varphi)]\,d\lambda$$
$$-\begin{cases} \int\limits_0^{1-\eta} S_{vh}\pi\lambda\,d\lambda & (\eta \le 1) \\ 0 & (\eta > 1), \end{cases} \tag{2.197b}$$

$$\delta_{2x}(r,R) = \frac{p_0(1-\nu')R^2}{2G_{rz0}\pi}\tilde{\delta}_{2x}^{(3)}(\eta,R), \tag{2.198a}$$

$$\tilde{\delta}_{2x}^{(3)}(\eta,R) = -\int\limits_{|\eta-1|}^{\eta+1} S_{vh}[\varphi - 0.5\sin(2\varphi)]\lambda\,d\lambda$$
$$-\begin{cases} \frac{1}{2}(1-\nu')S_{vh}\pi(1-\eta)^2 & (\eta \le 1) \\ 0 & (\eta > 1), \end{cases} \tag{2.198b}$$

$$\delta_{1z}(r,R) = \frac{p_0(1-\nu')R^2}{2G_{rz0}\pi}\tilde{\delta}_{1z}^{(3)}(\eta,R), \tag{2.199a}$$

$$\tilde{\delta}_{1z}^{(3)}(\eta,R) = \int\limits_{|\eta-1|}^{\eta+1} 2S_{vv}[\eta\varphi - \lambda\sin(\varphi)]\,d\lambda + \begin{cases} \int\limits_0^{1-\eta} 2S_{vv}\pi\eta\,d\lambda & (\eta \le 1) \\ 0 & (\eta > 1). \end{cases} \tag{2.199b}$$

The system of equations for unknowns q_j, q_{cj}, p_{0j} may be written in the following form ($i = 1,\ldots,N$ for each of the three groups of equations):

$$\sum_{j=1}^{N} C_{ij}^{(1)}\tilde{q}_j + \sum_{j=1}^{N} C_{ij}^{(2)}\tilde{q}_{0j} + \sum_{j=1}^{N} C_{ij}^{(3)}\tilde{p}_{0j} = 1,$$
$$\sum_{j=1}^{N} C_{ij}^{(4)}\tilde{q}_j + \sum_{j=1}^{N} C_{ij}^{(5)}\tilde{q}_{0j} + \sum_{j=1}^{N} C_{ij}^{(6)}\tilde{p}_{0j} = 1, \tag{2.200}$$
$$\sum_{j=1}^{N} C_{ij}^{(7)}\tilde{q}_j + \sum_{j=1}^{N} C_{ij}^{(8)}\tilde{q}_{0j} + \sum_{j=1}^{N} C_{ij}^{(9)}\tilde{p}_{0j} = 0,$$

where

$$\tilde{q}_j = \frac{q_j R}{2G_{rz0}\pi}, \tag{2.201a}$$

$$\tilde{q}_{0j} = \frac{q_{0j} R^3}{2G_{rz0}\pi}, \tag{2.201b}$$

$$\tilde{p}_{0j} = \frac{p_{0j}(1-\nu')R^2}{2G_{rz0}\pi}, \tag{2.201c}$$

$$C_{ij}^{(1)} = \overline{R}_j \tilde{\delta}_{1x}^{(1)}(\overline{r}_i/\overline{R}_j, R_j) - \overline{R}_{j-1}\tilde{\delta}_{1x}^{(1)}(\overline{r}_i/\overline{R}_{j-1}, R_{j-1}), \tag{2.202}$$

$$C_{ij}^{(2)} = \overline{R}_j^3 \tilde{\delta}_{1x}^{(2)}(\overline{r}_i/\overline{R}_j, R_j) - \overline{R}_{j-1}^3\tilde{\delta}_{1x}^{(2)}(\overline{r}_i/\overline{R}_{j-1}, R_{j-1}), \tag{2.203}$$

$$C_{ij}^{(3)} = \overline{R}_j^2 \tilde{\delta}_{1x}^{(3)}(\overline{r}_i/\overline{R}_j, R_j) - \overline{R}_{j-1}^2\tilde{\delta}_{1x}^{(3)}(\overline{r}_i/\overline{R}_{j-1}, R_{j-1}), \tag{2.204}$$

$$C_{ij}^{(4)} = \overline{R}_j \tilde{\delta}_{2x}^{(1)}(\overline{r}_i/\overline{R}_j, R_j) - \overline{R}_{j-1}\tilde{\delta}_{2x}^{(1)}(\overline{r}_i/\overline{R}_{j-1}, R_{j-1}), \tag{2.205}$$

$$C_{ij}^{(5)} = \overline{R}_j^3 \tilde{\delta}_{2x}^{(2)}(\overline{r}_i/\overline{R}_j, R_j) - \overline{R}_{j-1}^3\tilde{\delta}_{2x}^{(2)}(\overline{r}_i/\overline{R}_{j-1}, R_{j-1}), \tag{2.206}$$

$$C_{ij}^{(6)} = \overline{R}_j^2 \tilde{\delta}_{2x}^{(3)}(\overline{r}_i/\overline{R}_j, R_j) - \overline{R}_{j-1}^2\tilde{\delta}_{2x}^{(3)}(\overline{r}_i/\overline{R}_{j-1}, R_{j-1}), \tag{2.207}$$

$$C_{ij}^{(7)} = \overline{R}_j \tilde{\delta}_{1z}^{(1)}(\overline{r}_i/\overline{R}_j, R_j) - \overline{R}_{j-1}\tilde{\delta}_{1z}^{(1)}(\overline{r}_i/\overline{R}_{j-1}, R_{j-1}), \tag{2.208}$$

$$C_{ij}^{(8)} = \overline{R}_j^3 \tilde{\delta}_{1z}^{(2)}(\overline{r}_i/\overline{R}_j, R_j) - \overline{R}_{j-1}^3\tilde{\delta}_{1z}^{(2)}(\overline{r}_i/\overline{R}_{j-1}, R_{j-1}), \tag{2.209}$$

$$C_{ij}^{(9)} = \overline{R}_j^2 \tilde{\delta}_{1z}^{(3)}(\overline{r}_i/\overline{R}_j, R_j) - \overline{R}_{j-1}^2\tilde{\delta}_{1z}^{(3)}(\overline{r}_i/\overline{R}_{j-1}, R_{j-1}). \tag{2.210}$$

The first N equations (2.200) for $i = 1,\ldots,N$ express the fact of equality to unity for displacements of the points that belong to the X-axis along this axis; the next N equations are the same for the points that belong to the Y-axis; and the last N equations allow vertical displacements to vanish for the mean points of the ring-shaped elements.

Next, we determine the horizontal stiffness for the disk by summing the stresses q_j, employing an analogy with relationship (2.188):

$$K_x = \pi \sum_{j=1}^{N} q_j (R_j^2 - R_{j-1}^2) = 2\pi^2 G_{rz0} R \sum_{j=1}^{N} \tilde{q}_j (\overline{R}_j^2 - \overline{R}_{j-1}^2) . \tag{2.211}$$

Weakening of the contact conditions results in the simplification of equations. By neglecting the influence of vertical stresses on the horizontal displacements of the points in the contact area, we obtain the following system of equations:

$$\begin{aligned} &\sum_{j=1}^{N} C_{ij}^{(1)} \tilde{q}_j + \sum_{j=1}^{N} C_{ij}^{(2)} \tilde{q}_{0j} = 1 , \\ &\sum_{j=1}^{N} C_{ij}^{(4)} \tilde{q}_j + \sum_{j=1}^{N} C_{ij}^{(5)} \tilde{q}_{0j} = 1 . \end{aligned} \tag{2.212}$$

Considering only contact stresses that are uniformly distributed within the ring-shaped elements, and requiring that the contact conditions hold only for the points on the X-axis, we obtain

$$\sum_{j=1}^{N} C_{ij}^{(1)} \tilde{q}_j = 1 . \tag{2.213}$$

A statement of the problem similar to (2.213) has been applied to solve the static problem [36], and to solve the dynamic problem [61]; in both cases, the solution was constructed by using dual integral equations. Note that the application of equations (2.212) or (2.213) in static problems leads to identical results, since the load parallel to the X-axis and varying with radius according to the law

$$q_x = \frac{Q_0}{2\pi R \sqrt{R^2 - r^2}} , \tag{2.214}$$

ensures the equality of displacements along the X-axis for all points in the contact area and the vanishing of displacements along the Y-axis. In other words, the system (2.212) should yield practically zero values of q_{0j} and the same values of q_j as in the case of system (2.213). Indeed, calculations prove that this statement holds. Calculations also indicate that the parts of matrices $C_{ij}^{(1)}$ and $C_{ij}^{(4)}$, which are an outcome of the terms in the expressions for $\tilde{\delta}_{1x}^{(1)}$ and $\tilde{\delta}_{2x}^{(1)}$ containing $\sin(2\varphi)$, tend to zero as $N \to \infty$ when multiplied by vector q_j corresponding to distribution (2.214). Dropping these summands will make the expressions for $\tilde{\delta}_{1x}^{(1)}$ and $\tilde{\delta}_{2x}^{(1)}$ identical, and the elements of matrices $C_{ij}^{(1)}$ and $C_{ij}^{(4)}$ will become proportional to the corresponding values for the case of relaxed contact between the stamp and foundation under the action of a vertical force (see relationships (2.177a), (2.184), (2.190)), when the vertical load having the form (2.214) provides constancy of displacements over the contact area. Therefore, two statements of the problem, (2.212) and (2.213), yield identical results. The latter does not hold for dynamic problems.

First, we determine the horizontal stiffness by (2.211), with the help of equations (2.213). The value of sum $\sum_{j=1}^{N} \widetilde{q}_j(\overline{R}_j^2 - \overline{R}_{j-1}^2)$ needed for calculations may be found from the known value of $\sum_{j=1}^{N} \widetilde{p}_j(\overline{R}_j^2 - \overline{R}_{j-1}^2)$ in equations (2.188), (2.189), which equals $2/(S_{vv}\pi^2)$ for the case of relaxed contact ($\widetilde{K}_z = 1$). Keeping in mind that the matrices entering systems (2.190) and (2.213) (when dropping terms containing $\sin(2\varphi)$ in the expressions for normalized displacements) differ by multipliers $2S_{vv}$ and α_1 only, where

$$\alpha_1 = S_{hr} + (1-\nu')S_{h\vartheta} = \frac{1}{\sqrt{\widetilde{G}_{r\vartheta}}} + \frac{\widetilde{c}_{12} - \widetilde{c}_{11}}{\widetilde{D}}, \tag{2.215}$$

we obtain the following expression for the horizontal stiffness corresponding to equations (2.213) or (2.212) (the known value of $\sum_{j=1}^{N} \widetilde{p}_j(\overline{R}_j^2 - \overline{R}_{j-1}^2)$ is multiplied by the ratio of the mentioned above multipliers):

$$K_x = 2\pi^2 G_{rz0} R \sum_{j=1}^{N} \widetilde{q}_j(\overline{R}_j^2 - \overline{R}_{j-1}^2) = 2\pi^2 G_{rz0} R \frac{2}{\pi^2 S_{vv}} \frac{2S_{vv}}{\alpha_1} = \frac{8G_{rz0}R}{\alpha_1}. \tag{2.216}$$

For the isotropic half-space having shear modulus G and Poisson's ratio ν, we obtain $\alpha_1 = 2 - \nu$, resulting in the known formula [129]

$$K_{xi} = \frac{8GR}{2-\nu}. \tag{2.217}$$

Expression (2.216) leads to results which are identical to those obtained from the formulas in [36] for various relationships between the elastic constants. Ratio K_x / K_{xi} is shown in Fig. 2.15 for the particular case of relationships between the parameters of a transversely isotropic half-space, namely $\nu = \nu' = 1/3,\ \xi' = 2(1+\nu')$.

The horizontal stiffness in (2.211) corresponding to the solution of the contact problem when it is most completely set, i.e. to equations (2.200), may be written in the following form:

$$K_x = 2\pi^2 G_{rz0} R \sum_{j=1}^{N} \widetilde{q}_j(\overline{R}_j^2 - \overline{R}_{j-1}^2) = \overline{K}_x \frac{8G_{rz0}R}{\alpha_1}, \tag{2.218}$$

where the normalized stiffness $\overline{K}_x$, representing the ratio of the horizontal

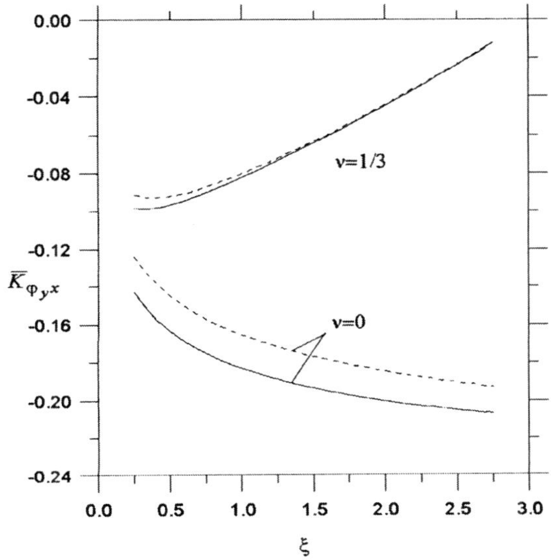

Fig. 2.17. Dependence of normalized side (horizontal rocking) stiffness on degree of anisotropy for exact (solid line) and approximate (dashed line) solution

stiffness to its value corresponding to the weakened contact condition, has the form

$$\overline{K}_x = \frac{\pi^2 \alpha_1}{4} \sum_{j=1}^{N} \widetilde{q}_j (\overline{R}_j^2 - \overline{R}_{j-1}^2). \tag{2.219}$$

In Fig. 2.16, $\overline{K}_x$ is shown at the same values of parameter relationships as those employed in previous sections. These results indicate that the static stiffness under weakened contact conditions is only slightly different from its value obtained by completely accounting for the conditions of the stiff contact. As shown below when studying dynamic stiffnesses for a circular disk, the role of the contact conditions is more significant in dynamic problems than in static ones.
In order to estimate the accuracy of calculations, consider the result, corresponding to the systems of equations (2.212) or (2.213), when $\overline{K}_x$ should be equal to unity. Calculations yield the following values of $\overline{K}_x$ at $N = 10, 20, 40$, respectively ($q = 0.9$): 0.9896, 0.9972, 0.99957.

When solving the problem when it is set by using equations (2.200), one should determine, in addition to $\widetilde{q}_j$, values $\widetilde{q}_{0j}$ and $\widetilde{p}_{0j}$. The latter determine the resultant moment with respect to the Y-axis, which should be applied to the disk together with the corresponding horizontal force, in order to provide the horizontal translational displacement of the disk. This moment $K_{x\varphi_y}$ representing the horizontal rocking stiffness may be calculated as follows:

$$K_{x\varphi_y} = 2\sum_{j=1}^{N}\int_0^{\pi}\int_{R_{j-1}}^{R_j} p_{0j} r^2 \cos^2(\vartheta) r \,\mathrm{d}r\,\mathrm{d}\vartheta$$

$$= \frac{\pi}{4}\sum_{j=1}^{N} p_{0j}(R_j^4 - R_{j-1}^4) = \frac{\pi^2 G_{rz0} R^2}{2(1-\nu')}\sum_{j=1}^{N}\widetilde{p}_{0j}(\overline{R}_j^4 - \overline{R}_{j-1}^4). \quad (2.220)$$

We introduce normalized stiffness $\overline{K}_{x\varphi_y}$ equal to the ratio of $K_{x\varphi_y}$ to the horizontal stiffness K_{xi} by (2.217) multiplied by R for the isotropic half-space with $G = G_{rz0}$ and $\nu = \nu'$,

$$\overline{K}_{x\varphi_y} = \frac{2-\nu'}{16(1-\nu')}\pi^2\sum_{j=1}^{N}\widetilde{p}_{0j}(\overline{R}_j^4 - \overline{R}_{j-1}^4). \quad (2.221)$$

In Fig. 2.17, values of $\overline{K}_{x\varphi_y}$ (solid lines) are shown versus ξ for the relations of parameters given in equations (2.26) with $\xi' = 2(1+\nu')$, $\nu = \nu'$, and for the values of Poisson's ratio, $\nu' = 0$ and $\nu' = 1/3$. Dashed lines correspond to the approximate solution that is discussed in the next subsection.

Rocking Stiffness under Conditions of Welded Contact

Let the disk be rotated by unit angle with respect to the Y-axis, and the horizontal displacements of the points in the contact area vanish. We employ the same matrices as in the previous case. The right-hand sides of the system of equations are modified: in the equations of the first two groups, the right sides vanish, while for the third group they equal $\overline{r}_i$:

$$\sum_{j=1}^{N} C_{ij}^{(1)}\widetilde{q}_j + \sum_{j=1}^{N} C_{ij}^{(2)}\widetilde{q}_{0j} + \sum_{j=1}^{N} C_{ij}^{(3)}\widetilde{p}_{0j} = 0,$$

$$\sum_{j=1}^{N} C_{ij}^{(4)}\widetilde{q}_j + \sum_{j=1}^{N} C_{ij}^{(5)}\widetilde{q}_{0j} + \sum_{j=1}^{N} C_{ij}^{(6)}\widetilde{p}_{0j} = 0, \quad (2.222)$$

$$\sum_{j=1}^{N} C_{ij}^{(7)}\widetilde{q}_j + \sum_{j=1}^{N} C_{ij}^{(8)}\widetilde{q}_{0j} + \sum_{j=1}^{N} C_{ij}^{(9)}\widetilde{p}_{0j} = \overline{r}_i.$$

Here, division by R is done on both sides of equations (2.222). The new dimensionless loads differ from those given in equations (2.201):

$$\widetilde{q}_j = \frac{q_j}{2G_{rz0}\pi}, \quad (2.223a)$$

$$\widetilde{q}_{0j} = \frac{q_{0j}R^2}{2G_{rz0}\pi}, \quad (2.223b)$$

$$\widetilde{p}_{0j} = \frac{p_{0j}(1-\nu')R}{2G_{rz0}\pi}. \tag{2.223c}$$

Analogously to the determination of $K_{x\varphi_y}$ by (2.220), we calculate the rocking stiffness by using the values of $\widetilde{p}_{0j}$:

$$K_{\varphi_y} = \frac{\pi^2 G_{rz0} R^3}{2(1-\nu')} \sum_{j=1}^{N} \widetilde{p}_{0j}(\overline{R}_j^4 - \overline{R}_{j-1}^4) = \overline{K}_{\varphi_y} \frac{8G_{rz0}R^3}{3S_{\nu\nu}(1-\nu')}, \tag{2.224}$$

where

$$\overline{K}_{\varphi_y} = \frac{3\pi^2 S_{\nu\nu}}{16} \sum_{j=1}^{N} \widetilde{p}_{0j}(\overline{R}_j^4 - \overline{R}_{j-1}^4). \tag{2.225}$$

The multiplier of normalized stiffness $\overline{K}_{\varphi_y}$ in expression (2.224) is equal to the rocking stiffness from (2.173), which corresponds to neglecting the influence of horizontal stresses on the vertical displacements of the points in the contact area. The results of calculations are presented in Fig. 2.16 with the same values of parameters as those employed in studying $\overline{K}_x$ and $\overline{K}_z$. The influence of completely accounting for the welded contact conditions is now more significant than in the case of the vertical and horizontal stiffnesses.

In order to estimate the accuracy of calculations, consider the system of equations corresponding to neglecting the influence of horizontal stresses on the rocking stiffness:

$$\sum_{j=1}^{N} C_{ij}^{(9)} \widetilde{p}_{0j} = \overline{r}_i \quad (i = 1,.., N).. \tag{2.226}$$

The value of $\overline{K}_{\varphi_y}$ corresponding to the solution of the given system must equal unity and be independent of the material properties. The solution based on the method of ring-shaped elements with $q = 0.9$ yields the following values: 0.9689, 0.9914, 0.99839, 0.99919 for $N = 10, 20, 40, 80$, respectively. These values are somewhat less accurate (at equal values of N) than the corresponding values considered above in the framework of studying vertical and horizontal stiffnesses.

The resultant horizontal force, which occurs due to rotation of the disk without horizontal displacements, represents the rocking horizontal stiffness, $K_{\varphi_y x}$. The latter is determined by summation of loads q_j, analogous to equation (2.211). According to Rayleigh's principle of reciprocity, this value should be equal to the stiffness $K_{x\varphi_y}$ that was calculated previously. The equality of values $K_{x\varphi_y}$ and $K_{\varphi_y x}$ calculated by using the numerical method above was found to hold with sufficiently high precision (a difference occurred in the 4th–5th digits only).

Note that the values of stiffnesses, $K_{x\varphi_y}$ or $K_{\varphi_y x}$, may be calculated approximately (under the weakening of contact conditions given above) by employing the known behavior of a distribution of contact stresses beneath the disk, and averaging displacements according to the principle of reciprocity. This approach yields the following approximate relationship for the side compliances $C_{x\varphi_y} = C_{\varphi_y x}$ of the disk resting on an isotropic half-space [67, 115]:

$$C_{x\varphi_y} = C_{\varphi_y x} = \frac{u}{M} = \frac{\varphi}{Q} = \frac{3(1-2\nu)}{16\pi G R^2}, \tag{2.227}$$

where u is the horizontal displacement of the disk due to the action of a moment M with respect to the Y-axis, and φ is the rotation angle for the disk subjected to the action of a horizontal force Q. In order to find the value of $C_{\varphi_y x}$, we determine the work done by the load having the form (2.214) on the horizontal displacements that occur due to the moment load, caused by rotation of the disk with respect to the Y-axis; this work should be divided by the value of Q. The idea of such averaging was suggested in [17]. When calculating the work by integration, Green's function w_{vh} is employed for horizontal displacements that occur on the surface of the half-space due to the action of the vertical unit force. These horizontal displacements for the isotropic and transversely isotropic half-spaces are proportional to each other. For the isotropic half-space:

$$w_{vh} = -\frac{1-2\nu}{4G\pi r}. \tag{2.228}$$

For the transversely isotropic half-space, in accordance with (2.1a), (2.24a):

$$w_{vh} = \frac{1-\nu'}{2G_{rz0}\pi r} S_{vh} = \frac{\tilde{c}_{21} - \tilde{c}_{21}}{2G_{rz0}\pi \tilde{D} r}. \tag{2.229}$$

Clearly, the expression for $C_{x\varphi_y}$ or $C_{\varphi_y x}$ in the case of the transversely isotropic half-space may be derived by multiplying the value on the right side of (2.227) by the ratio of Green's functions w_{vh} from (2.229) and (2.228):

$$C_{x\varphi_y} = C_{\varphi_y x} = -\frac{3(1-\nu')S_{vh}}{8\pi G_{rz0} R^2} = -\frac{3(\tilde{c}_{21} - \tilde{c}_{22})}{8\pi G_{rz0} R^2 \tilde{D}}. \tag{2.230}$$

Using the matrix of compliances, one can present the relationship between displacements and actions in the following form:

$$\begin{Bmatrix} \varphi_y \\ u_x \end{Bmatrix} = \begin{bmatrix} C_{\varphi_y} & C_{x\varphi_y} \\ C_{\varphi_y x} & C_x \end{bmatrix} \begin{Bmatrix} M \\ Q \end{Bmatrix}, \tag{2.231}$$

where the main compliances, C_{φ_y} and C_x, may be defined as inverse values of the corresponding stiffnesses calculated for the weakened contact conditions, i.e. values of K_{φ_y} by (2.173) and K_x by (2.216). An approximate value of the side stiffness may be found by inverting the compliance matrix:

$$K_{x\varphi_y} = K_{\varphi_y x} = -\frac{C_{\varphi_y x}}{C_{\varphi_y} C_x - C_{\varphi_y x}^2} = -\frac{1}{C_{\varphi_y x}} \frac{1}{\psi - 1}, \tag{2.232}$$

where

$$\psi = \frac{\pi^2 \alpha_1 S_{vv}}{3(1-\nu')S_{vh}^2}. \tag{2.233}$$

The normalized stiffness $\overline{K}_{x\varphi_y}$ corresponding to the approximate relationship (2.232) takes the form

$$\overline{K}_{x\varphi_y} = \frac{\pi(2-\nu')}{3(1-\nu')S_{vh}(\psi-1)}. \tag{2.234}$$

In Fig. 2.17, the stiffness $\overline{K}_{x\varphi_y}$ is shown for two values of Poisson's ratio (dashed lines). Thus, the approximate approach presented above provides a satisfactory estimation of the side stiffnesses. The error, as shown below, may be more significant in dynamic problems.

Torsional Stiffness

The solution of the corresponding static problem is given by the formula (2.175). Since the numerical method is also applicable for the dynamic problems and for the problems dealing with heterogeneous foundations, we present below the general equations for determination of the torsional stiffness. In the framework of the method of ring-shaped elements, we employ tangential loads having the form $q_{0j}r$, for which tangential displacements δ_{1y} are considered in section 1.7.7 (for the case of a circular loading area). The disk is rotated by unit angle with respect to the Z-axis; corresponding displacements of the mean points of ring-shaped elements are equal to the distances between these points and the Z-axis. Using relationships (2.53), (2.54), the displacements δ_{1y} may be written in the following form:

$$\delta_{1y}(r,R) = \frac{q_0 R^2}{2G_{rz0}\pi} \tilde{\delta}_{1y}(\eta, R), \tag{2.235a}$$

$$\widetilde{\delta}_{1y}(\eta, R) = \int\limits_{|\eta-1|}^{\eta+1} \left\{ S_{h\vartheta}(1-\nu')\left[\eta\varphi + \frac{\eta\sin(2\varphi)}{2} - 2\lambda\sin(\varphi)\right] + S_{hr}\eta\left[\varphi - \frac{\sin(2\varphi)}{2}\right]\right\} d\lambda$$

$$+\begin{cases} \int\limits_{0}^{1-\eta} \pi\eta[S_{h\vartheta}(1-\nu') + S_{hr}]\,d\lambda & (\eta \le 1) \\ 0 & (\eta > 1). \end{cases} \tag{2.235b}$$

Unknown loads acting on the ring-shaped elements are calculated from the following system of equations (both sides of the equations are divided by R):

$$\sum_{j=1}^{N} C_{ij}\widetilde{q}_{0j} = \overline{r}_i \qquad (i = 1,\dots,N\,), \tag{2.236}$$

where

$$\widetilde{q}_{0j} = \frac{q_{0j}R}{2G_{rz0}\pi}, \tag{2.237}$$

$$C_{ij} = \overline{R}_j^2\widetilde{\delta}_{1y}(\overline{r}_i/\overline{R}_j, R_j) - \overline{R}_{j-1}^2\widetilde{\delta}_{1y}(\overline{r}_i/\overline{R}_{j-1}, R_{j-1}). \tag{2.238}$$

The torsional stiffness is calculated analogously to relationship (2.220), by summing the moments of loads applied to the ring-shaped elements, with respect to the Z-axis:

$$K_{\varphi_z} = \frac{\pi}{2}\sum_{j=1}^{N} q_{0j}(R_j^4 - R_{j-1}^4) = \pi^2 G_{rz0}R^3\sum_{j=1}^{N}\widetilde{q}_{0j}(\overline{R}_j^4 - \overline{R}_{j-1}^4)$$

$$= \overline{K}_{\varphi_z}\frac{16\sqrt{G_{rz0}G_{r\vartheta 0}}\,R^3}{3}, \tag{2.239}$$

where the normalized torsional stiffness is introduced:

$$\overline{K}_{\varphi_z} = \frac{3\pi^2}{16\sqrt{\widetilde{G}_{r\vartheta}}}\sum_{j=1}^{N}\widetilde{q}_{0j}(\overline{R}_j^4 - \overline{R}_{j-1}^4). \tag{2.240}$$

The value of $\overline{K}_{\varphi_z}$ equals the ratio of the torsional stiffness to the value determined from (2.175), i.e. to the static torsional stiffness. Thus, in static problems, the value of $\overline{K}_{\varphi_z}$ should equal unity, irrespective of the values of the elastic parameters of the half-space, despite the fact that these parameters appear in the matrix elements in the system of equations (2.236). Calculations confirm that the independence holds up to a high precision, providing an efficient evaluation of the technique employed, and the Green's functions, which serve for the calculation of coefficients in system (2.236). The following are values of $\overline{K}_{\varphi_z}$ for various N and $q = 0.9$: $\overline{K}_{\varphi_z} = 0.9691, 0.9918, 0.9989, 0.9997$ for $N = 10$, 20, 40, 80, respectively.

2.7.2 Dynamic Stiffnesses for Circular Disk on Transversely Isotropic Half-Space

The method of ring-shaped elements considered in the previous subsection may be successfully applied to solve problems of steady-state harmonic vibrations. The difference between the latter and the static problems is as follows: values $S_{vv}, S_{vh}, S_{hr}, S_{h\vartheta}$ entering expressions (2.176a), (2.177a), (2.178a), etc, for normalized displacements (and matrices of corresponding systems of equations), are real constants in the static case, while in the dynamic case they are represented by complex functions dependent on the radii of loading circles and on the variable of integration in the integral representations of displacements. Thus, in normalized displacements denoted in the previous subsection by "~" the second argument, R, becomes significant. In dynamic problems, it is appropriate to use parameter $\tilde{r}$ from (2.19) as an independent variable for the normalized amplitudes of vibrations $S_{vv}, S_{vh}, S_{hr}, S_{h\vartheta}$, which corresponds to the value of parameter $\theta = 1$. Since in the final formulas constructed on the basis of the principle of superposition, argument r is replaced with λR in the Green's functions, where λ is an integration variable and R the radius of the circular loading area, the second arguments R_j and R_{j-1} given in the matrix elements in the systems of equations for the considered contact problems, mean that the argument $\tilde{r}$ in functions $S_{vv}, S_{vh}, S_{hr}, S_{h\vartheta}$ should be replaced with $\lambda\tilde{R}\overline{R}_j$ and $\lambda\tilde{R}\overline{R}_{j-1}$, respectively. Here, we use the notation analogous to (2.19a):

$$\tilde{R} = \frac{\omega R}{C_{rz}} = \omega R\sqrt{\frac{\rho}{G_{rz0}}} \ . \tag{2.241}$$

Calculations are performed for a series of given values of the parameter $\tilde{R}$. Note that it differs from the analogous parameter considered in connection with the torsional vibrations of a half-space (2.85b). In order to determine the matrix elements, first, the functions $S_{vv}(\tilde{r}), S_{vh}(\tilde{r}), S_{hr}(\tilde{r}), S_{h\vartheta}(\tilde{r})$ are calculated with a step which is sufficiently small to determine the intermediate values of these functions (required for integration with respect to λ) with sufficient precision by using local parabolic interpolation employing three points of the grid surrounding the considered point. In the calculations, the step of 0.05 was used. The real and imaginary parts of functions $S_{vv}(\tilde{r}), S_{vh}(\tilde{r}), S_{hr}(\tilde{r}), S_{h\vartheta}(\tilde{r})$, studied for some values of parameters of the half-space in sections 2.1 and 2.3, vary sufficiently slowly, and the step of 0.05 provides an acceptable accuracy (the differences between the results of interpolation and the direct calculation of the corresponding values occur in 5th digit only). The number N of ring-shaped elements was taken to be 40, while a geometric progression with a ratio $q = 0.9$ was used for dividing the elements.

Vertical Stiffness

In order to account for the welded contact, one should use equations (2.181), while for relaxed contact equations (2.190) are applied; the stiffness is determined by using values of $\tilde{p}_j$ in accordance with the formula (2.188):

$$K_z = \frac{2\pi^2 R G_{rz0}}{1-\nu'} \sum_{j=1}^{N} \tilde{p}_j (\overline{R}_j^2 - \overline{R}_{j-1}^2) . \tag{2.242}$$

The results of calculations at $\varepsilon = 0$, $\nu' = 1/3$, $\xi' = 2(1+\nu')$, $\nu = \nu'$, and for some values of parameter $\xi = E/E'$, are shown in Fig. 2.18, where the ratio of dynamic stiffness K_z to the static stiffness K_z^{st} (corresponding to $\omega = 0$) is presented. For the imaginary part, an additional division by the frequency parameter $\tilde{R}$ is made, yielding the damping coefficient, c_v :

$$c_v = \mathrm{Im}\left[\frac{K_z}{\tilde{R} K_z^{st}}\right] . \tag{2.243}$$

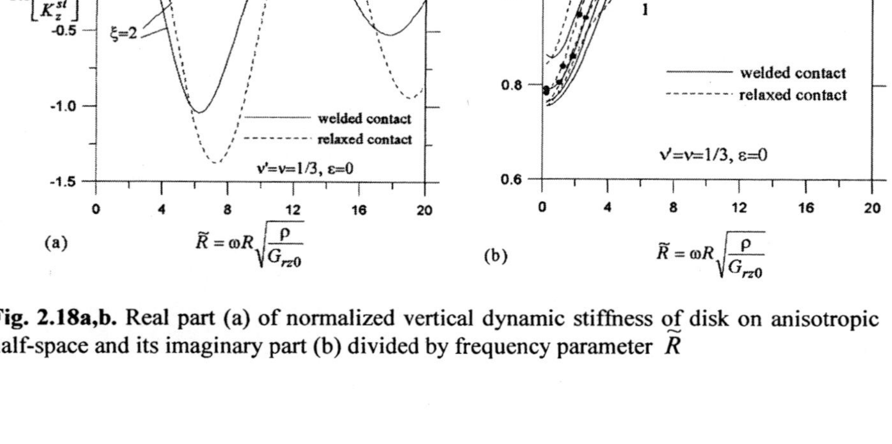

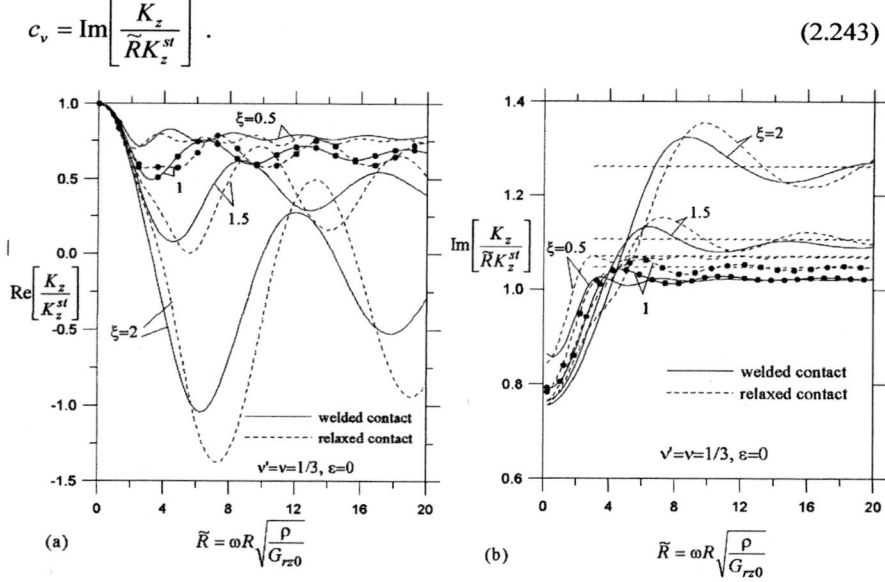

Fig. 2.18a,b. Real part (a) of normalized vertical dynamic stiffness of disk on anisotropic half-space and its imaginary part (b) divided by frequency parameter $\tilde{R}$

The case $\xi = 1$ corresponds to the isotropic half-space (the curves are marked with bold dots).

An estimate of the coefficient c_v corresponding to high frequencies (high values of $\widetilde{R}$), when the relationship between stresses and strains at each point of the contact area may be taken as that for a semi-infinite rod with excluded side deformations, gives

$$\sigma_z = -\frac{\partial u_z}{\partial t}\sqrt{\rho A_{zz}} . \tag{2.244}$$

According to relationships (1.2), (1.3), it is coefficient A_{zz} that should be taken as the deformation modulus of the rod when considering motion in the vertical direction. For the time-dependence law $\exp(\mathrm{i}\,\omega\, t)$, equation (2.244) results in the following relationship between the force amplitudes P_0 and vertical displacements of the disk, u_{z0}:

$$P_0 = \pi R^2\, \mathrm{i}\,\omega u_{z0}\sqrt{\rho A_{zz}} . \tag{2.245}$$

Relationships similar to (2.245) have been presented in numerous reports [5, 18, 26, 34, 35, 71], and various techniques were used to derive such equations. A transformation of (2.245) to the form that enables lateral deformations was proposed [26, 34, 35]; this issue is considered in section 3.6. When no internal friction exists in the material of a half-space ($\varepsilon = 0$), the dynamic stiffness expressed in terms of the coefficient of u_{z0} in equation (2.245) is a pure imaginary quantity. Using the expression for the static stiffness (2.188), one can evaluate the coefficient c_v corresponding to high values of parameter $\widetilde{R}$ (we consider the case of $\varepsilon = 0$):

$$c_{v\infty} = \frac{\pi(1-\nu')S_{vv}\sqrt{\widetilde{A}_{zz}}}{4\overline{K}_z} , \tag{2.246}$$

where S_{vv} represents the static value determined from (2.24b), $\overline{K}_z$ is the normalized static stiffness equal to unity in the case of relaxed contact, and calculated from (2.189) in the case of welded contact (see the graphs in Fig. 2.16). For the values of parameters corresponding to the graphs in Fig. 2.18, we have, respectively for $\xi = 0.5$, 1.0, 1.5, 2.0: $c_{v\infty} = 1.07026$, 1.0472, 1.10781, 1.26262 for the relaxed contact, and $c_{v\infty} = 1.0234$, 1.0250, 1.09719, 1.25815 for the welded contact. The horizontal dashed lines in Fig. 2.18b represent values of $c_{v\infty}$ for the relaxed contact. The rate of approaching limiting values increases as parameter ξ decreases. At $\xi = 0.5$ and $\widetilde{R} > 15$, the deviation of c_v from $c_{v\infty}$ does not exceed 0.1%; at $\xi = 2$ noticeable oscillations of c_v (with respect to its limiting value) occur in the considered range of varying $\widetilde{R}$. In the case of the welded contact, the rate of approaching the limiting value is somewhat higher than for the relaxed contact. Taking into account the internal friction of the material gives an additional multiplier of the order $1+\varepsilon^2/8$, by which $c_{v\infty}$ should be multiplied; this multiplication does not lead to any significant change in the value of $c_{v\infty}$. The calculations which take into account the energy dissipation in the material of the half-space show that an imaginary part of the normalized complex stiffness (i.e. value $c_v\widetilde{R}$) increases by a value of order ε compared with the case when no dissipation occurs.

The assumption (based on physical considerations) stating that the high-frequency asymptotics for the vertical dynamic stiffness may be calculated (for the compressible medium) by employing a uniform distribution of the contact normal stresses, indeed leads to the correct value of the principal term in the asymptotic representation of the complex stiffness. An attempt to construct the refined asymptotics for the case of an isotropic homogeneous elastic half-space has been done by determining the average vertical amplitudes over the circular area, where a uniformly distributed vertical load was applied as $\omega \to \infty$ [18]. As a result, in addition to the principal term in the asymptotic representation of the complex dynamic compliance (corresponding to (2.245)), the next term was found to be a real quantity. However, this technique yields a result for the real part of the stiffness, which is approximately 2/3 of the value, calculated from the numerical solution of the corresponding contact problem at high frequencies. Note that for the case of torsional vibrations of a circular stiff disk resting on an isotropic half-space, the corresponding high-frequency asymptotic value has been determined exactly [104, 110], and the second term in the asymptotic expansion above is $\pi/2$ of its value calculated by using the technique given in [18]. Apparently, such a coefficient should be introduced in the case of the other forms of vibrations. On the other hand, as seen from Fig. 2.18a, as the material properties approach those of the incompressible medium (with increasing parameter ξ), oscillations in the value of the real part of the dynamic stiffness become more significant, so that this value remains far from its limiting value, even at sufficiently high values of the frequency parameter $\widetilde{R}$. The differences between asymptotic values of the real parts of complex stiffnesses (or compliances), calculated from the exact solution of the problem and by employing the approach based on the assumption of a proportional relationship between loading and displacements over the entire area of the disk–half-space contact, occur due to violation of this proportionality in the vicinity of the line where the separation of boundary conditions takes place – the boundary of the contact area. Here the load undergoes a discontinuous change from relatively high values to zero.

Consider behavior of contact pressures (Fig. 2.19) under the weakened contact conditions at $\widetilde{R}$ = 20, while the rest of the parameters remain unchanged. The value of the dimensionless pressure $\widetilde{p}$, corresponding to quantities $\widetilde{p}_j$ obtained from the numerical solution of the contact problem, is presented for three values of parameter ξ; in the calculations, we take $N = 80$, $q = 0.95$. For $\xi = 0.5$ and $\xi = 1$

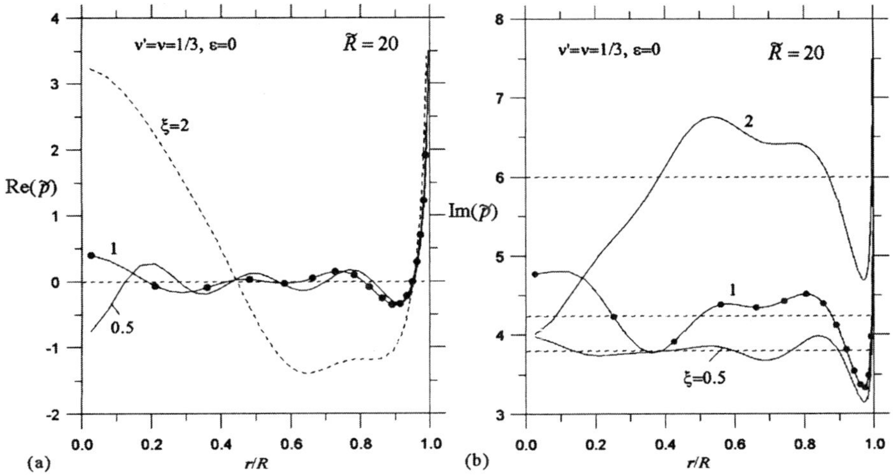

Fig. 2.19a,b. Real (a) and imaginary (b) parts of normalized vertical contact stresses for a number of values of anisotropy parameter ξ

(an isotropic case), the real part of the contact pressures is significantly smaller than the corresponding imaginary part over most of the contact area, but at $\xi = 2$, the considered values are comparable. When $r \to R$, both the real and imaginary parts tend to infinity in the narrow zone in the vicinity of the point $r = R$. In Fig. 2.19b, the horizontal dashed lines represent the average value of $\tilde{p}$ determined from relationships (2.244) and (2.245). Evidently, local deviations of the contact pressures from these average values may be rather significant, especially at $\xi = 2$, whereas the discordance between values of the resultant pressures, corresponding to the calculated values $\mathrm{Im}[\tilde{p}]$, and those found from relationship (2.245), is much less noticeable at high values of parameter $\tilde{R}$ (see Fig. 2.18b).

For the parameters considered in the calculations, the behavior of the complex dynamic stiffness whis increasing parameter ξ is similar to its behavior for the isotropic half-space with increasing Poisson's ratio. Note that parameter ξ approaching its value given in (2.27a) (i.e. 3 for the considered values of parameters) results in an unbounded growth of the coefficient A_{zz}; for the isotropic half-space, this situation takes place at $\nu = 0.5$. In Fig. 2.20, curves for the normalized dynamic stiffness are given for the anisotropic half-space with $\xi = 2$ and with the rest of parameters having the previously used values, and for the isotropic half-space with $G = G_{rz0}$, $\nu = 0.44$; relaxed contact is considered. Note that besides the normalized dynamic stiffnesses shown in Fig. 2.20, static stiffnesses determined from (2.168) and (2.171), for both the considered cases, are in close proximity (the ratio of static stiffness in the isotropic case to that in the anisotropic case equals 1.015).

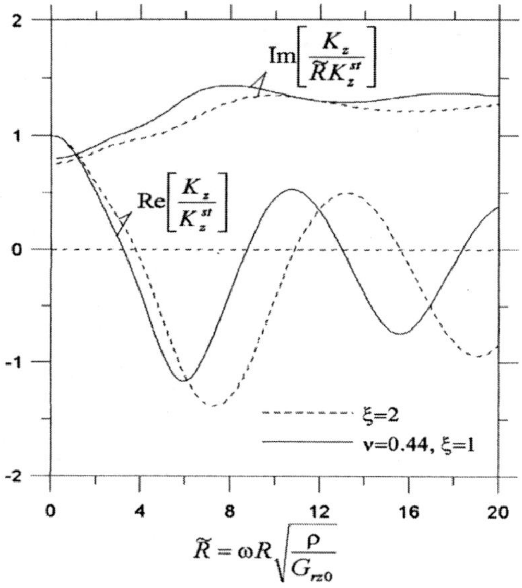

Fig. 2.20. Comparison of results corresponding to isotropic (solid line) and anisotropic (dashed line) half-spaces

Horizontal Stiffness

In the case of welded contact, one should use equations (2.200), and when neglecting the influence of normal stresses on the horizontal displacements of the points in the contact area, equations (2.212) are applied. The dynamic horizontal stiffness K_x is calculated by using values of $\widetilde{q}_j$, as we did in the static case:

$$K_x = 2\pi^2 G_{rz} R \sum_{j=1}^{N} \widetilde{q}_j (\overline{R}_j^2 - \overline{R}_{j-1}^2) . \tag{2.247}$$

Calculations were performed with values of parameters equal to those used for determining the vertical stiffness. The ratio of the real part of complex stiffness K_x to the corresponding static value K_x^{st} is shown in Fig. 2.21a; Fig. 2.21b represents the value c_x:

$$c_x = \mathrm{Im}\left[\frac{K_x}{\widetilde{R}K_x^{st}}\right] . \tag{2.248}$$

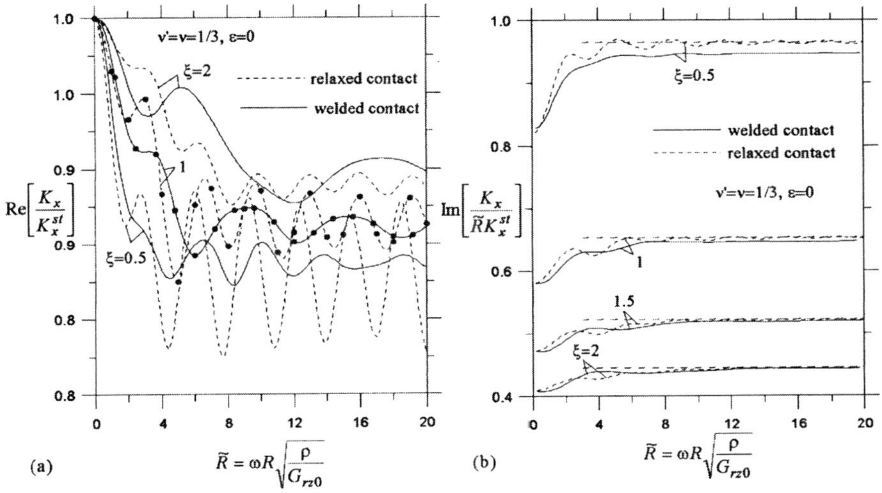

Fig. 2.21a,b. Real part (a) of normalized horizontal dynamic stiffness of disk on anisotropic half-space and its imaginary part (b) divided by frequency parameter $\widetilde{R}$

The limiting value $c_{x\infty}$ of the damping coefficient c_x corresponding to an unbounded increase of vibration frequency may be determined by using the following relationship, analogous to (2.244):

$$\tau_{zx} = -\frac{\partial u_x}{\partial t}\sqrt{\rho G_{rz}}\ . \tag{2.249}$$

This equation expresses the proportional dependence between the load applied to the surface of the half-space and the velocity along the line of action of the load; this relationship is the same as if the load were applied to the entire boundary of the half-space. The relationship between the amplitudes of the horizontal force Q_0 and horizontal displacements u_{x0} has the following form:

$$Q_0 = \pi R^2\, \mathrm{i}\omega\sqrt{\rho G_{rz}}\, u_{x0}\ . \tag{2.250}$$

Analogously to (2.246), the following expression holds for value $c_{x\infty}$:

$$c_{x\infty} = \frac{\pi\alpha_1}{8\overline{K}_x}, \tag{2.251}$$

where $\overline{K}_x$ is the normalized static stiffness, being equal to unity if the influence of vertical stresses in the contact area on the horizontal stiffness is neglected, otherwise its value is calculated from equation (2.219) (see Fig. 2.16). For the values of parameters used above we have, respectively, for $\xi = 0.5, 1.0, 1.5, 2.0$:

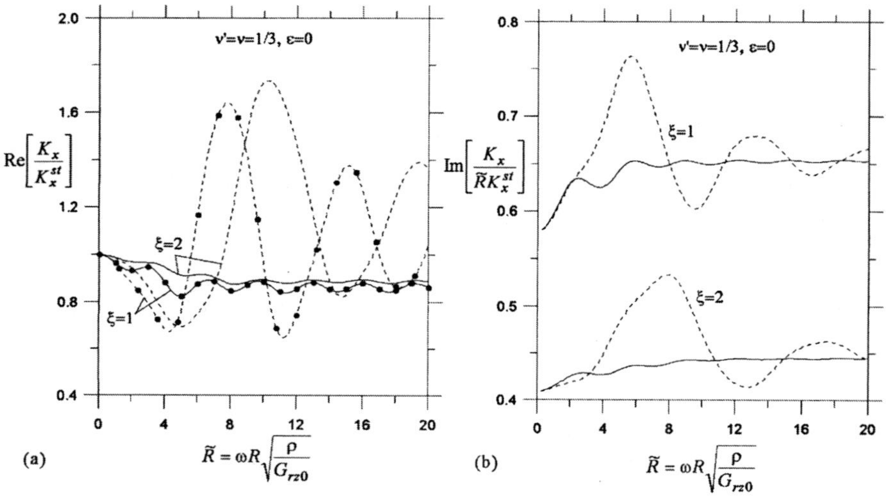

Fig. 2.22a,b. Comparison of dynamic horizontal stiffnesses for exact (solid lines) and approximate (dashed lines) accounting for relaxed contact conditions

$c_{x\infty}$ = 0.96579, 0.65450, 0.52289, 0.44640 when the vertical pressures are neglected over the contact area, and $c_{x\infty}$ = 0.94806, 0.64879, 0.52109, 0.44593 if the contact conditions are completely accounted for. The horizontal dashed lines in Fig. 2.21b represent the values of $c_{x\infty}$ in the first case. The damping coefficient tends to its corresponding limiting values rather quickly. As noted previously, employing (2.213), rather than (2.212), leads to the same values of static stiffness for the circular disk resting on a half-space. In dynamic problems, the results of calculations in the framework of the simplified approach (using equations (2.213)) may be appreciably different from those obtained by employing a more accurate determination of the contact conditions. In Fig. 2.22, this fact is illustrated for two values of parameter ξ (the rest of the parameters are taken equal to those used in the previous calculations); the dashed curves correspond to the application of equations (2.213), while the solid curves correspond to equations (2.212). An error due to employment of the approximate approach, corresponding to equations (2.213), becomes appreciable under dynamic loading conditions.

Horizontal rocking Stiffness

Formulas (2.220), (2.221) employed for constructing solutions of the static problems still hold in the dynamic case. The results of calculations for the previously used parameter values are given in Fig. 2.23. As parameter ξ increases, the horizontal rocking stiffness reaches appreciable values, substantially exceeding the corresponding values for the isotropic half-space.

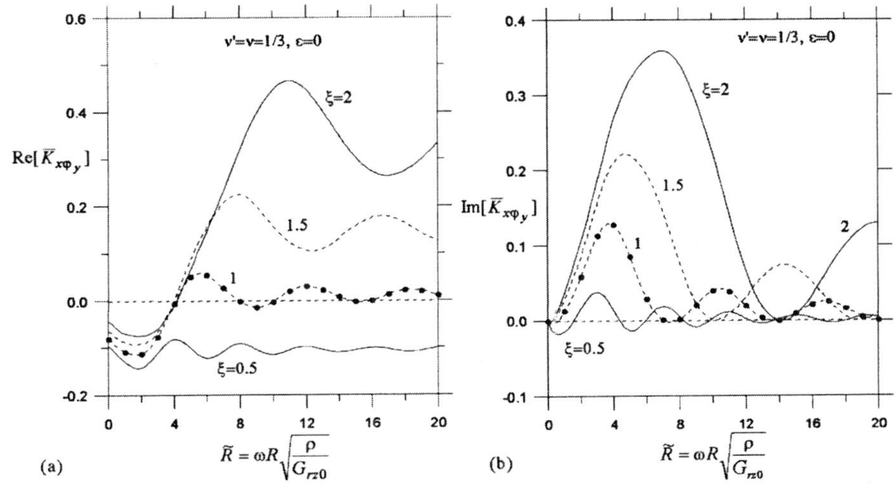

Fig. 2.23a,b. Real (a) and imaginary (b) parts of horizontal rocking stiffness of disk on anisotropic half-space

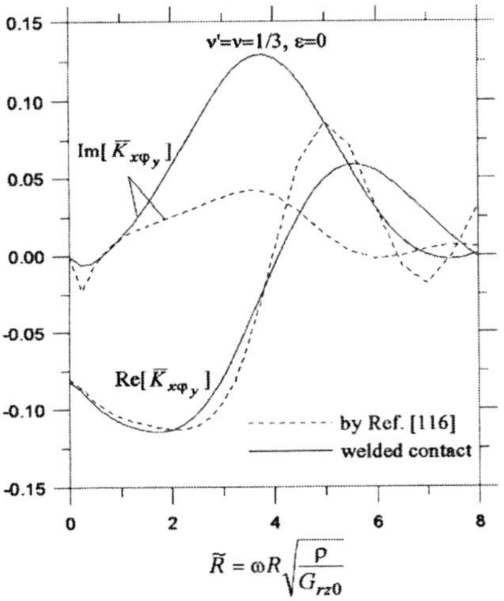

Fig. 2.24. Comparison of dynamic horizontal rocking stiffness for exact (solid lines) and approximate (dashed lines) solution of the contact problem ($\xi = 1$)

In static problems, we considered an approximate approach to determine the horizontal rocking stiffness (formula (2.234)). A similar technique has been employed elsewhere to construct solutions of dynamic problems [67,115], while the static distribution of contact stresses was used to average the corresponding displacements (in dynamics, this resulted in violation of the condition $C_{x\varphi_y} = C_{\varphi_y x}$) [67], or the averaging was performed by means of contact stresses corresponding to the solution of the dynamic contact problem under weakened contact conditions [115]. In Fig. 2.24, we present a comparison between the results of calculations for the isotropic half-space ($\xi = 1$) with $\nu = 1/3$ corresponding to the solution based on the latter reference, where a table of dynamic stiffnesses is given, and the solution, taking into account the welded contact between the disk and the half-space, i.e. based on equations (2.200). The discordance in the values of the imaginary part of the stiffness is rather significant. This approximate approach may cause a substantial error in the determination of the horizontal rocking stiffness in dynamic problems.

Rocking Stiffness

Equations (2.226) (for relaxed contact) or (2.222) (for welded contact) are used in order to determine unknowns p_{j0}, which enable calculation of the rocking stiffness, analogously to (2.224):

$$K_{\varphi_y} = \frac{\pi^2 G_{rz0} R^3}{2(1-\nu')} \sum_{j=1}^{N} \widetilde{p}_{0j} (\overline{R}_j^4 - \overline{R}_{j-1}^4) \ . \tag{2.252}$$

The results of calculations for the values of parameters employed previously when studying the vertical and horizontal stiffnesses are given in Fig. 2.25, where $\xi = 1$ corresponds to the isotropic half-space. Plots presenting the real part of the normalized stiffness, equal to the ratio of value K_{φ_y} from (2.252) to its static value $K_{\varphi_y}^{st}$ (corresponding to zero frequency), are shown in Fig. 2.25a. The damping coefficient, c_{φ_y}, equal to the imaginary part of the normalized stiffness divided by $\widetilde{R}$, is presented in Fig. 2.25b. The high-frequency asymptotics for c_{φ_y} may be determined following relationship (2.244), which provides a local connection between the load and the displacement rate. Calculation of the resultant moment of the load (with respect to the horizontal Y-axis) gives a relationship between the moment amplitudes M_{y0} and the disk rotation angle with respect to the Y-axis, φ_{y0}:

$$M_{y0} = \frac{i\omega\pi R^4}{4} \sqrt{A_{zz}\rho}\varphi_{y0} \ . \tag{2.253}$$

Hence, we obtain an asymptotic representation for the coefficient c_{φ_y} corresponding to high frequencies:

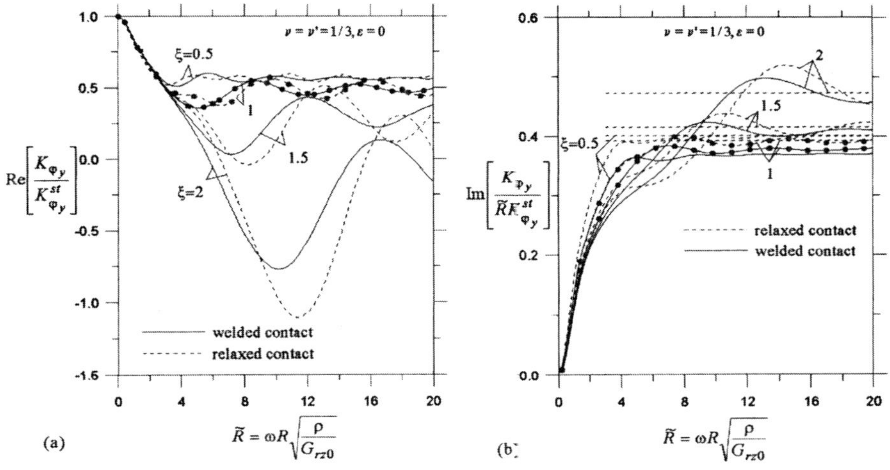

Fig. 2.25a,b. Real part (a) of normalized rocking dynamic stiffness of disk on anisotropic half-space and its imaginary part (b) divided by frequency parameter $\widetilde{R}$

$$c_{\varphi_y\infty} = \frac{3\pi(1-\nu')S_{\nu\nu}}{32\overline{K}_{\varphi_y}}\sqrt{\widetilde{A}_{zz}} \ . \tag{2.254}$$

Under the relaxed contact condition, when the normalized static stiffness $\overline{K}_{\varphi_y} = 1$, value of $c_{\varphi_y\infty}$ is 3/8 of the value $c_{\nu\infty}$ corresponding to the vertical vibrations of the disk. In Fig. 2.25b, horizontal dashed lines represent the coefficient $c_{\varphi_y\infty}$ for several values of parameter ξ.

Torsional Stiffness

As shown in section 2.4, the dynamic stiffness K_{φ_z} for the case of torsional vibrations of the transversely isotropic half-space may be derived by the following transformation of the expression for stiffness in the case of the isotropic half-space having shear modulus G: variable $a_0 = \omega R\sqrt{\rho/G_0}$ should be replaced with $\widetilde{R} = \omega R\sqrt{\rho/G_{r\vartheta 0}}$, while the stiffness values for the isotropic case should be multiplied by the ratio $\sqrt{G_{rz0}G_{r\vartheta 0}}/G_0$ (index 0 is used to denote the modulus of an ideally elastic body, when internal friction is neglected). Note that parameter $\widetilde{R}$ differs from that employed in the studies of other forms of vibrations. The normalized stiffness, being equal to the ratio $K_{\varphi_z}/K^{st}_{\varphi_z}$, where $K^{st}_{\varphi_z}$ is the static

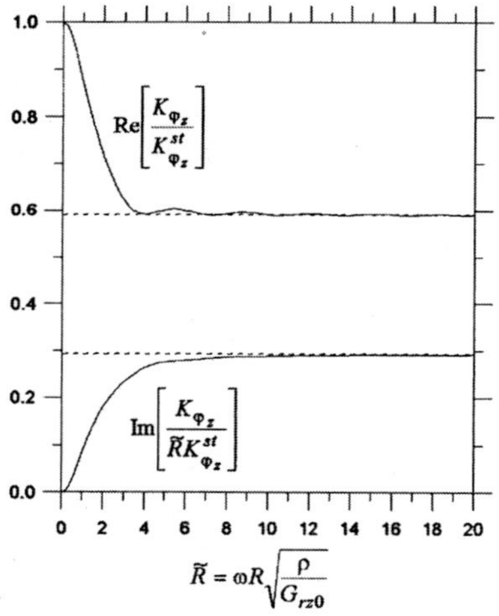

Fig. 2.26. Real part (a) of normalized torsional dynamic stiffness of disk on anisotropic half-space and its imaginary part (b) divided by frequency parameter $\widetilde{R}$

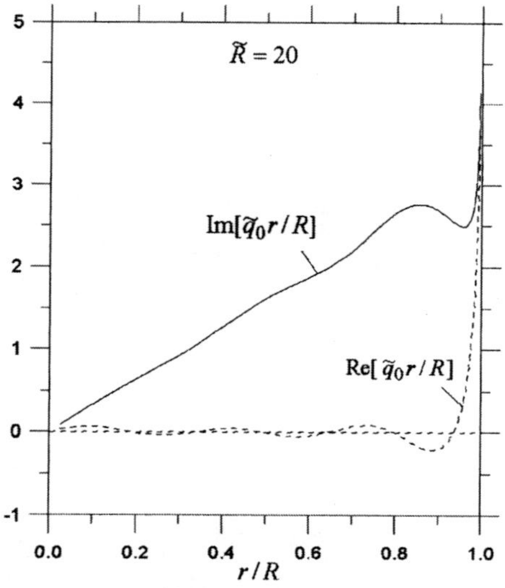

Fig. 2.27. High-frequency contact stresses for torsional vibrations of disk on half-space

stiffness determined by (2.175), depends in a similar way on the variable $\widetilde{R}$ for an arbitrary transversely isotropic half-space. Calculations performed using the technique presented in the previous subsection confirm this statement. The real part of the normalized torsional stiffness and coefficient $c_{\varphi_z} = \text{Im}[K_{\varphi_z} / K^{st}_{\varphi_z}]/\widetilde{R}$

are presented in Fig. 2.26. The corresponding limiting value $c_{\varphi_z\infty}$ may be found from the relationship between stresses and displacements analogous to (2.249):

$$\tau_{z\vartheta} = -\frac{\partial u_\vartheta}{\partial t}\sqrt{\rho G_{rz}}\ , \tag{2.255}$$

resulting in the following equation relating amplitudes M_{z0} of the torque and the disk rotation angle, φ_{z0}, with respect to the Z-axis:

$$M_{z0} = \frac{\mathrm{i}}{2}\pi\omega R^4 \varphi_{z0}\sqrt{\rho G_{rz}}\ . \tag{2.256}$$

This equation yields the following value of $c_{\varphi_z\infty}$:

$$c_{\varphi_z\infty} = \frac{3\pi}{32}\ , \tag{2.257}$$

which is, indeed, independent of the half-space properties. The result obtained agrees with the corresponding value given in the literature [104, 110]. According to the authors, the limiting value may be determined for the real part of the normalized complex stiffness:

$$\mathrm{Re}\left[\frac{K_{\varphi_z}}{K^{st}_{\varphi_z}}\right]_\infty = \frac{3\pi}{16}\ . \tag{2.258}$$

As noted previously, this value is $\pi/2$ greater than the result which stems from the approximate approach used by Bycroft [18]. In Fig. 2.26, the dashed horizontal lines represent the values determined from (2.257) and (2.258). Convergence to the limiting values is rather fast. Figure 2.27 illustrates how contact stresses are distributed over the contact area at high frequencies ($\widetilde{R} = 20$). Here, the graphs are constructed to represent quantity $\widetilde{q}_0 r/R$, which is proportional to the contact stresses, where $\widetilde{q}_0$ corresponds to the values $\widetilde{q}_{0_j}$ by (2.237) introduced in the numerical solution of the contact problem with a torsional action; ordinates are built at the mean points of the ring-shaped elements. Values of $N = 80$, $q = 0.95$ were employed. The assumption concerning the similarity of loads and displacements in the contact area is satisfied quite well, excluding a narrow zone in the vicinity of the contact area boundary.

2.8 Plane Problems for Transversely Isotropic Half-Space

In section 1.4.4, we derived general formulas for the amplitudes of vibrations under the action of loads distributed uniformly along an infinite line on the half-space surface. With known fundamental solutions $\widetilde{q}_j$, $\widetilde{w}_j$ $(j = 1,2)$ and

$\widetilde{p}_1$ constructed for a transversely isotropic half-space, one can perform an analysis (under the conditions of plane problems) of the amplitudes of vibrations of points that belong to the transversely isotropic half-space.

2.8.1 Action of Vertical Load Distributed Uniformly along Infinite Line on Half-Space Surface

According to expressions (1.91), the amplitudes of vibrations may be written as follows:

$$u_x(x,z)=\frac{p_0}{\pi G_{rz0}}S_{vx}, \tag{2.259a}$$

$$S_{vx}=\beta^2\int_0^{\infty}\sin(\widetilde{k}\widetilde{x})\frac{c_{21}\widetilde{q}_2(\widetilde{z},\widetilde{k})-c_{22}\widetilde{q}_1(\widetilde{z},\widetilde{k})}{D}\mathrm{d}\widetilde{k}, \tag{2.259b}$$

$$u_z(x,z)=\frac{p_0}{\pi G_{rz0}}S_{vz}, \tag{2.260a}$$

$$S_{vz}=\beta^2\int_0^{\infty}\cos(\widetilde{k}\widetilde{x})\frac{c_{21}\widetilde{w}_2(\widetilde{z},\widetilde{k})-c_{22}\widetilde{w}_1(\widetilde{z},\widetilde{k})}{D}\mathrm{d}\widetilde{k}, \tag{2.260b}$$

where, analogously to (2.2),

$$\widetilde{x}=\frac{x}{z_r}. \tag{2.261}$$

In the case of $\widetilde{z}>0$, the convergence of the integrals representing normalized amplitudes of vibrations, S_{vx} and S_{vz}, is provided by exponentially decreasing multipliers entering solutions $\widetilde{q}_j$, $\widetilde{w}_j$. Considering the amplitudes for the surface points ($\widetilde{z}=0$), we see that convergence of the integrals becomes rather slow – the multipliers of trigonometric functions have an order of decreasing $1/\widetilde{k}$. It is reasonable to apply the following presentation for the normalized amplitudes of vibrations at $\widetilde{z}=0$:

$$S_{vx}=\beta^2\int_0^{\infty}\frac{\sin(\widetilde{k}\widetilde{x})}{\widetilde{k}}\left[\widetilde{k}\frac{c_{21}\widetilde{q}_2(0,\widetilde{k})-c_{22}\widetilde{q}_1(0,\widetilde{k})}{D}-\Phi_{1\infty}\right]\mathrm{d}\widetilde{k}$$

$$+\beta^2\Phi_{1\infty}\int_0^{\infty}\frac{\sin(\widetilde{k}\widetilde{x})}{\widetilde{k}}\mathrm{d}\widetilde{k}, \tag{2.262a}$$

$$S_{vz}=\beta^2\int_0^{\infty}\frac{\cos(\widetilde{k}\widetilde{x})}{1+\widetilde{k}}\left[(1+\widetilde{k})\frac{c_{21}\widetilde{w}_2(0,\widetilde{k})-c_{22}\widetilde{w}_1(0,\widetilde{k})}{D}-\Phi_{2\infty}\right]\mathrm{d}\widetilde{k}$$

$$+\beta^2\Phi_{2\infty}\int_0^{\infty}\frac{\cos(\widetilde{k}\widetilde{x})}{1+\widetilde{k}}\mathrm{d}\widetilde{k}\,. \qquad (2.262b)$$

Here, quantities $\Phi_{1\infty}$ and $\Phi_{2\infty}$ represent limiting values as $\widetilde{k}\to\infty$ for the multipliers of $\sin(\widetilde{k}x)$ and $\cos(\widetilde{k}x)$, respectively, multiplied by $\widetilde{k}$, in the integrands in equations (2.259b), (2.260b) at $\widetilde{z}=0$. Clearly, these values correspond to the behavior of the integrands with vanishing frequency parameter θ (see section 2.1.1). Using the symbols adopted in section 2.1.1 and formulas (2.24) gives:

$$\Phi_{1\infty}=\frac{\widetilde{c}_{21}-\widetilde{c}_{22}}{\widetilde{D}}=(1-\nu')S_{vh}\,, \qquad (2.263a)$$

$$\Phi_{2\infty}=\frac{\widetilde{c}_{21}\widetilde{C}_{w2}-\widetilde{c}_{22}\widetilde{C}_{w1}}{\widetilde{D}}=(1-\nu')S_{vv}\,. \qquad (2.263b)$$

Note that the corresponding values may be determined by calculating the integrands in expressions (2.259b) and (2.260b) at very high values of argument $\widetilde{k}$ (e.g. $\widetilde{k}$ = 1000). The last integral in expression (2.262a) is equal to π/2, while the integral in expression (2.262b) may be expressed through the integral trigonometric functions Ci(u), Si(u) [1]:

$$\int_0^{\infty}\frac{\cos(\widetilde{x}\widetilde{k})}{1+\widetilde{k}}\mathrm{d}\widetilde{k}=g(\widetilde{x})=-\mathrm{Ci}(\widetilde{x})\cos(\widetilde{x})-[\mathrm{Si}(\widetilde{x})-0.5\pi]\sin(\widetilde{x})\,. \qquad (2.264)$$

The first of the integrals in expressions (2.262) converges faster than the initial integrals (2.259b) and (2.260b) at $\widetilde{z}=0$. Function $\mathrm{Ci}(\widetilde{x})$ contains logarithm $\ln(\widetilde{x})$, which determines its behavior at small values of argument $\widetilde{x}$, resulting in a logarithmic growth of S_{vz} when argument $\widetilde{x}$ tends to zero. Quantity S_{vz} may be presented in the following form ($\widetilde{z}=0$):

$$S_{vz}=-\beta^2\Phi_{2\infty}\ln(\widetilde{x})+\widetilde{S}_{vz}\,, \qquad (2.265)$$

where $\widetilde{S}_{vz}$ represents the bounded part of the normalized amplitudes. As for the value of S_{vx}, it is bounded.

As when calculating the amplitudes of vibrations under the action of the concentrated force, θ is taken equal to unity, so that the value of $\widetilde{x}$ may be determined from the following expression analogous to (2.19a):

$$\widetilde{x}=\frac{\omega x}{C_{rz}}=\omega x\sqrt{\frac{\rho}{G_{rz0}}}\,. \qquad (2.266)$$

The previously used expression (2.19b) still holds for $\widetilde{z}$. When the vibration frequency tends to zero, values of $\widetilde{z}$ and $\widetilde{x}$ also tend to zero, resulting in an

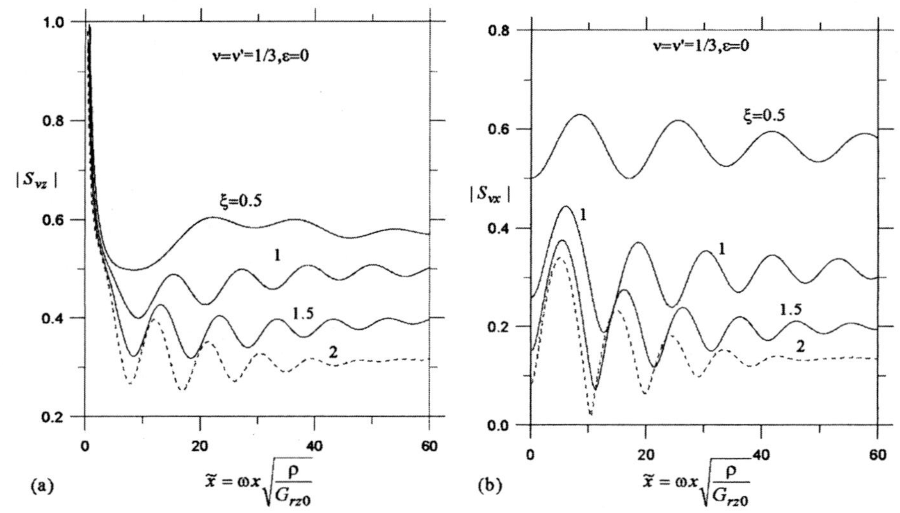

Fig. 2.28a,b. Absolute values of normalized vertical (a) and horizontal (b) amplitudes of vibrations of surface points of half-plane subjected to vertical load

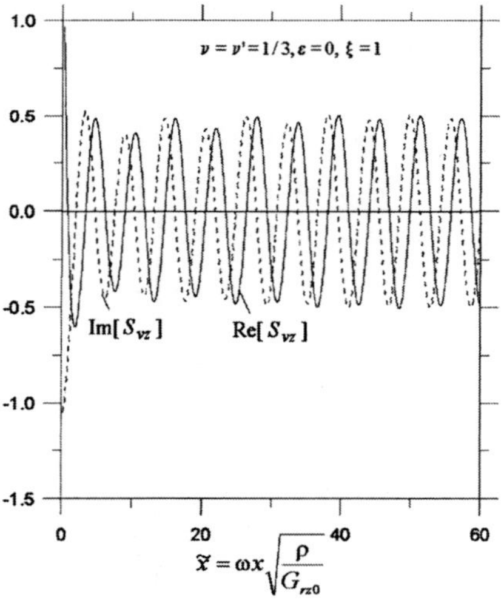

Fig. 2.29. Sample calculation illustrating fast convergence of solution to the result corresponding to Rayleigh wave of form $C\exp(-\mathrm{i}\widetilde{k}_R\widetilde{x}+\pi/2)$ at C=0.488, $\widetilde{k}_R = 1.07236$

unbounded growth of vertical displacements of all points in the half-space due to the term containing the logarithm.

In the process of calculation, we employ those considerations given in section 2.1.3 concerning the determination of amplitudes of vibrations that occur under the action of the concentrated force. The results of calculations corresponding to the values of parameters used for Fig. 2.4 are presented in Fig. 2.28. With increasing parameter $\widetilde{x}$, the magnitudes of the complex amplitudes tend to constant values, while the corresponding phase curves approach straight lines. This limiting behavior, representing Rayleigh waves, is due to the influence of integrand poles in equations (2.259b) and (2.260b). In contrast to the case of concentrated force (Figs. 2.4, 2.5), when the influence of Rayleigh waves results in an increase of order $\sqrt{\widetilde{r}}$ for the normalized amplitudes of vibrations (or decrease of order $1/\sqrt{\widetilde{r}}$ for the complete amplitudes of vibrations), under the conditions of the plane problem, amplitudes of vibrations for Rayleigh waves are constant. This conclusion stems from the structure of the integrals in expressions (2.259b) and (2.260b). In Fig. 2.29, the real and imaginary parts of S_{vz} are given at $\xi = 1$ (isotropic case), which are close at high values of $\widetilde{x}$ to the results obtained for the Rayleigh wave having the form $C\exp(-\mathrm{i}\widetilde{k}_R\widetilde{x}+\pi/2)$ with the constant amplitude C = 0.488, and $\widetilde{k}_R = 1.072357$ (see section 2.1.2). The value of the Rayleigh wave amplitude was found on the basis of the integrand residue in expression (2.260b) (at $\widetilde{z}=0$) with respect to the pole $\widetilde{k}=\widetilde{k}_R$.

2.8.2 Action of Horizontal Load Distributed Uniformly along Infinite Line on Half-Space Surface

Consider the vibrations of a point located in the plane (*X*, *Z*) under the action of a load that is parallel to the *X*-axis and distributed uniformly along an infinite line (which coincides with the *Y*-axis) on the half-space surface. For this case, the amplitudes of vibrations are presented in equations (1.98), (1.99). Next, we rewrite expressions for amplitudes using solutions given in section 2.1:

$$u_x(x,z) = \frac{q_0}{\pi G_{rz0}} S_{hx}\,, \tag{2.267a}$$

$$S_{hx} = \beta^2 \int_0^\infty \cos(\widetilde{k}\widetilde{x}) \frac{c_{12}\widetilde{q}_1(\widetilde{z},\widetilde{k}) - c_{11}\widetilde{q}_2(\widetilde{z},\widetilde{k})}{D} \mathrm{d}\widetilde{k}\,, \tag{2.267b}$$

$$u_z(x,z) = -\frac{q_0}{\pi G_{rz0}} S_{hz}\,, \tag{2.268a}$$

$$S_{hz} = \beta^2 \int_0^\infty \sin(\widetilde{k}\widetilde{x}) \frac{c_{12}\widetilde{w}_1(\widetilde{z},\widetilde{k}) - c_{11}\widetilde{w}_2(\widetilde{z},\widetilde{k})}{D} \mathrm{d}\widetilde{k}\,. \tag{2.268b}$$

When studying the amplitudes of vibrations for the points that belong to the half-space surface, we perform the transformations employed previously to study

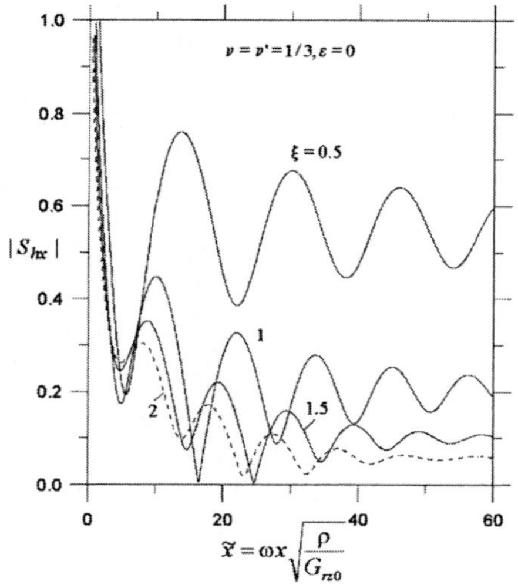

Fig. 2.30. Absolute values of normalized horizontal amplitudes of vibrations of surface points on half-plane subjected to horizontal load (in-plane vibrations)

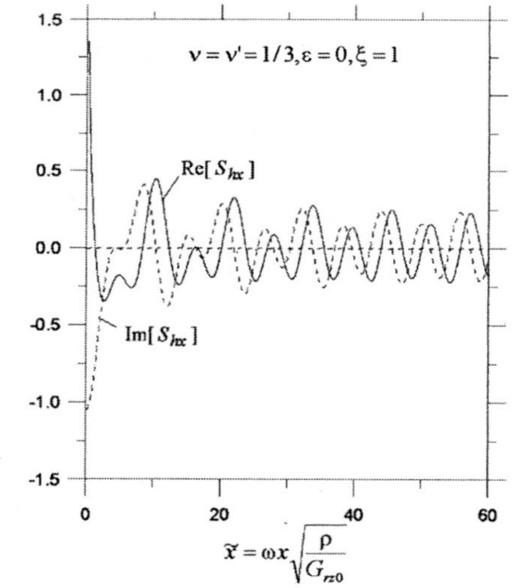

Fig. 2.31. Sample calculation illustrating slow convergence of solution to the result corresponding to Rayleigh wave

the action of the vertical load. Since the values S_{hz} and S_{vx} are interrelated due to the principle of reciprocity, in the following we shall consider S_{hx} only. In analogy with expression (2.262b)

$$S_{hx} = \beta^2 \int_0^\infty \frac{\cos(\tilde{k}\tilde{x})}{1+\tilde{k}} \left[(1+\tilde{k}) \frac{c_{12}\tilde{q}_1(0,\tilde{k}) - c_{11}\tilde{q}_2(0,\tilde{k})}{D} - \Phi_{3\infty} \right] \mathrm{d}\tilde{k}$$

$$+ \beta^2 \Phi_{3\infty} \int_0^\infty \frac{\cos(\tilde{k}\tilde{x})}{1+\tilde{k}} \mathrm{d}\tilde{k}\,, \tag{2.269}$$

where $\Phi_{3\infty}$ represents the limiting value as $\tilde{k} \to \infty$ of the multiplier of $\cos(\tilde{k}x)$ multiplied by $\tilde{k}$ in the integrand in equation (2.267b) (at $\tilde{z} = 0$). Similarly to $\Phi_{1\infty}$ in (2.263a), the expression for $\Phi_{3\infty}$ has the form

$$\Phi_{3\infty} = \frac{\tilde{c}_{12} - \tilde{c}_{11}}{\tilde{D}} = (1 - \nu')S_{h\vartheta}\,, \tag{2.270}$$

where relationship (2.72) is taken into account. The second integral in expression (2.269) is determined by employing relationship (2.264). In Fig. 2.30, the magnitudes of the normalized amplitudes of vibrations S_{hx} are plotted for the same values of parameters as those taken for Fig. 2.28. A comparison between the behavior of vertical amplitudes under action of the vertical load (Fig. 2.28a) and the horizontal amplitudes S_{hx} shows that in the latter case, approaching the constant amplitude of vibrations related to Rayleigh waves is slower. In Fig. 2.31, the real and imaginary parts of variable S_{hx} are plotted for the case $\xi = 1$; for the values considered, deviations from the sinusoidal character of variation are more appreciable than for the values shown in Fig. 2.29.

As in representation (2.265) for the amplitudes of vertical vibrations, one can separate out the term that contains the logarithmic singularity in expression (2.269):

$$S_{hx} = -\beta^2 \Phi_{3\infty} \ln(\tilde{x}) + \tilde{S}_{hx}\,. \tag{2.271}$$

Further, we consider anti-plane vibrations. Let a load q_0 be directed along the *Y*-axis and distributed uniformly along an infinite line (coinciding with the *Y*-axis) on the half-space surface. Employing expression (1.95) for the amplitudes of vibrations u_y, solution (2.57)–(2.59) and the symbols used when studying torsional vibrations of a half-space in section 2.4, we present the amplitudes of vibrations in the following form:

$$u_y = \frac{q_0}{\pi\sqrt{G_{rz0}G_{r\vartheta 0}}} S_y\,, \tag{2.272}$$

$$S_y = -\beta^2 \sqrt{G_{r\vartheta 0}} \int_0^\infty \frac{\tilde{p}_1(\tilde{z},\tilde{k})}{F} \cos(\tilde{k}\tilde{x})\,\mathrm{d}\tilde{k} = \beta^2 \int_0^\infty \exp(-\overline{\eta}_0 \tilde{z}_1) \frac{\cos(\tilde{k}\tilde{x})}{\overline{\eta}_0} \mathrm{d}\tilde{k}\,. \tag{2.273}$$

Application of expression (2.82) for parameter θ gives

$$\overline{\eta}_0 = \sqrt{\widetilde{k}^2 - \beta^2} \tag{2.274}$$

$$\widetilde{x} = \omega x \sqrt{\frac{\rho}{G_{r\vartheta 0}}} \tag{2.275}$$

$$\widetilde{z}_1 = \omega z \sqrt{\frac{\rho}{G_{rz0}}} \tag{2.276}$$

The integral in (2.273) may be expressed using cylindrical functions [43]:

$$S_y = \beta^2 K_0[\mathrm{i}\beta\sqrt{\widetilde{x}^2 + \widetilde{z}_1^2}] = -\frac{\beta^2 \pi \mathrm{i}}{2} H_0^{(2)}(\beta\sqrt{\widetilde{x}^2 + \widetilde{z}_1^2}) \tag{2.277}$$

As in the case of the torsional vibrations of a transversely isotropic half-space, the anisotropy affects the structure of the multiplier of quantity S_y in equation (2.272) and the structure of dimensionless coordinates $\widetilde{x}$ and $\widetilde{z}_1$ in (2.275) and (2.276). We see that in the anisotropic half-space, the velocities of propagation of the shear waves depend on directions in the plane (*X*, *Z*).

For the half-space surface ($\widetilde{z}_1 = 0$), the amplitudes of vibrations decrease as $1/\sqrt{\widetilde{x}}$ with increasing parameter $\widetilde{x}$, which is a result of the asymptotic behavior of cylindrical functions. This behavior is different from that of the surface amplitudes in the cases considered previously in the framework of the plane problem, when due to the influence of Rayleigh waves, the amplitudes of vibrations tended to a constant value with increasing $\widetilde{x}$.

As follows from relationship (2.277), logarithmic growth of the amplitudes of vibration takes place (when the vibration frequency tends to zero), similar to the case of values S_{vz} and S_{hx}. For the half-space surface, parameter S_y may be written in the following form (analogously to (2.265) and (2.271)):

$$S_y = -\beta^2 \ln(\widetilde{x}) + \widetilde{S}_y \tag{2.278}$$

Representations having the form of (2.265), (2.271), (2.278) are important for the numerical solution of contact problems, when the values of Green's functions required for integration are determined from interpolation over the values found at discrete points. Since functions $\widetilde{S}_{vz}(\widetilde{x})$, $\widetilde{S}_{hx}(\widetilde{x})$ and $\widetilde{S}_y$ vary sufficiently slowly, it is not necessary to apply an extremely small step for the discrete values, in order to carry out interpolation (a step of 0.05 may be used). The integrals for the logarithmic part of the normalized amplitudes of vibrations were determined in finite form.

2.8.3. Contact Problems for Strip Stamp

In order to study the harmonic vibrations of an infinite stiff strip stamp resting on the surface of a transversely isotropic half-space, we employ the approach used in section 2.6, where we considered vibrations of a circular stamp. Analogously to the ring-shaped elements, we take pairs of infinite strip elements located symmetrically with respect to the central line of the contact area considered. In order to construct matrices of the corresponding systems of equations for the contact problems, we dealt previously with expressions for the amplitudes of vibrations of points on the half-space surface that occur due to the action of some specific loads applied to the elements. For the individual elements, the results are derived from expressions related to the loading of an infinite continuous strip, analogous to the circular area in the 3-D case. The corresponding formulas are given in section 1.7.9. Let us rewrite them for the surface of the half-space taking into account separation of the logarithmic singular part in expressions for Green's functions; integrals for this singular part are taken in finite form. Note that parameter $\widetilde{x}$ determined by (2.266) serves as an argument for the normalized amplitudes of vibrations, and coordinate x in the formulas of section 1.7.9 is replaced by λR. Thus, we should replace the argument $\widetilde{x}$ in the corresponding integrals with $\lambda\widetilde{R}$, where the dimensionless half-width $\widetilde{R}$ of the strip is determined according to expression (2.241).

For the case of a vertical uniformly distributed load p_0 employment of expressions (2.259), (2.260) and (1.236), (1.237) gives

$$\delta_x(r,R) = \frac{p_0 R}{\pi G_{rz0}} \widetilde{\delta}_x(\eta,\widetilde{R}), \tag{2.279a}$$

$$\widetilde{\delta}_x(\eta,\widetilde{R}) = \int\limits_{|\eta-1|}^{\eta+1} S_{vx}(\lambda\widetilde{R})\,\mathrm{d}\lambda, \tag{2.279b}$$

$$\delta_z(r,R) = \frac{p_0 R}{\pi G_{rz0}} \widetilde{\delta}_z(\eta,\widetilde{R}), \tag{2.280a}$$

$$\widetilde{\delta}_z(\eta,\widetilde{R}) = \int\limits_{|\eta-1|}^{\eta+1} S_{vz}(\lambda\widetilde{R})\,\mathrm{d}\lambda + \begin{cases} 2\displaystyle\int\limits_0^{1-\eta} S_{vz}(\lambda\widetilde{R})\,\mathrm{d}\lambda & (r < R) \\ 0 & (r \ge R) \end{cases}$$

$$= \int\limits_{|\eta-1|}^{\eta+1} \widetilde{S}_{vz}(\lambda\widetilde{R})\,\mathrm{d}\lambda - \beta^2\Phi_{2\infty}\{(1+\eta)\ln[\widetilde{R}(1+\eta)] - (\eta-1)\ln[\widetilde{R}\,|\,\eta-1\,|] - 2\}$$

$$+ \begin{cases} 2\displaystyle\int\limits_0^{1-\eta} \widetilde{S}_{vz}(\lambda\widetilde{R})\,\mathrm{d}\lambda & (r < R) \\ 0 & (r \ge R). \end{cases} \tag{2.280b}$$

Analogous formulas hold for the case of horizontal uniformly distributed load q_0 acting in the direction parallel to the X-axis:

$$\delta_x(r,R) = \frac{q_0 R}{\pi G_{rz0}} \widetilde{\delta}_x^{(1)}(\eta, \widetilde{R}), \tag{2.281a}$$

$$\widetilde{\delta}_x^{(1)}(\eta, \widetilde{R}) = \int\limits_{|\eta-1|}^{\eta+1} S_{hx}(\lambda\widetilde{R})\, d\lambda + \begin{cases} 2\int\limits_0^{1-\eta} S_{hx}(\lambda\widetilde{R})\, d\lambda & (r < R) \\ 0 & (r \geq R) \end{cases}$$

$$= \int\limits_{|\eta-1|}^{\eta+1} \widetilde{S}_{hx}(\lambda\widetilde{R})\, d\lambda - \beta^2 \Phi_{3\infty}\{(1+\eta)\ln[\widetilde{R}(1+\eta)] - (\eta-1)\ln[\widetilde{R}\,|\,\eta-1\,|] - 2\}$$

$$+ \begin{cases} 2\int\limits_0^{1-\eta} \widetilde{S}_{hx}(\lambda\widetilde{R})\, d\lambda & (r < R) \\ 0 & (r \geq R), \end{cases} \tag{2.281b}$$

$$\delta_z(r,R) = \frac{q_0 R}{\pi G_{rz0}} \widetilde{\delta}_z^{(1)}(\eta, \widetilde{R}), \tag{2.282a}$$

$$\widetilde{\delta}_z^{(1)}(\eta, \widetilde{R}) = -\int\limits_{|\eta-1|}^{\eta+1} S_{vx}(\lambda\widetilde{R})\, d\lambda\ . \tag{2.282b}$$

In the case of a uniform load q_0 parallel to the Y-axis, we employ relationships (1.240) and (2.272), (2.273), (2.278),

$$\delta_y(r,R) = \frac{q_0 R}{\pi\sqrt{G_{rz0} G_{r\vartheta 0}}} \widetilde{\delta}_y(\eta, \widetilde{R}), \tag{2.283a}$$

$$\widetilde{\delta}_y(\eta, \widetilde{R}) = \int\limits_{|\eta-1|}^{\eta+1} S_y(\lambda\widetilde{R})\, d\lambda + \begin{cases} 2\int\limits_0^{1-\eta} S_y(\lambda\widetilde{R})\, d\lambda & (r < R) \\ 0 & (r \geq R) \end{cases}$$

$$= \int\limits_{|\eta-1|}^{\eta+1} \widetilde{S}_y(\lambda\widetilde{R})\, d\lambda - \beta^2\{(1+\eta)\ln[\widetilde{R}(1+\eta)] - (\eta-1)\ln[\widetilde{R}\,|\,\eta-1\,|] - 2\}$$

$$+\begin{cases} 2\int\limits_0^{1-\eta}\widetilde{S}_y(\lambda\widetilde{R})\,\mathrm{d}\lambda & (r<R) \\ 0 & (r\geq R). \end{cases} \tag{2.283b}$$

Note that when studying shear vibrations along the Y-axis, the definition (2.275) of variable $\widetilde{x}$ differs from that in other cases. In accordance with this fact, the value of $\widetilde{R}$ in relationships (2.283) should be determined from (2.85b), as in the case of torsional vibrations.

Consider the action of a horizontal load proportional to the distance from the central line of the loaded strip. Application of formulas (1.241), (1.242) and expressions (2.267), (2.268) gives

$$\delta_x(r,R)=\frac{q_0R^2}{\pi G_{rz0}}\widetilde{\delta}_x^{(2)}(\eta,\widetilde{R})\,, \tag{2.284a}$$

$$\widetilde{\delta}_x^{(2)}(\eta,\widetilde{R})=\int\limits_{|\eta-1|}^{\eta+1}(\eta-\lambda)S_{hx}(\lambda\widetilde{R})\,\mathrm{d}\lambda+\begin{cases} 2\eta\int\limits_0^{1-\eta}S_{hx}(\lambda\widetilde{R})\,\mathrm{d}\lambda & (r<R) \\ 0 & (r\geq R) \end{cases}$$

$$=\int\limits_{|\eta-1|}^{\eta+1}(\eta-\lambda)\widetilde{S}_{hx}(\lambda\widetilde{R})\,\mathrm{d}\lambda-\beta^2\Phi_{3\infty}\left[\frac{\eta^2-1}{2}\ln\frac{\eta+1}{|\eta-1|}-\eta\right]$$

$$+\begin{cases} 2\eta\int\limits_0^{1-\eta}\widetilde{S}_{hx}(\lambda\widetilde{R})\,\mathrm{d}\lambda & (r<R) \\ 0 & (r\geq R), \end{cases} \tag{2.284b}$$

$$\delta_z(r,R)=\frac{q_0R^2}{\pi G_{rz0}}\widetilde{\delta}_z^{(2)}(\eta,\widetilde{R}), \tag{2.285a}$$

$$\widetilde{\delta}_z^{(2)}(\eta,\widetilde{R})=-\int\limits_{|\eta-1|}^{\eta+1}(\eta-\lambda)S_{vx}(\lambda\widetilde{R})\,\mathrm{d}\lambda+\begin{cases} 2q\int\limits_0^{1-\eta}\lambda S_{vx}(\lambda\widetilde{R})\,\mathrm{d}\lambda & (r<R) \\ 0 & (r\geq R). \end{cases} \tag{2.285b}$$

Formulas related to the action of the antisymmetric vertical load applied to the infinite strip, located on the half-space surface, and varying in proportion to the distance between the point of application of the load and the central line of the strip, have similar structure, but they include another Green's functions.

Corresponding formulas are given in (1.243) and (1.244). Taking into account expressions (2.259), (2.260), (2.265), we obtain

$$\delta_x(r,R)=\frac{p_0R^2}{\pi G_{rz0}}\widetilde{\delta}_x^{(3)}(\eta,\widetilde{R}), \tag{2.286a}$$

$$\widetilde{\delta}_x^{(3)}(\eta,\widetilde{R})=\int\limits_{|\eta-1|}^{\eta+1}(\eta-\lambda)S_{vx}(\lambda\widetilde{R})\,\mathrm{d}\lambda-\begin{cases}2\int\limits_0^{1-\eta}\lambda S_{vx}(\lambda\widetilde{R})\,\mathrm{d}\lambda & (r<R)\\ 0 & (r\geq R),\end{cases} \tag{2.286b}$$

$$\delta_z(r,R)=\frac{p_0R^2}{\pi G_{rz0}}\widetilde{\delta}_z^{(3)}(\eta,\widetilde{R}), \tag{2.287a}$$

$$\widetilde{\delta}_z^{(3)}(\eta,\widetilde{R})=\int\limits_{|\eta-1|}^{\eta+1}(\eta-\lambda)S_{vz}(\lambda\widetilde{R})\,\mathrm{d}\lambda+\begin{cases}2\eta\int\limits_0^{1-\eta}S_{vz}(\lambda\widetilde{R})\,\mathrm{d}\lambda & (r<R)\\ 0 & (r\geq R)\end{cases}$$

$$=\int\limits_{|\eta-1|}^{\eta+1}(\eta-\lambda)\widetilde{S}_{vz}(\lambda\widetilde{R})\,\mathrm{d}\lambda-\beta^2\Phi_{2\infty}\left[\frac{\eta^2-1}{2}\ln\frac{\eta+1}{|\eta-1|}-\eta\right]$$

$$+\begin{cases}2\eta\int\limits_0^{1-\eta}\widetilde{S}_{vz}(\lambda\widetilde{R})\,\mathrm{d}\lambda & (r<R)\\ 0 & (r\geq R).\end{cases} \tag{2.287b}$$

Note that in the case of loads having zero vector sum, the displacements remain bounded when the vibration frequency tends to zero (see equations (2.284)–(2.287)), while in the case of uniformly distributed loads, an unbounded logarithmic growth of displacements occurs when parameter $\widetilde{R}$ tends to zero (see equations (2.280b), (2.281b) and (2.283b)).

Now, we come to the contact problems. Instead of ring-shaped elements applied in the case of a circular stamp, we shall use strip elements, and only the right-hand half of the contact area will be considered. This domain (having the width R) is divided into N elements with subsequently decreasing widths in the direction of the contact area boundary, following a geometrical progression with ratio q similar to the case of the circular stamp (considered in section 2.6). Let R_j $(j=0,...,N)$ denote the x-coordinates of the element's boundaries ($R_0=0,\ R_N=R$); $r_i=(R_i+R_{i-1})/2$ $(i=1,2,...,N)$ the x-coordinates of the mean points of the elements. Relative parameters (2.180) will be employed.

Vertical Vibrations of Strip Stamp

In order to account for the welded contact, we introduce $2N$ unknowns p_j, q_{0j} $(j = 1,...,N)$, where p_j represent normal pressures acting on the elements, being constant within each element, and $q_{0j}r$ is a horizontal load (parallel to the X-axis) for element j, varying in proportion with distance r from the central line of the contact area. The corresponding system of equations for unit vertical displacement of the stamp has the form of (2.181), with some changes in the meaning of the normalized unknowns and coefficients of the system of equations:

$$\widetilde{p}_j = \frac{p_j R}{\pi G_{rz0}}, \tag{2.288}$$

$$\widetilde{q}_{0j} = \frac{q_{0j} R^2}{\pi G_{rz0}}, \tag{2.289}$$

$$K_{ij} = \overline{R}_j \widetilde{\delta}_z(\overline{r}_i / \overline{R}_j, \widetilde{R}\overline{R}_j) - \overline{R}_{j-1} \widetilde{\delta}_z(\overline{r}_i / \overline{R}_{j-1}, \widetilde{R}\overline{R}_{j-1}), \tag{2.290}$$

$$L_{ij} = \overline{R}_j^2 \widetilde{\delta}_z^{(2)}(\overline{r}_i / \overline{R}_j, \widetilde{R}\overline{R}_j) - \overline{R}_{j-1}^2 \widetilde{\delta}_z^{(2)}(\overline{r}_i / \overline{R}_{j-1}, \widetilde{R}\overline{R}_{j-1}), \tag{2.291}$$

$$M_{ij} = \overline{R}_j \widetilde{\delta}_x(\overline{r}_i / \overline{R}_j, \widetilde{R}\overline{R}_j) - \overline{R}_{j-1} \widetilde{\delta}_x(\overline{r}_i / \overline{R}_{j-1}, \widetilde{R}\overline{R}_{j-1}), \tag{2.292}$$

$$N_{ij} = \overline{R}_j^2 \widetilde{\delta}_x^{(2)}(\overline{r}_i / \overline{R}_j, \widetilde{R}\overline{R}_j) - \overline{R}_{j-1}^2 \widetilde{\delta}_x^{(2)}(\overline{r}_i / \overline{R}_{j-1}, \widetilde{R}\overline{R}_{j-1}), \tag{2.293}$$

For determination of the vertical stiffness K_z, one should calculate the resultant vertical force (per unit length of the stamp), corresponding to pressures p_j:

$$K_z = 2\sum_{j=1}^{N} p_j (R_j - R_{j-1}) = 2\pi G_{rz0} \sum_{j=1}^{N} \widetilde{p}_j (\overline{R}_j - \overline{R}_{j-1}) = \overline{K}_z G_{rz0}, \tag{2.294}$$

where

$$\overline{K}_z = 2\pi \sum_{j=1}^{N} \widetilde{p}_j (\overline{R}_j - \overline{R}_{j-1}). \tag{2.295}$$

In the case of relaxed contact, the system of equations to be solved for values $\widetilde{p}_j$ takes the form of (2.190). The results of calculations for parameters, corresponding to those used in Fig. 2.18, are presented in Fig. 2.32; in Fig. 2.32b, the damping coefficient c_z, which is equal to the ratio of the imaginary part of the normalized complex stiffness $\widetilde{K}_z$ to $\widetilde{R}$, is shown. The limiting value (as $\widetilde{R} \to \infty$) for this coefficient corresponding to relationship (2.244) is

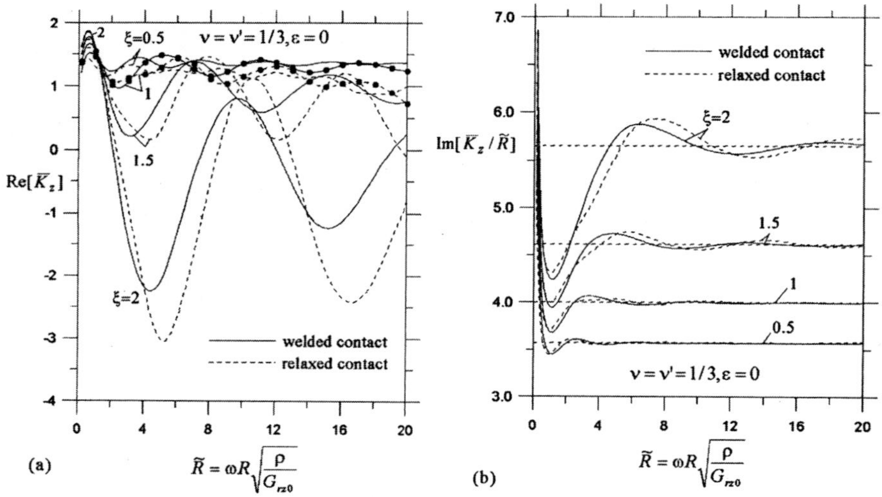

Fig. 2.32a,b. Normalized vertical dynamic stiffness of strip stamp on anisotropic half-space

$$c_{z\infty} = 2\sqrt{\widetilde{A}_{zz}} \,. \tag{2.296}$$

For the values of $\xi = 0.5, 1.0, 1.5, 2.0$, we obtain $c_{z\infty} = 3.5777, 4, 4.6188, 5.6569$, respectively. The dashed horizontal straight lines in Fig. 2.32b represent values of $c_{z\infty}$; with decreasing parameter ξ, the value of the damping coefficient tends to its limiting value faster (similar to the case of the circular stamp). The dynamic stiffness of the strip stamp tends to zero with decreasing frequency of vibrations, whereas in the case of the circular stamp, the static value of stiffness is different from zero. Since the dynamic stiffness for the plane problem decreases in inverse proportion to a logarithmic function with decreasing frequency of vibrations, as a result of dividing the imaginary part of the stiffness by frequency parameter $\widetilde{R}$, an unbounded growth of coefficient c_z takes place. The corresponding coefficient for the 3-D problem is bounded (see Fig. 2.18b).

Horizontal Vibrations of Strip Stamp

In order to study vibrations of the stamp along the *X*-axis under conditions of welded contact between the stamp and a half-space, we introduce $2N$ unknowns: horizontal loads q_j, distributed uniformly within each element (the results required for such loads are derived from relationships (2.281), (2.282), corresponding to uniform loading of a strip area), and loads having the form $p_{0j}\rho$, being parallel to the *Z*-axis and proportional to the *x*-coordinate ρ of the point of application of the load (the results required for such loads are derived by using relationships (2.286), (2.287), corresponding to loading of a continuous strip with an antisymmetric load). The system of equations for the unknowns represents a simplified version of the system (2.200) that determines unknown loads in the 3-D case:

$$\sum_{j=1}^{N} C_{ij}^{(1)}\widetilde{q}_j + \sum_{j=1}^{N} C_{ij}^{(2)}\widetilde{p}_{0j} = 1,$$

$$\sum_{j=1}^{N} C_{ij}^{(3)}\widetilde{q}_j + \sum_{j=1}^{N} C_{ij}^{(4)}\widetilde{p}_{0j} = 0, \tag{2.297}$$

where

$$\widetilde{q}_j = \frac{q_j R}{G_{rz0}\pi}, \tag{2.298a}$$

$$\widetilde{p}_{0j} = \frac{p_{0j} R^2}{G_{rz0}\pi}, \tag{2.298b}$$

$$C_{ij}^{(1)} = \overline{R}_j \widetilde{\delta}_x^{(1)}(\overline{r}_i / \overline{R}_j, \widetilde{RR}_j) - \overline{R}_{j-1} \widetilde{\delta}_x^{(1)}(\overline{r}_i / \overline{R}_{j-1}, \widetilde{RR}_{j-1}), \tag{2.299}$$

$$C_{ij}^{(2)} = \overline{R}_j^2 \widetilde{\delta}_x^{(3)}(\overline{r}_i / \overline{R}_j, \widetilde{RR}_j) - \overline{R}_{j-1}^2 \widetilde{\delta}_x^{(3)}(\overline{r}_i / \overline{R}_{j-1}, \widetilde{RR}_{j-1}), \tag{2.300}$$

$$C_{ij}^{(3)} = \overline{R}_j \widetilde{\delta}_z^{(1)}(\overline{r}_i / \overline{R}_j, \widetilde{RR}_j) - \overline{R}_{j-1} \widetilde{\delta}_z^{(1)}(\overline{r}_i / \overline{R}_{j-1}, \widetilde{RR}_{j-1}), \tag{2.301}$$

$$C_{ij}^{(4)} = \overline{R}_j^2 \widetilde{\delta}_z^{(3)}(\overline{r}_i / \overline{R}_j, \widetilde{RR}_j) - \overline{R}_{j-1}^2 \widetilde{\delta}_z^{(3)}(\overline{r}_i / \overline{R}_{j-1}, \widetilde{RR}_{j-1}). \tag{2.302}$$

The first N equations (2.297) for $i = 1,...,N$ express equality to unity of displacements of the central points of elements along the X-axis, and the next N equations make vertical displacements vanish for these points. If one neglects the influence of vertical stresses over the contact area on horizontal displacements of the contact points, the system reduces to N equations:

$$\sum_{j=1}^{N} C_{ij}^{(1)}\widetilde{q}_j = 1. \tag{2.303}$$

The results of calculations, corresponding to the welded contact (equations (2.297)), and to the relaxed contact (equations (2.303)), are shown in Fig. 2.33. The dynamic horizontal stiffness per unit length of stamp is calculated by using values of $\widetilde{q}_j$ in the following way:

$$K_x = 2\sum_{j=1}^{N} q_j (R_j - R_{j-1}) = 2\pi G_{rz0} \sum_{j=1}^{N} \widetilde{q}_j (\overline{R}_j - \overline{R}_{j-1}) = \overline{K}_x G_{rz0}, \tag{2.304}$$

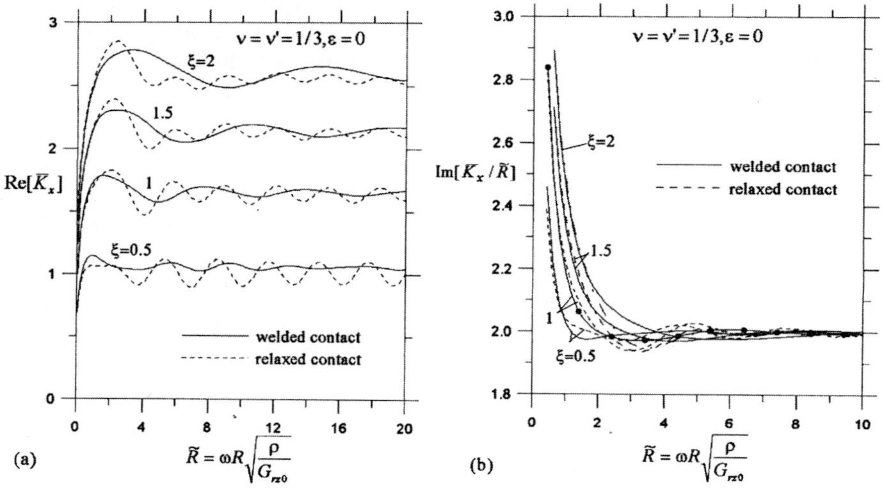

Fig. 2.33a,b. Normalized horizontal dynamic stiffness of strip stamp on anisotropic half-space

where

$$\overline{K}_x = 2\pi \sum_{j=1}^{N} \tilde{q}_j (\overline{R}_j - \overline{R}_{j-1}) . \tag{2.305}$$

The damping coefficient c_x shown in Fig. 2.33b equals the ratio of the imaginary part of normalized stiffness $\overline{K}_x$ to parameter $\widetilde{R}$. The limiting value of this coefficient $c_{x\infty}$ as $\widetilde{R} \to \infty$, determined from the relationship having the form (2.249), is equal to 2 (for $\varepsilon = 0$), being independent of an anisotropy parameter ξ, which influences the coefficient c_x at low and intermediate values of parameter $\widetilde{R}$. With decreasing $\widetilde{R}$, coefficient c_x grows unboundedly, while the real part of the dynamic stiffness tends to zero, analogously to the case of vertical vibrations of the stamp. Weakening of the contact conditions results in more significant oscillations of dynamic stiffness of the stamp with varying parameter $\widetilde{R}$.

Horizontal Rocking Stiffness

A moment arising due to horizontal displacement of the stamp without rotation determines the horizontal rocking stiffness $K_{x\varphi_y}$. Consider the moment with respect to the *Y*-axis of pressures $p_{0j}\rho$ determined by using equations (2.297):

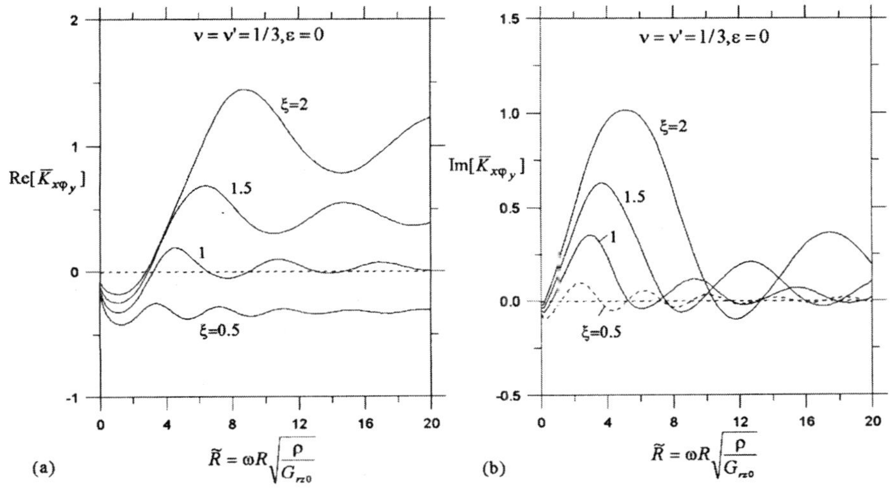

Fig. 2.34a,b. Normalized horizontal rocking dynamic stiffness of strip stamp on anisotropic half-space

$$K_{x\varphi_y} = \frac{2}{3}\sum_{j=1}^{N} p_{j0}(R_j^3 - R_{j-1}^3) = \frac{2}{3}\pi G_{rz0} R \sum_{j=1}^{N} \widetilde{p}_{j0}(\overline{R}_j^3 - \overline{R}_{j-1}^3) = \overline{K}_{x\varphi_y} G_{rz0} R\,, \tag{2.306}$$

where

$$\overline{K}_{x\varphi_y} = \frac{2}{3}\pi \sum_{j=1}^{N} \widetilde{p}_{j0}(\overline{R}_j^3 - \overline{R}_{j-1}^3)\,, \tag{2.307}$$

The results of calculations related to the value of $\overline{K}_{x\varphi_y}$ are presented in Fig. 2.34 for the values of parameters employed earlier. To a considerable extent, the curves obtained are similar to those plotted for the 3-D case, shown in Fig. 2.23. Note the appreciable influence of anisotropy on the behavior of the horizontal rocking dynamic stiffness.

Rocking Vibrations of Strip Stamp

When accounting for the influence of horizontal stresses arising in the contact area with a welded contact between the stamp and the half-space, on vertical displacements of the contact points, the system of equations for unknowns q_j and p_{0j} differs from system (2.297) in its right-hand sides only. In the case of rotation of the contact area with respect to the Y-axis with unit angle amplitude (without horizontal displacement), one should set the right sides of the first N equations equal to zero, and the next N equations equal to values r_i (i.e. to the x-coordinates of the mean points of the strip elements). Dividing both sides of the equations by R, we obtain

$$\sum_{j=1}^{N} C_{ij}^{(1)}\widetilde{q}_j + \sum_{j=1}^{N} C_{ij}^{(2)}\widetilde{p}_{0j} = 0,$$

$$\sum_{j=1}^{N} C_{ij}^{(3)}\widetilde{q}_j + \sum_{j=1}^{N} C_{ij}^{(4)}\widetilde{p}_{0j} = \overline{r}_i, \tag{2.308}$$

where

$$\widetilde{q}_j = \frac{q_j}{G_{rz0}\pi}, \tag{2.309a}$$

$$\widetilde{p}_{0j} = \frac{p_{0j}R}{G_{rz0}\pi}. \tag{2.309b}$$

Note that the definition of the dimensionless unknowns is modified compared with (2.298), due to division by R. The rocking stiffness per unit length of strip stamp is calculated analogously to (2.306):

$$K_{\varphi_y} = \frac{2}{3}\sum_{j=1}^{N} p_{j0}(R_j^3 - R_{j-1}^3) = \frac{2}{3}\pi G_{rz0}R^2 \sum_{j=1}^{N} \widetilde{p}_{j0}(\overline{R}_j^3 - \overline{R}_{j-1}^3) = \overline{K}_{\varphi_y} G_{rz0} R^2, \tag{2.310}$$

where

$$\overline{K}_{\varphi_y} = \frac{2}{3}\pi\sum_{j=1}^{N} \widetilde{p}_{j0}(\overline{R}_j^3 - \overline{R}_{j-1}^3), \tag{2.311}$$

For the case with weakened contact conditions, where only the vertical displacements of contact points are considered, we have the following system of equations:

$$\sum_{j=1}^{N} C_{ij}^{(4)}\widetilde{p}_{0j} = \overline{r}_i. \tag{2.312}$$

The results of calculations are given in Fig. 2.35, where, as earlier, the damping coefficient is introduced (Fig. 2.35b):

$$c_{\varphi_y} = \mathrm{Im}\left[\frac{\overline{K}_{\varphi_y}}{\widetilde{R}}\right]. \tag{2.313}$$

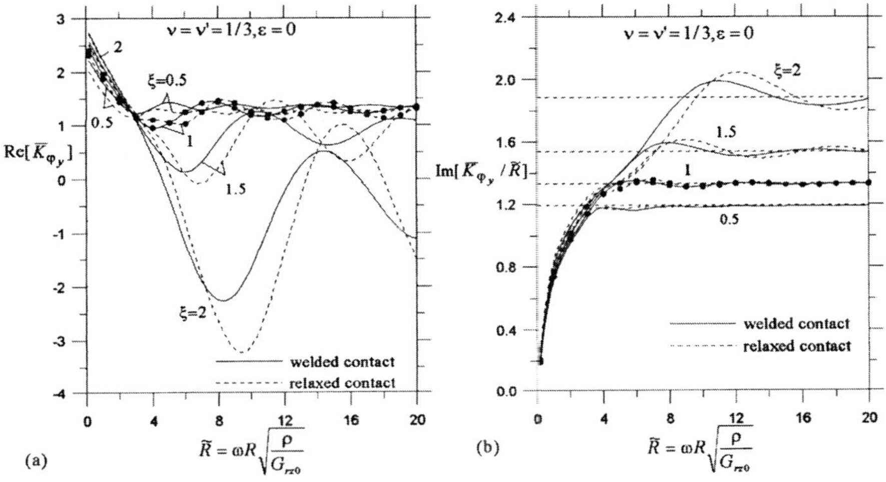

Fig. 2.35a,b. Normalized rocking dynamic stiffness of strip stamp on anisotropic half-space

The limiting value $c_{\varphi_y\infty}$ of this coefficient as $\widetilde{R} \to \infty$ may be determined by using the formula having the form (2.253), in which one should replace the moment of inertia of the contact area $\pi R^4/4$ with $2R^3/3$:

$$c_{\varphi_y\infty} = \frac{2}{3}\sqrt{\widetilde{A}_{zz}} = \frac{c_{z\infty}}{3}\,. \tag{2.314}$$

In Fig. 2.35b, the horizontal dashed lines represent the values of $c_{\varphi_y\infty}$.

The main differences from the case of translational vibrations of the stamp are in the boundedness of coefficient c_{φ_y}, and in the fact that the real part of the stiffness differs from zero at zero frequency of vibrations.

The rocking horizontal stiffness, defined as a horizontal force arising from rotation of the stamp without horizontal displacements, may be determined by using loads q_j analogously to (2.304):

$$K_{\varphi_y x} = 2\sum_{j=1}^{N} q_j (R_j - R_{j-1}) = 2\pi G_{rz0} R \sum_{j=1}^{N} \widetilde{q}_j (\overline{R}_j - \overline{R}_{j-1}) = \overline{K}_{\varphi_y x} G_{rz0} R\,, \tag{2.315}$$

where

$$\overline{K}_{\varphi_y x} = 2\pi \sum_{j=1}^{N} \widetilde{q}_j (\overline{R}_j - \overline{R}_{j-1}), \tag{2.316}$$

This value should be equal to $\overline{K}_{x\varphi_y}$, in accordance with the principle of reciprocity; this fact is confirmed with sufficiently high accuracy in our calculations.

Anti-Plane Vibrations

In order to solve this problem, we introduce horizontal loads q_j acting on the strip elements and directed parallel to the Y-axis. Employing relationships (2.283), we obtain the following system of equations:

$$\sum_{j=1}^{N} C_{ij} \widetilde{q}_j = 1, \tag{2.317}$$

where

$$\widetilde{q}_j = \frac{q_j R}{\pi \sqrt{G_{rz0} G_{r90}}}, \tag{2.318}$$

$$C_{ij} = \overline{R}_j \widetilde{\delta}_y (\overline{r}_i / \overline{R}_j, \widetilde{R}\overline{R}_j) - \overline{R}_{j-1} \widetilde{\delta}_y (\overline{r}_i / \overline{R}_{j-1}, \widetilde{R}\overline{R}_{j-1}). \tag{2.319}$$

Note that the definition of variable $\widetilde{R}$ differs from that used in previous cases: one should employ definition (2.85b) (see the remark following formulas (2.283)). Having determined the unknowns $\widetilde{q}_j$, we find the stiffness K_y:

$$K_y = 2\sum_{j=1}^{N} q_j (R_j - R_{j-1}) = 2\pi \sqrt{G_{rz0} G_{r90}} \sum_{j=1}^{N} \widetilde{q}_j (\overline{R}_j - \overline{R}_{j-1}) = \overline{K}_y \sqrt{G_{rz0} G_{r90}}, \tag{2.320}$$

where

$$\overline{K}_y = 2\pi \sum_{j=1}^{N} \widetilde{q}_j (\overline{R}_j - \overline{R}_{j-1}). \tag{2.321}$$

According to equations (2.277) and (2.283), when internal friction is neglected ($\varepsilon = 0$), normalized stiffness $\overline{K}_y$ depends on parameter $\widetilde{R}$ containing shear modulus G_{r90} only. The second shear modulus, G_{rz0}, enters into relationship (2.320). In the case considered, the influence of the half-space parameters is identical to their influence in the case of torsional vibrations of the circular stamp (Fig. 2.26). Further, we consider some results of calculations related to the normalized stiffness $\overline{K}_y$. Quantities $\mathrm{Re}[\overline{K}_y]$ and $c_y = \mathrm{Im}[\overline{K}_y / \widetilde{R}]$ are presented in

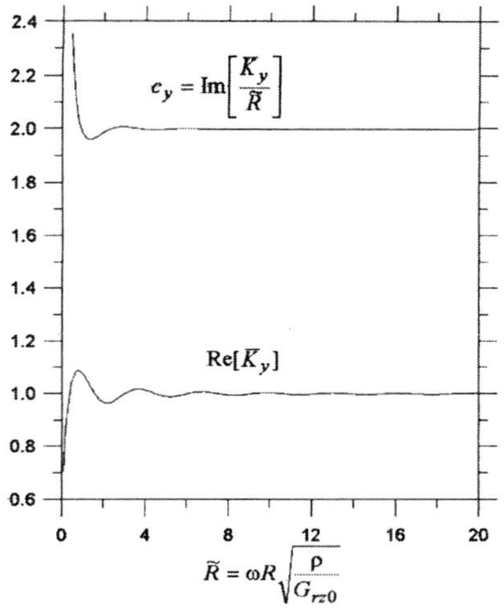

Fig. 2.36. Normalized horizontal dynamic stiffness of strip stamp on anisotropic half-space in the case of anti-plane vibrations

Fig. 2.36. The behavior of $\overline{K}_y$ is similar to that of $\overline{K}_x$ at $\xi = 0.5$ (see Fig. 2.33). The limiting value $c_{y\infty}$ of parameter c_y as $\widetilde{R} \to \infty$ determined from relationship (2.249) equals 2, as in the case of vibrations along the X-axis. The results of calculations indicate that the real part of the normalized stiffness tends to unity. Note that the same ratio of the limiting values occurs in the case of torsional vibrations of the circular stamp on the half-space.

Comparing results related to the 3-D case studied in section 2.7.2 and those corresponding to the plane problem shows their qualitative similarity at intermediate and high values of frequency parameter $\widetilde{R}$. The influence of anisotropy (parameter ξ) is identical in the plane and 3-D cases: with increasing parameter ξ, the variability of the dynamic stiffness (with varying frequency of vibration) grows and the imaginary part of the stiffness, responsible for emitting energy into the half-space, increases.

Chapter 3. Mechanics of Isotropic Half-Space with Shear Modulus Varying Linearly with Depth

In this chapter, we study dynamic and static problems stated for a half-space with shear modulus increasing linearly with depth. Further, we consider the case of an isotropic material, when the half-space, having a linearly varying shear modulus, enables a closed solution for the functions $\tilde{q}(\tilde{z},\tilde{k}), \tilde{w}(\tilde{z},\tilde{k}), \tilde{p}(\tilde{z},\tilde{k})$ to be found in the form of confluent hypergeometric functions. Dynamic problems for a half-space of this kind have been solved, in most cases, by studying free vibrations, i.e. the parameters of Rayleigh and Love waves have been determined [29, 106, 112–114]. Forced vibrations were a subject of concern in [6, 7, 78] for the case of an incompressible medium, and for arbitrary values of Poisson's ratio in [79].

We employ equations (1.130)–(1.133); instead of symbols G_{rz} and G_{rz0}, we use G and G_0, respectively. The internal friction is accounted for according to relationship (1.28), where indices r and z are dropped. For the reference length z_r, one can use the value of z_0 – the distance from the origin of coordinates to the half-space surface. Thus, the surface is located at $\tilde{z}=1$. Parameter θ defined in (1.32a) becomes

$$\theta = \omega z_0 \sqrt{\frac{\rho}{G_0}}\,, \tag{3.1}$$

where ρ represents the constant density of the half-space. Since the shear modulus varies linearly with depth, the normalized shear modulus, $\tilde{G}$, takes the form

$$\tilde{G} = \tilde{z}\,. \tag{3.2}$$

Parameter θ serves as a measure of heterogeneity of the half-space (at fixed values of shear modulus G_0 and of vibration frequency ω). With increasing z_0, the considered half-space approaches a homogeneous one.

3.1 Fundamental Solutions of System of Equations (1.130)–(1.133)

In order to construct fundamental solutions of the system of equations (1.130)–(1.133) for the isotropic linearly heterogeneous half-space, we perform some transformations of these equations. Assuming a constant density of material ($\widetilde{\rho}=1$), we differentiate equation (1.130) with respect to $\widetilde{z}$ and add the result to equation (1.131) multiplied by $\widetilde{k}$. This procedure yields

$$\frac{d^2\widetilde{\chi}}{d\widetilde{z}^2}+\left(\frac{\beta^2\theta^2}{\widetilde{G}(\widetilde{z})}-\widetilde{k}^2\right)\widetilde{\chi}-2\widetilde{k}\frac{d\widetilde{G}}{d\widetilde{z}}\frac{\tau^2\widetilde{e}}{\widetilde{G}(\widetilde{z})}-2\widetilde{k}\frac{d^2\widetilde{G}}{d\widetilde{z}^2}\widetilde{w}=0\,, \tag{3.3}$$

where parameter τ is given in relationship (1.134). Analogously, differentiation of equation (1.131) with respect to $\widetilde{z}$ with subsequent addition to equation (1.130) multiplied by $\widetilde{k}$ results in

$$\frac{d^2\widetilde{e}}{d\widetilde{z}^2}+\left(\frac{\beta^2\theta^2\tau^2}{\widetilde{G}(\widetilde{z})}-\widetilde{k}^2\right)\widetilde{e}-2\widetilde{k}\frac{d\widetilde{G}}{d\widetilde{z}}\frac{\widetilde{\chi}}{\widetilde{G}(\widetilde{z})}-2\widetilde{k}\frac{d^2\widetilde{G}}{d\widetilde{z}^2}\widetilde{q}=0\,. \tag{3.4}$$

With linear relationship (3.2), the system of equations (3.3), (3.4) contains functions $\widetilde{\chi}$ and $\widetilde{e}$ only. If, in addition, the medium is incompressible ($\tau=0$), then equation (3.3) becomes completely independent and reduces to Whittaker's equation. Substitution of the variable

$$\varsigma=2\widetilde{k}\widetilde{z} \tag{3.5}$$

enables the system of equations (3.3), (3.4) for the case $\widetilde{G}=\widetilde{z}$ to be rewritten in the following form:

$$\frac{d^2\widetilde{\chi}}{d\varsigma^2}-\left(\frac{1}{4}-\frac{\gamma}{\varsigma}\right)\widetilde{\chi}-\frac{\tau^2\widetilde{e}}{\varsigma}=0\,, \tag{3.6a}$$

$$\frac{d^2\widetilde{e}}{d\varsigma^2}-\left(\frac{1}{4}-\frac{\gamma\tau^2}{\varsigma}\right)\widetilde{e}-\frac{\widetilde{\chi}}{\varsigma}=0\,, \tag{3.6b}$$

where

$$\gamma=\frac{\beta^2\theta^2}{2\widetilde{k}}\,. \tag{3.7}$$

Then, the system of equations (1.130)–1.133) becomes

$$\frac{d\overline{\chi}}{d\varsigma}=\frac{\overline{e}}{2}+m(w-\gamma q)\,, \tag{3.8}$$

$$\frac{d\overline{e}}{d\varsigma}=\frac{\overline{\chi}}{2}+\overline{q}-\gamma\overline{w}\,, \tag{3.9}$$

$$\frac{d\overline{q}}{d\varsigma}=\frac{\overline{\chi}}{\varsigma}-\frac{\overline{w}}{2}\,, \tag{3.10}$$

$$\frac{\mathrm{d}\overline{w}}{\mathrm{d}\varsigma}=\frac{\tau^2\overline{e}}{\varsigma}-\frac{\overline{q}}{2}. \tag{3.11}$$

We seek the solution of system (3.6), decreasing with growing argument ς, in the following form:

$$\begin{aligned}\tilde{\chi}(\varsigma)&=A_\chi W_{\eta,1/2}(\varsigma)=A_\chi\varsigma\exp(-\varsigma/2)U(1-\eta,2,\varsigma),\\ \tilde{e}(\varsigma)&=A_e W_{\eta,1/2}(\varsigma)=A_e\varsigma\exp(-\varsigma/2)U(1-\eta,2,\varsigma),\end{aligned} \tag{3.12}$$

where Whittaker's function W and the confluent hypergeometric function U are used [1]; A_χ, A_e are arbitrary coefficients. In the framework of studying the vibrations of a half-space subjected to loads applied to its surface, representation (3.12) is sufficient to construct fundamental solutions having indices 1 and 2 in previous subsections. In order to solve other problems, e.g. those related to vibrations of a half-space under the action of loads applied in its depth, one should seek the solutions which increase with increasing argument ς (having indices 3 and 4). In order to construct these solutions, we represent the solution of system (3.6) in the following form:

$$\begin{aligned}\tilde{\chi}(\varsigma)&=A_\chi M_{\eta,1/2}(\varsigma)=A_\chi\varsigma\exp(-\varsigma/2)M(1-\eta,2,\varsigma),\\ \tilde{e}(\varsigma)&=A_e M_{\eta,1/2}(\varsigma)=A_e\varsigma\exp(-\varsigma/2)M(1-\eta,2,\varsigma),\end{aligned} \tag{3.13}$$

where Whittaker's function $M_{\eta,1/2}(\varsigma)$ and the confluent hypergeometric function $M(1-\eta,2,\varsigma)$ which increase with growth of argument ς are employed. Expressions (3.12) and (3.13) include parameter η, which may be determined by substituting expressions (3.12) and (3.13) into the system of differential equations (3.6). Using Whittaker's equation [1], we obtain the following system of equations to solve for coefficients A_χ, A_e:

$$\begin{aligned}&(\gamma-\eta)A_\chi-\tau^2A_e=0,\\ &-A_\chi+(\gamma\tau^2-\eta)A_e=0.\end{aligned} \tag{3.14}$$

In order to find a non-zero solution, we set the determinant of system (3.14) equal to zero:

$$\eta^2-\gamma(1+\tau^2)\eta+\tau^2(\gamma^2-1)=0. \tag{3.15}$$

The pair of solutions of equation (3.15) may be written as follows:

$$\eta_1=0.4[\gamma(1+\tau^2)+\sqrt{\gamma^2(1-\tau^2)+4\tau^2}],\eta_2=\frac{\tau^2(\gamma^2-1)}{\eta_1}. \tag{3.16}$$

In the static case ($\gamma = 0$), they reduce to

$$\eta_1 = \tau , \tag{3.17a}$$

$$\eta_2 = -\tau . \tag{3.17b}$$

Note that in the case of an incompressible medium ($\tau = 0$), this approach does not allow construction of linearly independent static solutions. In this case, one should employ very small values of τ rather than zero, or use equations which follow from (3.6), (3.8)–(3.11) at $\tau = 0$, $\gamma = 0$ (see [78]).

For the found values of η, coefficient A_χ can be determined through coefficient A_e from the second equation of (3.14):

$$A_\chi = (\gamma\tau^2 - \eta)A_e . \tag{3.18}$$

Coefficient A_e may be taken to be arbitrary, e.g. equal to unity. In order to determine functions $\widetilde{q}$ and $\widetilde{w}$, the found functions $\widetilde{\chi}$ and $\widetilde{e}$ from (3.12) or (3.13) are substituted into equations (3.10) and (3.11), and the method of variation of parameters employed:

$$\begin{pmatrix} \widetilde{q}(\varsigma) \\ \widetilde{w}(\varsigma) \end{pmatrix} = E_1(\varsigma)\begin{pmatrix} 1 \\ -1 \end{pmatrix}\exp(\varsigma/2) + E_2(\varsigma)\begin{pmatrix} 1 \\ 1 \end{pmatrix}\exp(-\varsigma/2) . \tag{3.19}$$

Here, we employ fundamental solutions of the system of differential equations corresponding to equations (3.10), (3.11), where terms containing $\widetilde{\chi}$ and $\widetilde{e}$ are dropped. Functions E_1 and E_2 are to be determined. Substitution of representation (3.19) into equations (3.10), (3.11) with functions $\widetilde{\chi}$ and $\widetilde{e}$ from (3.12) gives

$$\begin{aligned} \frac{\mathrm{d}E_1}{\mathrm{d}\varsigma} &= \frac{A_\chi - \tau^2 A_e}{2}\exp(-\varsigma)U(1-\eta,2,\varsigma) , \\ \frac{\mathrm{d}E_2}{\mathrm{d}\varsigma} &= \frac{A_\chi + \tau^2 A_e}{2}U(1-\eta,2,\varsigma) , \end{aligned} \tag{3.20}$$

Employing relationships between the confluent hypergeometric functions [1] and equation (3.18) yields

$$E_1 = -\frac{A_e}{2}(\gamma\tau^2 - \eta - \tau^2)\exp(-\varsigma)U(1-\eta,1,\varsigma) , \tag{3.21a}$$

$$E_2 = \frac{A_e}{2\eta}(\gamma\tau^2 - \eta + \tau^2)U(-\eta,1,\varsigma) . \tag{3.21b}$$

According to equation (3.19),

$$\tilde{q}(\varsigma) = \frac{A_e}{2}\exp(-\varsigma/2)\left[-(\gamma\tau^2 - \eta - \tau^2)U(1-\eta,1,\varsigma) + \frac{\gamma\tau^2 - \eta + \tau^2}{\eta}U(-\eta,1,\varsigma)\right]$$

$$= A_e \exp(-\varsigma/2)\left[(\eta - \gamma\tau^2)U(1-\eta,1,\varsigma) + \frac{\gamma\tau^2 - \eta + \tau^2}{2\eta}\varsigma U(1-\eta,2,\varsigma)\right],$$

(3.22a)

$$\tilde{w}(\varsigma) = \frac{A_e}{2}\exp(-\varsigma/2)\left[(\gamma\tau^2 - \eta - \tau^2)U(1-\eta,1,\varsigma) + \frac{\gamma\tau^2 - \eta + \tau^2}{\eta}U(-\eta,1,\varsigma)\right]$$

$$= A_e \exp(-\varsigma/2)\left[-\tau^2 U(1-\eta,1,\varsigma) + \frac{\gamma\tau^2 - \eta + \tau^2}{2\eta}\varsigma U(1-\eta,2,\varsigma)\right]. \quad (3.22b)$$

These results correspond to the solutions decreasing at infinity (having indices 1 and 2). To obtain other solutions, one should substitute expressions (3.19) into equations (3.10), (3.11) with functions $\tilde{\chi}$ and $\tilde{e}$ from (3.13). This results in relationships similar to (3.20) but including confluent functions M instead of U. Expressions for coefficients E_j will have a slightly different structure:

$$E_1 = -\frac{A_e}{2\eta}(\gamma\tau^2 - \eta - \tau^2)\exp(-\varsigma)M(1-\eta,1,\varsigma),$$

$$E_2 = -\frac{A_e}{2\eta}(\gamma\tau^2 - \eta + \tau^2)M(-\eta,1,\varsigma). \quad (3.23)$$

As follows from equation (3.19),

$$\tilde{q}(\varsigma) = \frac{A_e}{2\eta}\exp(-\varsigma/2)\left[-(\gamma\tau^2 - \eta - \tau^2)M(1-\eta,1,\varsigma) - (\gamma\tau^2 - \eta + \tau^2)M(-\eta,1,\varsigma)\right]$$

$$= \frac{A_e}{2\eta}\exp(-\varsigma/2)\left[2(\eta - \gamma\tau^2)M(1-\eta,1,\varsigma) + (\gamma\tau^2 - \eta + \tau^2)\varsigma M(1-\eta,2,\varsigma)\right],$$

(3.24a)

$$\tilde{w}(\varsigma) = \frac{A_e}{2\eta}\exp(-\varsigma/2)\left[(\gamma\tau^2 - \eta - \tau^2)M(1-\eta,1,\varsigma) - (\gamma\tau^2 - \eta + \tau^2)M(-\eta,1,\varsigma)\right]$$

$$= \frac{A_e}{2\eta}\exp(-\varsigma/2)\left[-2\tau^2 M(1-\eta,1,\varsigma) + (\gamma\tau^2 - \eta + \tau^2)\varsigma M(1-\eta,2,\varsigma)\right].$$

(3.24b)

By using $\eta = \eta_j\ (j = 1, 2)$, $A_e = 1$ and relationship (3.18), one can construct fundamental solutions having indices 1 and 2 from relationships (3.12), (3.22), while fundamental solutions having indices 3 and 4 are constructed from relationships (3.13), (3.24). Note that the confluent hypergeometric function $M(a,b,\varsigma)$ contains multiplier $\exp(\varsigma)$ in its asymptotic representation at high values of argument ς, so that solutions having indices 3 and 4 contain multiplier $\exp(\varsigma/2)$ in their asymptotic representation (see expressions (3.13) and (3.24)).

Consider fundamental solutions related to the function $\tilde{p}$ determined by equation (1.29). Taking into account the properties of isotropy, relationships $\tilde{\rho}=1$, $\tilde{G}=\tilde{z}$, and using the variable (3.5), we rewrite this equation as

$$\frac{d^2\tilde{p}}{d\varsigma^2}+\frac{1}{\varsigma}\frac{d\tilde{p}}{d\varsigma}-\left(\frac{1}{4}-\frac{\gamma}{\varsigma}\right)\tilde{p}=0\,. \tag{3.25}$$

This equation reduces to Whittaker's equation by means of substitution $\tilde{p}=\varsigma^{-1/2}f$:

$$\frac{d^2 f}{d\varsigma^2}+\left(-\frac{1}{4}+\frac{\gamma}{\varsigma}+\frac{1}{4\varsigma^2}\right)f=0\,. \tag{3.26}$$

Thus, solutions $\tilde{p}_1(\varsigma)$ (decreasing at infinity) and $\tilde{p}_2(\varsigma)$ (increasing) of equation (3.25) take the form

$$\tilde{p}_1(\varsigma)=\varsigma^{-1/2}W_{\gamma,0}(\varsigma)=\exp(-\varsigma/2)U(0.5-\gamma,1,\varsigma)\,, \tag{3.27a}$$

$$\tilde{p}_2(\varsigma)=\varsigma^{-1/2}M_{\gamma,0}(\varsigma)=\exp(-\varsigma/2)M(0.5-\gamma,1,\varsigma)\,. \tag{3.27b}$$

Expressions for the derivatives required to construct solutions of the considered problems may be written as follows [1]:

$$d_1(\varsigma)=\frac{d\tilde{p}_1}{d\tilde{z}}=2\tilde{k}\exp(-\varsigma/2)[0.5U(0.5-\gamma,1,\varsigma)-U(0.5-\gamma,2,\varsigma)]\,, \tag{3.28a}$$

$$d_2(\varsigma)=\frac{d\tilde{p}_2}{d\tilde{z}}=2\tilde{k}\exp(-\varsigma/2)[0.5M(0.5-\gamma,1,\varsigma)-(0.5+\gamma)M(0.5-\gamma,2,\varsigma)]\,. \tag{3.28b}$$

Taking into account the asymptotic behavior of function $M(a,b,\varsigma)$ with increasing argument ς, we note that the second solution contains multiplier $\exp(\varsigma/2)$ in its asymptotic representation for high values of ς.

In relationships (3.12), (3.13), (3.22), (3.24), (3.27), variable ς is given as a function argument on the left-hand side of the equations. However, one should keep in mind that an additional realization of the dependence of the functions on parameters $\tilde{k}$, τ, β, θ is due to parameter η, equal to η_1 or η_2.

3.2 Vibrations of Isotropic Linearly Heterogeneous Half-Space under Action of Vertical Force Applied to Half-Space Surface

Evidently, expressions (2.1) for the amplitudes of vibrations may be employed, into which one should substitute the fundamental solutions constructed in the previous section. Instead of G_{rz0} and ν', we shall use symbols G_0 and ν, respectively. Values of c_{1j} and c_{2j} are taken from (1.147), accounting for the fact that the half-space surface, $z = z_0$, corresponds to the value of variable $\varsigma = 2\tilde{k}$ (following (3.5); recall that $z_r = z_0$). In addition, we account for the presence of multiplier $\exp(-\varsigma/2)$ in the constructed fundamental solutions. Using expressions (1.147) and (1.56) for parameters c_{ij} and D, we take the corresponding exponential multipliers out of the brackets. Then, the expressions for the values of S_{vh} and S_{vv} become

$$S_{vh} = \frac{\beta^2\tilde{r}}{1-\nu}\int_0^\infty \exp[-\tilde{k}(\tilde{z}-1)]\frac{\overset{*}{c}_{21}\overset{*}{\tilde{q}}_2(\varsigma) - \overset{*}{c}_{22}\overset{*}{\tilde{q}}_1(\varsigma)}{D^*}\tilde{k}\,\mathrm{J}_1(\tilde{k}\tilde{r})\,\mathrm{d}\tilde{k}\,, \tag{3.29a}$$

$$S_{vv} = \frac{\beta^2\tilde{r}}{1-\nu}\int_0^\infty \exp[-\tilde{k}(\tilde{z}-1)]\frac{\overset{*}{c}_{21}\overset{*}{\tilde{w}}_2(\varsigma) - \overset{*}{c}_{22}\overset{*}{\tilde{w}}_1(\varsigma)}{D^*}\tilde{k}\,\mathrm{J}_0(\tilde{k}\tilde{r})\,\mathrm{d}\tilde{k}\,, \tag{3.29b}$$

$$\overset{*}{c}_{1j} = \overset{*}{\tilde{e}}_j(2\tilde{k}) - 2\tilde{k}\overset{*}{\tilde{q}}_j(2\tilde{k})\,, \tag{3.29c}$$

$$\overset{*}{c}_{2j} = \overset{*}{\tilde{\chi}}_j(2\tilde{k}) - 2\tilde{k}\overset{*}{\tilde{w}}_j(2\tilde{k})\,, \tag{3.29d}$$

$$D^* = \overset{*}{c}_{11}\overset{*}{c}_{22} - \overset{*}{c}_{11}\overset{*}{c}_{22}\,, \tag{3.29e}$$

where an asterisk means that in the expressions for the fundamental solutions determined by using equations (3.12), (3.18), (3.22) at $\eta = \eta_j\ (j = 1,2)$ and $A_e = 1$, multiplier $\exp(-\varsigma/2)$ is being dropped. When calculating the displacements of points located within the half-space ($\tilde{z} > 1$), integrals in expressions (3.29) converge relatively quickly due to their exponentially decreasing multipliers. For points on the half-space surface ($\tilde{z} = 1$), convergence is slow; it can be improved by employing the limiting values Φ_{vh}^∞ and Φ_{vv}^∞ as $\tilde{k} \to \infty$ of the multipliers of Bessel's functions in equations (3.29). Using asymptotic expansions for the confluent hypergeometric functions [1] gives

$$\Phi_{vh}^\infty = -\frac{1-2\nu}{2}\,, \tag{3.30a}$$

$$\Phi_{vv}^\infty = 1-\nu\,. \tag{3.30b}$$

Adding and subtracting the determined limiting values under the integral in relationships (3.29a) and (3.29b) yields ($\tilde{z}=1$)

$$S_{vh}=-\frac{1-2\nu}{2(1-\nu)}\beta^2+\frac{\beta^2\tilde{r}}{1-\nu}\int_0^\infty\left[\frac{\overset{*}{c}_{21}\overset{*}{\tilde{q}}_2(2\tilde{k})-\overset{*}{c}_{22}\overset{*}{\tilde{q}}_1(2\tilde{k})}{D^*}\tilde{k}-\Phi_{vh}^\infty\right]J_1(\tilde{k}\tilde{r})\,d\tilde{k}\,,$$

(3.31a)

$$S_{vv}=\beta^2+\frac{\beta^2\tilde{r}}{1-\nu}\int_0^\infty\left[\frac{\overset{*}{c}_{21}\overset{*}{\tilde{w}}_2(2\tilde{k})-\overset{*}{c}_{22}\overset{*}{\tilde{w}}_1(2\tilde{k})}{D^*}\tilde{k}-\Phi_{vv}^\infty\right]J_0(\tilde{k}\tilde{r})\,d\tilde{k}\,. \qquad (3.31b)$$

Various aspects concerning the calculation of integrals representing normalized vibration amplitudes S_{vh} and S_{vv} were discussed in section 2.1.3. In contrast to the case of the homogeneous half-space considered in Chap. 2, heterogeneity results in the appearance of multiple poles of the integrands in expressions (3.29), (3.31). The poles are located on the real axis of the complex plane $\tilde{k}$ in the case $\varepsilon=0$ ($\beta=1$), and, as follows from calculations, these poles are shifted downward when $\varepsilon>0$. Therefore, the contour of integration shown in Fig. 2.3 may be employed in this case, too. When seeking a solution of dynamic problems, we substitute the integration variable

$$\tilde{k}=s\theta \qquad (3.32)$$

and replace the variable $\tilde{r}$ with the following one:

$$a=\theta\tilde{r}=\omega r\sqrt{\frac{\rho}{G_0}}\,. \qquad (3.33)$$

In order to visualize the behavior of integrands in these integrals, graphs of the real and imaginary parts of the integrand $\Phi(s)$ in expression (3.29b) (excluding the multiplier expressed with Bessel's function) are shown in Fig. 3.1 at $\tilde{z}=1$, $s=\mathrm{Re}(s)+\mathrm{i}h$ (i.e. on the part of the contour similar to that shown in Fig. 2.3, which is parallel to the real axis). Here, $\varepsilon=0, \nu=1/3, \theta=10, h=0.01$. Note that in order to calculate the amplitudes of forced vibrations, we take for the variable s the value $h=0.05$. Points where drastic changes in the real part of $\Phi(s)$ occur (and points of the minimum of $\mathrm{Im}[\Phi(s)]$) correspond to the poles on the real axis s. With increasing parameter θ, the number of poles increases, while the pole having the highest abscissa (the principal pole) shifts to the right approaching the point $s=s_R$, which corresponds to Rayleigh waves for the homogeneous half-space ($s_R=1.0724$ at $\nu=1/3$). Under these conditions, a contribution of the rest of the poles to the value of the corresponding integrals becomes small, compared to the contribution of the principal pole. Next, we present the results of calculations of the normalized amplitudes of vibrations, S_{vh} and S_{vv}, for the half-space surface. Magnitudes $|S_{vh}|$ and phase angles δ_{vh} of the normalized

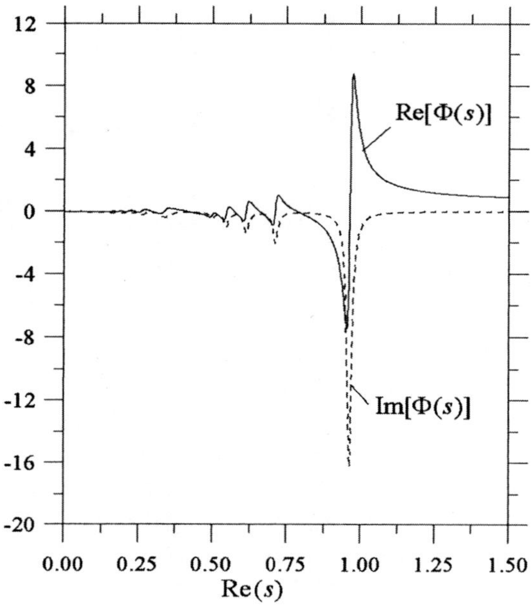

Fig. 3.1. Behavior of integrand on part of integration contour (Fig. 2.3) parallel to real axis ($s = \mathrm{Re}(s) + \mathrm{i}h$) at $\varepsilon = 0, \nu = 1/3, \theta = 10, h = 0.01$

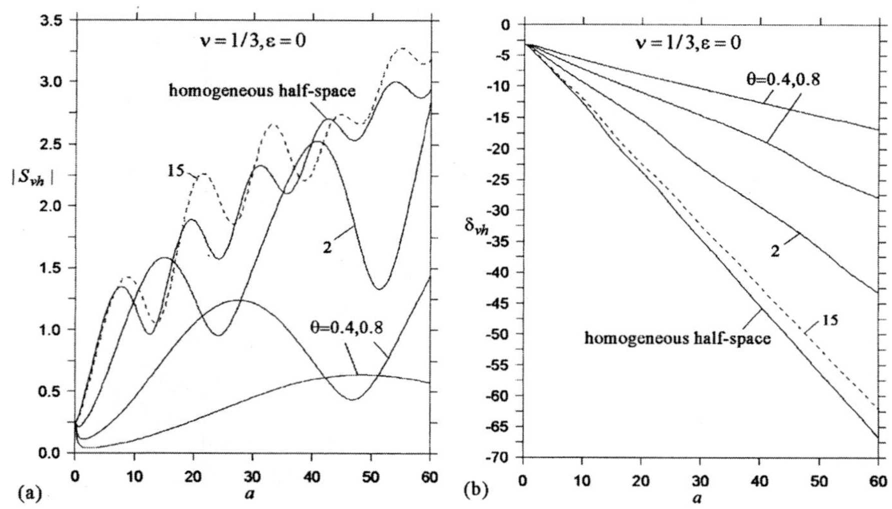

Fig. 3.2a,b. Absolute values (a) and phases (b) of normalized horizontal amplitudes of vibrations of linearly heterogeneous half-space subjected to vertical force

amplitudes S_{vh} are shown in Fig. 3.2, and $|S_{vv}|$ and δ_{vv} in Fig. 3.3. We use the following relationships:

$$S_{vh} = |S_{vh}| \exp(i\,\delta_{vh}), \tag{3.34}$$

$$S_{vv} = |S_{vv}| \exp(i\,\delta_{vv}). \tag{3.35}$$

Let parameter a approach zero when $r \neq 0$ and a nonzero value of parameter θ is fixed. This corresponds to the frequency of vibrations tending to zero and unbounded growth of variable z_0. Consequently, the corresponding solutions should tend to the static solution for the homogeneous half-space. Indeed, according to relationships (3.31), the initial values of S_{vv} and S_{vh} at $\varepsilon = 0$ are equal to the static values, namely 1 and $-0.5(1-2\nu)/(1-\nu)$, respectively. In Fig. 3.2a, all curves pass through the point (0, 0.25), and in Fig. 3.3a through the point (0, 1). Increasing of the normalized amplitudes of vibrations with increasing parameter a may be explained by including the influence of the poles of integrands in relationships (3.29). The poles are associated with a system of Rayleigh waves, among which the wave corresponding to the right side pole gives the most significant contribution to the final result (Fig. 3.1). The graphs for the homogeneous half-space and for the value $\theta = 15$ are similar. Note that an abrupt change of phase δ_{vv} occurs at $a \approx 40$ and $\theta = 0.4$. This fact contradicts a commonly accepted point of view concerning waves which move away with a definite velocity from the source of disturbance. Similar irregularities in phase

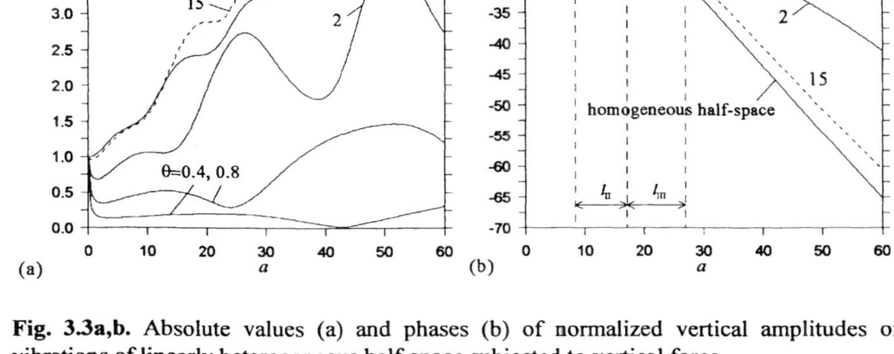

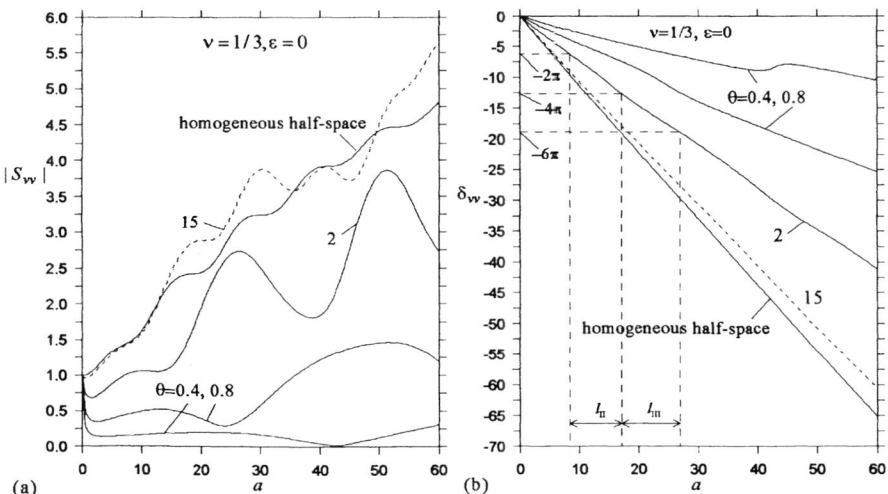

Fig. 3.3a,b. Absolute values (a) and phases (b) of normalized vertical amplitudes of vibrations of linearly heterogeneous half-space subjected to vertical force

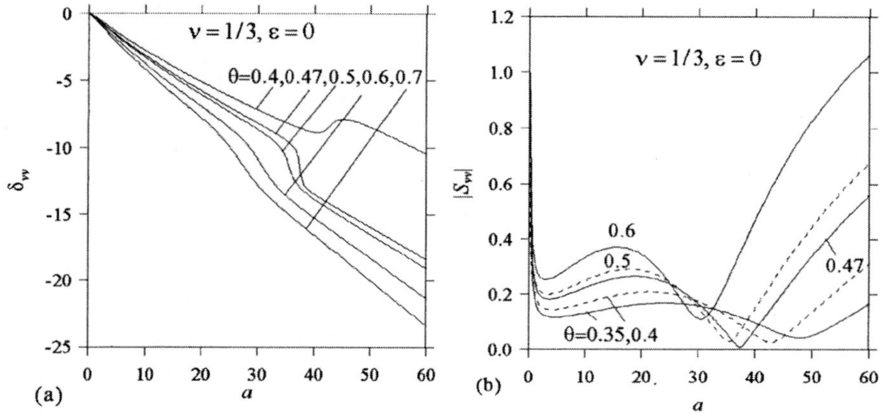

Fig. 3.4a,b. Phases (a) and absolute values (b) of normalized vertical amplitudes of vibrations for the values of parameter θ leading to abrupt change in vibrations amplitudes

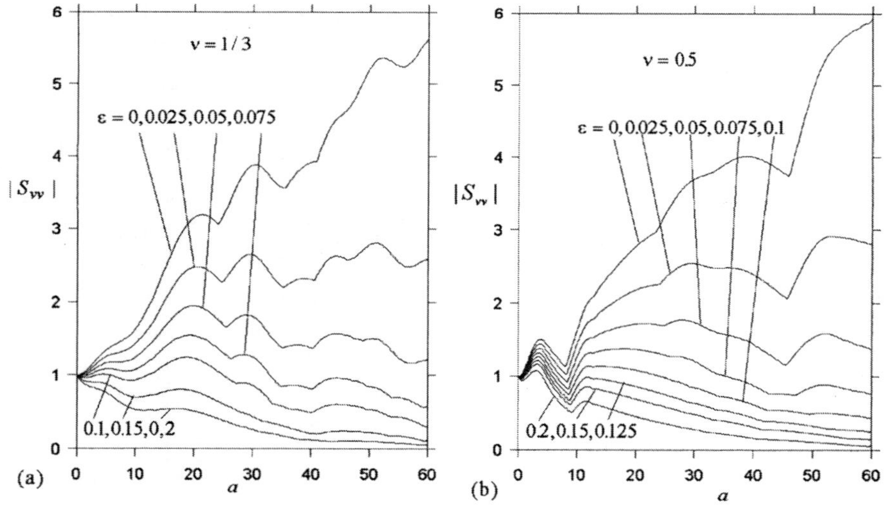

Fig. 3.5a,b. Envelope curves for absolute values of normalized vertical amplitudes of vibrations under action of vertical force at $\nu = 1/3$ (a) and $\nu = 0.5$ (b)

behavior, when the corresponding graphs differ significantly from straight lines, take place when the value of parameter θ provides amplitudes of vibrations which can approach zero in a narrow range of varying parameter a. This phenomenon is shown in Fig. 3.4, where the graphs are presented for values of parameter θ close to 0.47. The interaction of waves propagating along the surface of the half-space and in its zones located at depth (where the velocity of wave propagation is higher than in the vicinity of the surface) results in different values of the phase velocity on different parts of the straight line originating at the source of the disturbance. For values of parameter $\theta > 0.7$–0.8, the phase curves do not differ appreciably from the straight lines.

For the considered half-space, when parameters ε and ν are being fixed, the normalized amplitudes for points that belong to the half-space surface are determined by two parameters, θ and a, only (see (3.1) and (3.33)). By varying the vibration frequency, ω, and the distance, r, one can obtain any value of these parameters. It is useful to have envelope curves for the graphs $|S_{vv}(a)|$, corresponding to a large number of values of parameter θ. Such curves enable us to estimate the largest possible amplitudes of vibrations. In Fig. 3.5, envelope curves are plotted for two values of Poisson's ratio and for a series of values of the dissipative parameter ε.

3.3 Vibrations of Isotropic Linearly Heterogeneous Half-Space under Action of Horizontal Force Applied to Half-Space Surface

For the case of a horizontal concentrated force, we represent amplitudes of vibrations in the form (2.53)–(2.55), similar to that used in the study of vibrations of a homogeneous transversely isotropic half-space, where the shear modulus G_{rz0} is replaced with G_0, and ν' with ν. As in the derivation of formulas (3.29), we take into account the presence of the multiplier $\exp(-\varsigma/2)$ in the fundamental solutions constructed in section 3.1, and employ symbols introduced in formulas (3.29). Parts of the integrands, containing function $\tilde{p}_1$ and its derivative with respect to $\tilde{z}$ at $\tilde{z}=1$ (the value of F in H_ϑ and H_r is expressed through this derivative), are modified by taking the exponential multiplier out of the brackets. In so doing, relationships (3.27a), (3.28a) are employed.

Expressions for S_{hr}, $S_{h\vartheta}$ and S_{hz} may be written in the following form:

$$S_{hr} = \beta^2 \tilde{r} \int_0^\infty \tilde{k}\, e^{-\tilde{k}(\tilde{z}-1)} \{[\tilde{q}^*(\tilde{z},\tilde{k}) - \tilde{p}^*(\tilde{z},\tilde{k})] \frac{J_1(\tilde{k}\tilde{r})}{\tilde{k}\tilde{r}} - \tilde{q}^*(\tilde{z},\tilde{k}) J_0(\tilde{k}\tilde{r})\}\, d\tilde{k}\,, \tag{3.36a}$$

$$S_{h\vartheta} = -\frac{\beta^2 \tilde{r}}{1-\nu} \int_0^\infty \tilde{k}\, e^{-\tilde{k}(\tilde{z}-1)} \{[\tilde{q}^*(\tilde{z},\tilde{k}) - \tilde{p}^*(\tilde{z},\tilde{k})] \frac{J_1(\tilde{k}\tilde{r})}{\tilde{k}\tilde{r}} + \tilde{p}^*(\tilde{z},\tilde{k}) J_0(\tilde{k}\tilde{r})\}\, d\tilde{k}, \tag{3.36b}$$

$$S_{hz} = -\frac{\beta^2 \tilde{r}}{1-\nu} \int_0^{\infty} e^{-\tilde{k}(\tilde{z}-1)} \frac{\overset{*}{c}_{11}\tilde{w}_2^*(\tilde{z},\tilde{k}) - \overset{*}{c}_{12}\tilde{w}_1^*(\tilde{z},\tilde{k})}{D^*} \tilde{k}\, \mathrm{J}_1(\tilde{k}\tilde{r})\,\mathrm{d}\tilde{k}\,, \tag{3.36c}$$

where

$$\tilde{q}^*(\tilde{z},\tilde{k}) = \frac{\overset{*}{c}_{11}\tilde{q}_2^*(\varsigma) - \overset{*}{c}_{12}\tilde{q}_1^*(\varsigma)}{D^*}, \tag{3.37a}$$

$$\tilde{p}^*(\tilde{z},\tilde{k}) = \frac{U(0.5-\gamma,1,\varsigma)}{2\tilde{k}[0.5U(0.5-\gamma,1,2\tilde{k}) - U(0.5-\gamma,2,2\tilde{k})]}\,. \tag{3.37b}$$

While studying the vibrations of points that belong to the surface of the half-space ($\tilde{z}=1$), we shall consider only the values S_{hr} and $S_{h\vartheta}$. The value of S_{hz} is identical to S_{vh} due to the principle of reciprocity. In the case of $\tilde{z}=1$, it is reasonable to improve convergence of the integrals by using representations $\tilde{q}_l^*$ and $\tilde{p}_l^*$ as $\tilde{k} \to \infty$ for values $\tilde{q}^*$ and $\tilde{p}^*$, respectively:

$$\tilde{q}_l^* = -\frac{(1-\nu)}{\tilde{k}}\,, \tag{3.38a}$$

$$\tilde{p}_l^* = -\frac{1}{\tilde{k}}\,. \tag{3.38b}$$

These limits may be determined by applying known asymptotic expansions for confluent hypergeometric functions [1]. Adding and subtracting $\tilde{q}_l^*$ and $\tilde{p}_l^*$ under the integral in the expressions for S_{hr} and $S_{h\vartheta}$ gives ($\tilde{z}=1$, $\varsigma = 2\tilde{k}$):

$$S_{hr} = \beta^2 + \beta^2\tilde{r}\int_0^{\infty} \tilde{k}\{[\tilde{q}^* - \tilde{q}_l^* - \tilde{p}^* + \tilde{p}_l^*]\frac{\mathrm{J}_1(\tilde{k}\tilde{r})}{\tilde{k}\tilde{r}} - (\tilde{q}^* - \tilde{q}_l^*)\mathrm{J}_0(\tilde{k}\tilde{r})\}\,\mathrm{d}\tilde{k}\,, \tag{3.39a}$$

$$S_{h\vartheta} = \beta^2 - \frac{\beta^2\tilde{r}}{1-\nu}\int_0^{\infty} \tilde{k}\{[\tilde{q}^* - \tilde{q}_l^* - \tilde{p}^* + \tilde{p}_l^*]\frac{\mathrm{J}_1(\tilde{k}\tilde{r})}{\tilde{k}\tilde{r}} + (\tilde{p}^* - \tilde{p}_l^*)\mathrm{J}_0(\tilde{k}\tilde{r})\}\,\mathrm{d}\tilde{k}\,. \tag{3.39b}$$

We obtain the value β^2 for the normalized amplitudes as $\tilde{r} \to 0$. When variable $\tilde{r}$ tends to zero at fixed r and θ, the value of z_0 increases without bound, while the frequency of vibrations tends to zero. Hence, variables S_{hr} and $S_{h\vartheta}$ must tend to the static values for a homogeneous half-space (equal to unity when neglecting internal friction within the material of the half-space). The latter fact is in agreement with the result obtained from formulas (3.39).

Next, we present the results of calculations of the amplitudes of vibrations for points on the surface of the half-space. Considerations related to the integration given in the previous section remain valid. In Figs. 3.6 and 3.7, the magnitudes of

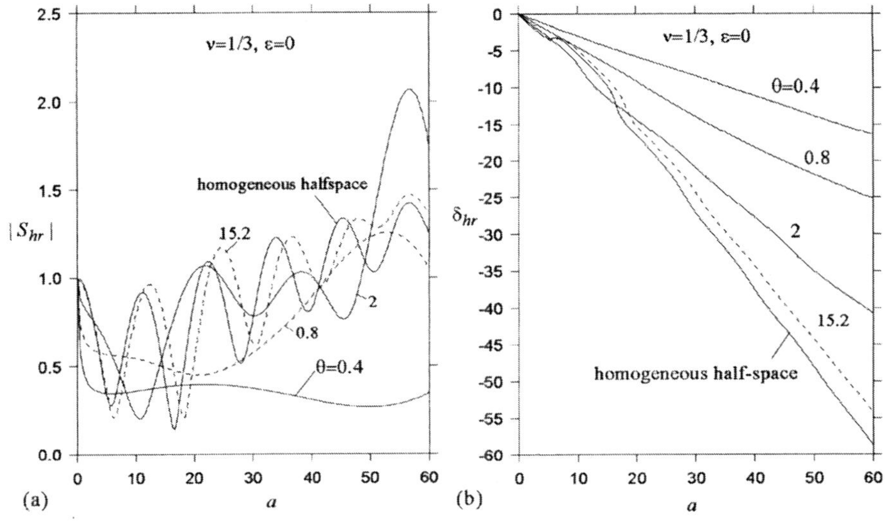

Fig. 3.6a,b. Absolute values (a) and phases (b) of normalized radial amplitudes of vibrations of linearly heterogeneous half-space subjected to horizontal force

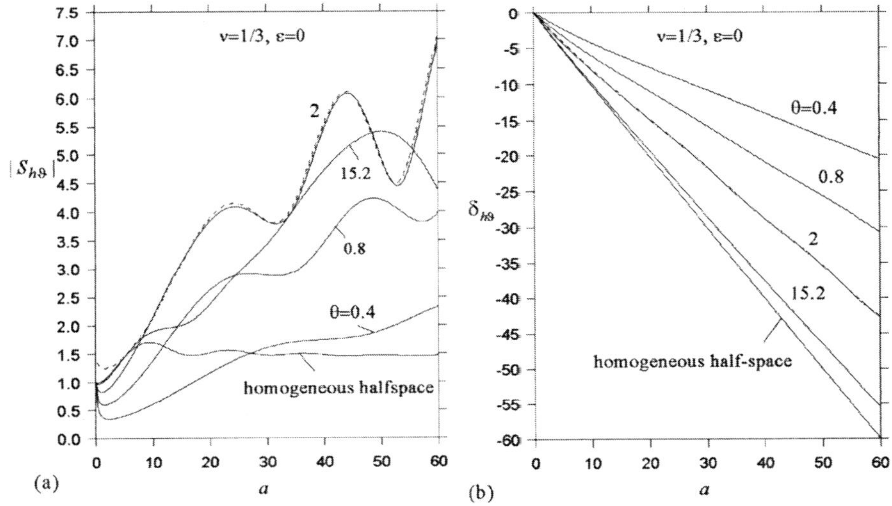

Fig. 3.7a,b. Absolute values (a) and phases (b) of normalized tangential amplitudes of vibrations of linearly heterogeneous half-space subjected to horizontal force

the normalized amplitudes, S_{hr} and $S_{h\vartheta}$, and their phases are presented as functions of parameter a, determined by (3.33). For the complex amplitudes, one should use relationships having the form of (3.34), (3.35). For the homogeneous half-space at high values of parameter θ, the magnitude $|S_{hr}|$ undergoes noticeable oscillations, resulting in a significant deviation of the corresponding phase curves from straight lines (at small values of parameter a). The growth of the normalized amplitudes $|S_{hr}|$ with increasing a is much slower than that of the amplitudes $|S_{h\vartheta}|$.

As follows from analysis of expressions (3.36b) or (3.39b) for $S_{h\vartheta}$, at high values of parameter a, the part of the integrand containing Bessel's function J_0, i.e. the term depending on function $\widetilde{p}^*$ only, predominates. The contribution of this part to the value of $S_{h\vartheta}$ has been denoted in expression (2.76) as $S'_{h\vartheta}$. In the considered case,

$$S'_{h\vartheta} = -\frac{\beta^2 \widetilde{r}}{1-\nu} \int_0^\infty \widetilde{k}\, e^{-\widetilde{k}(\widetilde{z}-1)}\, \widetilde{p}^*(\varsigma) J_0(\widetilde{k}\widetilde{r})\, d\widetilde{k} \,. \tag{3.40a}$$

For the surface of the half-space, according to (3.38b),

$$S'_{h\vartheta} = \frac{\beta^2}{1-\nu} - \frac{\beta^2 \widetilde{r}}{1-\nu} \int_0^\infty \widetilde{k}\, [\widetilde{p}^*(2\widetilde{k}) - \widetilde{p}_l^*] J_0(\widetilde{k}\widetilde{r})\, d\widetilde{k} \,. \tag{3.40b}$$

In order to demonstrate the efficiency of such an approximation, the corresponding graphs are plotted for $\theta = 2$ (the dashed curves in Fig. 3.7, where the initial point ordinate is equal to 1.5). According to expressions (2.77), in the case of a homogeneous half-space, the normalized amplitude of vibrations must tend to 1.5 (at $\nu = 1/3$) with increasing parameter a. This fact is verified by the corresponding graph in Fig. 3.7a. Evidently, the considered approximation provides rather high accuracy, starting with values of a of about 7. Note that the complete amplitude of vibrations, equal to the product of the normalized amplitude and static displacement (see relationship (2.54a)), is independent of Poisson's ratio for the approximate solution $S'_{h\vartheta}$. Hence, the complete amplitude of vibrations, $\hat{u}_\vartheta$, by the exact solution, is practically independent of Poisson's ratio at sufficiently high values of parameter a. Concerning the structure of the approximate solution, it is similar to the solution corresponding to the anti-plane vibrations which are determined by expression (1.95). The plane problem for the case of a medium consisting of a half-space and a layer having different shear modulus has characteristic free vibrations, first reported by Love (Love waves) [29]. For the considered heterogeneous half-space, similar free anti-plane vibrations may occur due to the poles in the integrand in (1.95), located on the real axis of the complex plane k (at $\varepsilon = 0$). For the approximate representation $S'_{h\vartheta}$ introduced for the normalized amplitudes, these poles are responsible for the growth of amplitudes with increasing parameter a, as seen in Fig. 3.7a; naturally, such a phenomenon may be treated as an influence of Love waves. Note that, unlike the amplitudes S_{hr}, the results corresponding to a homogeneous half-space and the heterogeneous half-space at $\theta = 15.2$ differ appreciably at $a > 10$.

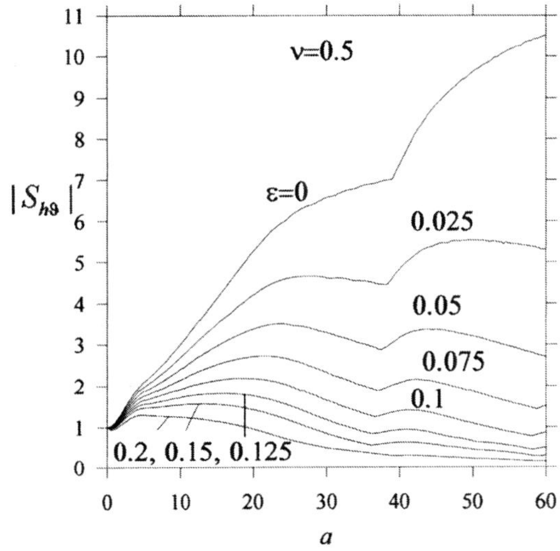

Fig. 3.8. Envelope curves for absolute values of normalized tangential amplitudes of vibrations under action of horizontal force

Following relationship (2.76), no poles exist for the considered approximation in the case of a homogeneous half-space, and as a result the normalized amplitudes do not increase with increase of parameter *a*.

The envelope curves for the magnitudes $|S_{h\vartheta}|$, corresponding to $\nu = 0.5$ and a number of values of ε, are shown in Fig. 3.8. Note that, starting with $a > 7$, application of the approximate representation (3.40a) is sufficiently accurate. This representation depends on Poisson's ratio because of the multiplier $1/(1-\nu)$ only, it reaches its peak value at $\nu = 0.5$.

The influence of Love waves on the amplitudes $\hat{u}_\vartheta$ may be appreciably more significant than the influence of Rayleigh waves on the amplitudes u_z (see relationship (2.1c)). Since the corresponding static displacements are identical (at $Q_0 = P_0$), it is sufficient to compare the normalized amplitudes S_{vv} and $S_{h\vartheta}$ only. Comparing the envelope curves in Figs. 3.5 and 3.8 shows that at $\nu = 0.5$ amplitudes S_{vv} are approximately half as great as $S_{h\vartheta}$.

3.4 Determining Properties of Linearly Heterogeneous Half-Space Using Characteristics of Surface Waves

The well-known fact that the heterogeneity of a half-space leads to the dependence of the propagation velocity of the surface waves on the vibration frequency has been extensively used determine the form and extent of heterogeneity from the dispersion properties of the surface waves [92, 96, 119]. Actually, the known technique of spectral analysis for surface waves (SASW) [84, 97, 105] also consists in employing the relationship between the wave propagation velocity and the vibrations frequency. This section presents a technique for determining the characteristics of the linearly heterogeneous half-space by using the properties of the surface waves (predominantly, the relationship between the wavelength and the frequency of vibrations).

3.4.1 Application of Solution Related to Vertical Vibrations of Half-Space Surface under Action of Vertical Force

Next, we consider how to use the characteristics of phase curves having the form shown in Fig. 3.3b, to determine the half-space properties, assuming that the shear modulus varies linearly with depth. Let a secant line for each curve δ_{vv} pass through the points corresponding to the values $\delta_{vv} = -2\pi$ and $\delta_{vv} = -4\pi$. On the surface of the half-space, the points corresponding to given phase values vibrate in phase with the source of vibrations. These points bound a zone, which may be called a "second wave" (according to the range of phase variation considered). The location of these points, and hence a length of the "second wave", may be found experimentally, when studying vibrations of the half-space surface generated by a small source vibrating in the vertical direction [96]. Denote as k_v the angle coefficient (reversed in sign) of the introduced secant line, so that the line equation in the considered range is taken in the form

$$\delta_{vv} = c_v - k_v a \,, \tag{3.41}$$

where c_v is a constant which has no influence on the wavelength. Parameter k_v may be called a dimensionless wave number (related to parameter a). A relationship between the value of k_v and the dimensional wave number K_v, which is related to distances r, follows from

$$k_v a = k_v \omega r (\rho / G_0)^{1/2} \,, \tag{3.42a}$$

$$K_v = k_v \omega (\rho / G_0)^{1/2} \,. \tag{3.42b}$$

Having performed calculations for numerous values of parameter θ, one can plot a graph describing the relationship between k_v and θ. Parallel with the "second wave" we shall consider the "third wave" which corresponds to variation of the phase δ_{vv} in the limits between -4π and -6π (further we drop the quotation marks). In Fig. 3.3b, dimensionless wavelengths (for the argument a) are presented for the second $l_a = l_{\text{II}}$ and the third $l_a = l_{\text{III}}$ waves at $\theta = 2$. The value of k_v equals the ratio of 2π to the dimensionless wavelength. In Fig. 3.9, parameter k_v is plotted for two values of Poisson's ratio, while the variable λ has the form

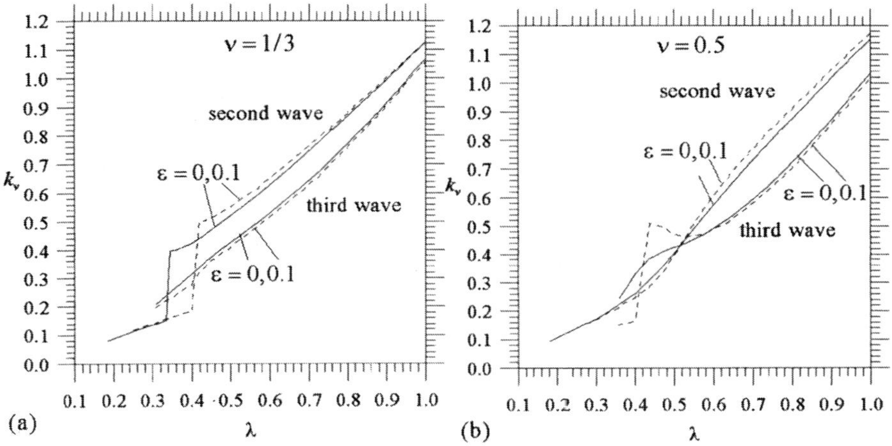

Fig. 3.9a,b. Dimensionless wave numbers k_v versus parameter $\lambda = \theta/(0.9+\theta)$ for $\nu = 1/3$ (a) and $\nu = 0.5$ (b)

$$\lambda = \frac{\theta}{\theta + 0.9} \tag{3.43}$$

transforming a set of various values of parameter θ into the interval [0,1]. The value $\lambda = 1$ ($z_0 \to \infty$, $\theta \to \infty$) corresponds to the homogeneous half-space. Discontinuities correspond to the values of θ for which the phase may change abruptly in the considered range of variations of *a*, analogous to the case of $\theta \approx 0.47$ shown in Fig. 3.4a. In order to determine the applicability of used relationships, one should carry out experiments at sufficiently high vibration frequencies, leading to values of parameter $\theta > 0.8$–1.

We see that the results obtained depend on the choice of which wave is considered, the second or the third one. This is due to deviation of the phase curves from straight lines. In addition, quantities k_v depend on Poisson's ratio, and, to a small extent, on the damping coefficient ε. Note that as $\lambda \to 1$ ($z_0 \to \infty$, $\theta \to \infty$), the value of k_v should approach the dimensionless wave number of Rayleigh waves in the corresponding homogeneous half-space. This wave number coincides with the ratio C_S / C_R of the velocity of shear waves to that of Rayleigh waves. This ratio equals 1.0468 for the material without internal damping for the incompressible half-space, and increases to 1.144 when Poisson's ratio vanishes. These data are in good agreement with the results obtained for the third wave corresponding to the homogeneous half-space; however, the results obtained for the second wave deviate from the values of C_S / C_R to a greater extent, since, in this case, the values of parameter *a* are still not sufficiently large to enable a good approximation for the complete solution by employing the part of the solution which corresponds to Rayleigh waves.

Consider the phase velocity C for the given vertical vibrations. Using the relationship (3.42b) for the wave numbers K_v and k_v, we have

$$C = \frac{\omega}{K_v} = \frac{\omega}{k_v \omega \sqrt{\rho / G_0}} = \frac{1}{k_v}\sqrt{\frac{G_0}{\rho}} \ . \tag{3.44}$$

Next, we consider a shear modulus for the homogeneous half-space G_h corresponding to the calculated velocities C of the surface waves. Employing (3.44), we obtain

$$G_h = k_{vh}^2 C^2 \rho = \frac{k_{vh}^2}{k_v^2} G_0 , \tag{3.45}$$

where k_{vh} is a dimensionless wave number for the homogeneous half-space; the value of k_{vh} is taken for the wave, having the same number as that for k_v. Note that at $\nu = 1/3$, $\varepsilon = 0$ we have k_{vh} = 1.1238 and k_{vh} = 1.066 respectively for the second and third waves (the value of k_{vh} for pure Rayleigh waves in a homogeneous half-space equals 1.0724). According to relationship (3.45), the ratio G_h / G_0 (or k_{vh}^2 / k_v^2) is a function of parameter θ, which may be related to the dimensionless wavelength $\tilde{l}$:

$$\tilde{l} = \frac{l}{z_0} = \frac{2\pi}{z_0 K_v} = \frac{2\pi}{z_0 k_v(\theta) \omega \sqrt{\rho / G_0}} = \frac{2\pi}{k_v(\theta)\theta} \ . \tag{3.46}$$

Thus, with a known relationship between k_v and parameter θ, one can express the ratio G_h / G_0 in terms of quantity $\tilde{l}$. In Fig. 3.10, this relationship is shown for the second and third waves. The curves obtained enable a good approximation by using the function

$$\frac{G_h}{G_0} = (1 + \gamma \tilde{l})^{\zeta} , \tag{3.47}$$

where $\gamma = 0.46$, $\zeta = 0.78$ for the second wave and $\gamma = 0.51$, $\zeta = 0.82$ for the third wave. Function (3.47) is represented by the dashed curves in Fig. 3.10. The curves in Fig. 3.10 correspond to the range of varying parameter θ: $0.8 < \theta < \infty$, where no discontinuities of quantity k_v exist.

The relationship connecting quantities G_h / G_0 and $\tilde{l}$ may be employed to determine the equivalent depth H_e, i.e. the depth of the given heterogeneous half-space, to which the shear modulus G_h should be referred. Equating quantities G_h / G_0 (as the function of $\tilde{l}$) to shear modulus at depth H_e divided by G_0, we obtain (see equation (3.2))

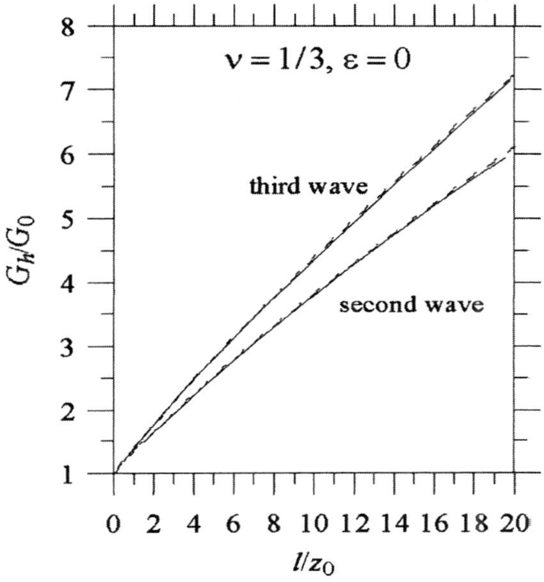

Fig. 3.10. Normalized shear modulus of homogeneous half-space, corresponding to velocities of surface waves under action of vertical force on heterogeneous half-space

$$\frac{G_h}{G_0} = 1 + \widetilde{H}_e, \tag{3.48}$$

where $\widetilde{H}_e = H_e / z_0$. Finally,

$$\frac{H_e}{l} = \left(\frac{G_h}{G_0} - 1 \right) \Big/ \widetilde{l} \ . \tag{3.49}$$

According to equation (3.49), the ratios H_e / l are determined as slopes of the secant lines, drawn for the curves presented in Fig. 3.10. In Fig. 3.11, the quantity H_e / l is shown as a function of the dimensionless wavelength $\widetilde{l}$ at $\nu = 1/3$ and $\varepsilon = 0$; the dashed lines correspond to the approximate relationship (3.47). A noticeable difference is observed between the results obtained for the second and third waves. The ratio H_e / l decreases with increasing wavelength, a fact that should be accounted for when interpreting the experimental results.

Now, we consider the determination of properties of the soil foundation by using experimentally obtained wavelengths l (corresponding to various frequencies) for the vertical amplitudes at the points on the foundation surface subjected to a vertical harmonic force. Suppose that the quantities C_1, l_1 and

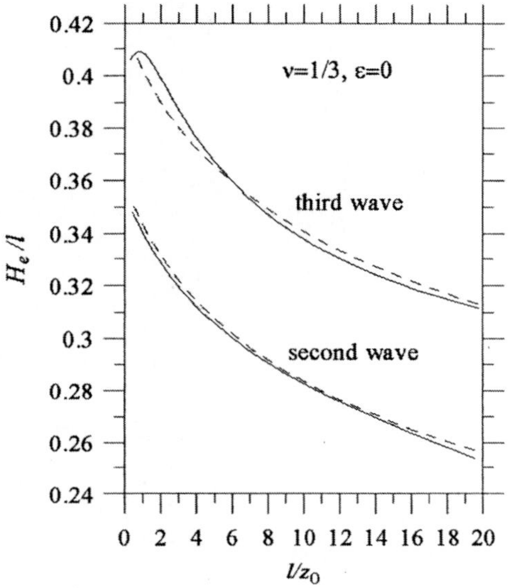

Fig. 3.11. Relationship between equivalent depth and surface wavelength under action of vertical force

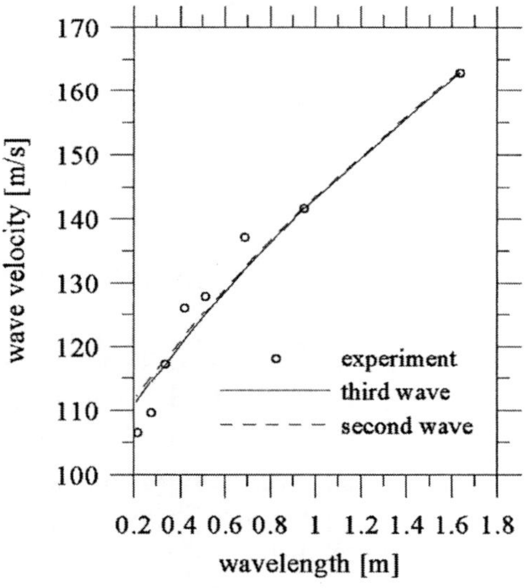

Fig. 3.12. Experimental data from [96] and theoretical relationships between velocities of surface waves and wavelength

C_2, l_2 are known for two values of vibration frequencies, f_1 and f_2. From equations (3.45) and (3.47)

$$\frac{C_2^2}{C_1^2} = \frac{(1+\gamma l_2 / z_0)^{\zeta}}{(1+\gamma l_1 / z_0)^{\zeta}}, \tag{3.50a}$$

$$z_0 = \gamma \frac{C_2^{2/\zeta} l_1 - C_1^{2/\zeta} l_2}{C_1^{2/\zeta} - C_2^{2/\zeta}} . \tag{3.50b}$$

With z_0 and l_1 known, the value of G_h / G_0 corresponding to l_1 may be determined by employing Fig. 3.10 or relationship (3.47). Next, we calculate the quantity k_v from equation (3.45):

$$k_v^2 = \frac{k_{vh}^2}{G_h / G_0} . \tag{3.51}$$

Further, parameter G_0 is calculated from equation (3.44), assuming that density ρ is known.

As an example, we consider experimental data obtained for silty fine sand [96] (p.114). Here, for a series of values of vibration frequency, the corresponding mean values of wavelengths and of wave propagation velocities are given. Assume that these mean values are related to the second or to the third wave. Let $\nu = 1/3, \varepsilon = 0$. Next, we apply the following values: $f_1 =$ 100 Hz, l_1=1.631 m (5.35 ft), $C_1 =$ 163.1 m/s (535 ft/s) and f_2=150 Hz, $l_2 =$ 0.945 m (3.1 ft), C_1=141.7 m/s (465 ft/s) [96] in formula (3.50b). The results of calculations are as follows: $z_0 =$ 0.294 m for the second wave, and $z_0 =$ 0.376 m for the third wave. According to (3.47), at $l = l_1$ the value of G_h / G_0 equals 2.685 and 2.604 for the second and third waves, respectively. Assuming $\rho = 1800\,\text{kg/m}^3$, we employ equations (3.51), (3.44) with respect to the second and third waves, resulting in: $G_0 =$ 22.51 MN/m^2 and $G_0 =$ 20.89 MN/m^2. By employing the obtained parameters of the half-space, one can determine theoretical values of G_h / G_0, and then the wave velocities from equation (3.45). A comparison between the experimental and these theoretical results is given in Fig. 3.12, where circles represent the experimental data, while the dashed (solid) curve represents the theoretical results for the second (third) wave. Though the parameters z_0 and G_0 for the second and third waves differ appreciably (resulting in substantially different half-spaces), the wave velocities are in close proximity. The model of the linearly heterogeneous half-space proves to be suitable for approximating these experimental data.

Consider the next example dealing with experimental data presented in [119]. In this report, the authors give the following relationship between the velocities of

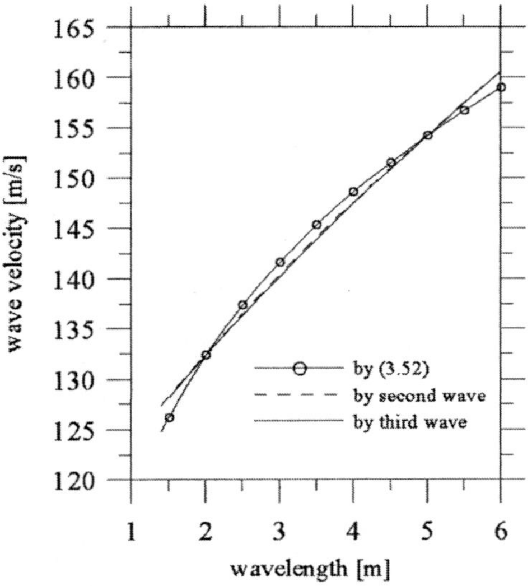

Fig. 3.13. Experimental from Ref. [119] and theoretical relationships between velocities of surface waves and wave length

the surface waves and the wavelengths, resulting from processing the experimental data,

$$C = 118 l^{1/6}, \tag{3.52}$$

where l is in meters and C in meters per second. Experiments were performed at lower frequencies than in the previous example; the range for wavelength was as follows: $1.4\text{ m} < l < 6.2\text{ m}$. Using relationship (3.52) with $l_1 = 5$ m and $l_2 = 2$ m, we obtain the corresponding velocities and then use the above procedure. We take $\rho = 1800\,\text{kg/m}^3$, $\nu = 1/3$, and we get $z_0 = 1.96\,\text{m}, G_0 = 29.53\ \text{MN/m}^2$ for the second wave, and $z_0 = 2.37\,\text{m}, G_0 = 26.76\,\text{MN/m}^2$ for the third wave. In Fig. 3.13, the relationship between velocities of the surface waves versus wavelengths is presented by using (3.52) (line with circles) as well as the above-described procedure for the second wave (dashed curve) and for the third wave (solid curve). Discrepancies between the results obtained from (3.52) and the theoretical ones based on the theory of the linearly heterogeneous half-space do not exceed 2%. However, the form of curves shown in Fig. 3.13 indicates that the actual growth of the half-space stiffness with depth is slower than predicted by the linear law.

3.4.2 Application of Solution Related to Horizontal Vibrations of Half-Space Surface under Action of Horizontal Force

Now, we consider the problem of employing the amplitudes of vibrations $\hat{u}_\vartheta$ (or $S_{h\vartheta}$) to determine the parameters of the linearly heterogeneous half-space. As shown in section 3.3, the properties of the corresponding surface waves are related to Love waves, being virtually independent of Poisson's ratio at $a > 7$. As previously, we shall use the terms the second, the third, etc., waves, writing k_θ for the angle coefficient (taken with the reverse sign), corresponding to the secant line. In Fig. 3.14, the behavior of k_θ is shown versus parameter λ calculated according to (3.43) for two values of damping coefficient ε. Poisson's ratio was taken to be equal to 0.5. However, the results obtained should be considered valid for arbitrary values of Poisson's ratio (intervals of variation of parameter a for the second etc. waves correspond to sufficiently high values of a, so that the phase curves may be assumed to be virtually independent of Poisson's ratio). The results of calculations indicate that for the second and the third waves, the influence of parameter ε on the quantities k_θ is negligible for $\varepsilon < 0.2$. In contrast to the case of vertical vibrations, considered above, the latter results depend on the wave number to a lesser extent. As before, the value of $\lambda = 1$ corresponds to the homogeneous half-space, with k_θ equal to unity for pure shear waves. For the second wave, calculations yield 0.9676 at $\varepsilon = 0$; the result tends to unity for the subsequent waves, since high values of parameter a lead to better accuracy of the "shear approximation" (3.40a) considered previously. The data presented in Fig. 3.14 indicate that the following relationship is suitable for practical applications:

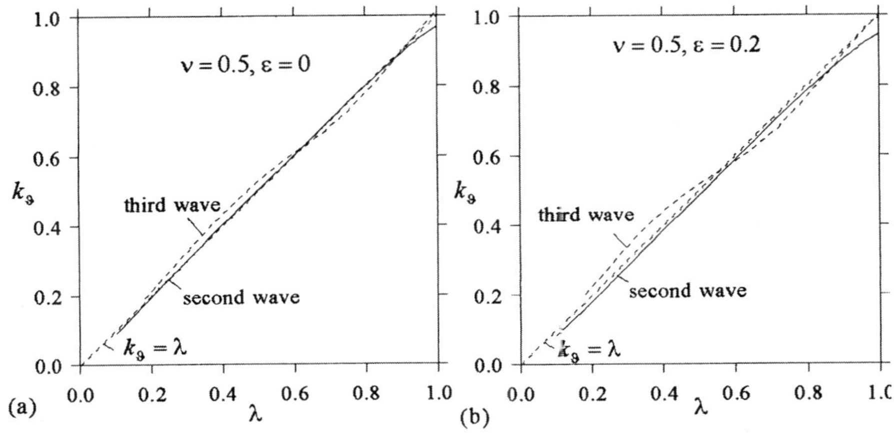

Fig. 3.14a,b. Dimensionless wave numbers k_ϑ versus parameter $\lambda = \theta/(0.9 + \theta)$ for $\varepsilon = 0$ (a) and $\varepsilon = 0.2$ (b)

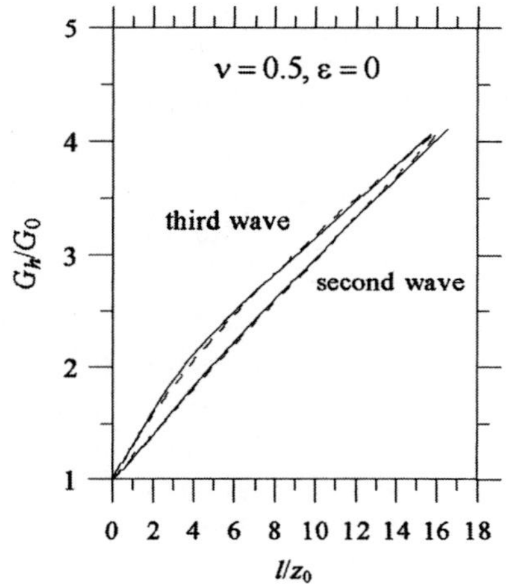

Fig. 3.15. Normalized shear modulus of homogeneous half-space, corresponding to velocities of surface waves under action of horizontal force

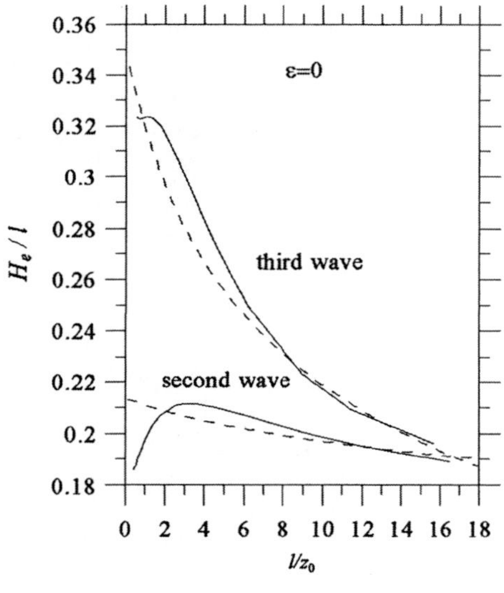

Fig. 3.16. Relationship between equivalent depth and surface wavelength under action of horizontal force

$$k_\vartheta = \lambda = \frac{\theta}{0.9 + \theta}. \tag{3.53}$$

Note that, in accordance with Fig. 3.14, this relationship is of sufficiently high accuracy for the second wave for $\lambda < 0.9$, and for the third wave for $\lambda > 0.9$.

When relationship $k_{\vartheta}(\theta)$ is available, determination of the parameters of the linearly heterogeneous half-space from the known (found experimentally) wavelengths l_1 and l_2 corresponding to the circular frequencies of vibrations ω_1 and ω_2 may be done in the same way as in the case of vertical vibrations. Now, a source of horizontal vibrations is being employed, and the amplitudes of vibrations are considered in the direction of the horizontal disturbing force for the surface points of the half-space, which belong to the line normal to this force and passing through the center of loading application domain. The variation of phase for these amplitudes with varying distance from the source corresponds to the variation of phase of quantity $S_{h\vartheta}$. Note that formulas (3.42)–(3.46) remain valid, with index ν replaced with ϑ. Certainly, a phase velocity C is related to the horizontal vibrations. The quantity $k_{h\vartheta}$ is related to the homogeneous half-space, its value being close to unity (at $\nu = 0.5$, $k_{h\vartheta} = 0.9676$ for the second wave and $k_{h\vartheta} = 1.01364$ for the third wave). The procedure for determining of the half-space parameters, analogous to that used for the analysis of the vertical amplitudes of vibrations, may be applied. Now, however, the dependence on Poisson's ratio is excluded. The curves presented in Fig. 3.15 express the quantity G_h / G_0 as a function of the dimensionless wavelength $\tilde{l}$ for the second and the third waves, analogous to graphs in Fig. 3.10. An approximation of the form (3.47) is presented in Fig. 3.15 (dashed lines) with $\gamma = 0.24, \zeta = 0.89$ for the second wave and $\gamma = 0.58, \zeta = 0.605$ for the third wave. An equivalent depth is determined by using formula (3.49). The results of calculations are presented in Fig. 3.16, where the dashed lines correspond to the above-mentioned approximation. For the considered waves, which are close to the shear waves, the equivalent depth is appreciably less than that calculated for the vertical vibrations under the action of a vertical force.

Applying the proximity of curves in Fig. 3.14 to the straight line with equation (3.53), we may propose, in addition, the following technique for determining the half-space parameters. Let equality (3.46) be rewritten twice (with the earlier index replacement) for two vibration conditions specified by quantities f_1, l_1, C_1 and f_2, l_2, C_2. Next, both sides of the relationships obtained are divided by each other, taking into account that the ratio of frequencies f_1 and f_2 equals the ratio of corresponding parameters θ_1 and θ_2. Thus, we obtain an equation with respect to θ_1, the solution of which, accounting for relationship (3.53), may be expressed in the form

$$\theta_1 = \frac{0.9[\zeta^2 - (l_1 / l_2)]}{\zeta[(l_1 / l_2) - \zeta]}, \tag{3.54}$$

where $\zeta = f_2 / f_1$. If the value of θ_1 was determined by using data corresponding to the second wave and the related value of parameter λ exceeds 0.9 (in this case, a noticeable error occurs in approximation (3.53) for the second wave), then it is advisable either to repeat the measurements at lower frequencies, or to employ measurements of wavelengths for the third wave at the same frequencies.

Following the calculation of parameter θ_1, we determine the corresponding value of k_9 by using formula (3.53) or curves in Fig. 3.14, and then z_0 from the relationship (3.46) (replacing k_v with k_9), applied for $l = l_1$. The shear modulus G_0 may be determined by using relationship (3.1) with known θ_1, z_0, ρ and ω_1.

3.5 Some Static Problems for Linearly Heterogeneous Half-Space

3.5.1 Displacements of Half-Space Surface under Action of Surface Concentrated Forces

Green's functions, related to displacements of the half-space, provide a relatively simple statement and solution of the problems dealing with the interactions of various constructions with deformable foundations. The model of the half-space with the shear modulus varying linearly with depth is rather attractive since in the framework of this model, one can vary the foundation properties – from those of the homogeneous half-space, which is "too good" in distributing the load, to the properties of Winkler's model which is locally deformable.

For the static case, the formulas have a structure similar to that of the corresponding relationships presented in previous sections for the case of harmonic vibrations. In static problems, one should take $\gamma = 0$, $\beta = 1$, $\eta_1 = \tau$, $\eta_2 = -\tau$, where η_1, η_2 represent the roots of equation (3.15), entering into the fundamental solutions. Moreover, the solution of the problems simplifies since no poles of the integrands exist. Consequently, integration over the real axis may be done with respect to the integration variable $\widetilde{k}$. Using relationships (2.1), (2.53), (2.54), we rewrite expressions for Green's functions shown in Fig. 1.6 for the case when both the points of application of the unit forces and points in which tpy displacements are determined are located on the surface of the half-spaceЖ

$$w_{vv} = \frac{1-\nu}{2G_0\pi r} S_{vv}(\widetilde{r}), \tag{3.55a}$$

$$w_{vh} = \frac{1-\nu}{2G_0\pi r} S_{vh}(\widetilde{r}), \tag{3.55b}$$

$$w_{hr} = \frac{1}{2G_0 \pi r} S_{hr}(\tilde{r}), \tag{3.55c}$$

$$w_{h\vartheta} = \frac{1-\nu}{2G_0 \pi r} S_{h\vartheta}(\tilde{r}). \tag{3.55d}$$

Green's function w_{hz} equals w_{vh} with the opposite sign. When writing the expression for $w_{h\vartheta}$, the following relationship was employed:

$$w_{h\vartheta} = -\hat{u}_{h\vartheta}. \tag{3.56}$$

The normalized displacements $S_{vv}(\tilde{r})$, $S_{vh}(\tilde{r})$, $S_{hr}(\tilde{r})$, $S_{h\vartheta}(\tilde{r})$ are determined according to (3.29), (3.36) with $\tilde{z}=1$, $\varsigma = 2\tilde{k}$:

$$S_{vv}(\tilde{r}) = \frac{\tilde{r}}{1-\nu} \int_0^\infty \frac{\overset{*}{c}_{21}\tilde{w}_2^*(2\tilde{k}) - \overset{*}{c}_{22}\tilde{w}_1^*(2\tilde{k})}{D^*} \tilde{k}\, \mathrm{J}_0(\tilde{k}\tilde{r})\, \mathrm{d}\tilde{k}, \tag{3.57a}$$

$$S_{vh}(\tilde{r}) = \frac{\tilde{r}}{1-\nu} \int_0^\infty \frac{\overset{*}{c}_{21}\tilde{q}_2^*(2\tilde{k}) - \overset{*}{c}_{22}\tilde{q}_1^*(2\tilde{k})}{D^*} \tilde{k}\, \mathrm{J}_1(\tilde{k}\tilde{r})\, \mathrm{d}\tilde{k}, \tag{3.57b}$$

$$S_{hr}(\tilde{r}) = \tilde{r} \int_0^\infty \tilde{k}\{[\tilde{q}^*(2\tilde{k}) - \tilde{p}^*(2\tilde{k})]\frac{\mathrm{J}_1(\tilde{k}\tilde{r})}{\tilde{k}\tilde{r}} - \tilde{q}^*(2\tilde{k})\mathrm{J}_0(\tilde{k}\tilde{r})\}\, \mathrm{d}\tilde{k}, \tag{3.57c}$$

$$S_{h\vartheta}(\tilde{r}) = -\frac{\tilde{r}}{1-\nu} \int_0^\infty \tilde{k}\{[\tilde{q}^*(2\tilde{k}) - \tilde{p}^*(2\tilde{k})]\frac{\mathrm{J}_1(\tilde{k}\tilde{r})}{\tilde{k}\tilde{r}} + \tilde{p}^*(2\tilde{k})\mathrm{J}_0(\tilde{k}\tilde{r})\}\, \mathrm{d}\tilde{k}, \tag{3.57d}$$

where

$$\tilde{q}^*(2\tilde{k}) = \frac{\overset{*}{c}_{11}\tilde{q}_2^*(2\tilde{k}) - \overset{*}{c}_{12}\tilde{q}_1^*(2\tilde{k})}{D^*}, \tag{3.58a}$$

$$\tilde{p}^*(2\tilde{k}) = \frac{U(0.5,1,2\tilde{k})}{2\tilde{k}[0.5U(0.5,1,2\tilde{k}) - U(0.5,2,2\tilde{k})]}, \tag{3.58b}$$

$$\overset{*}{c}_{1j} = \tilde{e}_j^*(2\tilde{k}) - 2\tilde{k}\tilde{q}_j^*(2\tilde{k}), \tag{3.59a}$$

$$\overset{*}{c}_{2j} = \tilde{\chi}_j^*(2\tilde{k}) - 2\tilde{k}\tilde{w}_j^*(2\tilde{k}), \tag{3.59b}$$

$$\tilde{\chi}_j^*(\varsigma) = -\eta_j \varsigma U(1-\eta_j,2,\varsigma), \tag{3.60a}$$

$$\tilde{e}_j^*(\varsigma) = \varsigma U(1-\eta_j,2,\varsigma), \tag{3.60b}$$

$$\tilde{q}_j^*(\varsigma) = \eta_j U(1-\eta_j,1,\varsigma) + \frac{\eta_j - 1}{2}\varsigma U(1-\eta_j,2,\varsigma), \tag{3.60c}$$

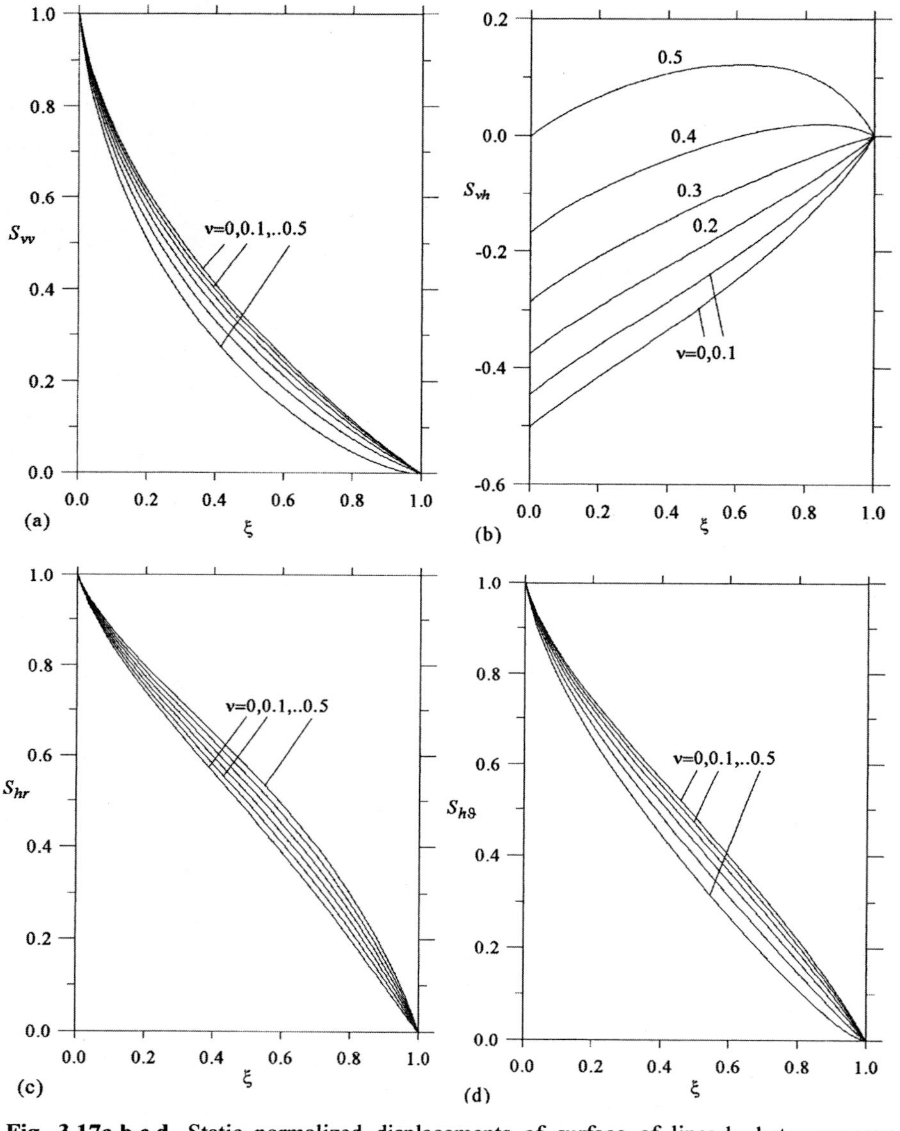

Fig. 3.17a,b,c,d. Static normalized displacements of surface of linearly heterogeneous half-space subjected to vertical ((a), (b)) and horizontal ((c), (d)) forces for various values of Poisson's ratio

$$\tilde{w}_j^*(\varsigma) = -\tau^2 U(1-\eta_j,1,\varsigma) + \frac{\eta_j - 1}{2}\varsigma U(1-\eta_j,2,\varsigma) . \tag{3.60d}$$

The previously determined limiting values given in relationships (3.30), (3.38), and equations of the form (3.31), (3.39), were employed in the present calculations. The results are shown in Fig. 3.17, where the quantity

$$\xi = \frac{\tilde{r}}{1+\tilde{r}} \tag{3.61}$$

serves as an argument. This value is equivalent to $\tilde{r}$ for small $\tilde{r}$, while quantity $1-\xi$ is equivalent to $1/\tilde{r}$ when $\tilde{r} \to \infty$.

Consider the behavior of the normalized displacements for high values of the variable $\tilde{r}$. In this case, the integrals in relationships (3.57) are determined by the values of their integrands (without the multiplier expressed through Bessel's function) at small values of their argument. This becomes obvious if one makes a change of the form $u = \tilde{k}\tilde{r}$ in the integration variable. Taking into account representations of the confluent hypergeometric functions at small values of the argument yields

$$S_{vv} \approx \frac{\tilde{r}}{2(1-\nu)} \int_0^{\infty} \frac{-\tilde{k}\tau^2 \ln(2\tilde{k})}{\tilde{k}+0.5} \mathrm{J}_0(\tilde{k}\tilde{r}) \mathrm{d}\tilde{k}, \tag{3.62a}$$

$$S_{vh} \approx \frac{\tilde{r}}{2(1-\nu)} \int_0^{\infty} \frac{b_1 \tilde{k}}{\tilde{k}+0.5} \mathrm{J}_1(\tilde{k}\tilde{r}) \mathrm{d}\tilde{k} = \frac{b_1 \pi \tilde{r}}{8(1-\nu)} [N_{-1}(\tilde{r}/2) - \mathbf{H}_{-1}(\tilde{r}/2)]$$

$$\approx \frac{2(1-\tau^2) b_1}{\tilde{r}}, \tag{3.62b}$$

$$S_{hr} \approx \frac{\tilde{r}}{2} \int_0^{\infty} \left[\left(\frac{\tilde{k}(b_2 + \ln(2\tilde{k}))}{\tilde{k}+0.5} - \frac{2\tilde{k}(b_3 + \ln(2\tilde{k})}{\tilde{k}+1} \right) \frac{\mathrm{J}_1(\tilde{k}\tilde{r})}{\tilde{k}\tilde{r}} \right.$$

$$\left. - \frac{\tilde{k}(b_2 + \ln(2\tilde{k}))}{\tilde{k}+0.5} \mathrm{J}_0(\tilde{k}\tilde{r}) \right] \mathrm{d}\tilde{k}, \tag{3.62c}$$

$$S_{h\vartheta} \approx -\frac{\tilde{r}}{2(1-\nu)} \int_0^{\infty} \left[\left(\frac{\tilde{k}(b_2 + \ln(2\tilde{k}))}{\tilde{k}+0.5} - \frac{2\tilde{k}(b_3 + \ln(2\tilde{k})}{\tilde{k}+1} \right) \frac{\mathrm{J}_1(\tilde{k}\tilde{r})}{\tilde{k}\tilde{r}} \right.$$

$$\left. + \frac{2\tilde{k}(b_3 + \ln(2\tilde{k}))}{\tilde{k}+1} \mathrm{J}_0(\tilde{k}\tilde{r}) \right] \mathrm{d}\tilde{k}, \tag{3.62d}$$

where the tabulated integral [43], containing Struve's function, **H**, and Neumann's function, *N*, is used in order to express function S_{vh}:

$$\int_0^{\infty} \frac{x^{\nu}\, \mathrm{J}_{\nu}(ax)}{x+k} \mathrm{d}x = \frac{\pi k^{\nu}}{2\cos(\nu\pi)} [\mathbf{H}_{-\nu}(ak) - N_{-\nu}(ak)] \tag{3.63}$$
$$-\frac{1}{2} < \mathrm{Re}(\nu) < \frac{3}{2}, \; a > 0, \; |\arg(k)| < \pi .$$

The constants b_j are expressed in terms of the psi-function Ψ:

$$b_1 = \frac{1}{2} + \frac{1}{2}\tau\left[\Psi(1-\tau) - \Psi(\tau) - \frac{1}{\tau}\right] = \frac{\pi\tau}{2\tan(\pi\tau)}, \tag{3.64a}$$

$$b_2 = \frac{1}{2}[\Psi(1-\tau) + \Psi(1+\tau) - 4\Psi(1) - 1], \tag{3.64b}$$

$$b_3 = \Psi(0.5) - 2\Psi(1) = -0.809079 . \tag{3.64c}$$

Parts of the integrals not containing the term $\ln(2\tilde{k})$ may be reduced to tabulated integrals similar to those used for representing S_{vh}; for the terms with $\ln(2\tilde{k})$, we apply integration by parts (including differentiation of these terms) with subsequent application of tabulated integrals. As a result, in addition to formula (3.62b), we obtain the following asymptotic representations for the normalized static displacements of the half-space surface corresponding to large values of $\tilde{r}$:

$$S_{vv} \approx \frac{\tau^2}{(1-\nu)\tilde{r}}, \tag{3.65a}$$

$$S_{hr} \approx \frac{1 - b_3 + b_2}{\tilde{r}}, \tag{3.65b}$$

$$S_{h\vartheta} \approx \frac{1 + b_3 - b_2}{(1-\nu)\tilde{r}}. \tag{3.65c}$$

Note that quantity S_{vv} in (3.65a) becomes zero in the case of an incompressible medium ($\tau = 0$); employing a more accurate representation for the corresponding integrand, one can prove that in this case, the first term in an asymptotic expansion has the form $2/\tilde{r}^2$. Consider approximate values of the normalized displacements by formulas (3.65), (3.62b), and their exact values by (3.57) for $\tilde{r} = 150$ and $\nu = 1/3$. Approximate values of $S_{vv}, S_{vh}, S_{hr}, S_{h\vartheta}$ equal 0.0025, 0.0, 0.01, 0.005, respectively; exact values equal 0.002499, –0.00003302, 0.009981, 0.0049997. Evidently, the asymptotic representation (3.62b) for the quantity S_{vh} vanishes at $\nu = 1/3$ (analogous to the case with S_{vv} at $\nu = 1/2$); a more accurate analysis indicates that in the case of S_{vh} at $\nu = 1/3$, the asymptotic expansion begins with a term of the form $-0.75/\tilde{r}^2$, which yields the value of –0.00003333. Note that the numerical evaluation of integrals in (3.57) provides rather high accuracy even for small values of the normalized displacements.

For a decreasing value of z_0 while keeping the ratio

$$m = G_0 / z_0 \tag{3.66}$$

fixed, we obtain a linearly heterogeneous half-space with zero modulus on its surface; in this case, the quantity $\tilde{r}$ increases without bound. Consider the formulas corresponding to the uniform distribution of loading over the circular area located on the surface of a half-space given in section 2.7.1. For the case of a compressible medium ($\nu < 0.5$, $\tau > 0$), as follows from (2.177), the asymptotic representation (3.65a) does not become zero, so that at $z_0 \to 0$ and $\eta > 1$ the result takes the form

$$\delta_z(r,R) = \frac{p_0 R(1-\nu)}{2\pi G_0} \int_{|\eta-1|}^{\eta+1} 2S_{\nu\nu}\left(\frac{\lambda R}{z_0}\right)\varphi \, d\lambda$$

$$\approx \frac{p_0 R(1-\nu)}{2\pi G_0} \int_{|\eta-1|}^{\eta+1} 2\frac{\tau^2 z_0}{(1-\nu)\lambda R}\varphi \, d\lambda = \frac{p_0 \tau^2}{\pi m} \int_{|\eta-1|}^{\eta+1} \frac{1}{\lambda}\varphi \, d\lambda \, , \tag{3.67}$$

where quantity φ depends on λ and η, following (1.212c); according to the formulas constructed by using the principle of superposition, variable r which enters into $S_{\nu\nu}$ is replaced with λR ($\tilde{r}$ is replaced with $\lambda R / z_0$). In the case when $\eta < 1$, according to (2.177a), one should add to the value by (3.67) the quantity

$$\frac{p_0 R(1-\nu)}{G_0} \int_0^{1-\eta} S_{\nu\nu}\left(\frac{\lambda R}{z_0}\right) d\lambda = \frac{p_0(1-\nu)}{m} \int_0^{(R-r)/z_0} S_{\nu\nu}(\tilde{r}) \, d\tilde{r} \, . \tag{3.68}$$

The last integral diverges as $z_0 \to 0$ in the case of representation having the form (3.65a) at high values of $\tilde{r}$ (for $\nu < 0.5$), i.e. displacements become unbounded over the loaded area. Outside the loaded area, as follows from (3.67), displacements are finite and decreasing with increasing η. A similar behavior takes place when determining horizontal displacements, which occur due to a horizontal load. In this case, the asymptotic representation of Green's functions, which is given in relationships (3.65b) and (3.65c), yields integrals of the form (3.67), (3.68). Note that for the incompressible medium, where, as noted earlier, quantity $S_{\nu\nu}$ is equivalent to $2/\tilde{r}^2$, we obtain instead of (3.67)

$$\delta_z(r,R) \approx \frac{p_0 R}{4\pi G_0} \int_{|\eta-1|}^{\eta+1} 2\frac{2z_0^2}{\lambda^2 R^2}\varphi \, d\lambda = \frac{p_0 z_0}{\pi m R} \int_{|\eta-1|}^{\eta+1} \frac{1}{\lambda^2}\varphi \, d\lambda \, . \tag{3.69}$$

This yields a zero result when $z_0 \to 0$. For $\eta < 1$, the integral in relationship (3.68) converges; displacements over the loaded area tend to the following constant value:

$$\delta_z = \frac{p_0}{2m}\int_0^\infty S_{vv}(\tilde{r})\,\mathrm{d}\tilde{r} = \frac{p_0}{2m}\int_0^1 S_{vv}(\xi)\frac{1}{(1-\xi)^2}\,\mathrm{d}\xi\ , \tag{3.70}$$

where variable ξ from (3.61) is employed. Thus, we obtain the known result [39]: an incompressible linearly heterogeneous half-space with zero shear modulus on its surface behaves (with respect to surface displacements) as Winkler's foundation. The coefficient of subgrade reaction equals $2m$ [39]; as a result, the integral in relationship (3.70) must be equal to unity. Indeed, numerical integration including the earlier mentioned behavior of S_{vv} at high values of $\tilde{r}$ (when $\xi \to 1$, the integrand in the integral tends to 2) yields unity with high accuracy (when dividing the interval $0 \le \xi \le 1$ into 50 equal segments, the calculation of the integral by Simpson's formula gives 1.00001).

Consider the behavior of the normalized displacements at small values of $\tilde{r}$. Obviously, the integrands in (3.57) should be considered at high values of the integration variable. Simple limiting representations were given previously in relationships (3.30) and (3.38) in connection with acceleration of convergence of the corresponding integrals. Below, these representations are refined in order to obtain a more accurate evaluation of the behavior of displacements at small values of $\tilde{r}$. Using the integral (3.63), we have

$$S_{vv} \approx \tilde{r}\int_0^\infty \frac{\tilde{k}}{\tilde{k}+b}\mathrm{J}_0(\tilde{k}\tilde{r})\,\mathrm{d}\tilde{k} = 1 - b\tilde{r}\frac{\pi}{2}[\mathbf{H}_0(\tilde{r}b) - N_0(\tilde{r}b)] \approx 1 + b\tilde{r}\ln(\tilde{r}), \tag{3.71a}$$

$$S_{vh} \approx \frac{\tilde{r}}{2}\int_0^\infty \frac{-2\tau^2\tilde{k} + (1-\tau^2)(1-2\tau^2)}{\tilde{k}+b}\mathrm{J}_1(\tilde{k}\tilde{r})\,\mathrm{d}\tilde{k} \approx -\tau^2 + \frac{1}{2}\tilde{r} + \frac{b}{4}\tilde{r}^2\ln(\tilde{r})\,, \tag{3.71b}$$

$$S_{hr} \approx \frac{\tilde{r}}{2}\int_0^\infty\left[\left(-\frac{\tilde{k}+1-\tau^2}{(\tilde{k}+b)(1-\tau^2)} + \frac{2\tilde{k}-1/4}{\tilde{k}+3/8}\right)\frac{\mathrm{J}_1(\tilde{k}\tilde{r})}{\tilde{k}\tilde{r}} + \frac{\tilde{k}+1-\tau^2}{(\tilde{k}+b)(1-\tau^2)}\mathrm{J}_0(\tilde{k}\tilde{r})\right]\mathrm{d}\tilde{k}$$

$$\approx 1 + \frac{b}{4(1-\tau^2)}\tilde{r}\ln(\tilde{r})\,, \tag{3.71c}$$

$$S_{h\vartheta} \approx -\tilde{r}(1-\tau^2)\int_0^\infty\left[\left(-\frac{\tilde{k}+1-\tau^2}{(\tilde{k}+b)(1-\tau^2)} + \frac{2\tilde{k}-1/4}{\tilde{k}+3/8}\right)\frac{\mathrm{J}_1(\tilde{k}\tilde{r})}{\tilde{k}\tilde{r}}\right.$$

$$\left. - \frac{2\tilde{k}-1/4}{\tilde{k}+3/8}\mathrm{J}_0(\tilde{k}\tilde{r})\right]\mathrm{d}\tilde{k} \approx 1 + \frac{b}{2}\tilde{r}\ln(\tilde{r}) \tag{3.71d}$$

where

$$b = 1.5 - \tau^2 . \tag{3.72}$$

In order to illustrate the accuracy of approximate formulas (3.71), we present the values calculated for the normalized displacements at $\tilde{r} = 0.005$, $\nu = 1/3$. Approximate values for $S_{vv}, S_{vh}, S_{hr}, S_{h\vartheta}$ equal 0.9669, –0.2475, 0.9890, 0.9834, respectively; exact values equal 0.9687, –0.2475, 0.9878, 0.9824. As $\tilde{r} \to 0$, quantities $S_{vv}, S_{vh}, S_{hr}, S_{h\vartheta}$ tend to their corresponding static values; and the first derivative of S_{vv}, $S_{hr}, S_{h\vartheta}$ and the second derivative of S_{vh} become unbounded.

Next, we construct an approximation for the normalized displacements, which enables us to perform calculations in a more efficient way than by using integral representations (3.57). An approximation using the least-squares technique could be applied to the functions themselves shown in Fig. 3.17. However, because of singularities in derivatives as $\tilde{r} \to 0$, and for the purpose of having a better approximation at high values of $\tilde{r}$, we preliminarily account for the behavior of the considered quantities at small and high values of $\tilde{r}$, determined earlier. Auxiliary functions, equivalent to the corresponding normalized displacements in the vicinity of the end points of the interval $0 \le \xi \le 1$ of varying ξ from (3.61), have the form

$$\tilde{S}_{vv} = [1 + b\xi \ln(\xi)](1-\xi^2)^2 + 2(1-\tau^2)\tau^2(1-\xi)\xi^2 , \tag{3.73a}$$

$$\tilde{S}_{vh} = \left[-\tau^2 + \frac{1}{2}\xi + \frac{b}{4}\xi^2 \ln(\xi) \right](1-\xi^2)^2 + 2(1-\tau^2)b_1(1-\xi)\xi^2 , \tag{3.73b}$$

$$\tilde{S}_{hr} = \left[1 + \frac{b}{4(1-\tau^2)}\xi \ln(\xi) \right](1-\xi^2)^2 + (1 - b_3 + b_2)(1-\xi)\xi^2 , \tag{3.73c}$$

$$\tilde{S}_{h\vartheta} = \left[1 + \frac{b}{2}\xi \ln(\xi) \right](1-\xi^2)^2 + 2(1-\tau^2)(1 + b_3 - b_2)(1-\xi)\xi^2 . \tag{3.73d}$$

We introduce the quantities

$$\delta_{vv} = \frac{S_{vv} - \tilde{S}_{vv}}{1-\xi} , \tag{3.74a}$$

$$\delta_{vh} = \frac{S_{vh} - \tilde{S}_{vh}}{\xi(1-\xi)} , \tag{3.74b}$$

$$\delta_{hr} = \frac{S_{hr} - \tilde{S}_{hr}}{1-\xi} , \tag{3.74c}$$

$$\delta_{h\vartheta} = \frac{S_{h\vartheta} - \tilde{S}_{h\vartheta}}{1-\xi} . \tag{3.74d}$$

These quantities enable a good approximation by using polynomials. Calculations show that it is reasonable to take variable ξ for δ_{vh} and variable

$$\xi_1 = \frac{\tilde{r}}{2+\tilde{r}} = \frac{\xi}{2-\xi} \tag{3.75}$$

for the rest of the values in (3.74) as arguments of the polynomials. The least-squares technique was applied to the quantities (3.74) with 11 values of Poisson's ratio, $\nu = 0, 0.05, \ldots, 0.5$, for the degree of the polynomials equal to 7. Next, the least-squares technique was applied once again to each coefficient of the polynomials considered as a function ν. As a result, the coefficients were obtained as polynomials of ν (also of 7th degree). The following approximate expressions are obtained for quantities (3.74):

$$\delta_{vv} \approx P_{vv}(\xi_1) = \sum_{i=0}^{7} c_{ivv}\xi_1^i, \tag{3.76a}$$

where

$$c_{0vv} = 0.000288 + 0.000292\,\nu + 0.000024\,\nu^2 + 0.000648\,\nu^3 - 0.005032\,\nu^4 + 0.015438\,\nu^5 - 0.026678\,\nu^6 + 0.017978\,\nu^7,$$

$$c_{1vv} = 0.139254 + 0.943450\,\nu + 0.909272\,\nu^2 + 0.997539\,\nu^3 + 0.061070\,\nu^4 + 3.981760\,\nu^5 - 5.671376\,\nu^6 + 6.564648\,\nu^7,$$

$$c_{2vv} = -5.093887 - 4.487810\,\nu - 3.353558\,\nu^2 + 3.393978\,\nu^3 - 33.310154\,\nu^4 + 123.55918\,\nu^5 - 223.364158\,\nu^6 + 186.082413\,\nu^7,$$

$$c_{3vv} = 11.496923 + 4.807464\,\nu + 4.476579\,\nu^2 - 14.205996\,\nu^3 + 105.109565\,\nu^4 - 383.605368\,\nu^5 + 687.332467\,\nu^6 - 556.647456\,\nu^7,$$

$$c_{4vv} = -10.940509 + 2.213314\,\nu - 3.725624\,\nu^2 + 33.183434\,\nu^3 - 241.842308\,\nu^4 + 845.108815\,\nu^5 - 1512.908946\,\nu^6 + 1181.321645\,\nu^7,$$

$$c_{5vv} = 6.420134 - 9.321105\,\nu + 4.119596\,\nu^2 - 48.409530\,\nu^3 + 371.087358\,\nu^4 - 1265.396912\,\nu^5 + 2270.477559\,\nu^6 - 1741.745776\,\nu^7,$$

$$c_{6vv} = -2.619294 + 8.676073\,\nu - 4.137793\,\nu^2 + 38.169155\,\nu^3 - 309.508061\,\nu^4 + 1044.039534\,\nu^5 - 1879.987846\,\nu^6 + 1433.940633\,\nu^7,$$

$$c_{7vv} = 0.597031 - 2.831377\,\nu + 1.711461\,\nu^2 - 13.129495\,\nu^3 + 108.404510\,\nu^4 - 367.692159\,\nu^5 + 664.120112\,\nu^6 - 509.516076\,\nu^7; \tag{3.76b}$$

$$\delta_{vh} \approx P_{vh}(\xi) = \sum_{i=0}^{7} c_{ivh}\xi^{i} \ , \tag{3.77a}$$

where

$$c_{0vh} = -\,0.001177 + 0.000457\,\nu + 0.000693\,\nu^{2} - 0.002363\,\nu^{3} + 0.017237\,\nu^{4} - 0.063319\,\nu^{5} + 0.112844\,\nu^{6} - 0.093101\,\nu^{7},$$

$$c_{1vh} = -\,0.114987 - 0.962011\,\nu - 1.251783\,\nu^{2} - 2.255396\,\nu^{3} + 3.022420\,\nu^{4} - 20.310734\,\nu^{5} + 33.600109\,\nu^{6} - 32.489736\,\nu^{7},$$

$$c_{2vh} = -\,0.046389 + 3.747733\,\nu + 4.021385\,\nu^{2} - 1.013133\,\nu^{3} + 30.096655\,\nu^{4} - 95.747420\,\nu^{5} + 172.707459\,\nu^{6} - 132.815569\,\nu^{7},$$

$$c_{3vh} = 0.775366 - 17.961415\,\nu - 19.445183\,\nu^{2} + 17.680289\,\nu^{3} - 212.913063\,\nu^{4} + 724.132612\,\nu^{5} - 1290.739684\,\nu^{6} + 1017.021970\,\nu^{7},$$

$$c_{4vh} = -\,6.515719 + 52.680266\,\nu + 56.064210\,\nu^{2} - 65.207248\,\nu^{3} + 684.244870\,\nu^{4} - 2378.149311\,\nu^{5} + 4236.993389\,\nu^{6} - 3375.066362\,\nu^{7},$$

$$c_{5vh} = 15.563835 - 82.702445\,\nu - 85.355538\,\nu^{2} + 118.296409\,\nu^{3} - 1129.671218\,\nu^{4} + 4002.690858\,\nu^{5} - 7135.682236\,\nu^{6} + - 5746.046483\,\nu^{7},$$

$$c_{6vh} = -\,15.035042 + 66.442723\,\nu + 65.846577\,\nu^{2} - 104.472178\,\nu^{3} + 928.445783\,\nu^{4} - 3346.804462\,\nu^{5} + 5971.098181\,\nu^{6} - 4855.819523\,\nu^{7},$$

$$c_{7vh} = 5.372597 - 21.242742\,\nu - 19.875782\,\nu^{2} + 36.976344\,\nu^{3} - 303.222122\,\nu^{4} + 1114.201623\,\nu^{5} - 1987.984392\,\nu^{6} + 1633.133960\,\nu^{7};$$

(3.77b)

$$\delta_{hr} \approx P_{hr}(\xi_1) = \sum_{i=0}^{7} c_{ihr}\xi_1^{i} \ , \tag{3.78a}$$

where

$$c_{0hr} = 0.000303 + 0.000842\nu + 0.000291\nu^{2} + 0.000844\nu^{3} - 0.003080\nu^{4} + 0.009179\nu^{5} - 0.012129\nu^{6} + 0.006360\nu^{7},$$

$$c_{1hr} = -0.583588 + 0.085565\nu - 0.034099\nu^{2} - 0.069823\nu^{3} + 0.197368\nu^{4} - 0.599582\nu^{5} + 0.734181\nu^{6} - 0.367573\nu^{7},$$

$$c_{2hr} = -1.201163 - 4.200003\nu + 0.012831\nu^2 + 0.499341\nu^3 - 2.750364\nu^4 + 6.070838\nu^5 - 6.371420\nu^6 + 0.985308\nu^7,$$

$$c_{3hr} = 5.988020 + 4.093030\nu - 2.148194\nu^2 - 5.100414\nu^3 + 16.542979\nu^4 - 42.617670\nu^5 + 44.884284\nu^6 - 13.222967\nu^7,$$

$$c_{4hr} = -5.832984 + 12.000917\nu + 8.049432\nu^2 + 16.960555\nu^3 - 49.946523\nu^4 + 131.439171\nu^5 - 135.835508\nu^6 + 40.509898\nu^7,$$

$$c_{5hr} = 0.384864 - 33.769889\nu - 13.224451\nu^2 - 27.124057\nu^3 + 78.290007\nu^4 - 203.930489\nu^5 + 205.297778\nu^6 - 55.248841\nu^7,$$

$$c_{6hr} = 3.055265 + 33.775092\nu + 10.249095\nu^2 + 20.942851\nu^3 - 60.647274\nu^4 + 154.433197\nu^5 - 150.330550\nu^6 + 33.243549\nu^7,$$

$$c_{7hr} = -1.810423 - 11.983565\nu - 2.903543\nu^2 - 6.108766\nu^3 + 18.319200\nu^4 - 44.801664\nu^5 + 41.619914\nu^6 - 5.889694\nu^7; \quad (3.78b)$$

$$\delta_{h\vartheta} \approx P_{h\vartheta}(\xi_1) = \sum_{i=0}^{7} c_{ih\vartheta}\xi_1^i, \quad (3.79a)$$

where

$$c_{0h\vartheta} = 0.000261 - 0.000111\nu - 0.000234\nu^2 - 0.000768\nu^3 + 0.002035\nu^4 - 0.009557\nu^5 + 0.015671\nu^6 - 0.013527\nu^7,$$

$$c_{1h\vartheta} = -0.579077 + 0.129488\nu + 0.132872\nu^2 + 0.204734\nu^3 - 0.267435\nu^4 + 1.637869\nu^5 - 2.59218\nu^6 + 2.346601\nu^7,$$

$$c_{2h\vartheta} = -2.938552 + 0.788853\nu + 0.936981\nu^2 + 1.525651\nu^3 - 1.698312\nu^4 + 13.34448\nu^5 - 22.89941\nu^6 + 23.44393\nu^7,$$

$$c_{3h\vartheta} = 9.912372 - 2.260316\nu - 1.596999\nu^2 + 0.089071\nu^3 - 6.834943\nu^4 + 19.95824\nu^5 - 31.74231\nu^6 + 20.50947\nu^7,$$

$$c_{4h\vartheta} = -11.90488 + 1.488211\nu - 2.058863\nu^2 - 12.53976\nu^3 + 36.54814\nu^4 - 170.9144\nu^5 + 280.8795\nu^6 - 244.5578\nu^7,$$

$$c_{5h\vartheta} = 7.84316 + 2.065757\nu + 8.55048\nu^2 + 27.71553\nu^3 - 63.38808\nu^4 + 327.3412\nu^5 - 538.3505\nu^6 + 486.1915\nu^7,$$

$$c_{6h\vartheta} = -2.500945 - 4.051516\nu - 9.581091\nu^2 - 25.62767\nu^3 + 50.86606\nu^4 - 279.7694\nu^5 + 459.2169\nu^6 - 423.0173\nu^7,$$

$$c_{7h\vartheta} = 0.167876 + 1.839337\nu + 3.616189\nu^2 + 8.631172\nu^3 - 15.22254\nu^4 + 88.38577\nu^5 - 144.4839\nu^6 + 135.0566\nu^7. \tag{3.79b}$$

Finally, the approximate formulas for the normalized displacements may be written as

$$S_{vv} \approx \widetilde{S}_{vv} + (1-\xi)P_{vv}(\xi_1), \tag{3.80a}$$

$$S_{vh} \approx \widetilde{S}_{vh} + (1-\xi)\xi P_{vh}(\xi), \tag{3.80b}$$

$$S_{hr} \approx \widetilde{S}_{hr} + (1-\xi)P_{hr}(\xi_1), \tag{3.80c}$$

$$S_{h\vartheta} \approx \widetilde{S}_{h\vartheta} + (1-\xi)P_{h\vartheta}(\xi_1). \tag{3.80d}$$

This approximation provides a sufficiently high accuracy: curves plotted on the basis of formulas (3.80) merge with those shown in Fig. 3.17. In Table 3.1, exact values of the normalized displacements (in the upper line), and their values calculated from the approximate formulas (3.80), are given for several values of $\widetilde{r}$ at $\nu = 1/3$.

Table 3.1. Exact and approximate values of normalized displacements

	$\widetilde{r}$ =0.01	0.5	2	10	100
S_{vv}	0.94603	0.41297	0.16164	0.036958	0.003749
	0.94592	0.41293	0.16164	0.036961	0.003749
S_{vh}	− 0.24522	− 0.13275	− 0.048319	− 0.005560	− 0.000074
	− 0.24522	− 0.13273	− 0.048293	− 0.005545	− 0.000076
S_{hr}	0.97846	0.67152	0.39835	0.13219	0.014961
	0.97872	0.67170	0.39846	0.13222	0.014935
$S_{h\vartheta}$	0.96894	0.56671	0.27788	0.072806	0.007507
	0.96891	0.56674	0.27790	0.072806	0.007498

A relative error may appear to be appreciable in those cases when the normalized displacements are extremely small – corresponding to high values of $\widetilde{r}$ with respect to S_{vv} at $\nu = 0.5$, and with respect to S_{vh} at $\nu = 1/3$. In these cases, the asymptotic representation takes the form $2/\widetilde{r}^2$ and $-0.75/\widetilde{r}^2$, respectively, while the asymptotics (3.65a) and (3.62b), employed previously for constructing approximate formulas (3.80), vanish for these values of ν. Note that the values for the last column of the table, calculated from the asymptotic formulas which correspond to high values of $\tilde{r}$ with the above-mentioned refinement for S_{vh}, equal 0.00375, −0.000075, 0.015, 0.0075, for $S_{vv}, S_{vh}, S_{hr}, S_{h\vartheta}$, respectively. Therefore, sufficiently high accuracy is provided. For $\tilde{r} \geq 100$, one should use these asymptotic formulas rather than expressions (3.80).

3.5.2 Static Stiffnesses for Circular Disk Resting on Isotropic Linearly Heterogeneous Half-Space

The previously determined Green's functions are further applied in order to calculate the stiffnesses for a thin circular disk resting on the surface of a half-space. One can employ, for the considered heterogeneous half-space, equations derived for the corresponding contact problems and general expressions for the displacements given in section 2.6.1; in the formulas for the displacements, G_{rz0} and ν' are replaced with G_0 and ν, respectively. In the equations similar to (2.177a), the quantity $\lambda R / z_0$ serves as an argument instead of an initially used argument $\tilde{r}$ (in formulas related to the circular loaded area, constructed by using the principle of superposition in section 1.7, argument r in Green's function is replaced with λR). Consequently, the second arguments, R_j and R_{j-1}, specified in the elements of matrices in the systems of equations constructed for the contact problems, mean that the argument $\tilde{r}$ in functions $S_{vv}, S_{vh}, S_{hr}, S_{h\vartheta}$ must be replaced with $\lambda \tilde{R} \overline{R}_j$ and $\lambda \tilde{R} \overline{R}_{j-1}$, respectively, where notation (2.180b) is used and

$$\tilde{R} = \frac{R}{z_0}. \tag{3.81}$$

In the calculations, the number of ring-shaped elements $N = 40$, and the common ratio of the geometric progression, according to which the width of elements decreases as they approach the contact area boundary, is taken as $q = 0.9$.

Vertical Stiffness

When considering welded contact, one should employ equations (2.181), and in the case of relaxed contact, equations (2.190). Calculations were done for three values of Poisson's ratio, taken for relaxed contact (Fig. 3.18a) and for welded contact (Fig. 3.18b). The normalized vertical stiffness $\overline{K}_z$, shown in Fig. 3.18, is defined as

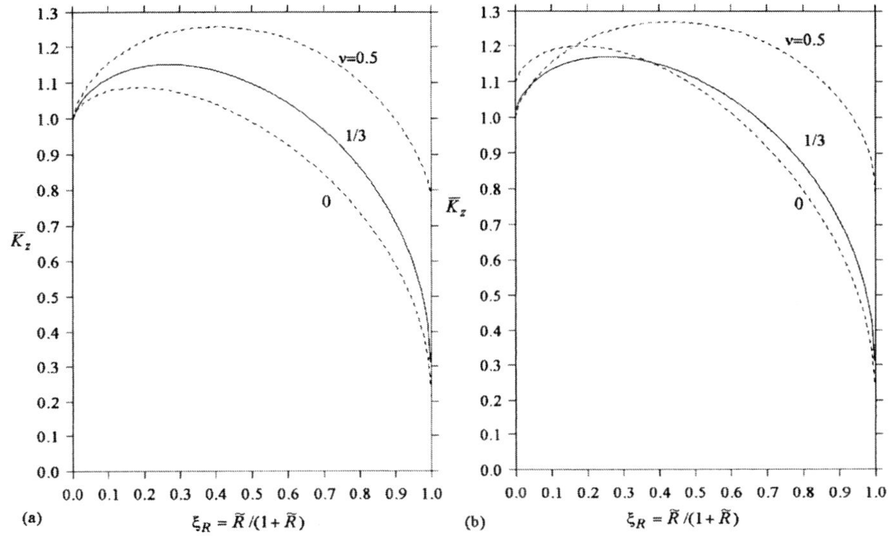

Fig. 3.18a,b. Normalized vertical static stiffness of disk on linearly heterogeneous half-space for (a) relaxed and (b) welded contact

$$\overline{K}_z = \frac{K_z}{k_z^{\text{hom}}}, \tag{3.82a}$$

$$k_z^{\text{hom}} = \frac{4G_\mu R}{1-\nu}, \tag{3.82b}$$

$$G_\mu = G_0(1+\mu\widetilde{R}), \tag{3.82c}$$

where K_z is the vertical stiffness, calculated analogously to (2.188),

$$K_z = \frac{2\pi^2 R G_0}{1-\nu}\sum_{j=1}^{N}\widetilde{p}_j(\overline{R}_j^2 - \overline{R}_{j-1}^2). \tag{3.83}$$

Quantity k_z^{hom} represents stiffness in the case of relaxed contact between the disk and foundation, corresponding to the homogeneous half-space having shear modulus G_μ which, as follows from (3.82c), equals the shear modulus of the given half-space taken at depth μR; for the present case, we take $\mu = 1$. This normalization technique results in an insignificant deviation of the quantity $\overline{K}_z$ from unity on the larger part of the interval of variation of

$$\xi_R = \frac{\widetilde{R}}{1+\widetilde{R}} \ . \tag{3.84}$$

As follows from relationships (3.83),

$$\overline{K}_z = \frac{\pi^2}{2(1+\mu\widetilde{R})} \sum_{j=1}^{N} \widetilde{p}_j (\overline{R}_j^2 - \overline{R}_{j-1}^2) \ . \tag{3.85}$$

In the case with $\nu = 0.5$ and $\xi_R \to 1$ $(\widetilde{R} \to \infty,\ z_0 \to 0)$, when, according to the previous subsection, the half-space behaves similarly to Winkler's foundation with the coefficient of subgrade reaction equal to $2m$ (see (3.66)), we have

$$K_z = 2m\pi R^2 , \tag{3.86a}$$

$$k_z^{\text{hom}} = 8mR^2 , \tag{3.86b}$$

$$\overline{K}_z = \frac{\pi}{4} . \tag{3.86c}$$

For $\nu < 0.5$, the limiting value of the stiffness, when ξ_R tends to unity, equals zero. This fact corresponds to unbounded values of displacements of the points in the loaded area in the case of a zero value of the modulus at the half-space surface (see the previous subsection).

The value $\xi_R = 0$ $(z_0 \to \infty,\ \widetilde{R} = 0)$ corresponds to the homogeneous half-space, for which the normalized stiffness should equal unity in the case of relaxed contact. Calculations, carried out with $N = 40, q = 0.9$, yield values 0.9993–0.9995 (for the considered values of Poisson's ratio).

The method employed to determine stiffness for the disk resting on the half-space requires calculation of Green's functions for a large number of argument values. On the other hand, using the approximate formulas constructed in the previous subsection enables us to speed up calculations considerably.

Horizontal Stiffness

In what follows, the approach presented in section 2.6.1 is modified according to another type of half-space. First, we consider relaxed contact, so that equations (2.212) can be employed. Using an analogy with (3.82) gives

$$\overline{K}_x = \frac{K_x}{k_x^{\text{hom}}} , \tag{3.87a}$$

$$k_x^{\text{hom}} = \frac{8G_\mu R}{2-\nu} , \tag{3.87b}$$

where K_x is the horizontal stiffness calculated analogously to (2.216):

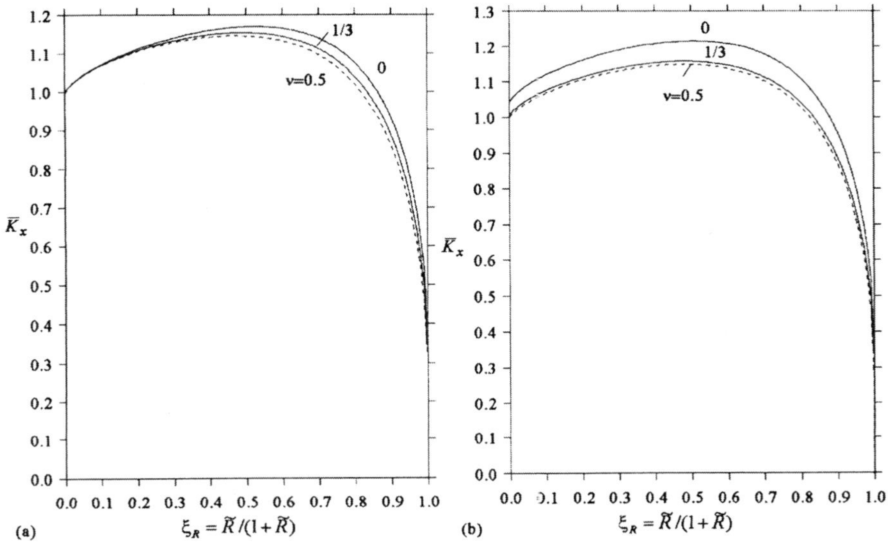

Fig. 3.19a,b. Normalized horizontal static stiffness of disk on linearly heterogeneous half-space for (a) relaxed and (b) welded contact

$$K_x = 2\pi^2 R G_0 \sum_{j=1}^{N} \widetilde{q}_j (\overline{R}_j^2 - \overline{R}_{j-1}^2) . \tag{3.88}$$

When determining the stiffness for the "equivalent" homogeneous half-space, k_x^{hom}, it is reasonable to take the quantity μ in expression (3.82c) for G_μ to be equal to 0.35; in this case, the normalized stiffness is close to unity on the larger part of the interval of variation of the parameter ξ_R. The normalized stiffness, $\overline{K}_x$, takes the form

$$\overline{K}_x = \frac{\pi^2(2-\nu)}{4(1+\mu\widetilde{R})} \sum_{j=1}^{N} \widetilde{q}_j (\overline{R}_j^2 - \overline{R}_{j-1}^2) . \tag{3.89}$$

The quantity $\overline{K}_x$ is presented in Fig. 3.19a for three different values of Poisson's ratio. Now, the stiffness vanishes when $\xi_R \to 1$ for the case of the incompressible half-space, too. At $\xi_R = 0$, values of 0.9993–0.9994 are obtained instead of unity. Note that in this case, the normalized stiffness depends on Poisson's ratio to a lesser extent than the normalized stiffness $\overline{K}_z$.

When solving the problem in its complete statement context (with welded contact), i.e. by taking into account the influence of normal stresses in the contact area on the horizontal displacements of points of this area, we employ the system

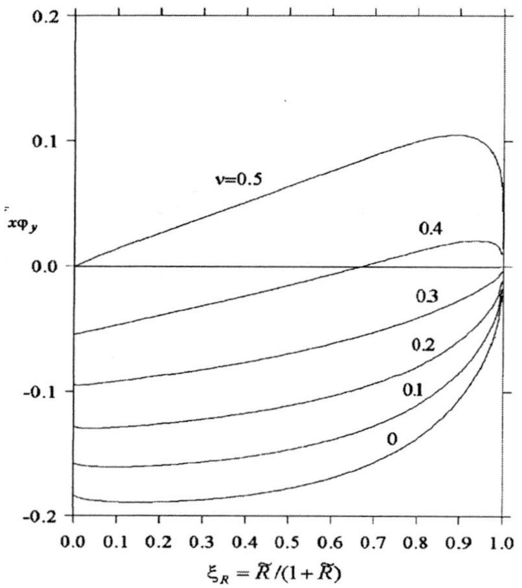

Fig. 3.20. Normalized horizontal rocking static stiffness of disk on linearly heterogeneous half-space for various values of Poisson's ratio

of equations (2.200). The results of the calculations are shown in Fig. 3.19b. As in the case of vertical displacements of the stamp, the results corresponding to the relaxed contact conditions are only slightly different from those obtained for the case with welded contact.

The resultant moment with respect to the *Y*-axis, which appears due to normal pressures arising with the translation displacement of the disk along the *X*-axis, determines the horizontal rocking stiffness, $K_{x\varphi_y}$, which may be calculated from relationship (2.220):

$$K_{x\varphi_y} = \frac{\pi^2 R^2 G_0}{2(1-\nu)} \sum_{j=1}^{N} \widetilde{p}_{0j} (\overline{R}_j^4 - \overline{R}_{j-1}^4). \tag{3.90}$$

We introduce the corresponding normalized stiffness $\overline{K}_{x\varphi_y}$ as the ratio of $K_{x\varphi_y}$ to Rk_x^{hom}:

$$\overline{K}_{x\varphi_y} = \frac{\pi^2 (2-\nu)}{16(1-\nu)(1+\mu\widetilde{R})} \sum_{j=1}^{N} \widetilde{p}_{0j} (\overline{R}_j^4 - \overline{R}_{j-1}^4). \tag{3.91}$$

In Fig. 3.20, the graphs represent the quantity $\overline{K}_{x\varphi_y}$ for the series of values of Poisson's ratio.

Rocking Stiffness

First, we consider a simplified statement of the problem, neglecting the influence of the horizontal contact stresses on the vertical displacements of points within the contact area. Knowing the solution for the system of equations (2.226), we employ the obtained quantities $\widetilde{p}_{0j}$ to calculate the rocking stiffness, analogously to relationship (2.224),

$$K_{\varphi_y} = \frac{\pi^2 R^3 G_0}{2(1-\nu)} \sum_{j=1}^{N} \widetilde{p}_{0j} (\overline{R}_j^4 - \overline{R}_{j-1}^4) \ . \tag{3.92}$$

As previously, we introduce the normalized stiffness $\overline{K}_{\varphi_y}$ as a ratio of K_{φ_y} to the stiffness of a homogeneous half-space, $k_{\varphi_y}^{\text{hom}}$,

$$k_{\varphi_y}^{\text{hom}} = \frac{8 G_\mu R^3}{3(1-\nu)}, \tag{3.93}$$

resulting in

$$\overline{K}_{\varphi_y} = \frac{K_{\varphi_y}}{k_{\varphi_y}^{\text{hom}}} = \frac{3\pi^2}{16(1+\mu\widetilde{R})} \sum_{j=1}^{N} \widetilde{p}_{0j} (\overline{R}_j^4 - \overline{R}_{j-1}^4) \,. \tag{3.94}$$

Similar to the case of the horizontal stiffness, it is reasonable to take $\mu = 0.35$ in the expression (3.82c) for G_μ. The normalized rocking stiffness under relaxed contact conditions is presented in Fig. 3.21a. As noted previously, at $\nu = 0.5$ and $\xi_R \to 1$ the half-space becomes similar to Winkler's foundation having the stiffness coefficient $2m$, resulting in the following relationships:

$$K_{\varphi_y} = 2m \frac{\pi R^4}{4}, \tag{3.95a}$$

$$k_{\varphi_y}^{\text{hom}} = \frac{16 m \mu R^4}{3}, \tag{3.95b}$$

$$\overline{K}_{\varphi_y} = \frac{3\pi}{32\mu} = 0.8125 \,. \tag{3.95c}$$

When $\xi_R \to 1$, the normalized stiffness tends to the value in (3.95c) for the case with $\nu = 0.5$ (Fig. 3.21). For another limiting case, $\xi_R = 0$, the exact value of the normalized stiffness equals unity (under relaxed contact conditions), and calculations yield the value 0.999.

The problem, in its complete context, is solved by using the system of equations (2.222). The results of calculations are presented in Fig. 3.21b; the

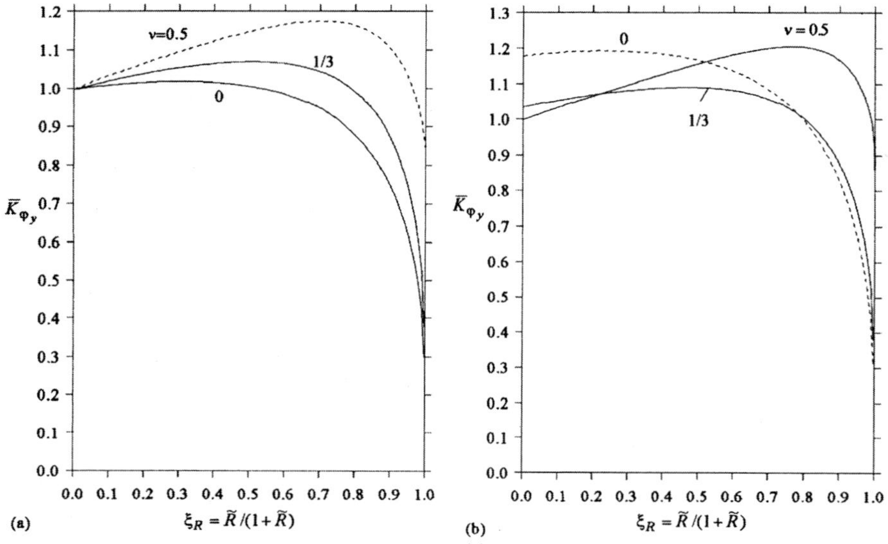

Fig. 3.21a,b. Normalized static rocking stiffness of disk on linearly heterogeneous half-space for (a) relaxed and (b) welded contact

curves shown here are similar to those representing the vertical stiffness of the disk.

When determining the resultant horizontal force arising due to rotation of the disk with respect to the *Y*-axis without horizontal displacements we come to the concept of the rocking horizontal stiffness, $K_{\varphi_y x}$, which should equal the quantity $K_{x\varphi_y}$ considered earlier, in accordance with the principle of reciprocity (Fig. 3.20). This fact is confirmed by calculations to a high precision, which serves as an acceptable validation of the Green's functions used and the numerical method employed.

Torsional Stiffness

Here, we again use the approach and equations given in the corresponding representation in section 2.6.1, employing Green's functions for the considered linearly heterogeneous half-space. Having solved the system of equations (2.236), one can determine the torsional stiffness from the relationship analogous to (2.239),

$$K_{\varphi_z} = \pi^2 G_0 R^3 \sum_{j=1}^{N} \widetilde{q}_{0j} (\overline{R}_j^4 - \overline{R}_{j-1}^4) . \tag{3.96}$$

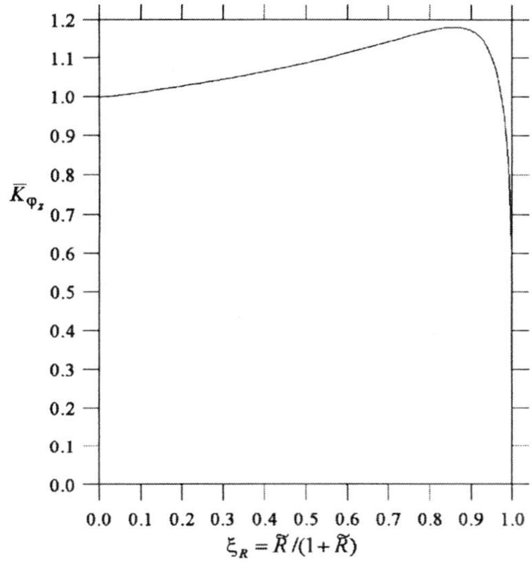

Fig. 3.22. Normalized torsional static stiffness of disk on linearly heterogeneous half-space

We introduce the normalized stiffness as the ratio of K_{φ_z} to the torsional stiffness $k_{\varphi_z}^{\text{hom}}$ of the "equivalent" homogeneous half-space,

$$k_{\varphi_z}^{\text{hom}} = \frac{16 G_\mu R^3}{3} . \tag{3.97}$$

We obtain

$$\overline{K}_{\varphi_z} = \frac{K_{\varphi_z}}{k_{\varphi_z}^{\text{hom}}} = \frac{3\pi^2}{16(1+\mu\widetilde{R})} \sum_{j=1}^{N} \widetilde{q}_{0j} (\overline{R}_j^4 - \overline{R}_{j-1}^4) . \tag{3.98}$$

In order to reach a close proximity between the results corresponding to the present half-space and to the "equivalent" homogeneous half-space, we take $\mu = 0.1$. Although the coefficients in the system of equations of the form (2.236) contain Poisson's ratio, the final result should be independent of Poisson's ratio, since the problem may be solved completely by using the function $p(z,k)$ satisfying equation (1.24). This fact is confirmed to a high precision in the calculations performed for various values of Poisson's ratio. The results of calculations are shown in Fig. 3.22. When parameter ξ_R tends to unity, the torsional stiffness tends to zero, corresponding to the behavior of the normalized

Green's functions at high values of argument $\tilde{r}$ (see section 3.5.1). For $\xi_R = 0$, the corresponding value of $\overline{K}_{\varphi_y}$ equals 0.9986 instead of its exact value 1.

In conclusion, note that the shear moduli of the "equivalent" half-spaces, determined by coefficient μ, are in good agreement with the corresponding values proposed elsewhere [35]. Following this report, the value of μ for the square stamp equals 1, 0.5, 0.3333, 0.1, for the vertical, horizontal, rocking and torsional stiffnesses, respectively; an appreciable difference occurs only for the horizontal stiffness, for which we obtain 0.35 instead of 0.5. Note that, as follows from Figs. 3.18–3.22, results corresponding to the "equivalent" half-space may deviate substantially from their exact values at high values of parameter $\tilde{R}$, with ξ_R being close to unity. Indeed, in these cases, the values of the normalized stiffness appear to be appreciably smaller than unity, so that for the given half-space, the stiffness is much smaller than for the "equivalent" homogeneous half-space. Moreover, as follows from the graphs shown in Fig. 3.20, the concept of "equivalent" homogeneous half-space is inapplicable with respect to the stiffneses, $K_{x\varphi_y}$ and $K_{\varphi_y x}$.

3.6 Dynamic Stiffness of Circular Disk Resting on Linearly Heterogeneous Half-Space

In order to determine the dynamic stiffness of the disk resting on a heterogeneous half-space, we employ the method given in section 2.6. The calculations are performed on the basis of Green's functions which reduce to the corresponding normalized amplitudes of vibrations $S_{vv}, S_{vh}, S_{hr}, S_{h\vartheta}$ of the half-space surface subjected to the action of a concentrated forces. These quantities are considered in sections 3.2, 3.3 where the parameter a determined by (3.33) serves as a main argument; the quantity θ from (3.1) is used as a parameter accounting for the degree of heterogeneity (and for the vibration frequency). The calculations are performed for a series of values of the parameter $\tilde{R}$ from (3.81) with corresponding values of the parameter a_R:

$$a_R = \omega R \sqrt{\frac{\rho}{G_0}} = \tilde{R}\theta \,, \tag{3.99}$$

If the functions $S_{vv}(a), S_{vh}(a), S_{hr}(a), S_{h\vartheta}(a)$ are found for a fixed value of parameter θ, then one can determine the value of a_R for each $\tilde{R}$ by using (3.99). Recall that in the formulas for the displacements constructed by employing the principle of superposition, the argument r in the Green's functions is replaced with λR, where R is a radius of the loaded area; the arguments R_j and R_{j-1} in the elements of matrices in the systems of equations (e.g. having the form (2.181)) indicate that in the integrals which determine the displacements the argument a in the normalized amplitudes is replaced with $\lambda a_R \overline{R}_j$ and $\lambda a_R \overline{R}_{j-1}$, respectively. Evidently, in contrast to the static problems, in order to obtain a sufficiently complete representation of the behavior of the dynamic stiffness versus parameter a_R, one should determine the functions $S_{vv}(a), S_{va}(a), S_{hr}(a), S_{h\vartheta}(a)$ for a large number of values of the parameter θ. The results of calculations presented below are performed with the Poisson's ratio value $\nu = 1/3$ for a number of values of the parameter $\widetilde{R}$, while the dissipative parameter ε is taken to be equal to zero. As in the previous calculations the values $N = 40$, $q = 0.9$ are used.

Vertical Stiffness

Analogous to relationship (2.242), an expression for the vertical stiffness may be written as follows

$$K_z = \frac{2\pi^2 R G_0}{1-\nu} \sum_{j=1}^{N} \widetilde{p}_j (\overline{R}_j^2 - \overline{R}_{j-1}^2) \ , \tag{3.100}$$

where the quantities $\widetilde{p}_j$ are determined from the system of equations having the form (2.181) under welded contact conditions and from the system (2.190) under relaxed contact conditions. The results of calculations performed for $\nu = 1/3$ are presented in Fig. 3.23 (relaxed contact) and in Fig. 3.24 (welded contact) for a number of values of parameter $\widetilde{R}$ determined by (3.81). The graphs are plotted for the normalized dynamic stiffness, being equal to the ratio of the parameter K_z from (3.100) to the corresponding static value K_z^{st}; the static stiffnesses are considered in section 3.5.2 (see the graphs in Fig. 3.18). As an argument of the considered functions, we use the parameter a_μ determined by the shear modulus G_μ of the "equivalent" homogeneous half-space (3.82c),

$$a_\mu = \omega R \sqrt{\frac{\rho}{G_\mu}} = \omega R \sqrt{\frac{\rho}{G_0(1+\mu\widetilde{R})}} \ . \tag{3.101}$$

Each peak of the broken lines shown in Figs. 3.23 and 3.24 corresponds to the calculation performed with a specific value of the parameter θ; i.e. it corresponds to the employment of the functions $S_{vv}(a), S_{vh}(a), S_{hr}(a), S_{h\vartheta}(a)$ for this value of θ. These functions are calculated with a very small step on an initial interval of varying the parameter a (with the step of 0.002 on the interval $0 < a < 1$). Furthermore, the step is increased (the step of 0.05 was taken for $a > 1$).

The parameter a_μ given in relationship (3.101) is analogous to a_R from (3.99), but with the modulus of the "equivalent" homogeneous half-space, G_μ, determined by (3.82c), instead of the shear modulus G_0 corresponding to the

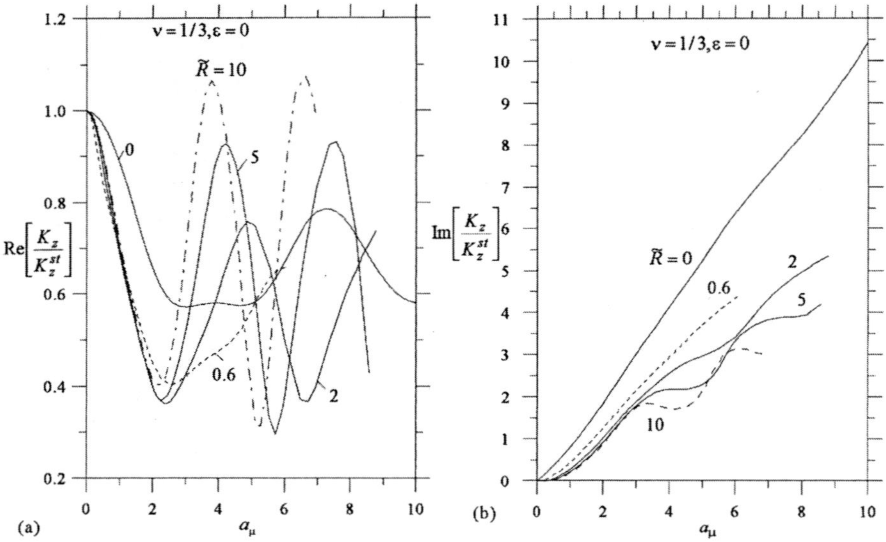

Fig. 3.23a,b. Real (a) and imaginary (b) parts of normalized dynamic vertical stiffness of disk on linearly heterogeneous half-space under relaxed contact conditions ($\nu = 1/3$)

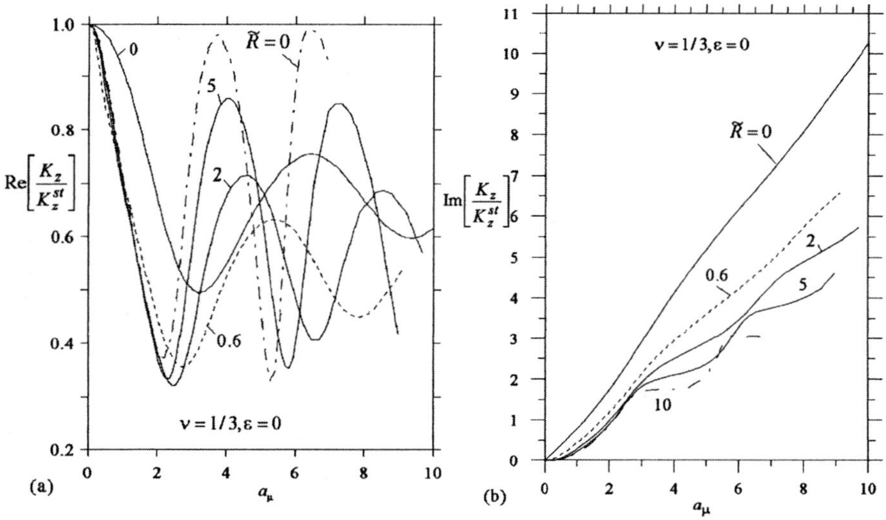

Fig. 3.24a,b. Real (a) and imaginary (b) parts of normalized dynamic vertical stiffness of disk on linearly heterogeneous half-space under welded contact conditions ($\nu = 1/3$)

surface of the half-space (in the considered case of vertical vibrations $\mu = 1$). Such a choice results in a proximity of the normalized dynamic stiffnesses corresponding to different values of $\widetilde{R}$, on a sufficiently wide interval of varying of the parameter a_μ. This is because the considered heterogeneous half-spaces are reduced to the same "equivalent" homogeneous half-space which was introduced when studying the static contact problems. Actually, for each value of $\widetilde{R}$, an abscissa axis corresponding to this value is used.

Note that the normalized stiffness varies with parameter $\widetilde{R}$ in a non-uniform fashion: at small values of a_μ, in going from a zero value of $\widetilde{R}$ corresponding to the homogeneous half-space to small values of $\widetilde{R}$, normalized stiffnesses vary relatively quickly, and in the following, the variability becomes less significant (the curves corresponding to rather different values of $\widetilde{R}$ lay in a close proximity). The graphs shown in Figs. 3.23b, 3.24b represent the quantity proportional to the damping ratio. One can see that heterogeneity causes a significant decreasing of this parameter. As follows from the comparison of the graphs shown in Figs. 3.23, 3.24, the influence of the contact conditions (welded or relaxed contact) on the value of the dynamic stiffness appears to be rather weak.

Analogous with representation given in section 2.6.2 we consider the asymptotics for the imaginary part of dynamic stiffness corresponding to high frequencies when the force of interaction between the disk and the half-space should be determined by the properties of the half-space in a vicinity of its surface [35]. The replacement of the quantity A_{zz} with G_0/τ^2 in relationship (2.245), where τ^2 is expressed according to (1.134), yields

$$P_0 \approx \frac{\pi R^2}{\tau} \mathrm{i}\,\omega\sqrt{\rho G_0}\, u_{z0} = \mathrm{i}\,\pi R^2 \omega\rho C_p u_{z0} = \mathrm{i}\frac{\pi}{\tau} G_0 R a_R u_{z0}, \tag{3.102}$$

where C_p denotes the propagation velocity of the tension-compression waves in the corresponding homogeneous half-space. The coefficient of amplitude u_{z0} in (3.102) represents the dynamic stiffness $K_{z\infty}$ corresponding to the unbounded increase of the vibration frequency,

$$K_{z\infty} = \mathrm{i}\,\pi R^2 \omega\rho C_p = \mathrm{i}\frac{\pi}{\tau} G_0 R a_R. \tag{3.103}$$

In Fig. 3.25, the ratio of the imaginary parts of K_z to $K_{z\infty}$ versus frequency parameter a_R is presented for the above-considered values of parameters and for relaxed contact. One can see that this ratio tends to unity confirming the physically clear concept that at high frequencies the heterogeneity has no effect on the interaction between a stamp and a half-space. A rate of approaching the limiting value decreases with increasing the parameter $\widetilde{R}$. In the cases of a homogeneous

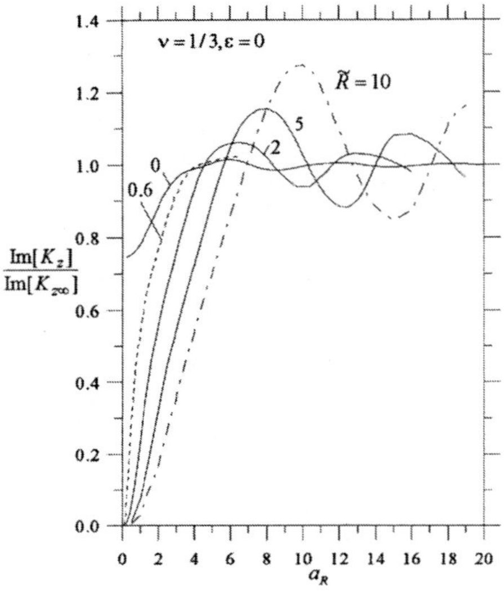

Fig. 3.25. Ratio of imaginary part of vertical stiffness of disk on linearly heterogeneous half-space ($\nu = 1/3$) to corresponding limiting value for $\omega \to \infty$ (relaxed contact)

half-space ($\widetilde{R} = 0$) and a heterogeneous half-space with small values of $\widetilde{R}$, it is advisable to use the limiting value $\mathrm{Im}[K_{z\infty}]$ for the imaginary part of the stiffness starting with $a_R \approx 2$, whereas for values of $\widetilde{R}$ equal to 5 and 10, the values of functions in Fig. 3.25 differ from unity substantially at $a_R = 2$: their values are equal to 0.3302 and 0.1661, respectively. The heterogeneity results in appreciable oscillations of the normalized stiffness presented in Fig. 3.25 about its limiting value. Even at small values of the parameter $\widetilde{R}$, the imaginary parts of the dynamic stiffness may differ significantly from the values corresponding to the homogeneous half-space at low frequencies of vibrations. The neglecting heterogeneity (even though the shear modulus increases with depth very slowly ($z_0 >> 1$)) may result in a significant overestimation of the parameter which represents the dissipative properties of the half-space and, correspondingly, in an underestimation of the amplitudes of forced vibrations of structures resting on soil foundations.

Expression (3.103) for the dynamic stiffness corresponding to high values of the frequency parameter a_R has no meaning in the case of $\nu = 0.5$ $(\tau = 0)$. In this case, the solution of the problem yields results which are significantly different from those obtained for the compressible medium. The curves corresponding to the case of welded contact are shown in Fig. 3.26. For relaxed contact, curves have a similar form [78]. First of all, the oscillations of the real part of the dynamic stiffness disappear; the absolute value of the real part increases without

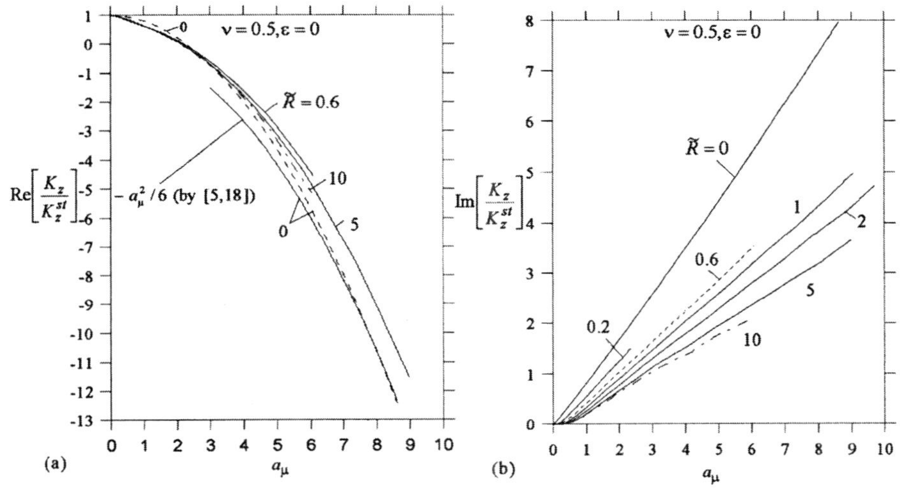

Fig. 3.26a,b. Real (a) and imaginary (b) parts of normalized dynamic vertical stiffness of disk on linearly heterogeneous half-space under welded contact conditions ($\nu = 1/2$)

bound following approximately the parabolic law, with increasing the vibration frequency. The imaginary part increases slower, according to the law that is close to the linear law (at high values of a_μ). These conclusions are in good agreement with the results reported elsewhere [5, 18]. The graph shown in Fig. 3.26a represents the high-frequency approximation for the normalized real part of the vertical stiffness as given in [5, 18].

In some reports (see [26, 34, 35]), the authors propose a correction for asymptotic representation (3.103) based on the approximation recommended by Lysmer [71, 96] for small values of parameter a_R :

$$K_{zL} = \mathrm{i}\frac{3.4 G_0 R a_R}{1-\nu}. \tag{3.104}$$

This formula corresponds to the replacement of the velocity C_p in expression (3.103) with the so called Lysmer's analog "wave velocity" [34, 35], C_{La} ,

$$C_{\mathrm{La}} = \frac{3.4}{\pi(1-\nu)}\sqrt{\frac{G_0}{\rho}}. \tag{3.105}$$

The ratios $K_{zL}/K_{z\infty}$ and C_{La}/C_p are identical. The calculations as well as physical considerations show that the quantity (3.103) rather than (3.104) serves as the correct high-frequency asymptotic representation for the stiffness at $\nu < 0.5$. It is another matter that the stiffness approaches its limiting value for more and

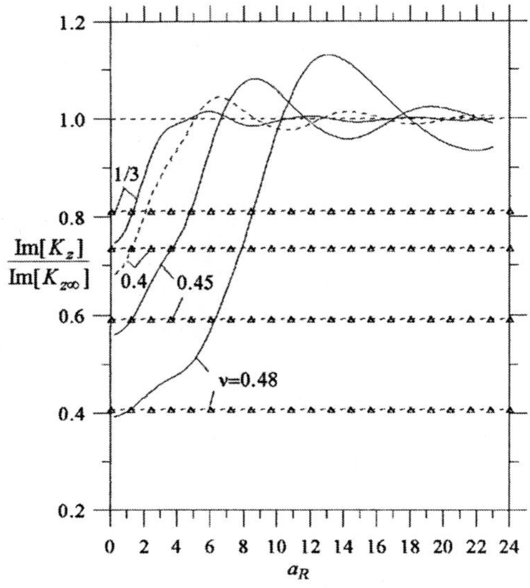

Fig. 3.27. Influence of Poisson's ratio on behavior of imaginary part of vertical stiffness of disk on homogeneous half-space for relaxed contact. Lines denoted with triangles represent ratio of K_{zL} from (3.104) to $K_{z\infty}$ from (3.103)

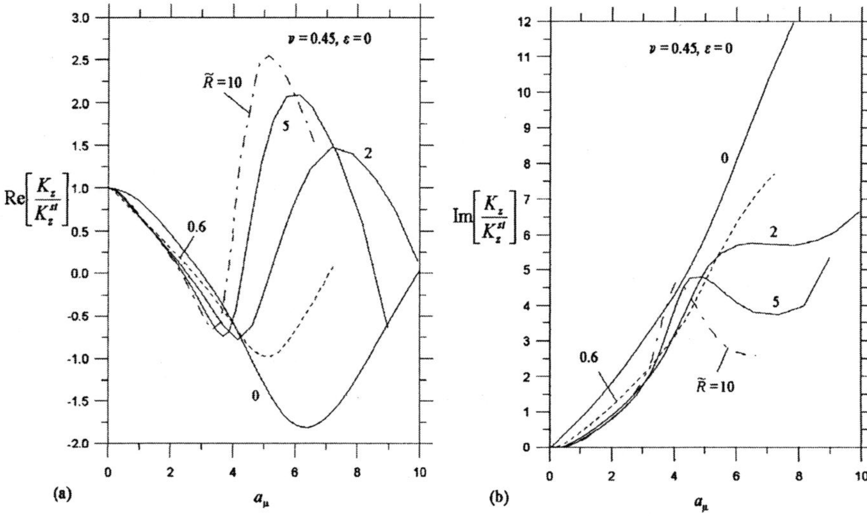

Fig. 3.28a,b. Real (a) and imaginary (b) parts of normalized vertical dynamic stiffness of disk on linearly heterogeneous half-space under relaxed contact conditions ($\nu = 0.45$)

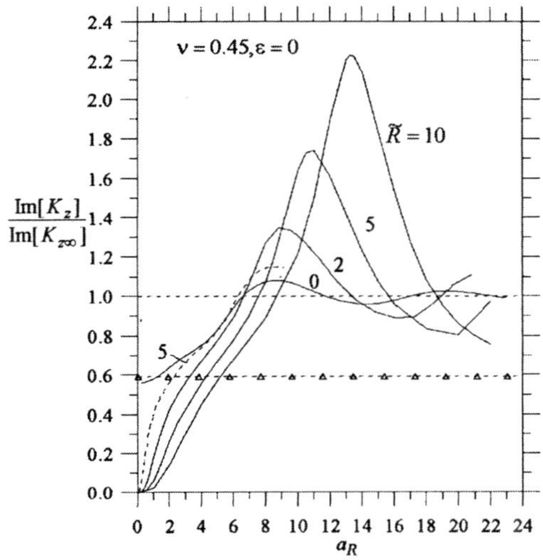

Fig. 3.29. Ratio of imaginary part of vertical stiffness of disk on linearly heterogeneous half-space under relaxed contact conditions ($\nu = 0.45$) to corresponding limiting value for $\omega \to \infty$

more values of the frequency parameter a_R when Poisson's ratio becomes close to 0.5. The results of calculations performed for the homogeneous half-space for a number of values of Poisson's ratio are presented in Fig. 3.27. The straight lines marked with triangles in Fig. 3.27 represent the ratio of the quantity K_{zL} from (3.104) to $K_{z\infty}$ from (3.103). One can see that the adoption of K_{zL} for the high-frequency asymptotics for the imaginary part of the dynamic stiffness leads to an appreciable error when Poisson's ratio approaches 0.5. In addition, the graphs show that the quantity K_{zL} provides an acceptable approximation for the imaginary part of the dynamic stiffness at small values of the parameter a_R as stated by Lysmer.

It is interesting to take a look at the results of calculations of the stiffness of the disk resting on the heterogeneous half-space for the values of Poisson's ratio close to 0.5. In Fig. 3.28, values of the vertical stiffness are presented for the case of relaxed contact at $\nu = 0.45$. The ratio $\mathrm{Im}[K_z]/\mathrm{Im}[K_{z\infty}]$ is shown in Fig. 3.29. The deviations of the imaginary part of the stiffness from the corresponding limiting values become significant on the wide interval of variation of the frequency parameter, especially, at high values of the parameter $\widetilde{R}$. The straight line marked with triangles represents the ratio of K_{zL} from (3.104) to $K_{z\infty}$ from (3.103).

Horizontal Stiffness

The horizontal dynamic stiffness is determined from the equation analogous to (2.247):

$$K_x = 2\pi^2 R G_0 \sum_{j=1}^{N} \widetilde{q}_j (\overline{R}_j^2 - \overline{R}_{j-1}^2), \tag{3.106}$$

where the quantities $\widetilde{q}_j$ (together with the quantities $\widetilde{q}_{0j}$ and $\widetilde{p}_{0j}$) satisfy the system of equations (2.200) under welded contact conditions and the system of equations (2.212) when neglecting the influence of vertical stresses on the horizontal displacements of points in the contact area (relaxed contact). The results of calculations corresponding to these cases are in close proximity to each other; in practice, it is sufficient to employ the relaxed contact conditions. However, the complete accounting for the contact conditions is required when one should determine moments which occur due to the vertical contact stresses during the horizontal translation motion of the disk. In Fig. 3.30, the results of calculations are presented at $\nu = 1/3$ (the complete including the contact conditions). Here, the graphs are plotted for the normalized dynamic stiffness equal to the ratio of K_x from (3.100) to the corresponding static value K_x^{st} (see the graphs in Fig. 3.19). As above, the quantity a_μ from (3.101) is taken for the argument of the considered functions at $\mu = 0.35$ (such a value was given in section 3.5.2 to determine the static horizontal stiffness). In contrast to the case of vertical vibrations, the horizontal stiffness varies more smoothly with varying the frequency parameter a_μ.

Next we consider an asymptotic representation for the dynamic stiffness corresponding to high frequencies. Analogously to relationship (2.250), the following asymptotic expression may be written for the amplitude of the resultant horizontal force Q_0:

$$Q_0 \approx \pi R^2 \, \mathrm{i}\,\omega \sqrt{\rho G_0}\, u_{x0} . \tag{3.107}$$

The coefficient of the amplitude of horizontal vibrations of the disk, u_{x0}, represents the horizontal stiffness for $\omega \to \infty$,

$$K_{x\infty} = \pi R^2 \, \mathrm{i}\,\omega \sqrt{\rho G_0} = \mathrm{i}\,\pi R G_0 a_R . \tag{3.108}$$

The ratio of $\mathrm{Im}[K_x]$ to $\mathrm{Im}[K_{x\infty}]$ is shown in Fig. 3.31. As in the case of vertical vibrations, approaching the limiting value becomes slower with increasing the degree of heterogeneity (the parameter $\widetilde{R}$). Notice that the value $a_R \approx 1.5$ [35], taken as a lower bound of an interval where $\mathrm{Im}[K_x]$ may be replaced with $\mathrm{Im}[K_{x\infty}]$, is clearly insufficient at high values of $\widetilde{R}$. In the graphs (Fig. 3.31), the

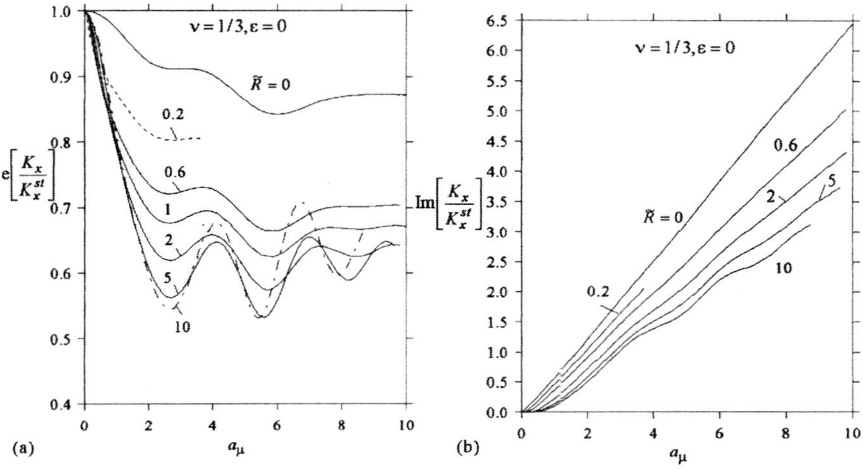

Fig. 3.30a,b. Real (a) and imaginary (b) parts of normalized horizontal dynamic stiffness of disk on linearly heterogeneous half-space under welded contact conditions

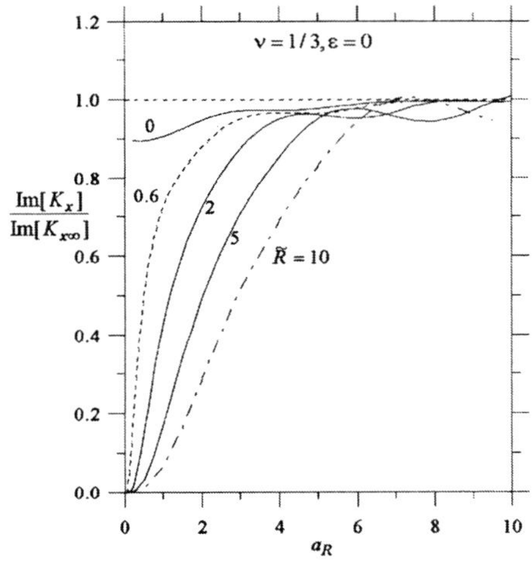

Fig. 3.31. Ratio of imaginary part of horizontal stiffness of disk on linearly heterogeneous half-space to corresponding limiting value for $\omega \to \infty$

ordinates for $a_R = 1.5$ are rather distant from unity, being equal to 0.82, 0.61, 0.35, 0.17 for $\widetilde{R} = 0.6, 2, 5, 10$, respectively.

Next, we examine the horizontal rocking stiffness, $K_{x\varphi_y}$ (side dynamic stiffness), representing the value of the moment which should be applied to the

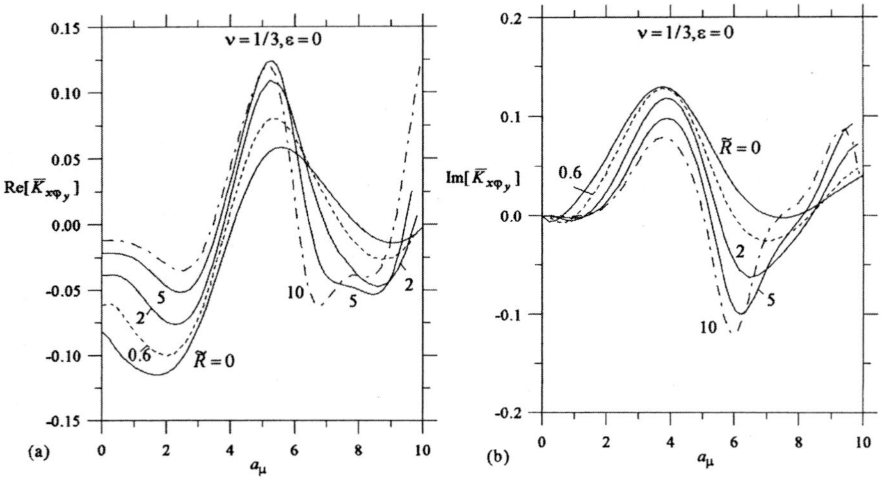

Fig. 3.32a,b. Real (a) and imaginary (b) parts of normalized horizontal rocking stiffness of disk on linearly heterogeneous half-space ($\nu = 1/3$)

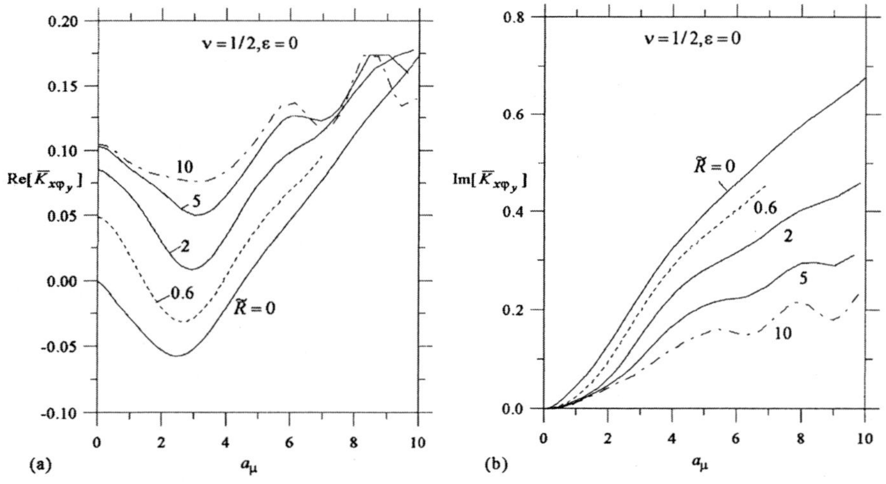

Fig. 3.33a,b. Real (a) and imaginary (b) parts of normalized horizontal rocking stiffness of disk on linearly heterogeneous half-space ($\nu = 0.5$)

disk (together with the horizontal force), in order to provide its horizontal translational vibrations. Expression for $K_{x\varphi_y}$ given in relationship (3.90) is still valid as well as the previously employed normalization (3.91). In order to determine the horizontal rocking stiffness, one should use the complete system of equations having form (2.200). In Fig. 3.32, the quantity $\overline{K}_{x\varphi_y}$ is shown for a number of values of the parameter $\widetilde{R}$ and at $\nu = 1/3$. Once again, the parameter a_μ stands for the argument at $\mu = 0.35$. At low and medium frequencies ($a_\mu < 3$), the heterogeneity leads to decreasing of the absolute values of $\overline{K}_{x\varphi_y}$. For the incompressible medium ($\nu = 1/2$), the behavior of the side stiffness differs substantially from its behavior at $\nu = 1/3$ (see the graphs in Fig. 3.33).

Rocking Stiffness

The rocking stiffness determined by the resultant moment of the contact pressures about the *Y*-axis may be calculated by using the formula analogous to (2.252):

$$K_{\varphi_y} = \frac{\pi^2 G_0 R^3}{2(1-\nu)} \sum_{j=1}^{N} \widetilde{p}_{0j} (\overline{R}_j^4 - \overline{R}_{j-1}^4), \tag{3.109}$$

where the quantities $\widetilde{p}_{0j}$ determined by (2.223c) (ν' is replaced with ν and G_{rz0} with G_0) satisfy the system of equations (2.222) in the case of welded contact and the system of equations (2.226) when neglecting the influence of the horizontal contact stresses on the rocking stiffness (relaxed contact). In Fig. 3.34, the ratio of the stiffness under welded contact conditions to its corresponding static value is presented at $\nu = 1/3$ with the value of argument a_μ used previously ($\mu = 0.35$). The static values are shown in Fig. 3.21. The behavior of the real part of the normalized stiffness differs from its behavior in the cases with the vertical and horizontal vibrations when at low frequencies the corresponding values are smaller than those calculated for a homogeneous half-space. In the case of rotational vibrations, an increase occurs (at low frequencies) in going from a homogeneous to the heterogeneous half-space. Notice that this increase would be even more obvious, if one used the common argument a_R for all the curves instead of a_μ. The imaginary parts of the normalized stiffness are appreciably smaller than the corresponding values in the cases of vertical and horizontal vibrations.

The high-frequency asymptotics for the dynamic stiffness can be given using relationship (2.253) between the amplitude of resultant moment about the *Y*-axis and the amplitude of variation of the rotation angle (in the considered isotropic case, the parameter A_{zz} is replaced with G_0/τ^2),

$$M_{y0} \approx \frac{\pi R^4}{4\tau} \mathrm{i}\omega\sqrt{\rho G_0}\,\varphi_{y0} = \frac{\pi R^4}{4} \mathrm{i}\omega\rho C_p\,\varphi_{y0} = \mathrm{i}\frac{\pi}{4\tau} G_0 R^3 a_R \varphi_{y0}. \tag{3.110}$$

From (3.110) the expression for the stiffness corresponding to unlimited increase in the vibration frequency becomes

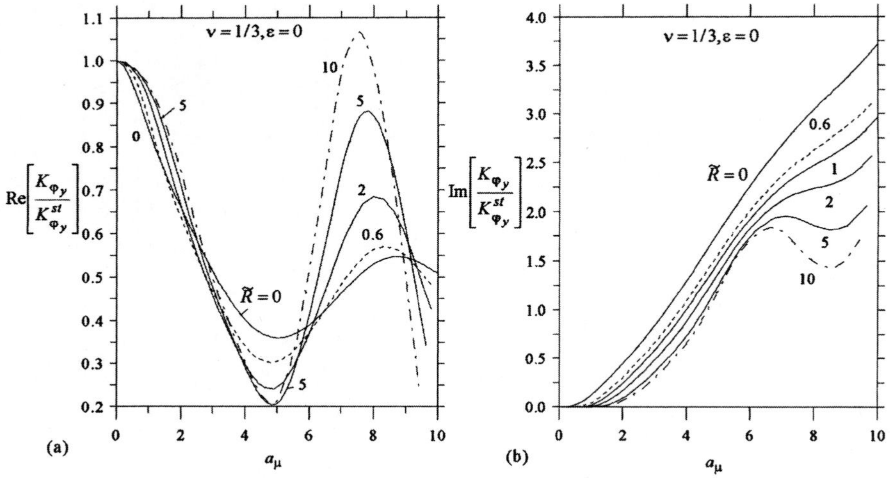

Fig. 3.34a,b. Real (a) and imaginary (b) parts of normalized dynamic rocking stiffness of disk on linearly heterogeneous half-space under welded contact conditions ($\nu = 1/3$)

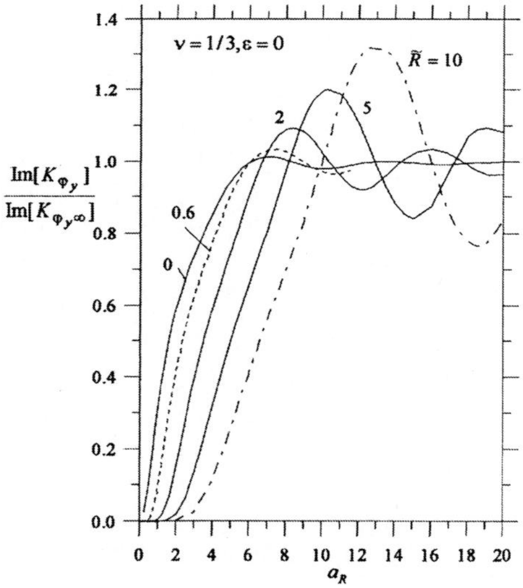

Fig. 3.35. Ratio of imaginary part of rocking stiffness of disk to corresponding limiting value for $\omega \to \infty$

$$K_{\varphi_y \infty} = \frac{\pi R^4}{4} \mathrm{i}\,\omega \rho C_p = \mathrm{i} \frac{\pi}{4\tau} G_0 R^3 a_R \,. \tag{3.111}$$

In Fig. 3.35, the ratio of $\mathrm{Im}[K_{\varphi_y}]$ from (3.109) to $\mathrm{Im}[K_{\varphi_y\infty}]$ is given for the data used in the previous figure. These curves are similar to those shown in Fig. 3.25. The rate of approaching the limiting value being equal to unity is now much lower than in the case of vertical vibrations of the disk.

For the welded contact conditions, the results of calculations for the incompressible medium ($\nu = 0.5$) are presented in Fig. 3.36. One can see that the graphs are similar to those related to the vertical vibrations of the disk (Fig. 3.26), but the absolute values of the ordinates increase now much slower at high values of the parameter a_μ.

As in the case of vertical vibrations, the discontinuous behavior of the dynamic stiffness occurs when going from the case with $\nu = 0.5$ to the case of the compressible medium. The results of calculations related to the value $\nu = 0.45$ (relaxed contact) are shown in Fig. 3.37. For $a_\mu < 3$, the curves are only slightly different from those plotted for $\nu = 0.5$. As a_μ increases the difference becomes rather significant: at $\nu = 0.45$ the real parts oscillate about some constant value, while at $\nu = 0.5$ an increase according to the parabolic law occurs; at $\nu = 0.45$ imaginary parts become much larger than in the case of $\nu = 0.5$.

Consider high-frequency asymptotics of the dynamic rocking stiffness for $\nu = 0.45$ (Fig. 3.38). The curves shown are analogous to those presented in Fig. 3.35, however, as parameter $\tilde{R}$ increases the imaginary parts of the stiffness approach their limiting values at a slower rate. Let us suppose that quantity C_p in the asymptotic expression (3.110) for the dynamic stiffness is replaced with the value C_{La} from (3.105) as recommended in [34, 35]. This replacement yields an

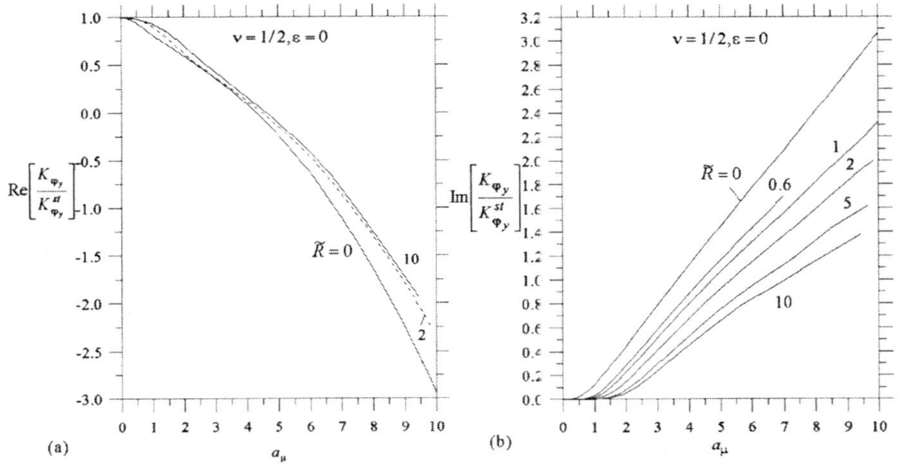

Fig. 3.36a,b. Real (a) and imaginary (b) parts of normalized dynamic rocking stiffness of disk on linearly heterogeneous half-space under welded contact conditions ($\nu = 0.5$)

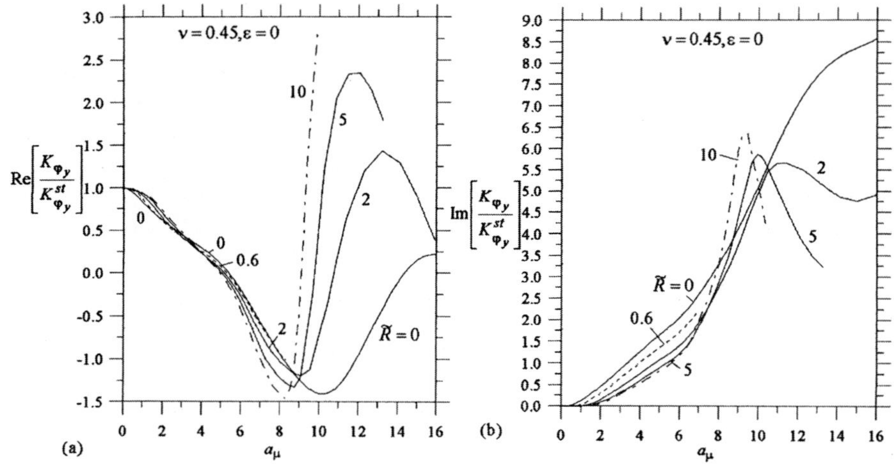

Fig. 3.37a,b. Real (a) and imaginary (b) parts of normalized dynamic rocking stiffness of disk on linearly heterogeneous half-space under relaxed contact conditions ($\nu = 0.45$)

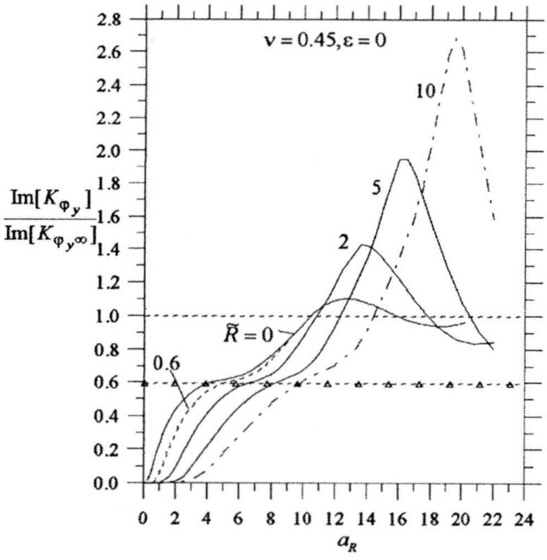

Fig. 3.38. Ratio of imaginary part of rocking stiffness of disk on linearly heterogeneous half-space ($\nu = 0.45$) to corresponding limiting value for $\omega \to \infty$

underestimated value for the limiting rocking stiffness ($K_{\varphi_y L}$). The ratio of $K_{\varphi_y L}$ to $K_{\varphi_y \infty}$ is shown in Fig. 3.38 (the straight line marked with triangles).

When determining the resultant horizontal force which appears due to rotation of the disk about horizontal the Y-axis, we obtain the rocking horizontal stiffness, that should be equal to the horizontal rocking stiffness, $K_{x\varphi_y}$, according to the Rayleigh's principle of reciprocity. In the calculations this equality is confirmed to a high precision providing a good verification for the employed technique of solving the contact problems and for the Green's functions used.

Torsional Stiffness

In the following, we use the technique presented in section 2.7.1. The equation for the torsional stiffness can be written in the form analogous to relationship (2.239):

$$K_{\varphi_z} = \frac{\pi}{2}\sum_{j=1}^{N} q_{0j}(R_j^4 - R_{j-1}^4) = \pi^2 G_0 R^3 \sum_{j=1}^{N} \tilde{q}_{0j}(\overline{R}_j^4 - \overline{R}_{j-1}^4), \tag{3.112}$$

where the quantities $\tilde{q}_{0j}$ are determined by (2.237) (G_{rz0} is replaced with G_0) using the tangential loads q_{0j} applied to the ring-shaped elements. We consider the normalized stiffnesses equal to the ratios of quantities K_{φ_z} to the corresponding static values of stiffness studied in the previous subsection. As previously, the parameter a_μ stands for the argument; in the considered case, μ = 0.1. The results of calculations independent of Poisson's ratio are presented in Fig. 3.39. Note that the graph corresponding to the homogeneous half-space shown in Fig. 3.39a ($\widetilde{R} = 0$) should approach the horizontal line with ordinate 0.589 from (2.258), as a_μ increases (the dashed line). As the parameter $\widetilde{R}$ increases, the corresponding limiting values decrease, approaching the value 0.53 and, in addition, the oscillations about this limiting value increase.

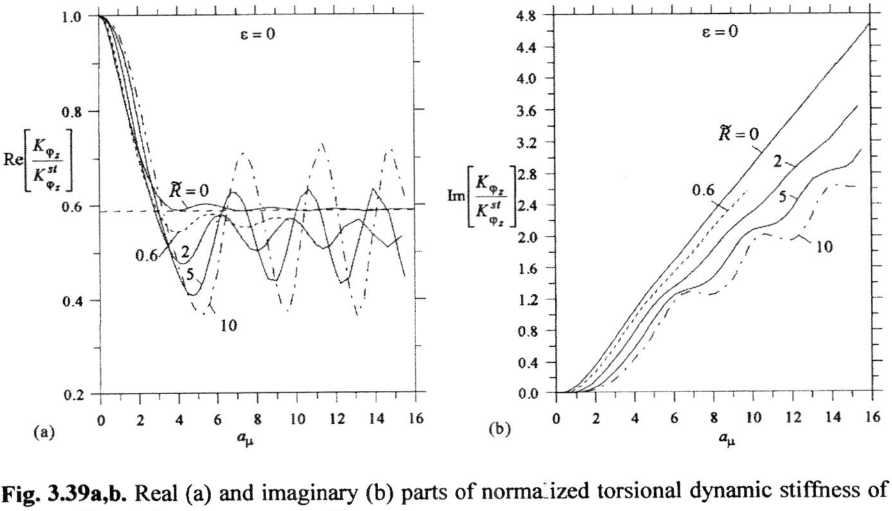

Fig. 3.39a,b. Real (a) and imaginary (b) parts of normalized torsional dynamic stiffness of disk on linearly heterogeneous half-space

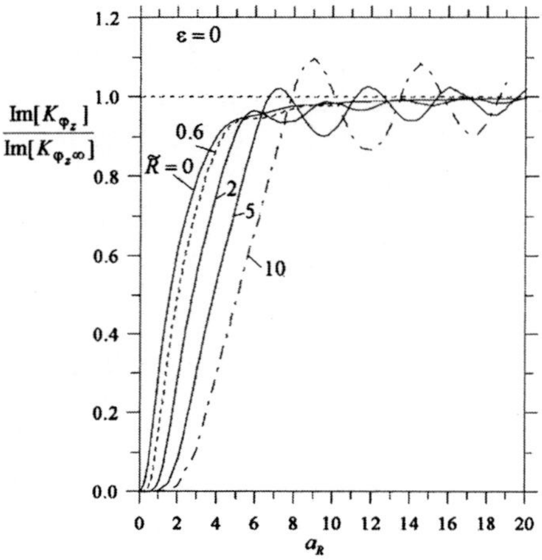

Fig. 3.40. Ratio of imaginary part of torsional stiffness of disk on linearly heterogeneous half-space to corresponding limiting value for $\omega \to \infty$

An asymptotic representation for the imaginary part of the dynamic stiffness corresponding to high frequencies of vibrations can be determined using relationship (2.256):

$$K_{\varphi_z \infty} = \frac{\mathrm{i}}{2}\pi R^4 \omega \sqrt{\rho G_0} = \frac{\mathrm{i}}{2}\pi R^3 G_0 a_R . \qquad (3.113)$$

In Fig. 3.40, the graphs represent the ratio $\mathrm{Im}[K_{\varphi_z}]/\mathrm{Im}[K_{\varphi_z \infty}]$. As in the case with other forms of disk vibrations, the rate of approaching the limiting value (being equal to unity) decreases as the parameter $\widetilde{R}$ increases. Note that the value $a_R \approx 2.5$, above which the limiting representation of the imaginary part of stiffness becomes applicable according to recommendation in [35], is obviously insufficient at high values of the parameter $\widetilde{R}$.

3.7 Vibrations of Linearly Heterogeneous Half-Space Subjected to Force Applied within Half-Space

3.7.1 Action of Vertical Force

We employ the solution of the problem in the form given in section 1.6. Using equation (1.157) with accounting for the isotropy of the half-space yields

$$u_r = \frac{P_0}{G(z_0)\pi R}\int_0^{\infty} \mathrm{J}_1(\tilde{k}\tilde{R})\mathrm{J}_1(\tilde{k}\tilde{r})[A_1^*(\tilde{k})\tilde{q}_1(\tilde{z},\tilde{k}) + A_2^*(\tilde{k})\tilde{q}_2(\tilde{z},\tilde{k}) + \tilde{q}_a^*(\tilde{z},\tilde{k})]\mathrm{d}\tilde{k}\ , \tag{3.114a}$$

$$u_z = \frac{P_0}{G(z_0)\pi R}\int_0^{\infty} \mathrm{J}_1(\tilde{k}\tilde{R})\mathrm{J}_0(\tilde{k}\tilde{r})[A_1^*(\tilde{k})\tilde{w}_1(\tilde{z},\tilde{k}) - A_2^*(\tilde{k})\tilde{w}_2(\tilde{z},\tilde{k}) + \tilde{w}_a^*(\tilde{z},\tilde{k})]\mathrm{d}\tilde{k}, \tag{3.114b}$$

where the value z_r is replaced with z_0 in order to determine the dimensionless quantities marked with "~" (see relationships (1.27), (1.158)). The particular solutions (denoted by index a) are determined using equation (1.153) where b_e is replaced with unity,

$$\begin{Bmatrix} \tilde{\chi}_a^* \\ \tilde{e}_a^* \\ \tilde{q}_a^* \\ \tilde{w}_a^* \end{Bmatrix} = \sum_{j=1}^{4} B_j \begin{Bmatrix} \tilde{\chi}_j \\ \tilde{e}_j \\ \tilde{q}_j \\ \tilde{w}_j \end{Bmatrix}, \tag{3.115}$$

where the coefficients B_j satisfy the system of equations (1.154). The coefficients A_j^* are calculated from the system of equations similar to (1.53):

$$\begin{aligned} c_{11}A_1^* + c_{12}A_2^* &= d_1, \\ c_{21}A_1^* + c_{22}A_2^* &= d_2\,, \end{aligned} \tag{3.116}$$

where the right-hand sides of the equations are determined from relationships (1.156) (by taking into account isotropy):

$$d_1 = -[\tilde{e}_a^*(1,\tilde{k}) - 2\tilde{k}\tilde{q}_a^*(1,\tilde{k})] = -\sum_{j=1}^{4} c_{1j}B_j\ , \tag{3.117a}$$

$$d_2 = -[\tilde{\chi}_a^*(1,\tilde{k}) - 2\tilde{k}\tilde{w}_a^*(1,\tilde{k})] = -\sum_{j=1}^{4} c_{2j}B_j\ . \tag{3.117b}$$

Here, we take into account that $\tilde{z}_0 = 1$ and an asterisk mean that in the particular solutions b_e is replaced with unity. The coefficients c_{ij} in equations (3.116), (3.117) are calculated according to relationships (1.147).

The fundamental solutions entering relationships (3.114), (3.115) which were constructed in section 3.1 contain the multiplier $\exp(-\varsigma/2)$ for $j = 1,2$, and $\exp(\varsigma/2)$ (in the asymptotic representation for high values of ς) for $j = 3,4$. It is advisable to modify the coefficients B_j from the system of equations (1.154) by multiplying by $\exp(-\varsigma_1/2)$ (for $j = 1,2$) and by $\exp(\varsigma_1/2)$ (for $j = 3,4$):

$$B_1^* = B_1 \exp(-\varsigma_1/2), \tag{3.118a}$$

$$B_2^* = B_2 \exp(-\varsigma_1/2), \tag{3.118b}$$

$$B_3^* = B_3 \exp(\varsigma_1/2), \tag{3.118c}$$

$$B_4^* = B_4 \exp(\varsigma_1/2). \tag{3.118d}$$

Instead of system (1.154) we have

$$\sum_{j=1}^{4} B_j^* \begin{pmatrix} \widetilde{\chi}_j^*(\varsigma_1) \\ \widetilde{e}_j^*(\varsigma_1) \\ \widetilde{q}_j^*(\varsigma_1) \\ \widetilde{w}_j^*(\varsigma_1) \end{pmatrix} = \begin{pmatrix} 0 \\ 1 \\ 0 \\ 0 \end{pmatrix}, \tag{3.119}$$

where $\varsigma_1 = 2kz_1 = 2\widetilde{k}\widetilde{z}_1$ and the fundamental solutions, marked with an asterisk, represent the corresponding solutions with the omitted multiplier $\exp(-\varsigma/2)$ for $j = 1,2$ (see equations (3.12), (3.22)), and with the multiplier $\exp(-\varsigma)$ instead of $\exp(-\varsigma/2)$ for $j = 3,4$ (see equations (3.13), (3.24)). Note that the quantity ς stands for the argument of the fundamental solutions; parameter k or $\widetilde{k}$ included in the fundamental solutions is not mentioned. According to the asymptotics of confluent hypergeometric functions at high values of the argument, coefficients in the system of equations (3.119) do not contain terms with exponentially increasing or decreasing argument ς_1. This is important when performing calculations. Furthermore, in order to clarify the behavior of the integrands in relationships (3.114), we join the terms having identical fundamental solutions ($z < z_1$):

$$\begin{aligned} A_1^*\widetilde{q}_1 + A_2^*\widetilde{q}_2 + \widetilde{q}_a^* &= B_1^{\wedge}\widetilde{q}_1 + B_2^{\wedge}\widetilde{q}_2 + B_3\widetilde{q}_3 + B_4\widetilde{q}_4 \\ &= B_1^{\wedge}\widetilde{q}_1 + B_2^{\wedge}\widetilde{q}_2 + B_3^* e^{-\varsigma_1/2}\,\widetilde{q}_3 + B_4^* e^{-\varsigma_1/2}\,\widetilde{q}_4, \\ A_1^*\widetilde{w}_1 + A_2^*\widetilde{w}_2 + \widetilde{w}_a^* &= B_1^{\wedge}\widetilde{w}_1 + B_2^{\wedge}\widetilde{w}_2 + B_3\widetilde{w}_3 + B_4\widetilde{w}_4 \\ &= B_1^{\wedge}\widetilde{w}_1 + B_2^{\wedge}\widetilde{w}_2 + B_3^* e^{-\varsigma_1/2}\,\widetilde{w}_3 + B_4^* e^{-\varsigma_1/2}\,\widetilde{w}_4, \end{aligned} \tag{3.120}$$

where

$$B_j^{\wedge} = A_j^* + B_j \qquad (j = 1,2). \tag{3.121}$$

Using equations (3.116), (3.117) and taking the exponents out of brackets, one can see that the coefficients $B_j^{\wedge}$ satisfy the following system of equations:

$$\overset{*}{c}_{11}B_1^{\wedge} + \overset{*}{c}_{12}B_2^{\wedge} = \mathrm{e}^{(\varsigma_0 - \varsigma_1/2)}\, d_1^{\wedge},$$

$$\overset{*}{c}_{21}B_1^{\wedge} + \overset{*}{c}_{22}B_2^{\wedge} = \mathrm{e}^{(\varsigma_0 - \varsigma_1/2)}\, d_2^{\wedge}, \qquad (3.122)$$

where

$$\overset{*}{c}_{1j} = \widetilde{e}_j^{\,*}(\varsigma_0) - 2\widetilde{k}\widetilde{q}_j^{\,*}(\varsigma_0), \qquad (3.123a)$$

$$\overset{*}{c}_{2j} = \widetilde{\chi}_j^{\,*}(\varsigma_0) - 2\widetilde{k}\widetilde{w}_j^{\,*}(\varsigma_0), \qquad (3.123b)$$

$$d_1^{\wedge} = -\sum_{j=3}^{4} \overset{*}{B}_j \overset{*}{c}_{1j}, \qquad (3.124a)$$

$$d_2^{\wedge} = -\sum_{j=3}^{4} \overset{*}{B}_j \overset{*}{c}_{2j}. \qquad (3.124b)$$

Notice that $\varsigma_0 = 2\widetilde{k}$. In equations (3.123), the fundamental solutions marked with the asterisk have the meaning given after equation (3.119). As follows from equations (3.122),

$$B_1^{\wedge} = \mathrm{e}^{(\varsigma_0 - \varsigma_1/2)} \frac{\overset{*}{c}_{22}d_1^{\wedge} - \overset{*}{c}_{12}d_2^{\wedge}}{D^*}, \qquad (3.125a)$$

$$B_2^{\wedge} = \mathrm{e}^{(\varsigma_0 - \varsigma_1/2)} \frac{\overset{*}{c}_{11}d_2^{\wedge} - \overset{*}{c}_{21}d_1^{\wedge}}{D^*}, \qquad (3.125b)$$

$$D^* = \overset{*}{c}_{11}\overset{*}{c}_{22} - \overset{*}{c}_{12}\overset{*}{c}_{21}. \qquad (3.125c)$$

For $z \geq z_1$, parts of the expression between the brackets in the solution (3.114) have the form

$$\begin{aligned} \overset{*}{A}_1\widetilde{q}_1 + \overset{*}{A}_2\widetilde{q}_2 &= B_1^{\wedge}\widetilde{q}_1 + B_2^{\wedge}\widetilde{q}_2 - B_1\widetilde{q}_1 - B_2\widetilde{q}_2 \\ &= B_1^{\wedge}\widetilde{q}_1 + B_2^{\wedge}\widetilde{q}_2 - \overset{*}{B}_1 \mathrm{e}^{\varsigma_1/2}\widetilde{q}_1 - \overset{*}{B}_2 \mathrm{e}^{\varsigma_1/2}\widetilde{q}_2, \end{aligned} \qquad (3.126a)$$

$$\begin{aligned} \overset{*}{A}_1\widetilde{w}_1 + \overset{*}{A}_2\widetilde{w}_2 &= B_1^{\wedge}\widetilde{w}_1 + B_2^{\wedge}\widetilde{w}_2 - B_1\widetilde{w}_1 - B_2\widetilde{w}_2 \\ &= B_1^{\wedge}\widetilde{w}_1 + B_2^{\wedge}\widetilde{w}_2 - \overset{*}{B}_1 \mathrm{e}^{\varsigma_1/2}\widetilde{w}_1 - \overset{*}{B}_2 \mathrm{e}^{\varsigma_1/2}\widetilde{w}_2. \end{aligned} \qquad (3.126b)$$

Using the performed transformations we can rewrite the solution (3.114) as

$$u_r = \frac{P_0}{\pi R G(z_0)} \int_0^{\infty} \mathrm{J}_1(\widetilde{k}\widetilde{R})\mathrm{J}_1(\widetilde{k}\widetilde{r})\widetilde{q}\,\mathrm{d}\widetilde{k}, \qquad (3.127a)$$

$$\widetilde{q} = \frac{\mathrm{e}^{\widetilde{k}(2-\widetilde{z}-\widetilde{z}_1)}}{D^*}[(\overset{*}{c}_{22}d_1^{\wedge} - \overset{*}{c}_{12}d_2^{\wedge})\widetilde{q}_1^{\,*} + (\overset{*}{c}_{11}d_2^{\wedge} - \overset{*}{c}_{21}d_1^{\wedge})\widetilde{q}_2^{\,*}]$$

$$+\mathrm{e}^{\tilde{k}(\tilde{z}-\tilde{z}_1)}(B_3^*\tilde{q}_3^* + B_4^*\tilde{q}_4^*) \quad (z \le z_1), \tag{3.127b}$$

$$\tilde{q} = \frac{\mathrm{e}^{\tilde{k}(2-\tilde{z}-\tilde{z}_1)}}{D^*}[(c_{22}^* d_1^{\wedge} - c_{12}^* d_2^{\wedge})\tilde{q}_1^* + (c_{11}^* d_2^{\wedge} - c_{21}^* d_1^{\wedge})\tilde{q}_2^*]$$

$$-\mathrm{e}^{\tilde{k}(\tilde{z}_1-\tilde{z})}(B_1^*\tilde{q}_1^* + B_2^*\tilde{q}_2^*) \quad (z \ge z_1), \tag{3.127c}$$

$$u_z = \frac{P_0}{\pi R G(z_0)} \int_0^{\infty} \mathrm{J}_1(\tilde{k}\tilde{R})\,\mathrm{J}_0(\tilde{k}\tilde{r})\tilde{w}\,\mathrm{d}\tilde{k}\,, \tag{3.128a}$$

$$\tilde{w} = \frac{\mathrm{e}^{\tilde{k}(2-\tilde{z}-\tilde{z}_1)}}{D^*}[(c_{22}^* d_1^{\wedge} - c_{12}^* d_2^{\wedge})\tilde{w}_1^* + (c_{11}^* d_2^{\wedge} - c_{21}^* d_1^{\wedge})\tilde{w}_2^*]$$

$$+\mathrm{e}^{\tilde{k}(\tilde{z}-\tilde{z}_1)}(B_3^*\tilde{w}_3^* + B_4^*\tilde{w}_4^*) \quad (z \le z_1), \tag{3.128b}$$

$$\tilde{w} = \frac{\mathrm{e}^{\tilde{k}(2-\tilde{z}-\tilde{z}_1)}}{D^*}[(c_{22}^* d_1^{\wedge} - c_{12}^* d_2^{\wedge})\tilde{w}_1^* + (c_{11}^* d_2^{\wedge} - c_{21}^* d_1^{\wedge})\tilde{w}_2^*]$$

$$-\mathrm{e}^{\tilde{k}(\tilde{z}_1-\tilde{z})}(B_1^*\tilde{w}_1^* + B_2^*\tilde{w}_2^*) \quad (z \ge z_1). \tag{3.128c}$$

The results obtained indicate that in the case of $z \ne z_1$, the integrands contain multipliers which decrease exponentially as the variable of integration, $\tilde{k}$, increases. As a result, the convergence of the integrals is sufficiently fast. In the case of $z = z_1$, the terms containing D^* still decrease exponentially (for $z_1 > z_0$), whereas the rest of the terms in $\tilde{q}$ and $\tilde{w}$ has the following asymptotic representation at high values of $\tilde{k}$:

$$-[B_1^*\tilde{q}_1^*(\varsigma_1) + B_2^*\tilde{q}_2^*(\varsigma_1)] = [B_3^*\tilde{q}_3^*(\varsigma_1) + B_4^*\tilde{q}_4^*(\varsigma_1)] \approx \frac{1-3\tau^2}{8\tilde{k}^2\tilde{z}_1^2}, \tag{3.129a}$$

$$-[B_1^*\tilde{w}_1^*(\varsigma_1) + B_2^*\tilde{w}_2^*(\varsigma_1)] = [B_3^*\tilde{w}_3^*(\varsigma_1) + B_4^*\tilde{w}_4^*(\varsigma_1)] \approx \frac{1+\tau^2}{4\tilde{k}\tilde{z}_1}. \tag{3.129b}$$

These relationships can be obtained by using asymptotic representations for the confluent hypergeometric functions and equations (3.119). As seen, in the case of $z = z_1$ the convergence of the integrals is rather slow, and it is advisable to speed it up by employing the representations (3.129).

Further, we consider the action of a concentrated vertical force ($R \to 0$) when the amplitudes of vibrations can be written as

$$u_r = \frac{P_0(1-\nu)}{2\pi G_0 r} S_{vh}\,, \tag{3.130a}$$

$$S_{vh} = \frac{\beta^2 \tilde{r}}{1-\nu} \int_0^\infty \mathrm{J}_1(\tilde{k}\tilde{r})\tilde{k}\tilde{q}\,\mathrm{d}\tilde{k}\,, \tag{3.130b}$$

$$u_z = \frac{P_0(1-\nu)}{2\pi G_0 r} S_{vv}\,, \tag{3.131a}$$

$$S_{vv} = \frac{\beta^2 \tilde{r}}{1-\nu} \int_0^\infty \mathrm{J}_0(\tilde{k}\tilde{r})\tilde{k}\tilde{w}\,\mathrm{d}\tilde{k}\,, \tag{3.131b}$$

where the coefficients of the normalized amplitudes of vibrations S_{vh}, S_{vv} represent the static vertical displacements on the surface of a homogeneous half-space in the case of the vertical force applied to the surface. Considering the displacements in the horizontal plane where the force is applied, we shall employ the limiting values (3.129) by adding them to and subtracting them from the quantities $\tilde{q}$ and $\tilde{w}$, respectively. Using tabulated integrals gives ($z = z_1$)

$$S_{vh} = \frac{\beta^2 \tilde{r}(1-3\tau^2)}{8(1-\nu)\tilde{z}_1^2} + \frac{\beta^2 \tilde{r}}{1-\nu} \int_0^\infty \mathrm{J}_1(\tilde{k}\tilde{r})\tilde{k}\left[\tilde{q} - \frac{1-3\tau^2}{8\tilde{k}^2\tilde{z}_1^2}\right]\mathrm{d}\tilde{k}\,, \tag{3.132}$$

$$S_{vv} = \frac{\beta^2(1+\tau^2)}{4(1-\nu)\tilde{z}_1} + \frac{\beta^2 \tilde{r}}{1-\nu} \int_0^\infty \mathrm{J}_0(\tilde{k}\tilde{r})\tilde{k}\left[\tilde{w} - \frac{1+\tau^2}{4\tilde{k}\tilde{z}_1}\right]\mathrm{d}\tilde{k}\,. \tag{3.133}$$

The first summands in the right side of these equations represent values of the normalized amplitudes as $\tilde{r} \to 0$ (with the parameters θ and $\tilde{z}_1$ kept fixed).

In the case when the force is applied to the surface of the half-space, limiting values for the quantities $\tilde{q}$ and $\tilde{w}$ differ from those in (3.129). For $\tilde{z} = \tilde{z}_1 = 1$, considering all terms in relationships (3.127b), (3.128b), we obtain instead of (3.129):

$$\tilde{q} \approx -\frac{\tau^2}{2(1-\tau^2)\tilde{k}} = -\frac{1-2\nu}{2\tilde{k}}\,, \tag{3.134a}$$

$$\tilde{w} \approx \frac{1}{2(1-\tau^2)\tilde{k}} = \frac{1-\nu}{\tilde{k}}\,. \tag{3.134b}$$

These results are in agreement with relationships (3.30). Equations (3.134a) and (3.134b) lead to the values $-\tau^2\beta^2$ and β^2 for S_{vh} and S_{vv}, respectively, at $\tilde{r} = 0$. Thus, a discontinuity of the initial values S_{vh} and S_{vv} occurs in going from the case $\tilde{z}_1 = 1$ to the case $\tilde{z}_1 > 1$. Notice that such a behavior is a feature of the known Mindlin's solution corresponding to the static action of a force within a homogeneous half-space [75, 91]. The principal terms (when $r \to 0$) in the equations for the displacements of points in the horizontal plane in which the force

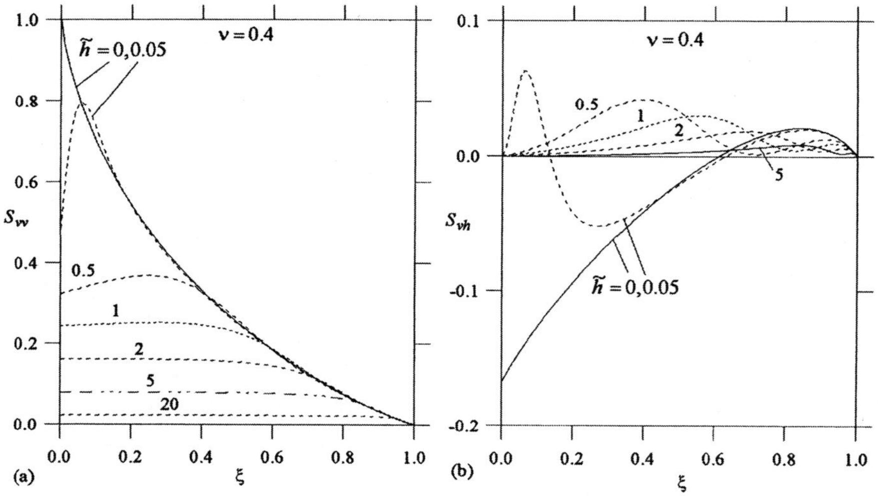

Fig. 3.41a,b. Normalized static vertical (a) and horizontal (b) displacements of points of linearly heterogeneous half-space subjected to vertical force applied within half-space ($\tilde{z} = 1 + \tilde{h}$)

acts are different for the force applied within the half-space and at the half-space surface.

Let us consider the results of calculations. When the solving static problems we take $\beta = 1$ and integrate over the real axis in the complex plane of the integration parameter $\tilde{k}$. In order to solve the dynamic problems, one should apply the contour shown in Fig. 2.3 which is considered in section 3.2 and in section 2.3.1. The following results correspond to the static action of a vertical force. In Fig. 3.41, the behavior of the normalized static displacements S_{vh} and S_{vv} in the horizontal planes containing the point of force application ($z = z_1$), is shown at ν=0.4 for a number of values of the dimensionless depth, $\tilde{h}$, calculated as follows:

$$\tilde{h} = \frac{z_1 - z_0}{z_0} = \tilde{z}_1 - 1.$$

The variable ξ determined from (3.61) is used as an argument. As the dimensionless distance $\tilde{r}$ increases displacements become identical to those corresponding to the force applied to the surface of the half-space (in the case of horizontal displacements, the identity takes place only at extremely high values of $\tilde{r}$). Even at very small values of the parameter $\tilde{h}$, there is a certain interval of varying of parameter $\tilde{r}$ in which the penetration of the point of force application significantly effects the displacements in the half-space. Note that, starting with

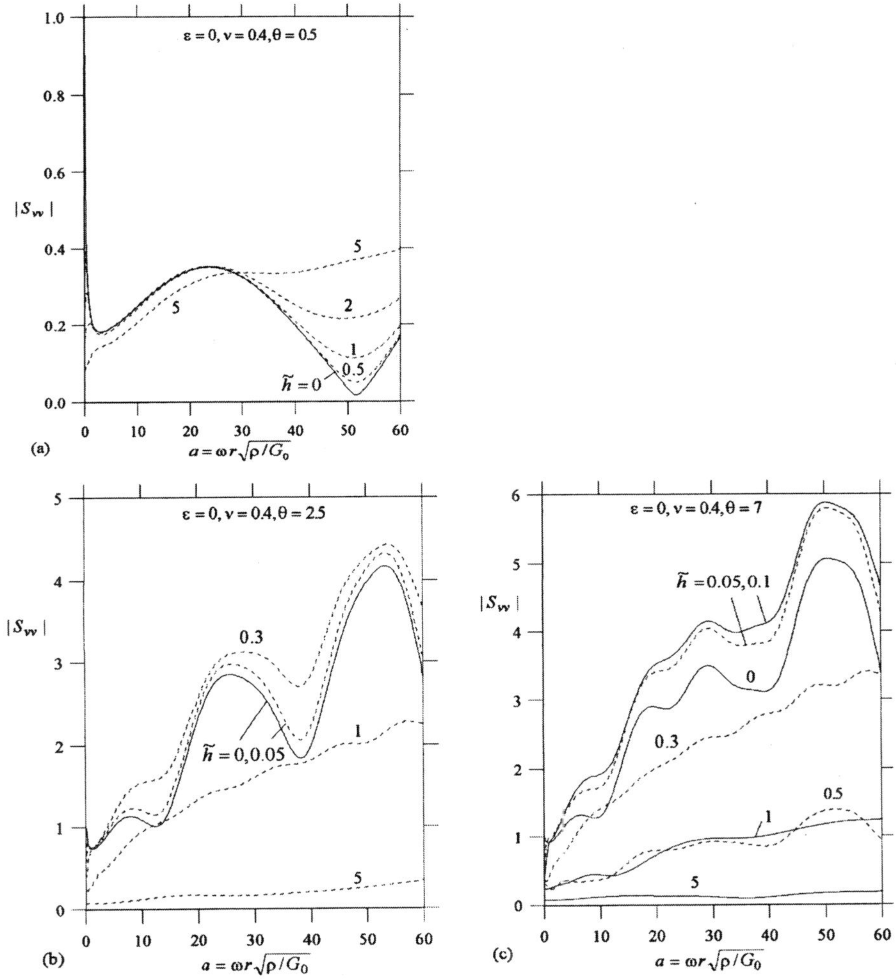

Fig. 3.42a,b,c. Absolute values of normalized vertical amplitudes of vibrations under dynamic action of vertical force applied within linearly heterogeneous half-space for three values of parameter θ ($\tilde{z} = 1 + \tilde{h}$).

the value $\tilde{h} \approx 1$, the normalized vertical displacements become virtually constant on a large interval of $\tilde{r}$ variation. Taking into account relationships (3.130a) and (3.131a) one can see that on this interval, a qualitative similarity takes place between the plotted curves and the curves which express the displacements of the surface of a homogeneous half-space under the action of the surface concentrated force.

When constructing solutions of dynamic problems, in integrals, the substitution $s = \tilde{k} / \theta$ and the parameter a defined in (3.33) are employed. In Fig. 3.42, the normalized vertical amplitudes (absolute values of the complex normalized amplitudes) in the planes $z = z_1$ are shown for the case when no dissipation occurs (ε=0) and ν = 0.4; three values of parameter θ are taken, namely, θ = 0.5, 2.5, 7. As in the case with the static action of the force, an abrupt change in the values of the amplitudes of vibration takes place at small values of the parameter a in going from the case $\tilde{z}_1 = 1$ (in this case, an initial value of the normalized amplitude equals unity) to the case $\tilde{z}_1 > 1$. Increasing of the normalized amplitudes with increasing a is due to the influence of the Rayleigh-type waves. Note that at small values of $\tilde{h}$, the amplitudes corresponding to $\tilde{h} > 0$ may exceed the amplitudes for the case with $\tilde{h} = 0$.

3.7.2 Action of Horizontal Force

The solution represented in relationships (1.163)–(1.165) may be transformed by using a technique similar to that employed in the case of the vertical force. In the consideration of the quantities $\tilde{q}$ and $\tilde{w}$, we use the modified coefficients B_j^*, as in equations (3.118). The system of equations for the coefficients B_j^* corresponding to system (3.119) has the form

$$\sum_{j=1}^{4} B_j^* \begin{pmatrix} \tilde{\chi}_j^*(\varsigma_1) \\ \tilde{e}_j^*(\varsigma_1) \\ \tilde{q}_j^*(\varsigma_1) \\ \tilde{w}_j^*(\varsigma_1) \end{pmatrix} = \begin{pmatrix} -1 \\ 0 \\ 0 \\ 0 \end{pmatrix}. \tag{3.135}$$

The rest of transformations related to functions $\tilde{q}$ and $\tilde{w}$ are identical to the transformations performed above for the case with the vertical force. Regarding the additional function p, for which the corresponding fundamental solutions and derivatives are given in relationships (3.27) and (3.28), we introduce the following modified coefficients C_j^* instead of C_j used in equations (1.116), (1.117)

$$C_1^* = C_1 \,\mathrm{e}^{-\varsigma_1/2}, \tag{3.136a}$$

$$C_2^* = C_2 \,\mathrm{e}^{\varsigma_1/2}. \tag{3.136b}$$

Instead of the system of equations (1.117), we have

$$\tilde{p}_1^*(\varsigma_1)C_1^* + \tilde{p}_2^*(\varsigma_1)C_2^* = 0, \tag{3.137a}$$

$$d_1^*(\varsigma_1)C_1^* + d_2^*(\varsigma_1)C_2^* = -1, \tag{3.137b}$$

where $\widetilde{p}_1^*(\varsigma)$ and $d_1^*(\varsigma)$ are determined according to relationships (3.27a), (3.28a), where the multiplier $\exp(-\varsigma/2)$ was dropped; $\widetilde{p}_2^*(\varsigma)$ and $d_2^*(\varsigma)$ are calculated from (3.27b), (3.28b) where the multiplier $\exp(-\varsigma/2)$ is replaced with $\exp(-\varsigma)$:

$$\widetilde{p}_1^*(\varsigma) = U(0.5-\gamma,1,\varsigma)\,, \tag{3.138a}$$

$$\widetilde{p}_2^*(\varsigma) = \exp(-\varsigma)M(0.5-\gamma,1,\varsigma)\,, \tag{3.138b}$$

$$d_1^*(\varsigma) = 2\widetilde{k}[0.5U(0.5-\gamma,1,\varsigma) - U(0.5-\gamma,2,\varsigma)]\,, \tag{3.138c}$$

$$d_2^*(\varsigma) = 2\widetilde{k}\exp(-\varsigma)[0.5M(0.5-\gamma,1,\varsigma) - (0.5+\gamma)M(0.5-\gamma,2,\varsigma)]\,. \tag{3.138d}$$

These functions do not increase or decrease exponentially, as the argument ς increases. Furthermore, we transform the expression for function $\widetilde{p}$ entering equations (1.164). Using equation (1.119) gives ($z \le z_1$)

$$\widetilde{p} = C^*\widetilde{p}_1 + C_1\widetilde{p}_1 + C_2\widetilde{p}_2 = C_1^{\wedge}\widetilde{p}_1 + C_2^*\,\mathrm{e}^{-\varsigma_1/2}\,\widetilde{p}_2\,, \tag{3.139a}$$

where

$$C_1^{\wedge} = C^* + C_1 = -C_2^*\,\mathrm{e}^{\varsigma_0-\varsigma_1/2}\,\frac{d_2^*(\varsigma_0)}{d_1^*(\varsigma_0)}\,. \tag{3.139b}$$

For $z \ge z_1$,

$$\widetilde{p} = C^*\widetilde{p}_1 = C_1^{\wedge}\widetilde{p}_1 - C_1\widetilde{p}_1 = C_1^{\wedge}\widetilde{p}_1 - C_1^*\,\mathrm{e}^{\varsigma_1/2}\,\widetilde{p}_1\,. \tag{3.140}$$

For the linearly heterogeneous half-space, the solution (1.164) may be presented in the following form:

$$\hat{u}_r = \frac{Q_0}{\pi R G(z_0)}\int_0^{\infty} \mathrm{J}_1(\widetilde{k}\widetilde{R})\left[\left(\widetilde{q} - \frac{\widetilde{p}}{\widetilde{z}_1}\right)\frac{\mathrm{J}_1(\widetilde{k}\widetilde{r})}{\widetilde{k}\widetilde{r}} - \widetilde{q}\,\mathrm{J}_0(\widetilde{k}\widetilde{r})\right]\mathrm{d}\widetilde{k}\,, \tag{3.141a}$$

$$\hat{u}_\vartheta = \frac{Q_0}{\pi R G(z_0)}\int_0^{\infty} \mathrm{J}_1(\widetilde{k}\widetilde{R})\left[\left(\widetilde{q} - \frac{\widetilde{p}}{\widetilde{z}_1}\right)\frac{\mathrm{J}_1(\widetilde{k}\widetilde{r})}{\widetilde{k}\widetilde{r}} + \frac{\widetilde{p}}{\widetilde{z}_1}\mathrm{J}_0(\widetilde{k}\widetilde{r})\right]\mathrm{d}\widetilde{k}\,, \tag{3.141b}$$

$$\hat{u}_z = \frac{Q_0}{\pi R G(z_0)}\int_0^{\infty} \mathrm{J}_1(\widetilde{k}\widetilde{R})\,\mathrm{J}_1(\widetilde{k}\widetilde{r})\widetilde{w}\,\mathrm{d}\widetilde{k}\,, \tag{3.141c}$$

where function $\widetilde{p}$ is determined according to equations (3.139)–(3.140). Using the previously determined quantities $\widetilde{p}_1^*(\varsigma)$, $d_1^*(\varsigma)$, $\widetilde{p}_2^*(\varsigma)$, $d_2^*(\varsigma)$ this function may be written as

$$\widetilde{p} = \begin{cases} -C_2^{*}\,\mathrm{e}^{\widetilde{k}(2-\widetilde{z}_1-\widetilde{z})}\,\dfrac{d_2^{*}(2\widetilde{k})}{d_1^{*}(2\widetilde{k})}\,\overline{p}_1^{*} + C_2^{*}\,\mathrm{e}^{\widetilde{k}(\widetilde{z}-\widetilde{z}_1)}\,\overline{p}_2^{*} & (z \le z_1) \\ -C_2^{*}\,\mathrm{e}^{\widetilde{k}(2-\widetilde{z}_1-\widetilde{z})}\,\dfrac{d_2^{*}(2\widetilde{k})}{d_1^{*}(2\widetilde{k})}\,\overline{p}_1^{*} - C_1^{*}\,\mathrm{e}^{\widetilde{k}(\widetilde{z}_1-\widetilde{z})}\,\overline{p}_1^{*} & (z \ge z_1). \end{cases} \tag{3.142}$$

For the functions $\widetilde{q}$ and $\widetilde{w}$, equations (3.127b), (3.128b) are still valid, but now the coefficients B_j^{*} satisfy the system of equations (3.135). Relationships (3.123)–(3.125) still hold.

Analogously to the case with the action of a vertical force the integrands in the constructed solutions decrease exponentially at $z \neq z_1$.

Next, we consider the action of a horizontal concentrated force. Note that the amplitudes $\hat{u}_z$ may be expressed in terms of the amplitudes u_r in relationship (3.130a) by using the principle of reciprocity. The quantities $\hat{u}_r, \hat{u}_\vartheta$ may be written as

$$\hat{u}_r = \frac{Q_0}{2\pi G_0 r} S_{hr}, \tag{3.143a}$$

$$S_{hr} = \beta^2 \widetilde{r} \int_0^{\infty} \widetilde{k} \left[\left(\widetilde{q} - \frac{\widetilde{p}}{\widetilde{z}_1} \right) \frac{\mathrm{J}_1(\widetilde{k}\widetilde{r})}{\widetilde{k}\widetilde{r}} - \widetilde{q}\,\mathrm{J}_0(\widetilde{k}\widetilde{r}) \right] \mathrm{d}\widetilde{k}\,, \tag{3.143b}$$

$$\hat{u}_\vartheta = -\frac{Q_0(1-\nu)}{2\pi G_0 r} S_{h\vartheta}, \tag{3.144a}$$

$$S_{h\vartheta} = -\frac{\beta^2 \widetilde{r}}{1-\nu} \int_0^{\infty} \widetilde{k} \left[\left(\widetilde{q} - \frac{\widetilde{p}}{\widetilde{z}_1} \right) \frac{\mathrm{J}_1(\widetilde{k}\widetilde{r})}{\widetilde{k}\widetilde{r}} + \frac{\widetilde{p}}{\widetilde{z}_1} \mathrm{J}_0(\widetilde{k}\widetilde{r}) \right] \mathrm{d}\widetilde{k}\,, \tag{3.144b}$$

where the coefficients of the normalized amplitudes S_{hr} and $S_{h\vartheta}$ in (3.143a) and (3.144a) represent the corresponding static displacements at the surface of a homogeneous half-space, which occur due to the force applied to the surface. Let us consider the amplitudes of vibrations in the horizontal plane where the force is applied ($z = z_1$). Determination of the limiting values at $\widetilde{k} \to \infty$ for the functions constituting relationships (3.143b), (3.144b) yields

$$\widetilde{q}_l = -\frac{1+\tau^2}{4\widetilde{k}\widetilde{z}_1}, \tag{3.145a}$$

$$\widetilde{p}_l = -\frac{1}{2\widetilde{k}}. \tag{3.145b}$$

In the case $\widetilde{z}_1 = 1$, we have as $\widetilde{k} \to \infty$

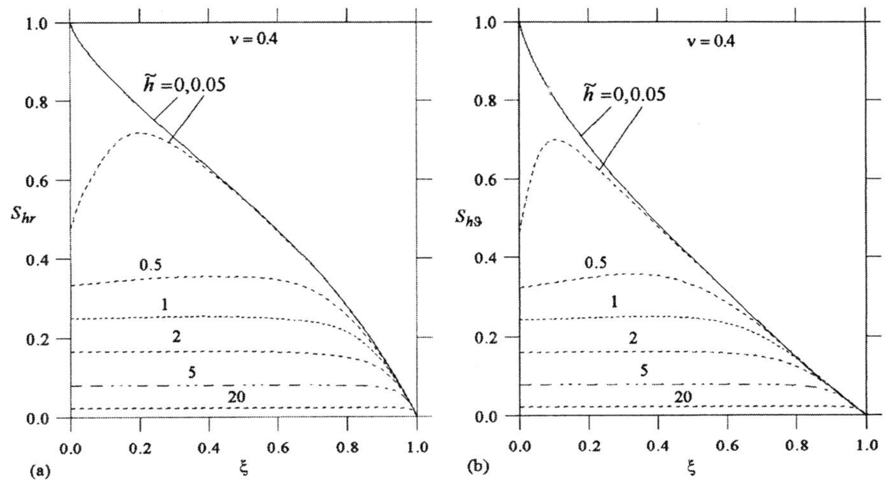

Fig. 3.43a,b. Normalized static radial (a) and tangential (b) displacements of points of linearly heterogeneous half-space subjected to horizontal force applied within half-space ($\tilde{z} = 1 + \tilde{h}$)

$$\tilde{q}_l = -\frac{1}{2(1-\tau^2)\tilde{k}}, \tag{3.146a}$$

$$p_l = -\frac{1}{\tilde{k}}. \tag{3.146b}$$

Analogously to (3.132), (3.133), one may write ($\tilde{z}_1 > 1$, $z = z_1$):

$$S_{hr} = \frac{\beta^2}{2\tilde{z}_1} + \beta^2 \tilde{r} \int_0^\infty \tilde{k} \left[\left(\tilde{q} - \frac{\tilde{p}}{\tilde{z}_1} - \tilde{q}_l + \frac{\tilde{p}_l}{\tilde{z}_1} \right) \frac{J_1(\tilde{k}\tilde{r})}{\tilde{k}\tilde{r}} - (\tilde{q} - \tilde{q}_l) J_0(\tilde{k}\tilde{r}) \right] d\tilde{k} \tag{3.147a}$$

$$S_{h\vartheta} = \frac{\beta^2(1+\tau^2)}{4(1-\nu)\tilde{z}_1}$$

$$- \frac{\beta^2 \tilde{r}}{1-\nu} \int_0^\infty \tilde{k} \left[\left(\tilde{q} - \frac{\tilde{p}}{\tilde{z}_1} - \tilde{q}_l + \frac{\tilde{p}_l}{\tilde{z}_1} \right) \frac{J_1(\tilde{k}\tilde{r})}{\tilde{k}\tilde{r}} + \left(\frac{\tilde{p}}{\tilde{z}_1} - \frac{\tilde{p}_l}{\tilde{z}_1} \right) J_0(\tilde{k}\tilde{r}) \right] d\tilde{k}\,. \tag{3.147b}$$

In these expressions, the first term represents the normalized amplitudes at $\tilde{r} = 0$. Note that in the case with the force applied to the half-space surface, these quantities equal β^2.

Some results of calculations for the static case are shown in Fig. 3.43 where the quantities S_{hr} and $S_{h\vartheta}$ are shown for the points located at the same depth h as

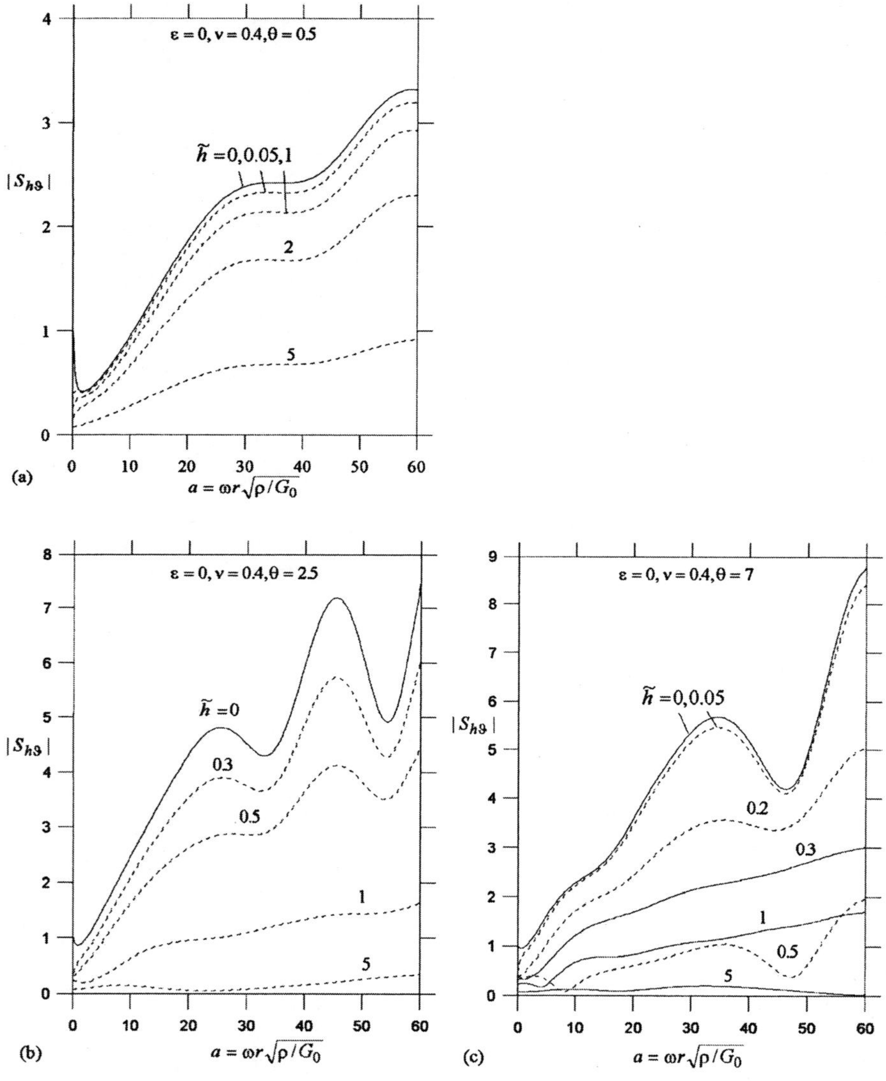

Fig. 3.44a,b,c. Absolute values of normalized tangential amplitudes of vibrations under dynamic action of horizontal force applied within linearly heterogeneous half-space for three values of parameter θ ($\tilde{z} = 1 + \tilde{h}$).

that of the point of force application. The plotted curves are similar to the curves corresponding to the action of vertical force (Fig. 3.41).

The magnitudes of the normalized amplitudes of vibrations $|S_{h\vartheta}|$ for a number of values of parameters are shown in Fig. 3.44. In contrast to the case with the

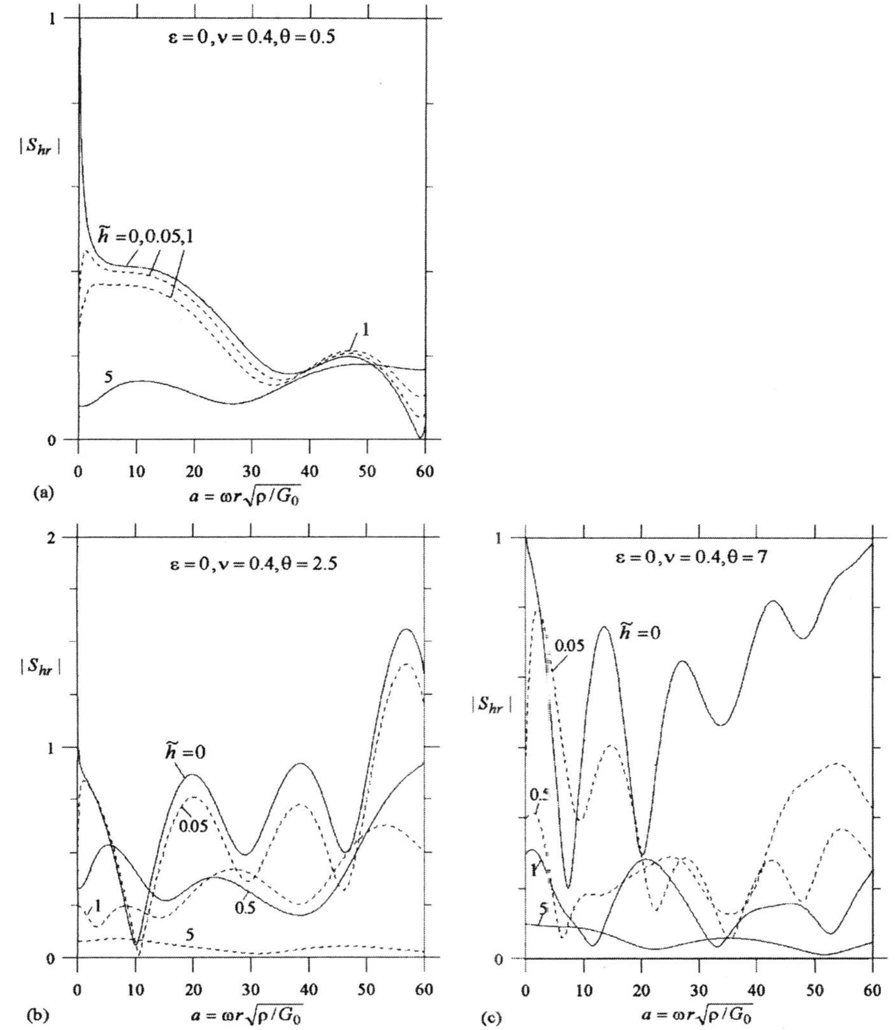

Fig. 3.45a,b,c. Absolute values of normalized radial amplitudes of vibrations under dynamic action of horizontal force applied within linearly heterogeneous half-space for three values of parameter θ ($\widetilde{z} = 1 + \widetilde{h}$).

vertical force, values of $|S_{h9}|$ for $\widetilde{z}_1 > 1$ ($\widetilde{h} > 0$) are always smaller than those for $\widetilde{z}_1 = 1$ ($\widetilde{h} = 0$). Note that the considered horizontal amplitudes significantly exceed the vertical amplitudes caused by the vertical force applied at the same depth. This is especially noticeable at the value of the parameter θ =0.5. As follows from calculations values of $|S_{hr}|$ shown in Fig. 3.45 for the values of parameters used in the calculations of $|S_{h\vartheta}|$ are appreciably smaller than the values of $|S_{h\vartheta}|$. As noted previously, increase in the normalized amplitudes $|S_{h\vartheta}|$ is related to the influence of the Love-type waves.

3.8 Plane Problems for Linearly Heterogeneous Half-space

3.8.1 Action of Vertical Load Distributed Uniformly over Infinite Straight Line on Half-Space Surface

In the following, we employ the general form of the solution given in section 1.4.4. Taking into account the form of the fundamental solutions constructed in section 3.1 we obtain the following equations for amplitudes of vibrations of the half-space subjected to the vertical load of intensity p_0 which is distributed uniformly over an infinite line located on the surface of the half-space (Fig. 1.4a):

$$u_x(x,z) = \frac{p_0}{\pi G_0} S_{vx}, \tag{3.148a}$$

$$S_{vx} = \beta^2 \int_0^\infty e^{-\tilde{k}(\tilde{z}-1)} \frac{\overset{*}{c}_{21}\overset{*}{\tilde{q}}_2(\varsigma) - \overset{*}{c}_{22}\overset{*}{\tilde{q}}_1(\varsigma)}{D^*} \sin(\tilde{k}\tilde{x})\, d\tilde{k}, \tag{3.148b}$$

$$u_z(x,z) = \frac{p_0}{\pi G_0} S_{vz}, \tag{3.149a}$$

$$S_{vz} = \beta^2 \int_0^\infty e^{-\tilde{k}(\tilde{z}-1)} \frac{\overset{*}{c}_{21}\overset{*}{\tilde{w}}_2(\varsigma) - \overset{*}{c}_{22}\overset{*}{\tilde{w}}_1(\varsigma)}{D^*} \cos(\tilde{k}\tilde{x})\, d\tilde{k}, \tag{3.149b}$$

where analogously to (2.261)

$$\tilde{x} = \frac{x}{z_r} = \frac{x}{z_0}. \tag{3.150}$$

Here, the structure of the normalized amplitudes of vibrations is similar to that given in relationships (3.29a) and (3.29b); nomenclature used in these relationships is adopted.

We consider amplitudes of vibrations of the points that belong to the surface of the half-space. It is reasonable to accelerate the convergence of the integrals by accounting for the behavior of the integrands at high values of the variable of integration. Using the limiting values obtained in relationships (3.30a), (3.30b) gives

$$\Phi_{vx}^{\infty} = \frac{\Phi_{vh}^{\infty}}{\widetilde{k}} = -\frac{1-2\nu}{2\widetilde{k}}, \tag{3.151a}$$

$$\Phi_{vz}^{\infty} = \frac{\Phi_{vv}^{\infty}}{\widetilde{k}} = \frac{1-\nu}{\widetilde{k}}, \tag{3.151b}$$

where the asymptotic representations Φ_{vx}^{∞} and Φ_{vz}^{∞} are introduced for the multipliers of $\sin(\widetilde{k}\widetilde{x})$ and $\cos(\widetilde{k}\widetilde{x})$ under the integral in expressions (3.148b) and (3.149b) ($\widetilde{z}=1$). The transformation of the integrals in (3.148b), (3.149b) is performed analogously to the transformations resulting in relationships (2.262) and (3.31),

$$S_{vx} = \beta^2 \int_0^{\infty} \left[\frac{\overset{*}{c}_{21}\overset{*}{\widetilde{q}}_2(2\widetilde{k}) - \overset{*}{c}_{22}\overset{*}{\widetilde{q}}_1(2\widetilde{k})}{D^*} \widetilde{k} - \Phi_{vh}^{\infty} \right] \frac{\sin(\widetilde{k}\widetilde{r})}{\widetilde{k}} \mathrm{d}\widetilde{k} + \beta^2 \frac{\Phi_{vh}^{\infty}\pi}{2}, \tag{3.152a}$$

$$S_{vz} = \beta^2 \int_0^{\infty} \frac{\cos(\widetilde{k}\widetilde{x})}{1+\widetilde{k}} \left[(1+\widetilde{k}) \frac{\overset{*}{c}_{21}\overset{*}{\widetilde{w}}_2(2\widetilde{k}) - \overset{*}{c}_{22}\overset{*}{\widetilde{w}}_1(2\widetilde{k})}{D^*} - \Phi_{vv}^{\infty} \right] \mathrm{d}\widetilde{k}$$

$$+ \beta^2 \Phi_{vv}^{\infty} \int_0^{\infty} \frac{\cos(\widetilde{k}\widetilde{x})}{1+\widetilde{k}} \mathrm{d}\widetilde{k}. \tag{3.152b}$$

According to equation (2.264) the last integral in relationship (3.152b) yields for small values of $\widetilde{x}$ the term having a logarithmic singularity. Analogously to expression (2.265) we have

$$S_{vz} = -\beta^2 \Phi_{vv}^{\infty} \ln(\widetilde{x}) + \widetilde{S}_{vz} = -(1-\nu)\beta^2 \ln(\widetilde{x}) + \widetilde{S}_{vz}, \tag{3.153}$$

where $\widetilde{S}_{vz}$ represents a part of the normalized amplitude of vibrations which is bounded for finite values of $\widetilde{x}$. When solving the dynamic problems the substitution (3.32) is applied and similar to (3.33) one should use the following variable instead of the variable $\widetilde{x}$:

$$a_x = \theta\widetilde{x} = \omega x \sqrt{\frac{\rho}{G_0}}. \tag{3.154}$$

Let us consider the results of calculations related to the values $\nu = 1/3, \varepsilon = 0$ ($\beta = 1$). In Fig. 3.46, absolute values of the normalized amplitudes S_{vx} and S_{vz} are presented for a number of values of the parameter θ. According to relationship (3.152a) all the curves shown in Fig. 3.46a pass through the initial point having the ordinate -0.2618. Following relationship (3.153) the curves representing $|S_{vz}|$ (Fig. 3.46b) have the logarithmic singularity at $a_x \to 0$

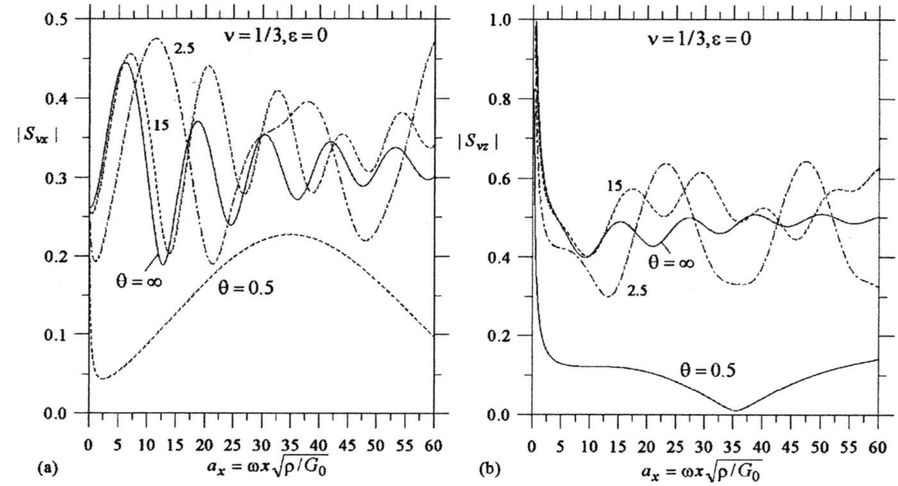

Fig. 3.46a,b. Absolute values of normalized horizontal (a) and vertical (b) amplitudes of vibrations of surface points of linearly heterogeneous half-plane under action of vertical load

$$S_{vz} \approx -(1-\nu)\beta^2 \ln(\frac{a_x}{\theta})$$

Note that at small values of the parameter a_x for the value $\theta = 15$, the results are close to those obtained for the homogeneous half-space which corresponds to going to the limit $z_0 \to \infty, \theta \to \infty$. As noted previously in studying vibrations of the transversely isotropic homogeneous half-space (section 2.8.1), in plane problems, the amplitudes of vibration tend to a constant values (in the case with the material where no internal friction exists), as the distance to the line of the load application increases, whereas in the corresponding 3-D problems, as the distance r to the point of the force application increases, amplitudes decrease as $1/\sqrt{r}$. A similar phenomenon which occurs due to the influence of the poles in the corresponding integrands is inherent in the considered problem of vibrations of the linearly heterogeneous half-plane.

3.8.2 Action of Horizontal Load Distributed Uniformly over Infinite Straight Line on Half-Space Surface

We consider the vibrations of points of the half-space located in the plane *(X, Z)* under action of a load which is directed parallel to the *X*-axis and distributed uniformly over an infinite straight line (parallel to the *Y*-axis) on the half-space surface (in-plane vibrations). Employing the general expressions for amplitudes of vibrations given in equations (1.98), (1.99) yields

$$u_x(x,z) = \frac{q_0}{\pi G_0} S_{hx}, \tag{3.155a}$$

$$S_{hx} = \beta^2 \int_0^\infty e^{-\widetilde{k}(\widetilde{z}-1)} \frac{c_{12}^* \widetilde{q}_1^*(\varsigma) - c_{11}^* \widetilde{q}_2^*(\varsigma)}{D^*} \cos(\widetilde{k}\widetilde{x}) \mathrm{d}\widetilde{k}, \tag{3.155b}$$

$$u_z(x,z) = -\frac{q_0}{\pi G_0} S_{hz}, \tag{3.156a}$$

$$S_{hz} = \beta^2 \int_0^\infty e^{-\widetilde{k}(\widetilde{z}-1)} \frac{c_{12}^* \widetilde{w}_1^*(\varsigma) - c_{11}^* \widetilde{w}_2^*(\varsigma)}{D^*} \sin(\widetilde{k}\widetilde{x}) \mathrm{d}\widetilde{k}, \tag{3.156b}$$

where the nomenclature used in equations (3.29) is employed. According to the Rayleigh's principle of reciprocity, the amplitudes of vertical vibrations by (3.156a) which occur due to the action of the horizontal load are directly related to the amplitudes of horizontal vibrations under action of the vertical load considered in the previous subsection. We examine further the quantity S_{hx}. For the half-space surface, we apply the transformation improving convergence of the integral in relationship (3.155b). Employing the asymptotic representation given in (3.38a) we have ($\widetilde{z} = 1$)

$$S_{hx} = \beta^2 \int_0^\infty \frac{\cos(\widetilde{k}\widetilde{x})}{1+\widetilde{k}} \left[(1+\widetilde{k}) \frac{c_{12}^* \widetilde{q}_1^*(2\widetilde{k}) - c_{11}^* \widetilde{q}_2^*(2\widetilde{k})}{D^*} - (1-\nu) \right] \mathrm{d}\widetilde{k}$$

$$+ \beta^2 (1-\nu) \int_0^\infty \frac{\cos(\widetilde{k}\widetilde{x})}{1+\widetilde{k}} \mathrm{d}\widetilde{k}. \tag{3.157}$$

Analogously to relationship (3.153),

$$S_{hx} = -\beta^2 (1-\nu) \ln(\widetilde{x}) + \widetilde{S}_{hx} = -(1-\nu)\beta^2 \ln(\frac{a_x}{\theta}) + \widetilde{S}_{hx}. \tag{3.158}$$

In Fig. 3.47, the graphs are plotted for the quantity S_{hx} at a number of values of the parameter θ with the previously used values of ν and ε. The case of the homogeneous half-space corresponds to the going to the limit $\theta \to \infty$.

Next, we consider the purely shear vibrations of a half-space due to a uniformly distributed load acting over the infinite line on the half-space surface (Fig. 1.4b) (anti-plane vibrations). The equation (1.95) determining the amplitudes of vibrations along the Y-axis is employed. Taking into account the expression (1.82) and the relationships (3.27a), (3.28a) we present the solution in the form:

$$u_y(x,z) = \frac{q_0}{\pi G_0} S_y, \tag{3.159a}$$

$$S_y = -\beta^2 \int_0^\infty e^{-\widetilde{k}(\widetilde{z}-1)} \widetilde{p}^*(\widetilde{z},\widetilde{k}) \cos(\widetilde{k}\widetilde{x}) \mathrm{d}\widetilde{k}. \tag{3.159b}$$

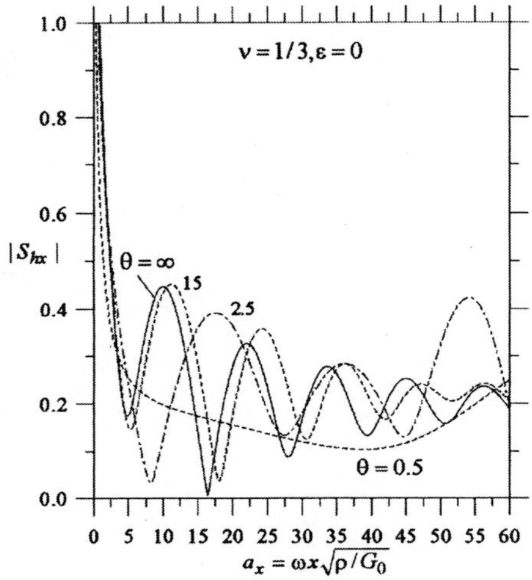

Fig. 3.47. Absolute values of normalized horizontal amplitudes of in-plane vibrations of surface points of linearly heterogeneous half-plane under action of horizontal load

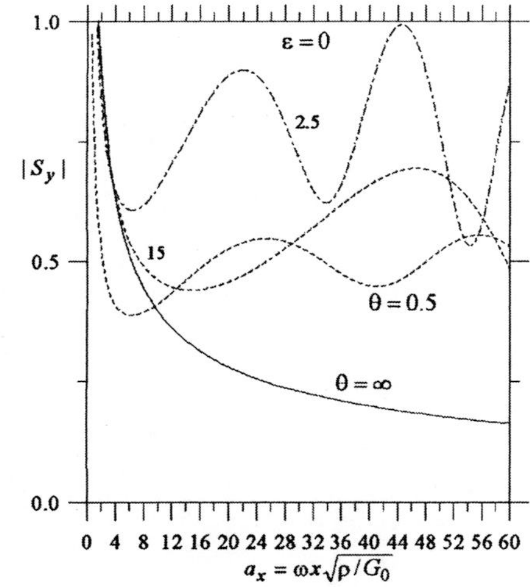

Fig. 3.48. Absolute values of normalized horizontal amplitudes of anti-plane vibrations of surface points of linearly heterogeneous half-plane under action of horizontal load

The quantity $\widetilde{p}^*$ is given in the relationship (3.37b). Considering the amplitudes of vibrations on the half-space surface ($\widetilde{z} = 1$), we can improve convergence of the integral in (3.159b) using the asymptotic representation for the quantity $\widetilde{p}^*$ given in (3.38b):

$$S_y = -\beta^2 \int_0^\infty \frac{\cos(\widetilde{k}\widetilde{x})}{1+\widetilde{k}}[(1+\widetilde{k})\widetilde{p}^*(1,\widetilde{k})+1]\mathrm{d}\widetilde{k} + \beta^2 \int_0^\infty \frac{\cos(\widetilde{k}\widetilde{x})}{1+\widetilde{k}}\mathrm{d}\widetilde{k}, \tag{3.160}$$

where according to (3.37b)

$$\widetilde{p}^*(1,\widetilde{k}) = \frac{U(0.5-\gamma,1,2\widetilde{k})}{2\widetilde{k}[0.5U(0.5-\gamma,1,2\widetilde{k}) - U(0.5-\gamma,2,2\widetilde{k})]}. \tag{3.161}$$

As in the case of the quantities S_{vz} and S_{hx} in the present case, the amplitudes of vibrations contain a logarithmic singularity as $\widetilde{x} \to 0$. Analogously to the relationships (3.158), (2.278) we present the quantity S_y in the following form:

$$S_y = -\beta^2 \ln(\widetilde{x}) + \widetilde{S}_y = -\beta^2 \ln(\frac{a_x}{\theta}) + \widetilde{S}_y. \tag{3.162}$$

In Fig. 3.48, the behavior of the absolute value of the normalized complex amplitude S_y is shown for a number of values of the parameter θ. According to (2.277) the quantity S_y for the surface points of the homogeneous half-plane ($\theta \to \infty$) has the form

$$S_y = -\frac{\beta^2 \pi \mathrm{i}}{2} H_0^{(2)}(\beta\widetilde{x}). \tag{3.163}$$

One can see that according to (3.163) the magnitude of the normalized amplitude decreases as $1/\sqrt{\widetilde{x}}$, whereas in the case of the inhomogeneous half-space, the amplitude values oscillate about a constant value. This difference in behavior may be explained by accounting for the influence of the poles of the function $\widetilde{p}^*$ which exist in the heterogeneous case and are absent in the case of the homogeneous half-space for the considered anti-plane vibrations. In the case with the anti-plane shear vibrations in the heterogeneous half-space, Love waves are related to the poles of function $\widetilde{p}^*$. For other forms of the plane vibrations, the integrands have singularities in the form of poles (Rayleigh poles) both for the homogeneous and heterogeneous half-spaces, and the difference in the behavior of the amplitudes of vibrations is less significant than in the case of the anti-plane vibrations.

3.8.3 Static Surface Green's Functions in Plane Problems for Linearly Heterogeneous Half-Space

The Green's functions for the half-space surface in the 3-D case were considered in section 3.5.1. Below similar transformations are applied for the plane problems. In the formulas for the amplitudes of vibrations given above, we take the intensity of the load distributed over an infinite line on the half-space surface to be equale to unity; $\beta=1$, $\widetilde{z}=1$ ($\varsigma=2\widetilde{k}$). Similar to relationships (3.55) we have

$$w_{vz}=\frac{1}{G_0\pi}S_{vz}(\widetilde{x}), \tag{3.164a}$$

$$w_{vx}=\frac{1}{G_0\pi}S_{vx}(\widetilde{x}), \tag{3. 164b}$$

$$w_{hx}=\frac{1}{G_0\pi}S_{hx}(\widetilde{x}), \tag{3. 164c}$$

$$w_y=\frac{1}{G_0\pi}S_y(\widetilde{r}). \tag{3.164d}$$

These relationships correspond to the formulas (3.149), (3.148), (3.155) and (3.159). Green's function w_{hz} equals function w_{vx} taken with the opposite sign. Below the results of calculations of the normalized displacements are presented for a number of values of Poisson's ratio (Fig. 3.49). Quantity ξ analogous to the parameter defined in relationship (3.61) is used as an argument,

$$\xi=\frac{\widetilde{x}}{1+\widetilde{x}}, \tag{3.165}$$

which is equivalent to $\widetilde{x}$ for small values of $\widetilde{x}$ while the quantity $1-\xi$ is equivalent to $1/\widetilde{x}$ for $\widetilde{x}\to\infty$. In order to construct a highly accurate approximation for considered Green's functions, it is advisable to account for behavior of the displacements for small and high values of the dimensionless distance $\widetilde{x}$. Actually, the behavior at small values of $\widetilde{x}$ has been determined in the previous considerations. According to relationships (3.153), (3.158), (3.162) the quantities S_{vz}, S_{hx}, S_y are equivalent to the logarithmic functions while the quantity S_{vx} is close to a constant value (following equation (3.152a)),

$$S_{vz}\approx-(1-\nu)\ln(\widetilde{x}), \tag{3.166a}$$

$$S_{vx}\approx-\frac{(1-2\nu)\pi}{4}=-\frac{\tau^2\pi(1-\nu)}{2}, \tag{3.166b}$$

$$S_{hx}\approx-(1-\nu)\ln(\widetilde{x}), \tag{3.166c}$$

$$S_y\approx-\ln(\widetilde{x}). \tag{3.166d}$$

A more detailed examination of the quantity S_{vx} shows that the next term in the representation for S_{vx} for $\widetilde{x}\to 0$ has an infinite derivative. Prior to derivation of a more accurate subsequent approximation for the normalized displacements S_{vx} by

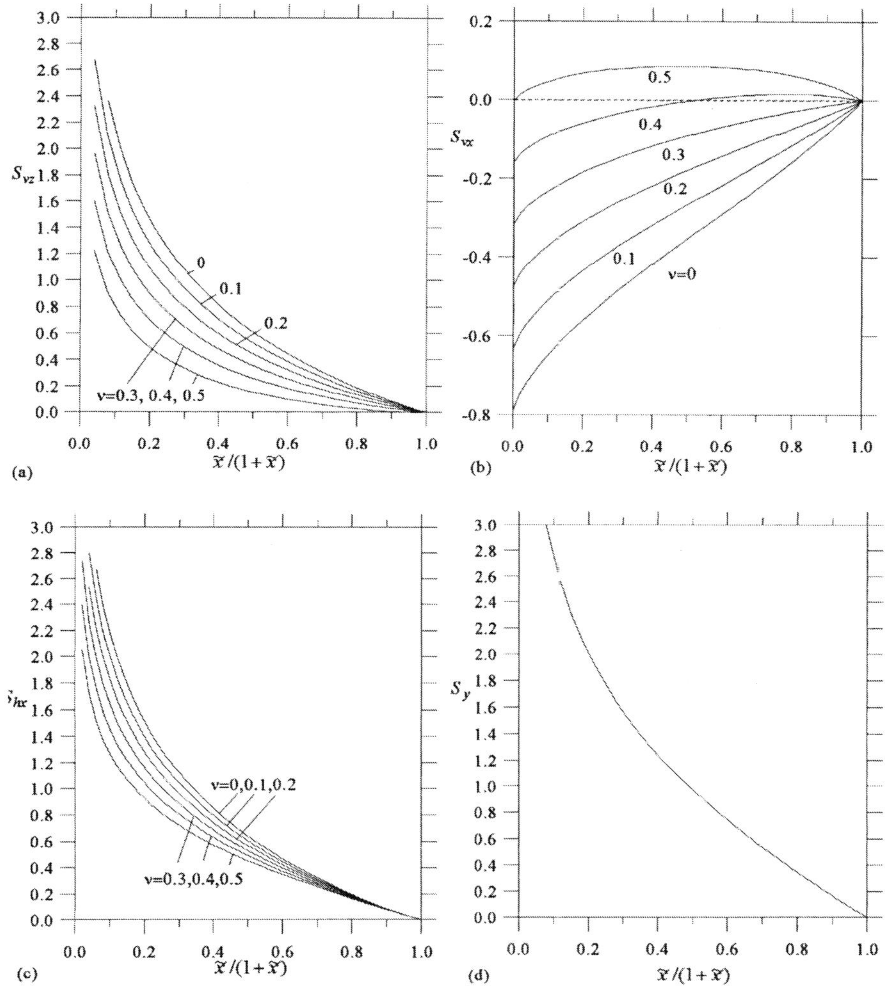

Fig. 3.49a,b,c,d. Static normalized surface displacements of linearly heterogeneous half-plane subjected to vertical ((a), (b)) and horizontal ((c), (d)) loads for different values of Poisson's ratio

using polynomials, it is advisable to determine this term. Employing relationship (3.148b) in the static case and taking into account the asymptotic expansions for the confluent hypergeometric functions at high values of their argument, give (analogously to (3.71b))

$$S_{vx} \approx \frac{(1-\nu)}{2}\int_0^{\infty}\frac{-2\tau^2\tilde{k}+(1-\tau^2)(1-2\tau^2)}{\tilde{k}(\tilde{k}+b)}\sin(\tilde{k}\tilde{x})\,\mathrm{d}\tilde{k}$$

$$= -(1-\nu)\tau^2 \int_0^{\infty} \frac{\sin(\tilde{k}\tilde{x})}{\tilde{k}+b} \mathrm{d}\tilde{k} + \frac{1-2\tau^2}{4} \int_0^{\infty} \frac{\sin(\tilde{k}\tilde{x})}{\tilde{k}(\tilde{k}+b)} \mathrm{d}\tilde{k} . \tag{3.167}$$

The first integral is expressed by the sine and cosine integral functions [1]:

$$\int_0^{\infty} \frac{\sin(\tilde{k}\tilde{x})}{\tilde{k}+b} \mathrm{d}\tilde{k} = f(b\tilde{x}) = \mathrm{Ci}(b\tilde{x})\sin(b\tilde{x}) - [\mathrm{Si}(b\tilde{x}) - \frac{\pi}{2}]\cos(b\tilde{x}) \approx \frac{\pi}{2} + b\tilde{x}\ln(\tilde{x}), \tag{3.168}$$

and the second integral is transformed as follows:

$$\int_0^{\infty} \frac{\sin(\tilde{k}\tilde{x})}{\tilde{k}(\tilde{k}+b)} \mathrm{d}\tilde{k} = \frac{1}{b} \int_0^{\infty} [\frac{1}{\tilde{k}} - \frac{1}{\tilde{k}+b}]\sin(\tilde{k}\tilde{x})\mathrm{d}\tilde{k} = \frac{\pi}{2b} - \frac{1}{b} f(b\tilde{x}) \approx -\tilde{x}\ln(\tilde{x}). \tag{3.169}$$

A refined (compared to (3.166b)) representation for S_{vx} corresponding to the small values of $\tilde{x}$ becomes

$$S_{vx} \approx -(1-\nu)[\frac{\tau^2\pi}{2} + \frac{\tilde{x}}{2}\ln(\tilde{x})] . \tag{3.170}$$

Furthermore, we determine the behavior of the considered Green's functions for high values of the argument $\tilde{x}$. Analogously to derivation of relationships (3.62) the representations are employed for the integrands in the corresponding integrals when the integration variable tends to zero:

$$S_{vz} \approx \frac{-\tau^2}{2} \int_0^{\infty} \frac{\ln(2\tilde{k})}{\tilde{k}+0.5} \cos(\tilde{k}\tilde{x})\mathrm{d}\tilde{k} , \tag{3.171a}$$

$$S_{vx} \approx \frac{1}{2} \int_0^{\infty} \frac{b_1}{\tilde{k}+0.5} \sin(\tilde{k}\tilde{x})\mathrm{d}\tilde{k} = \frac{b_1}{2} f(0.5\tilde{x}) \approx \frac{b_1}{\tilde{x}} , \tag{3.171b}$$

$$S_{hx} \approx \frac{1}{2} \int_0^{\infty} [-\frac{b_2 + \ln(2\tilde{k})}{\tilde{k}+0.5}]\cos(\tilde{k}\tilde{x})\mathrm{d}\tilde{k} , \tag{3.171c}$$

$$S_y \approx -\int_0^{\infty} \frac{(b_3 + \ln(2\tilde{k}))}{\tilde{k}+1} \cos(\tilde{k}\tilde{x})]\mathrm{d}\tilde{k} . \tag{3.171d}$$

The constant values b_1, b_2, b_3 are defined in relationships (3.64). The integrals containing the logarithm are transformed using integration by parts (the cosine is integrated and the rest is differentiated),

$$\int_0^{\infty} \frac{\ln(A\tilde{k})}{\tilde{k}+B} \cos(\tilde{k}\tilde{x})\mathrm{d}\tilde{k} = \int_0^{\infty} \frac{\sin(\tilde{k}\tilde{x})}{\tilde{x}\tilde{k}(\tilde{k}+B)} \mathrm{d}\tilde{k} - \int_0^{\infty} \frac{\sin(\tilde{k}\tilde{x})\ln(A\tilde{k})}{\tilde{x}(\tilde{k}+B)^2} \mathrm{d}\tilde{k} , \tag{3.172}$$

where A and B are constants. The first integral is similar to the integral in (3.169); as variable $\tilde{x}$ tends to infinity the integral becomes equivalent to $\pi/(2\tilde{x}B)$. The subsequent integration by parts shows that, as $\tilde{x}$ increases the second integral in (3.172) decreases faster than $1/\tilde{x}$. This is also true with respect to the integrals resulting from the terms b_2 and b_3 in (3.171c) and (3.171d). Thus, we obtain the following asymptotic representations for the normalized displacements when $\tilde{x} \to \infty$ (in addition to (3.171b)):

$$S_{vz} \approx \frac{\pi\tau^2}{2\tilde{x}}, \tag{3.173a}$$

$$S_{hx} \approx \frac{\pi}{2\tilde{x}}, \tag{3.173b}$$

$$S_y \approx \frac{\pi}{2\tilde{x}}. \tag{3.173c}$$

Note that the behavior of the normalized displacements at high values of $\tilde{x}$ is similar to behavior of the normalized displacements in the 3-D case for high values of the dimensionless distance $\tilde{r}$ (see relationships (3.65), (3.62b)). However, in this case, in order to determine the complete displacements one should multiply the normalized displacements by the corresponding displacements of the homogeneous half-space decreasing as $1/r$, whereas in the plane problems, the normalized displacements are multiplied by a constant value (see (3.164)). The approximate values of the normalized displacements in (3.173), (3.171b) and their exact values at $\tilde{r} = 150$ and $\nu = 1/3$ are shown below. The approximate values S_{vz}, S_{vx}, S_{hx}, S_y equal 0.002618, 0.0, 0.01047, 0.01047, respectively; the exact values equal 0.002631, – 0.0000174, 0.01046, 0.01050, respectively. Similarly to the 3-D case the asymptotic representation for S_{vx} becomes zero at $\nu = 1/3$.

Furthermore, we construct auxiliary functions by using the variable ξ defined in (3.165), which are equivalent to the corresponding normalized displacements in the vicinity of the bound points of the interval $0 \le \xi \le 1$:

$$\tilde{S}_{vz} = -(1-\nu)\ln(\xi)(1-\xi^2) + \frac{\pi\tau^2}{2}(1-\xi)\xi, \tag{3.174a}$$

$$\tilde{S}_{vx} = -\frac{(1-\nu)}{2}[\pi\tau^2 + \xi\ln(\xi)](1-\xi^2)^2 + b_1(1-\xi)\xi, \tag{3.174b}$$

$$\tilde{S}_{hx} = -(1-\nu)\ln(\xi)(1-\xi^2) + \frac{\pi}{2}(1-\xi)\xi, \tag{3.174c}$$

$$\tilde{S}_y = -\ln(\xi)(1-\xi^2) + \frac{\pi}{2}(1-\xi)\xi. \tag{3.174d}$$

Next, we employ the approximation using polynomials for the following quantities:

$$\delta_{vz} = \frac{S_{vz} - \tilde{S}_{vz}}{1-\xi}, \tag{3.175a}$$

$$\delta_{vx} = \frac{S_{vx} - \tilde{S}_{vx}}{1-\xi}, \tag{3.175b}$$

$$\delta_{hx} = \frac{S_{hx} - \tilde{S}_{hx}}{1-\xi}, \tag{3.175c}$$

$$\delta_y = \frac{S_y - \widetilde{S}_y}{1-\xi}. \tag{3.175d}$$

As follows from the test calculations the approximation accuracy, when using the least-squares technique for the quantities given in relationships (3.175) becomes better if an argument ξ_1 rather than ξ is taken in polynomials

$$\xi_1 = \frac{\widetilde{x}}{2+\widetilde{x}} = \frac{\xi}{2-\xi}. \tag{3.176}$$

The subsequent transformations are identical to those performed when the corresponding approximations were constructed for the 3-D case (section 3.5.1). For 11 values of Poisson's ratio, namely, $\nu = 0, 0.05, ..., 0.5$, the least-squares technique is applied to the quantities (3.175) with the degree of the polynomials equal to 7. Then, for each polynomial coefficient considered as a function of ν, the least-squares technique is applied again. As a result the coefficients are represented as polynomials with the argument ν (of 7th order). Regarding the quantity δ_y, which is independent of Poisson's ratio, its least-squares approximation was constructed directly. The following approximate expressions hold for the quantities (3.175), (3.176):

$$\delta_{vz} \approx P_{vz}(\xi_1) = \sum_{i=0}^{7} c_{ivz}\xi_1^i, \tag{3.177a}$$

where

$$c_{0vz} = -0.2309053 - 0.2462105\,\nu - 0.12366\,\nu^2 - 0.1354968\,\nu^3 + 0.02560879\,\nu^4 - 0.6133123\,\nu^5 + 0.9121424\,\nu^6 - 0.9627193\,\nu^7,$$

$$c_{1vz} = -0.7598588 + 1.211222\,\nu + 1.334338\,\nu^2 + 1.727907\,\nu^3 - 1.101648\,\nu^4 + 10.84507\,\nu^5 - 17.15911\,\nu^6 + 17.28026\,\nu^7,$$

$$c_{2vz} = 5.227069 - 6.432647\,\nu - 2.259866\,\nu^2 - 2.418379\,\nu^3 - 1.202416\,\nu^4 - 7.22941\,\nu^5 + 9.781098\,\nu^6 - 14.1773\,\nu^7,$$

$$c_{3vz} = -14.04158 + 17.40426\,\nu + 4.463105\,\nu^2 + 3.331193\,\nu^3 + 12.58146\,\nu^4 - 21.65629\,\nu^5 + 44.13008\,\nu^6 - 21.34131\,\nu^7,$$

$$c_{4vz} = 24.09984 - 28.39316\,\nu - 10.62686\,\nu^2 - 6.555002\,\nu^3 - 38.23384\,\nu^4 + 84.77811\,\nu^5 - 165.7191\,\nu^6 + 103.1088\,\nu^7,$$

$$c_{5vz} = -25.69252 + 28.24279\,\nu + 15.5989\,\nu^2 + 8.470591\,\nu^3 + 61.20759\,\nu^4 - 148.9279\,\nu^5 + 288.7424\,\nu^6 - 194.3473\,\nu^7,$$

$$c_{6vz} = 15.11583 - 14.9972\,\nu - 12.34048\,\nu^2 - 5.964781\,\nu^3 - 50.23585\,\nu^4 + 129.8757\,\nu^5 - 250.6361\,\nu^6 + 176.9361\,\nu^7,$$

$$c_{7vz} = -3.717039 + 3.210334\,\nu + 3.953405\,\nu^2 + 1.542513\,\nu^3 + 16.95922\,\nu^4 - 47.07969\,\nu^5 + 89.96021\,\nu^6 - 66.50961\,\nu^7; \tag{3.177b}$$

$$\delta_{vx} \approx P_{vx}(\xi_1) = \sum_{i=0}^{7} c_{ivx}\xi_1^i, \tag{3.178a}$$

where

$$c_{0vx} = -0.00020784 + 0.0002678\,\nu + 0.00025942\,\nu^2 + 0.00030858\,\nu^3 - 0.00026228\,\nu^4 + 0.00147553\,\nu^5 - 0.00175299\,\nu^6 + 0.00115704\,\nu^7,$$

$$c_{1vx} = 1.957949 - 5.905537\,\nu - 0.6776701\,\nu^2 - 0.7528748\,\nu^3 + 0.2422821\,\nu^4 - 3.586585\,\nu^5 + 5.303484\,\nu^6 - 5.331816\,\nu^7,$$

$$c_{2vx} = -2.133654 + 10.40352\,\nu + 1.165314\,\nu^2 + 1.399326\,\nu^3 - 1.168683\,\nu^4 + 7.812749\,\nu^5 - 11.03439\,\nu^6 + 9.140028\,\nu^7,$$

$$c_{3vx} = -0.2816862 - 6.515782\,\nu - 3.418828\,\nu^2 - 4.07887\,\nu^3 + 4.135938\,\nu^4 - 21.98289\,\nu^5 + 28.52761\,\nu^6 - 19.29029\,\nu^7,$$

$$c_{4vx} = -0.5505098 + 4.362211\,\nu + 9.818035\,\nu^2 + 10.51558\,\nu^3 - 6.461445\,\nu^4 + 42.77645\,\nu^5 - 48.99612\,\nu^6 + 30.18147\,\nu^7,$$

$$c_{5vx} = 2.924715 - 5.254723\,\nu - 15.65826\,\nu^2 - 15.08218\,\nu^3 + 2.553617\,\nu^4 - 40.28532\,\nu^5 + 34.05979\,\nu^6 - 16.05324\,\nu^7,$$

$$c_{6vx} = -3.007154 + 4.918458\,\nu + 12.84897\,\nu^2 + 11.16033\,\nu^3 + 3.249457\,\nu^4 + 13.57842\,\nu^5 + 2.538079\,\nu^6 - 9.106691\,\nu^7,$$

$$c_{7vx} = 1.090418 - 2.008245\,\nu - 4.077148\,\nu^2 - 3.160768\,\nu^3 - 2.551183\,\nu^4 + 1.689435\,\nu^5 - 10.40112\,\nu^6 + 10.4639\,\nu^7; \tag{3.178b}$$

$$\delta_{hx} \approx P_{hx}(\xi_1) = \sum_{i=0}^{7} c_{ihx}\xi_1^i, \tag{3.179a}$$

where

$$c_{0hx} = 0.5241327 - 0.1349591\,\nu - 0.0206017\,\nu^2 - 0.01956359\,\nu^3 - 0.00850399\,\nu^4 - 0.05240186\,\nu^5 + 0.07157362\,\nu^6 - 0.09726016\,\nu^7,$$

$$c_{1hx} = -2.269002 - 0.213405\,\nu - 0.07550746\,\nu^2 - 0.1085163\,\nu^3 + 0.2125443\,\nu^4 - 0.8835387\,\nu^5 + 1.270464\,\nu^6 - 0.8618935\,\nu^7,$$

$$c_{2hx} = 8.021226 - 3.67188\,\nu + 0.4335535\,\nu^2 + 0.6754332\,\nu^3 - 1.799069\,\nu^4 + 6.347884\,\nu^5 - 8.899052\,\nu^6 + 5.115448\,\nu^7,$$

$$c_{3hx} = -19.32593 + 10.01339\,\nu - 2.729424\,\nu^2 - 3.492173\,\nu^3 + 6.977836\,\nu^4 - 26.45739\,\nu^5 + 35.56952\,\nu^6 - 19.98382\,\nu^7,$$

$$c_{4hx} = 32.55194 - 8.995831\,\nu + 8.024314\,\nu^2 + 9.140126\,\nu^3 - 14.44389\,\nu^4 + 58.15694\,\nu^5 - 74.44257\,\nu^6 + 39.66467\,\nu^7,$$

$$c_{5hx} = -35.30487 - 1.143539\,\nu - 12.26408\,\nu^2 - 12.79233\,\nu^3 + 16.13421\,\nu^4 - 68.84525\,\nu^5 + 82.74138\,\nu^6 - 39.88025\,\nu^7,$$

$$c_{6hx} = 21.48348 + 8.271392\,\nu + 9.327955\,\nu^2 + 9.02779\,\nu^3 - 8.986012\,\nu^4 + 40.70534\,\nu^5 - 44.6814\,\nu^6 + 17.38102\,\nu^7,$$

$$c_{7hx} = -5.680841 - 4.124108\,\nu - 2.695541\,\nu^2 - 2.429905\,\nu^3 + 1.911661\,\nu^4 - 8.965522\,\nu^5 + 8.361438\,\nu^6 - 1.331586\,\nu^7; \qquad (3.179b)$$

$$\delta_y \approx P_y(\xi_1) = 0.4534537 - 2.413347\,\xi_1 + 8.406338\,\xi_1^2 - 20.94925\,\xi_1^3 + 36.71775\,\xi_1^4 - 40.86615\,\xi_1^5 + 25.23488\,\xi_1^6 - 6.582561\,\xi_1^7. \qquad (3.180)$$

The expressions for the static displacements of the surface points of the linearly heterogeneous half-space, under conditions of the plane deformation, may be rewritten in the following form convenient for computations:

$$w_{vz} \approx \frac{1}{G_0\pi}[\widetilde{S}_{vz}(\widetilde{x}) + (1-\xi)\sum_{i=0}^{7} c_{ivz}\xi_1^i], \qquad (3.181a)$$

$$w_{vx} \approx \frac{1}{G_0\pi}[\widetilde{S}_{vx}(\widetilde{x}) + (1-\xi)\sum_{i=0}^{7} c_{ivx}\xi_1^i], \qquad (3.181b)$$

$$w_{hx} \approx \frac{1}{G_0\pi}[\widetilde{S}_{hx}(\widetilde{x}) + (1-\xi)\sum_{i=0}^{7} c_{ihx}\xi_1^i], \qquad (3.181c)$$

$$w_y \approx \frac{1}{G_0\pi}[\widetilde{S}_y(\widetilde{x}) + (1-\xi)P_y(\xi_1)]. \qquad (3.181d)$$

The accuracy of these formulas is as high as that reached by numeric evaluations of the initial integral representations.

Chapter 4. Mechanics of Transversely Isotropic Half-Space with Stiffness Varying Exponentially with Depth

In this chapter, we consider some dynamic and static problems for a half-space having elastic characteristics containing exponential functions, whereas all elements of matrix (1.3c) are assumed to vary similarly. For the isotropic case, this assumption means that Poisson's ratio is considered to be constant.

In section 4.1, we consider a law of parameter variation in the form of a of second-order polynomial whose argument is an exponential with a negative power containing the depth of the considered half-space point [81]. This law provides boundedness of the half-space stiffness at infinite depth; it is a generalization of the relationship (having the form of a first-order polynomial) for the shear modulus applied in numerous reports [94, 95, 116–121]. Introducing the second-order term into the polynomial results in a wider variety of half-space properties. Further, in comparison with these reports, the analysis presented includes the case of the action of horizontal loads, and the case of transverse isotropy.

In section 4.2, we introduce another model of a half-space, which is described by using the exponential functions: an exponentially decreasing term which affects the behavior of the elastic parameters of the anisotropic half-space in the vicinity of its surface is added to an exponential increasing unboundedly with depth.

The common approach employed to solve the considered problems is given in [94, 95]: following substitution of the independent variable, which removes exponential functions, the solutions of equations having the form (1.29)–(1.31) are expanded into the generalized power series.

4.1 Vibrations of Transversely Isotropic Half-Space Having Elastic Coefficients Bounded at Infinite Depth

4.1.1 Variation of Elastic Parameters of Half-Space with Depth

In this subsection, we consider the vibrations of a transversely isotropic half-space possessing properties that are a generalization of those of the isotropic half-space considered in [94, 95, 116–121]. In these works, the shear modulus is assumed to vary by the following law:

$$G(z) = G_\infty - (G_\infty - G_0)\mathrm{e}^{-\alpha z} \ . \tag{4.1}$$

An origin of coordinates is assumed to be located on the half-space surface, G_0 represents the shear modulus on the half-space surface, G_∞ is the limiting value for the shear modulus as $z \to \infty$, and α is a parameter. In [94, 95], the forced vibrations problem was solved for the case with the action of a vertical force, where the problem formulation in stresses was adopted. In [116–121], the formulation in displacements was employed, and, besides forced vibrations under the action of a vertical force and forced anti-plane vibrations, free vibrations were considered.

Generalizing relationship (4.1), applied to the shear modulus G_{rz} of the transversely isotropic half-space, gives

$$G_{rz} = G_{rz}(z_0)\widetilde{G} \ , \tag{4.2a}$$

$$\widetilde{G} = \widetilde{G}_\infty + A\mathrm{e}^{-\widetilde{z}} + L\mathrm{e}^{-2\widetilde{z}} \ , \tag{4.2b}$$

where $z_0 = 0$; $\widetilde{G}_\infty > 0$, A, L are constant values. Values $\widetilde{z}$ and $\widetilde{G}$ are given in relationships (1.27). Clearly, $\widetilde{G} \to \widetilde{G}_\infty$ as $\widetilde{z} \to \infty$. To satisfy the condition $\widetilde{G}(0) = 1$, we take

$$A = 1 - \widetilde{G}_\infty - L \ . \tag{4.3}$$

The behavior of the shear modulus G_{rz} versus depth is determined by the following parameters: $\widetilde{G}_\infty, L, z_r$ (the reference length, z_r, enters the definition of the dimensionless length $\widetilde{z}$). For the case of the isotropic half-space, generalization (4.2) was considered in [80]. At $L = 0$, $z_r = 1/\alpha$, we obtain relationship (4.1); an additional parameter L provides a more diverse behavior of the shear modulus with varying coordinate z.

The relationship identical to (4.2) is also taken for the rest of the elastic parameters. For the normalized parameters introduced in relationships (1.27) we have

$$\begin{aligned} \widetilde{G}_{r\vartheta} &= \widetilde{G}_{r\vartheta 0}\widetilde{G} \ , \\ \widetilde{A}_{rr} &= \widetilde{A}_{rr0}\widetilde{G} \ , \\ \widetilde{A}_{rz} &= \widetilde{A}_{rz0}\widetilde{G} \ , \\ \widetilde{A}_{rr} &= \widetilde{A}_{rr0}\widetilde{G} \ , \\ \widetilde{A}_{zz} &= \widetilde{A}_{zz0}\widetilde{G} \ , \end{aligned} \tag{4.4}$$

where the values containing 0 in their index are taken to be constant; they represent the corresponding initial values. As follows from relationships (4.4), the ratios of the elastic parameters of the anisotropic medium to the shear modulus G_{rz} is independent of the coordinate z. Recall that the normalized elastic coefficients marked with a tilde in (1.27) were taken to be real.

Similar to the technique used in [94, 95, 116–121], we employ the following substitution of the variable in the equations of motion (1.29)–(1.31):

$$\xi = e^{-\tilde{z}} . \tag{4.5}$$

This substitution transforms the half-space $0 \le z < \infty$ to the interval $1 \ge \xi > 0$. The normalized shear modulus takes the form

$$\tilde{G} = \tilde{G}_\infty + A\xi + L\xi^2 . \tag{4.6}$$

The expression for the initial derivative of the quantity $\tilde{G}$ may be written as

$$p_0 = \left.\frac{d\tilde{G}}{d\tilde{z}}\right|_{\tilde{z}=0} = -\xi \left.\frac{d\tilde{G}}{d\xi}\right|_{\xi=1} = -1 + G_\infty - L . \tag{4.7}$$

We seek solutions of equations (1.29)–(1.31) (taking into account (4.2), (4.4), (4.5)) in the form of a power series of the variable ξ. In order to provide convergence of such a series, the function $\tilde{G}(\xi)$ entering the coefficient of the second derivative in the equations should be non-zero over the circle $|\xi|<1$ in the complex plane of variable ξ. Following the form of function (4.6), the necessary conditions, under which the roots of equation $\tilde{G} = 0$ are located outside the circle $|\xi| \le 1$, may be written as

$$\left|\frac{L}{\tilde{G}_\infty}\right| < 1 , \tag{4.8a}$$

$$\left.\tilde{G}\right|_{\xi=-1} > 0 , \tag{4.8b}$$

or

$$-\tilde{G}_\infty < L < \tilde{G}_\infty , \tag{4.9a}$$

$$0.5 - \tilde{G}_\infty < L . \tag{4.9b}$$

The necessary and sufficient condition of location of the roots of $\tilde{G}$ outside the circle $|\xi| \le 1$ has the form

$$0.5 - \tilde{G}_\infty < L < \tilde{G}_\infty . \tag{4.10}$$

When this condition is satisfied, an absolute value of the product of the roots is greater than unity, so that the existence of complex roots inside the circle

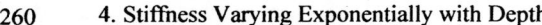

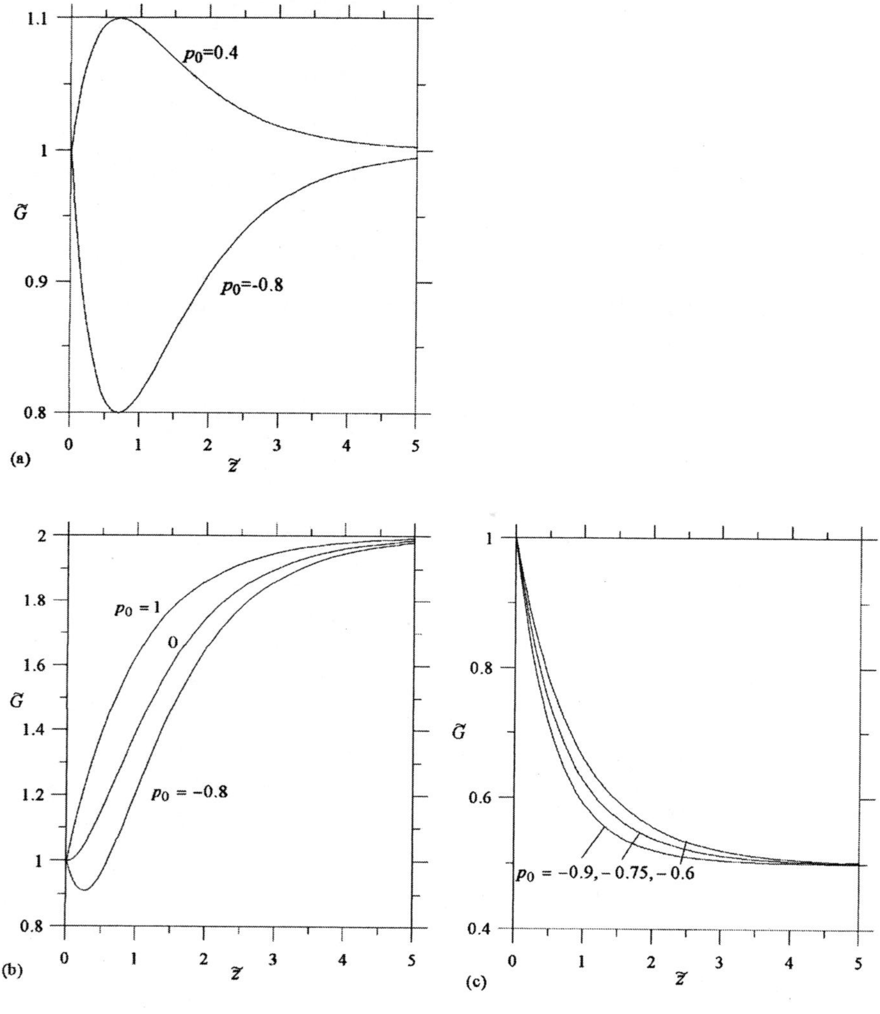

Fig. 4.1a,b,c. Variation of normalized shear modulus with depth for three values of $\widetilde{G}_\infty$ (ratio of half-space stiffness at infinite depth to surface stiffness): $\widetilde{G}_\infty = 1$ (a), 2 (b), 0.5 (c)

$|\xi| \le 1$ and the existence of two roots in the interval $-1 \le \xi \le 1$ are excluded. The case with a single root existing in this interval is excluded due to the positiveness of the function $\widetilde{G}$ at the end points of the interval. Note that for the practical application of condition (4.10), the following inequality should hold:

$$\widetilde{G}_\infty > 0.5 - \widetilde{G}_\infty \tag{4.11a}$$

or

$$\widetilde{G}_\infty > 0.25\,. \tag{4.11b}$$

A condition for the initial derivative p_0, corresponding to inequalities (4.10), may be written as

$$-1 < p_0 < 2\widetilde{G}_\infty - 1.5\ . \tag{4.12a}$$

At $\widetilde{z} = 0$ ($\xi = 1$), the value of the derivative may be negative even at $\widetilde{G}_\infty > 1$. Following the above presentation, the suggested relationship (4.2) enables us to obtain diverse forms of varying the shear modulus with depth. Consider some examples of such a variation. In Fig. 4.1a, the case with $\widetilde{G}_\infty = 1$ is presented for two values of p_0: $p_0 = 0.4$ and $p_0 = -0.8$; note that at $p_0 = 0$, the half-space becomes homogeneous (according to relationships (4.7), (4.3), we have $L = 0,\ A = 0$). In Fig. 4.1b, graphs are plotted for the case with $\widetilde{G}_\infty = 2$; the value $p_0 = 1$ results in $L = 0,\ A = -1$, i.e. we obtain the relationship used in [94, 95, 116–121]. A case with shear modulus decreasing with depth is shown in Fig. 4.1c. At $\widetilde{G}_\infty = 0.5$, as follows from (4.12), we obtain the following permitted interval of variation of the initial derivative p_0:

$$-1 < p_0 < -0.5\,. \tag{4.12b}$$

4.1.2 Construction of Fundamental Solutions

Following the substitution of variable (4.5), equations (1.29)–(1.31) become (assuming constant density, i.e. $\widetilde{\rho} = 1$)

$$\xi^2[\widetilde{G}_\infty + A\xi + L\xi^2]\frac{d^2\widetilde{p}}{d\xi^2} + (\widetilde{G}_\infty + 2A\xi + 3L\xi^2)\xi\frac{d\widetilde{p}}{d\xi}$$
$$+[\beta^2\theta^2 - \widetilde{k}^2\widetilde{G}_{r\theta 0}(\widetilde{G}_\infty + A\xi + L\xi^2)]\widetilde{p} = 0\,, \tag{4.13}$$

$$\xi^2[\widetilde{G}_\infty + A\xi + L\xi^2]\frac{d^2\widetilde{q}}{d\xi^2} + [\widetilde{G}_\infty + 2A\xi + 3L\xi^2]\xi\frac{d\widetilde{q}}{d\xi} + [\beta^2\theta^2 - \widetilde{k}^2\widetilde{A}_{rr0}(\widetilde{G}_\infty + A\xi$$
$$+ L\xi^2)]\widetilde{q} + \widetilde{k}\xi(\widetilde{A}_{rz0} + 1)[\widetilde{G}_\infty + A\xi + L\xi^2]\frac{d\widetilde{w}}{d\xi} + \widetilde{k}\xi[A + 2L\xi]\widetilde{w} = 0\,, \tag{4.14}$$

$$\widetilde{A}_{zz0}\xi^2[\widetilde{G}_\infty + A\xi + L\xi^2]\frac{d^2\widetilde{w}}{d\xi^2} + \widetilde{A}_{zz0}\xi[\widetilde{G}_\infty + 2A\xi + 3L\xi^2]\frac{d\widetilde{w}}{d\xi} + [\beta^2\theta^2 - \widetilde{k}^2(\widetilde{G}_\infty$$
$$+ A\xi + L\xi^2)]\widetilde{w} - \widetilde{k}\xi(\widetilde{A}_{rz0} + 1)[\widetilde{G}_\infty + A\xi + L\xi^2]\frac{d\widetilde{q}}{d\xi} - \widetilde{k}\widetilde{A}_{rz0}\xi[A + 2L\xi]\widetilde{q} = 0. \tag{4.15}$$

Dimensionless length z_r, which enters the law of varying normalized shear modulus $\widetilde{G}$, is also used in the definition of parameter θ introduced in (1.32a).

We seek solutions of equations (4.13)–(4.15) in the form of a power series

$$\widetilde{p} = \xi^{\mu} \sum_{n=0}^{\infty} c_n \xi^n , \tag{4.16}$$

$$\widetilde{q} = \xi^{m} \sum_{n=0}^{\infty} a_n \xi^n , \tag{4.17}$$

$$w = \xi^{m} \sum_{n=0}^{\infty} b_n \xi^n , \tag{4.18}$$

where the values μ, m and series coefficients c_n, a_n, b_n are determined by substituting these series into equations (4.13)–(4.15). Substituting series (4.16) into equation (4.13) and setting the sum of ξ^{μ}-containing terms equal to zero gives

$$(\widetilde{G}_{\infty}\mu^2 + \beta^2\theta^2 - \widetilde{k}^2\widetilde{G}_{r\vartheta 0}\widetilde{G}_{\infty})c_0 = 0 \ . \tag{4.19}$$

Requiring that the coefficient c_0 should differ from zero yields the following equation for μ:

$$\mu^2 + \beta^2\theta_{\infty}^2 - \widetilde{k}^2\widetilde{G}_{r\vartheta 0} = 0 , \tag{4.20}$$

where

$$\theta_{\infty} = \frac{\theta}{\sqrt{\widetilde{G}_{\infty}}} = \omega z_r \sqrt{\frac{\rho}{\widetilde{G}_{rz\infty}}} , \tag{4.21a}$$

$$\widetilde{G}_{rz\infty} = \widetilde{G}_{\infty}\widetilde{G}_{rz0} \, . \tag{4.21b}$$

The roots μ_1 and μ_2 may be written as follows:

$$\mu_j = -(-1)^j \sqrt{\widetilde{k}^2\widetilde{G}_{r\vartheta 0} - \beta^2\theta_{\infty}^2} , \tag{4.22}$$

where the principal value of the radical is used. Taking coefficient c_0 equal to unity, we determine the subsequent coefficients by considering the $\xi^{\mu+n}$-containing terms ($n = 1,2,\dots$) in equation (4.13). As a result, we arrive at the following recurrent relationship:

$$c_n = \frac{R_n}{\widetilde{G}_{\infty}[\widetilde{k}^2\widetilde{G}_{r\vartheta 0} - \beta^2\theta_{\infty}^2 - (\mu+n)^2]} = -\frac{R_n}{\widetilde{G}_{\infty}(2\mu n + n^2)} , \tag{4.23}$$

where

$$R_n = c_{n-1}A[(\mu+n-1)(\mu+n) - \widetilde{k}^2\widetilde{G}_{r\vartheta 0}] + c_{n-2}L[(\mu+n-2)(\mu+n) - \widetilde{k}^2\widetilde{G}_{r\vartheta 0}]\,. \tag{4.24}$$

When calculating R_1, we take $c_{-1}=0$. It can be shown that, when condition (4.10) is satisfied, convergence of the considered series is provided. Having determined the series coefficients at $\mu=\mu_j$ ($j=1,2$), we obtain two fundamental solutions of equation (4.13). The first solution, corresponding to the root μ_1, has a positive real part and tends to zero when $\xi \to 0$ ($\widetilde{z} \to \infty$); this solution is used when solving problems dealing with the half-space subjected to the forces applied to its surface. The second solution is employed when solving problems of vibrations of layers, or in the case of loads applied within the half-space.

The fundamental solutions of system (4.14), (4.15) are constructed in a similar way. Substitution of series (4.17), (4.18) and consideration of ξ^m-containing terms gives the system of equations for the coefficients a_0, b_0:

$$a_0[m^2 + \beta^2\theta_\infty^2 - \widetilde{k}^2\widetilde{A}_{rr0}] + b_0\widetilde{k}(\widetilde{A}_{rz0}+1)m = 0,$$

$$-a_0\widetilde{k}(\widetilde{A}_{rz0}+1)m + b_0[\widetilde{A}_{zz0}m^2 + \beta^2\theta_\infty^2 - \widetilde{k}^2] = 0\,. \tag{4.25}$$

This system is identical to the system of equations (2.6) for the homogeneous half-space. The requirement of the existence of a non-zero solution of this system of equations results in the vanishing of the system determinant, which gives an equation for *m*. This equation coincides with (2.7) after replacing θ with θ_∞ and adding 0 in the indices of the elastic coefficients,

$$\widetilde{A}_{zz0}m^4 + Bm^2 + C = 0\,, \tag{4.26}$$

where

$$B = \widetilde{k}^2\widetilde{B} + \beta^2\theta_\infty^2(1+\widetilde{A}_{zz0})\,, \tag{4.27a}$$

$$C = (\widetilde{k}^2 - \beta^2\theta_\infty^2)(\widetilde{A}_{rr0}\widetilde{k}^2 - \beta^2\theta_\infty^2)\,, \tag{4.27b}$$

$$\widetilde{B} = 2\widetilde{A}_{rz0} + \widetilde{A}_{rz0}^2 - \widetilde{A}_{zz0}\widetilde{A}_{rr0}\,. \tag{4.27c}$$

Expressions for the roots m_1 and m_2 that have a positive real part, i.e. provide a decrease in the corresponding solutions of the form (4.17), (4.18) when $\widetilde{z} \to \infty$ ($\xi \to 0$) (for sufficiently high values of parameter $\widetilde{k}$), are analogous to expressions (2.9),

$$m_1 = \sqrt{\frac{-B + \sqrt{B^2 - 4\widetilde{A}_{zz0}C}}{2\widetilde{A}_{zz0}}}\,, \tag{4.28a}$$

$$m_2 = \sqrt{\frac{-B - \sqrt{B^2 - 4\widetilde{A}_{zz0}C}}{2\widetilde{A}_{zz0}}} \,. \tag{4.28b}$$

Roots $m_3 = -m_1$ and $m_4 = -m_2$ have a negative real part (at sufficiently high values of parameter $\widetilde{k}$); they are used when solving problems for a layer, or in those cases when the acting loads are applied within the half-space. Clearly, at high values of $\widetilde{k}$, quantities m_j become proportional to $\widetilde{k}$.

Multiplier ξ^m in solution (4.17), (4.18) for the first two roots m_j results (for $\widetilde{z} > 0$) in an exponential decrease of solutions, as $\widetilde{k}$ increases; for the roots m_3, m_4, an exponential increase takes place.

For each of four values $m = m_j$, setting the determinant of the system of equations (4.25) to zero, we take $a_0 = 1$ and calculate the corresponding value of b_0 from the first equation (4.25):

$$b_0 = -\frac{m^2 + \beta^2\theta_\infty^2 - \widetilde{k}^2\widetilde{A}_{rr0}}{\widetilde{k}(\widetilde{A}_{rz0} + 1)m} \,. \tag{4.29}$$

In order to calculate the subsequent coefficients a_n, b_n ($n = 1,2,\ldots$), we consider ξ^{m+n}-containing terms in equations (4.14), (4.15) (following the substituting into these equations of series (4.17), (4.18)). Setting the sum of such terms equal to zero in each equation, we obtain the following recurrent relationships for the coefficients a_n, b_n:

$$a_n[(m+n)^2 + \beta^2\theta_\infty^2 - \widetilde{k}^2\widetilde{A}_{rr0}] + b_n\widetilde{k}(\widetilde{A}_{rz0} + 1)(n+m) = -\frac{P_n}{\widetilde{G}_\infty},$$

$$-a_n\widetilde{k}(\widetilde{A}_{rz0} + 1)(n+m) + b_n[\widetilde{A}_{zz0}(m+n)^2 + \beta^2\theta_\infty^2 - \widetilde{k}^2] = -\frac{Q_n}{\widetilde{G}_\infty}, \tag{4.30}$$

$$\begin{aligned} P_n &= a_{n-1}A[(m+n-1)(m+n) - \widetilde{k}^2\widetilde{A}_{rr0}] + a_{n-2}L[(m+n-2)(m+n) - \widetilde{k}^2\widetilde{A}_{rr0}] \\ &\quad + b_{n-1}\widetilde{k}A[(\widetilde{A}_{rz0} + 1)(n+m-1) + 1] + b_{n-2}L\widetilde{k}[(\widetilde{A}_{rz0} + 1)(n+m-2) + 2], \end{aligned} \tag{4.31a}$$

$$\begin{aligned} Q_n &= -a_{n-1}\widetilde{k}A[(\widetilde{A}_{rz0} + 1)(n+m-1) + \widetilde{A}_{rz0}] - a_{n-2}\widetilde{k}L[(\widetilde{A}_{rz0} + 1)(n+m-2) \\ &\quad + 2\widetilde{A}_{rz0}] + b_{n-1}A[\widetilde{A}_{zz0}(m+n-1)(m+n) - \widetilde{k}^2] \\ &\quad + b_{n-2}L[\widetilde{A}_{zz0}(m+n-2)(m+n) - \widetilde{k}^2]. \end{aligned} \tag{4.31b}$$

Here, coefficients a_{-1} and b_{-1} are assumed to vanish.

Having determined the required number of series coefficients for each value of $m = m_j$, one can consider the construction of the four fundamental solutions of the system of differential equations as complete. The series convergence slows down when parameter θ increases, and when an absolute value of the roots of equation $\widetilde{G}(\xi) = 0$ approaches unity. The required number of terms in the series is determined from test calculations.

At $\widetilde{G}_\infty = 1,\ L = 0,\ A = 0$, we have the case of an homogeneous anisotropic half-space. In this case, the only term remaining in the series is that corresponding to the zero index. We obtain the fundamental solutions constructed in Chap. 2.

Considering an isotropic half-space, one should account for relationships (1.5). This gives

$$\widetilde{G}_{r\vartheta 0} = 1, \tag{4.32a}$$

$$\widetilde{A}_{zz0} = \widetilde{A}_{rr0} = \frac{\lambda + 2G}{G} = \frac{1}{\tau^2}, \tag{4.32b}$$

$$\widetilde{A}_{rz0} = \widetilde{A}_{r\vartheta 0} = \frac{\lambda}{G} = \frac{1}{\tau^2} - 2. \tag{4.32c}$$

Analogously to relationships (2.15), we have for the isotropic half-space

$$m_1 = \sqrt{\widetilde{k}^2 - \beta^2\theta_\infty^2}, \tag{4.33a}$$

$$m_2 = \sqrt{\widetilde{k}^2 - \beta^2\theta_\infty^2 / \widetilde{A}_{zz0}} = \sqrt{\widetilde{k}^2 - \beta^2\theta_\infty^2\tau^2}. \tag{4.33b}$$

4.1.3 Vibrations of Half-Space under Action of Vertical Force Applied to Half-Space Surface

After constructing of the corresponding fundamental solutions, one can use relationships (2.1) directly. Quantities c_{ij} determined from the fundamental solutions for the half-space surface, according to (1.54), may be written as follows:

$$c_{1j} = \widetilde{A}_{rz0}\widetilde{k}\widetilde{q}_j(1,\widetilde{k}) - \widetilde{A}_{zz0}\frac{\mathrm{d}\,\widetilde{w}_j(1,\widetilde{k})}{\mathrm{d}\xi}, \tag{4.34a}$$

$$c_{2j} = -\frac{\mathrm{d}\,\widetilde{q}_j(1,\widetilde{k})}{\mathrm{d}\xi} - \widetilde{k}\widetilde{w}_j(1,\widetilde{k}), \tag{4.34b}$$

where the given value of the argument ξ, equal to unity, corresponds to the half-space surface; in formulas (2.1), we use ξ instead of argument $\widetilde{z}$.

In dynamic problems, instead of variable $\widetilde{r}$, we use variable a having the form (3.33):

$$a = \theta\tilde{r} = \omega r\sqrt{\frac{\rho}{G_{rz0}}} \ . \tag{4.35}$$

In the integrals having the form (2.1b), (2.1d), replacement (3.32) is performed,

$$\tilde{k} = s\theta \ . \tag{4.36}$$

Next, we consider singularities of the integrand in the case when no energy dissipation occurs ($\varepsilon = 0$). At small values of parameter θ, a single pole exists

$$s = s_R \approx s_{R\infty} \tag{4.37}$$

close to Rayleigh's pole for the homogeneous half-space possessing the properties of the given half-space at infinite depth. Since the replacement of variable (4.36) corresponds to the half-space properties on its surface, we have

$$s_{R\infty} = \frac{\tilde{k}_R}{\sqrt{\tilde{G}_\infty}} , \tag{4.38}$$

where the value $\tilde{k}_R$ represents the corresponding Rayleigh's pole for the homogeneous half-space considered in section 2.1.2. As the frequency of vibrations (parameter θ) increases, the pole s_R shifts, approaching the value $\tilde{k}_R$, i.e. the pole for the homogeneous half-space having the properties of the given half-space on its surface. In the case $\tilde{G}_\infty > 1$, as parameter θ increases, additional poles $s_i < \tilde{k}_R$ appear; for the case of an isotropic material, these poles are bounded below:

$$s_i > \frac{1}{\sqrt{\tilde{G}_\infty}} \ . \tag{4.39}$$

At $\tilde{G}_\infty \le 1$, a single pole exists; its abscissa decreases as the value of θ increases, from the value $s_{R\infty}$ to the value $\tilde{k}_R$.

Considerations regarding calculations, given in section 3.2, also hold in the present case; the integration contour shown in Fig. 2.3 is employed. The above information concerning the interval of location of the poles is used, when organizing calculations. A smaller numerical integration step is selected on the part of the contour where the poles are located.

Next, examples of calculations are presented for the case of an isotropic half-space. We take $\varepsilon = 0, \nu = 1/3, \tilde{G}_\infty = 2$. In expression (4.6), we take $L = 1$, $A = -2$, resulting in the value $p_0 = 0$. These parameters satisfy the given requirements. In Fig. 4.2, the normalized amplitudes of vibrations S_{vh} and S_{vv} for the half-space surface are shown for a number of values θ. Due to the influence of Rayleigh's poles, the normalized amplitudes of vibrations increase, analogously to

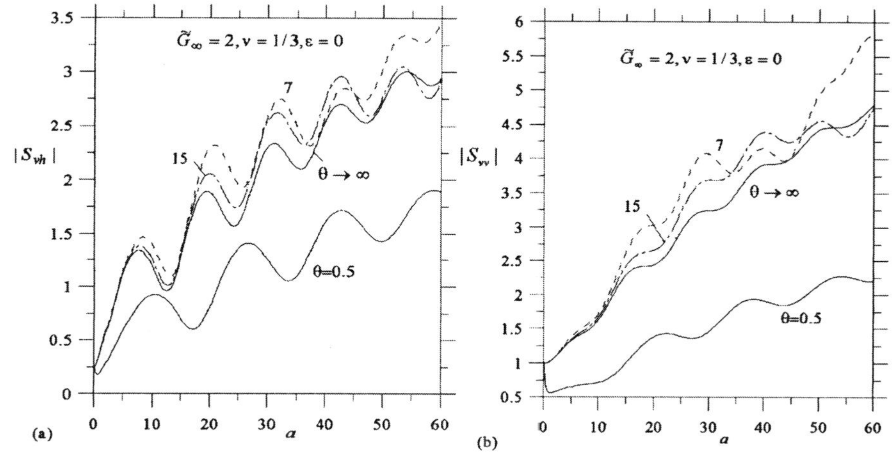

Fig. 4.2a,b. Absolute values of normalized horizontal (a) and vertical (b) amplitudes of vibrations of half-space surface points under action of vertical force, $\widetilde{G}_\infty = 2$

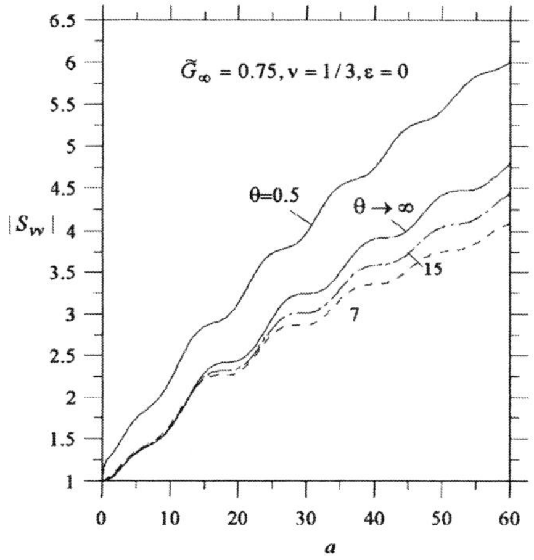

Fig. 4.3. Absolute values of normalized vertical amplitudes of vibrations of half-space surface points under action of vertical force in the case of stiffness decreasing with depth, $\widetilde{G}_\infty = 0.75$

the case of the linearly heterogeneous half-space studied in the previous chapter. The results for the homogeneous half-space correspond to the unbounded growth of parameter θ. The process of approaching zero of parameter a (while parameter θ is kept constant) may be considered as a process of decreasing frequency of vibrations with simultaneous growth of the reference length z_r, which enters definition (1.32a) for parameter θ. Therefore, the solution should approach the static solution for a homogeneous half-space, which takes the form (for the isotropic case)

$$S_{vh} = -\tau^2 = -\frac{1-2\nu}{2(1-\nu)}, \tag{4.40a}$$

$$S_{vv} = 1. \tag{4.40b}$$

The following example calculation is related to the case of an isotropic half-space having stiffness decreasing with depth. Let $\widetilde{G}_\infty = 0.75$, $A = 0.25$, $L = 0$, whereas the remainder of the parameters are the same, as in the previous case. Graphs representing varying magnitudes of the normalized amplitude S_{vv} are shown in Fig. 4.3. The difference between results corresponding to the values $\widetilde{G}_\infty = 0.75$ and $\widetilde{G}_\infty = 2$ is rather significant for small values of parameter θ. In the first case, at $\theta = 0.5$, an abrupt increase in $|S_{vv}|$ occurs at small values of a, whereas in the latter case, an abrupt decrease takes place. In order to explain such behavior, one should account for the difference between static stiffnesses of the considered half-spaces. At high values of θ, the results of calculations in both cases appear to be close to the results corresponding to the homogeneous half-space.

4.1.4 Vibrations of Half-Space under Action of Horizontal Force Applied to Half-Space Surface

One can employ the general form of solution given in expressions (2.53)–(2.55). Fundamental solutions constructed for the heterogeneous anisotropic half-space should be substituted into these expressions. Here, we take into account expressions (4.34) for the coefficients c_{ij} and expression for the quantity F entering functions $\widetilde{H}_r$, $\widetilde{H}_\vartheta$ ($z_0 = 0$),

$$F = \frac{\mathrm{d}\, p_1(z_0, k)}{\mathrm{d}z} = \frac{\mathrm{d}\,\widetilde{p}_1(\widetilde{z}_0, \widetilde{k})}{\mathrm{d}\widetilde{z}} = -\frac{\mathrm{d}\,\widetilde{p}_1(1, \widetilde{k})}{\mathrm{d}\xi}\,. \tag{4.41}$$

In Fig. 4.4, the results of calculations of the magnitudes of the normalized vibration amplitudes $S_{h\vartheta}$ and S_{hr} for the half-space surface in the isotropic case are presented for the values of parameters employed previously when studying the action of the vertical force (Fig. 4.2). As a increases, the normalized amplitudes $|S_{hr}|$ increase much slower than the amplitudes $|S_{h\vartheta}|$. Consider the possibility of approximating values $S_{h\vartheta}$ by using the term entering the integrand in relationship (2.54b), which contains Bessel's function J_0. Analogously to relationship (2.76),

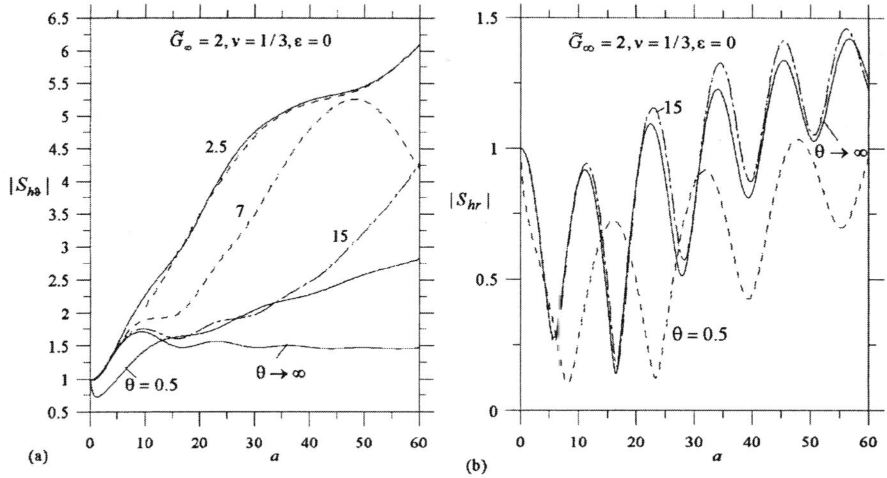

Fig. 4.4a,b. Absolute values of normalized tangential (a) and radial (b) amplitudes of vibrations of half-space surface points under action of horizontal force

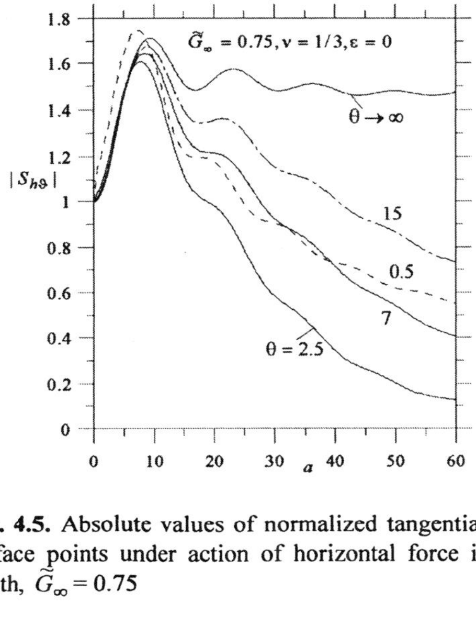

Fig. 4.5. Absolute values of normalized tangential amplitudes of vibrations of half-space surface points under action of horizontal force in the case of stiffness decreasing with depth, $\widetilde{G}_\infty = 0.75$

$$S_{h\vartheta} \approx S'_{h\vartheta} = -\frac{\beta^2 \widetilde{r}}{1-\nu'} \int_0^\infty \frac{\widetilde{k}\widetilde{p}_1(\widetilde{z},\widetilde{k})}{F} \mathrm{J}_0(\widetilde{k}\widetilde{r})\,\mathrm{d}\widetilde{k}\ . \tag{4.42}$$

This term becomes increasingly dominant in the complete solution, as parameter a increases. In Fig. 4.4a, the dashed curve corresponding to $\theta = 2.5$ represents the magnitude of $S'_{h\vartheta}$ for the isotropic half-space. As noted in the previous chapter, the behavior of $S'_{h\vartheta}$ expresses the half-space properties related to Love-type waves. For a homogeneous half-space, the poles of the integrand entering the expression for $S'_{h\vartheta}$ are absent; however, according to (2.76), a singularity having the form $(\tilde{k}^2 - 1/\tilde{G}_{r\vartheta})^{-1/2}$ exists (at $\beta = 1$). As a result, the normalized amplitude of vibrations on the half-space surface tends to the constant value (2.77a), as parameter a increases. If $\tilde{G}_\infty < 1$ (i.e. in the cases when the stiffness of the half-space at infinite depth is smaller than on its surface) the integrand in (4.42) is bounded. This leads to a decrease of the normalized amplitude at high values of a. In Fig. 4.5, values $|S_{h\vartheta}|$ are presented for the case with $\tilde{G}_\infty = 0.75$, $A = 0.25$, $L = 0$, whereas the rest of the parameters are as in the previous example. One can see that the case of the homogeneous half-space discriminates between the case with the normalized amplitudes, increasing due to the influence of Love's poles when $\tilde{G}_\infty > 1$, and the case with decreasing amplitudes when $\tilde{G}_\infty < 1$.

4.2 Vibrations of Transversely Isotropic Half-Space with Stiffness Increasing without Bounds

4.2.1 Varying of Elastic Parameters of Half-Space with Depth

In the present section, we consider vibrations of a half-space with a stiffness that increases exponentially with depth without bounds. A simple exponential relationship is generalized, and an including the anisotropic properties of the half-space is given.

Consider the following relationship between the shear modulus G_{rz} of the anisotropic half-space and coordinate z (the origin of coordinates is located on the half-space surface):

$$G_{rz} = G_{rz}(z_0)\tilde{G}\,, \tag{4.43a}$$

$$\tilde{G} = A\mathrm{e}^{\tilde{z}} + (1-A)\mathrm{e}^{-N\tilde{z}}\,, \tag{4.43b}$$

where $z_0 = 0$; A is a positive constant, N is a positive integer. Quantities $\tilde{z}$ and $\tilde{G}$ are given in relationships (1.27). The behavior of the shear modulus G_{rz} with depth is determined by three parameters, namely A, N, z_r (reference length z_r enters the definition of dimensionless length $\tilde{z}$). At $A = 1$, we obtain a simple exponential relationship, while the additional parameter N (at $A \neq 1$) provides a more diverse behavior of the shear modulus with varying z. A relationship identical to (4.43) is adopted for the remainder of the elastic parameters. For the normalized parameters introduced in relationships (1.27), we apply relationships (4.4). According to the latter, the ratios of elastic parameters of the anisotropic medium to the shear modulus G_{rz} are independent of the coordinate z. Recall that the normalized elastic parameters of the medium, marked with a tilde, were taken to be real.

In the equations of motion (1.29)–(1.31), we perform substitution of the variable (4.5), which transforms the half-space $0 \le z < \infty$ to the interval $1 \ge \xi > 0$. The normalized shear modulus takes the form

$$\tilde{G}(\xi) = \frac{A}{\xi} + (1 - A)\xi^N . \tag{4.44}$$

In the following, we seek solutions of equations (1.29)–(1.31) (after substitution (4.5)) in the form of power series (4.16)–(4.18) of the variable ξ. In order to provide convergence of such a series, function $\tilde{G}(\xi)$, entering the coefficient to the second derivative in considered equations, should remain non-zero over the circle $|\xi|<1$ of the complex variable ξ. This yields the inequality

$$\left|\frac{A}{A-1}\right| > 1 \tag{4.45}$$

or

$$A > 0.5 . \tag{4.46}$$

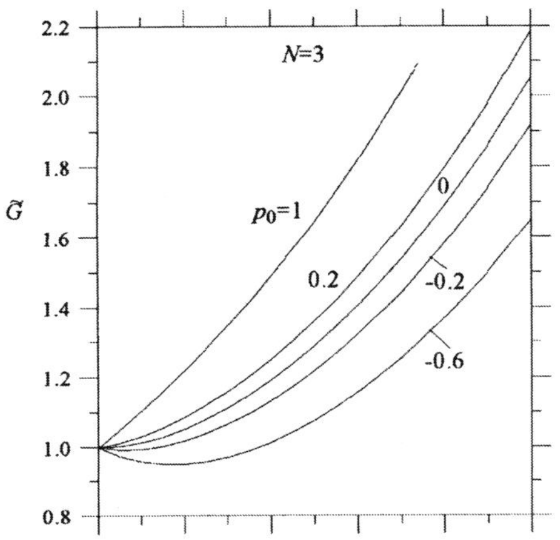

Fig. 4.6. Variation of normalized shear modulus with depth for various values of initial derivative p_0

An initial derivative of the normalized shear modulus may be written as

$$p_0 = \left.\frac{\mathrm{d}\widetilde{G}}{\mathrm{d}\widetilde{z}}\right|_{\widetilde{z}=0} = -\xi\left.\frac{\mathrm{d}\widetilde{G}}{\mathrm{d}\xi}\right|_{\xi=1} = A(1+N)-N\,. \tag{4.47}$$

Condition (4.46) becomes:

$$p_0 > 0.5(1-N)\,. \tag{4.48}$$

The initial derivative may be negative at $N = 2, 3, \ldots$. Following these considerations, relationship (4.43) provides a more diverse behavior of the shear modulus of the half-space with depth than the simple exponential law. A family of curves, representing the function $\widetilde{G}(\widetilde{z})$ at $N=3$ for a number of values p_0, is shown in Fig. 4.6. Value $p_0 = 1$ yields $A=1$, i.e. the conventional exponential relationship is obtained.

4.2.2 Construction of Fundamental Solutions

Following the substitution of variable (4.5), equations (1.29)–(1.31) take the form (assuming constant density, i.e. $\widetilde{\rho}=1$)

$$\xi^2[A+(1-A)\xi^{N+1}]\frac{\mathrm{d}^2\widetilde{p}}{\mathrm{d}\xi^2}+(N+1)(1-A)\xi^{N+2}\frac{\mathrm{d}\widetilde{p}}{\mathrm{d}\xi}+[\beta^2\theta^2\xi$$

$$-\widetilde{k}^2\widetilde{G}_{r\vartheta 0}(A+(1-A)\xi^{N+1})]\widetilde{p}=0, \tag{4.49}$$

$$\xi^2[A+(1-A)\xi^{N+1}]\frac{\mathrm{d}^2\widetilde{q}}{\mathrm{d}\xi^2}+(N+1)(1-A)\xi^{N+2}\frac{\mathrm{d}\widetilde{q}}{\mathrm{d}\xi}+[\beta^2\theta^2\xi-\widetilde{k}^2\widetilde{A}_{rr0}(A$$

$$+(1-A)\xi^{N+1})]\widetilde{q}+\widetilde{k}\xi(\widetilde{A}_{rz0}+1)[A+(1-A)\xi^{N+1}]\frac{\mathrm{d}\widetilde{w}}{\mathrm{d}\xi}-\widetilde{k}[A-(1-A)N\xi^{N+1}]\widetilde{w}=0, \tag{4.50}$$

$$\widetilde{A}_{zz0}\xi^2[A+(1-A)\xi^{N+1}]\frac{\mathrm{d}^2\widetilde{w}}{\mathrm{d}\xi^2}+(N+1)\widetilde{A}_{zz0}(1-A)\xi^{N+2}\frac{\mathrm{d}\widetilde{w}}{\mathrm{d}\xi}+[\beta^2\theta^2\xi-\widetilde{k}^2(A$$

$$+(1-A)\xi^{N+1})]\widetilde{w}-\widetilde{k}\xi(\widetilde{A}_{rz0}+1)[A+(1-A)\xi^{N+1}]\frac{\mathrm{d}\widetilde{q}}{\mathrm{d}\xi}$$

$$+\widetilde{k}A_{rz0}[A-(1-A)N\xi^{N+1}]\widetilde{q}=0\,. \qquad 4.51)$$

We seek the solution of these equations in the form of the power series (4.16)–(4.18). Following the substitution of series (4.16) into equation (4.49) and subsequently setting the sum of ξ^μ-containing terms equal to zero, we arrive at the following equation:

$$c_0[\mu(\mu-1)-\widetilde{k}^2\widetilde{C}_{r\vartheta 0}]=0\,. \tag{4.52}$$

The requirement that coefficient c_0 should differ from zero results in the equation for μ:

$$\mu^2 - \mu - \tilde{k}^2 \tilde{C}_{r\vartheta 0} = 0 . \tag{4.53}$$

From here,

$$\mu_j = 0.5[1 - (-1)^j \sqrt{1 + 4\tilde{k}^2 \tilde{C}_{r\vartheta 0}}\,] . \tag{4.54}$$

The solution corresponding to the root μ_1 tends to zero as $\xi \to 0$ ($\tilde{z} \to \infty$); the solution corresponding to the second root increases as $\tilde{z} \to \infty$. For both solutions, we assume $c_0 = 1$. Consideration of $\xi^{\mu+n}$-containing terms in equation (4.49) after the substitution of series (4.16) yields the following recurrent relationship enabling to determine the subsequent coefficients c_1, c_2,...:

$$c_n = \frac{R_n}{A[(n+\mu)(n+\mu-1) - \tilde{k}^2 \tilde{C}_{r\vartheta 0}]} = \frac{R_n}{A[n^2 + n(2\mu - 1)]} , \tag{4.55}$$

where

$$R_n = -\beta^2 \theta^2 c_{n-1} + (1 - A)[\tilde{k}^2 \tilde{C}_{r\vartheta 0} - (n - N - 1 + \mu)(n - 1 + \mu)] c_{n-1-N} . \tag{4.56}$$

Coefficients having negative indices are considered equal to zero. Substituting the value $\mu = \mu_j$ ($j = 1, 2$) into relationships (4.55), (4.56) gives the series coefficients for the considered fundamental solutions. It can be shown that absolute convergence of the series is provided when condition (4.46) is satisfied.

In the case with $A = 1$, when the simple exponential relationship holds, the solution may be written in the form

$$\tilde{p} = \xi^{\mu} \sum_{n=0}^{\infty} c_n (\beta^2 \theta^2)^n \xi^n , \tag{4.57}$$

whereas in relationship (4.55)

$$R_s = -c_{s-1} \tag{4.58}$$

and $\mu = \mu_1, \mu_2$. The corresponding static solutions are ($A = 1$)

$$\tilde{p}_j^{st} = \xi^{\mu_j} . \tag{4.59}$$

Notice that in the case $A = 1$, equation (4.49) has the form which enables us to construct a solution in the form of Bessel's functions,

$$\xi^2 \frac{d^2 \tilde{p}}{d\xi^2} + [\beta^2 \theta^2 \xi - \tilde{k}^2 \tilde{G}_{r\vartheta 0}] \tilde{p} = 0 . \tag{4.60}$$

According to [1], fundamental solutions of this equation have the form

$$p_1(\xi) = \xi^{1/2} \, \mathrm{J}_{\sqrt{1+4\widetilde{k}^2\widetilde{G}_{r\vartheta 0}}}(2\beta\theta\xi^{1/2}), \tag{4.61a}$$

$$p_2(\xi) = \xi^{1/2} \, \mathrm{J}_{-\sqrt{1+4\widetilde{k}^2\widetilde{G}_{r\vartheta 0}}}(2\beta\theta\xi^{1/2}). \tag{4.61b}$$

These representations are in agreement with solution (4.57).

Next, we construct the fundamental solutions of the system of differential equations (4.50) and (4.51). Following the substitution of series (4.17), (4.18) into these equations and consideration of ξ^m-containing terms, we obtain the system of equations to solve for the coefficients a_0, b_0:

$$\begin{aligned} &[m(m-1) - \widetilde{k}^2 \widetilde{A}_{rr0}]a_0 + \widetilde{k}[m(\widetilde{A}_{rz0}+1) - 1]b_0 = 0, \\ &-\widetilde{k}[m(\widetilde{A}_{rz0}+1) - \widetilde{A}_{rz0}]a_0 + [m(m-1)\widetilde{A}_{zz0} - \widetilde{k}^2]b_0 = 0. \end{aligned} \tag{4.62}$$

The requirement for the existence of a non-zero solution of this system of equations results in the system determinant being zero. The following equation serves for calculation of m:

$$\widetilde{A}_{zz0} g^2 + bg + c = 0, \tag{4.63}$$

where

$$g = m(m-1), \tag{4.64}$$

$$b = \widetilde{k}^2 (2\widetilde{A}_{rz0} + \widetilde{A}_{rz0}^2 - \widetilde{A}_{zz0}\widetilde{A}_{rr0}), \tag{4.65a}$$

$$c = \widetilde{k}^4 \widetilde{A}_{rr0} + \widetilde{k}^2 \widetilde{A}_{rz0}. \tag{4.65b}$$

The roots g_1 and g_2 of equation (4.63) take the form

$$g_j = \frac{-b - (-1)^j \sqrt{b^2 - 4\widetilde{A}_{zz0}c}}{2\widetilde{A}_{zz0}}, \tag{4.66}$$

where principal values of the radicals are implied. In the case with $b < 0$, both roots have a positive real part; at $b \geq 0$, we have two complex conjugate roots (the negativity of the radicand for this case may be proved in the same way as in relationships (2.13)). Calculation of the corresponding values of m gives

$$m_j = 0.5 + (0.25 + g_j)^{1/2} \quad (j = 1,2), \tag{4.67a}$$

$$m_j = 0.5 - (0.25 + g_{j-2})^{1/2} \quad (j = 3,4). \tag{4.67b}$$

The first two roots having positive real parts will be used to construct solutions of the problems dealing with half-space vibrations under loads applied to its surface; corresponding to these roots the solutions tend to zero as $\xi \to 0$ ($\widetilde{z} \to \infty$). The roots m_3, m_4 have a negative real part (at sufficiently high values of parameter $\tilde{k}$); they are used to solve problems for a layer, or in those cases when the acting loads are applied within the half-space. One can see that at high values of parameter $\tilde{k}$, quantities m_j are proportional to $\tilde{k}$. For the first two roots m_j, multiplier ξ^m in relationships (4.17), (4.18) results (for $\tilde{z} > 0$) in exponentially decreasing solutions as $\tilde{k}$ increases; for the roots m_3, m_4, exponential growth takes place. For each of the four values m_j that make the determinant of system (4.62) zero, we adopt $a_0 = 1$ and for each $m = m_j$ the corresponding value b_0 is determined from the first equation in system (4.62),

$$b_0 = \frac{m(m-1) - \tilde{k}^2 \tilde{A}_{rr0}}{\tilde{k}[1 - m(\tilde{A}_{rz0} + 1)]} . \tag{4.68}$$

In order to calculate the subsequent coefficients a_n, b_n ($n = 1,2,\ldots$), we consider ξ^{m+n}-containing terms entering equations (4.50), (4.51) (following the substitution of series (4.17), (4.18) into these equations). Setting the sum of such terms in each equation equal to zero yields recurrent relationships for the coefficients a_n, b_n:

$$\begin{aligned} &[(n+m)(n+m-1) - \tilde{k}^2 \tilde{A}_{rr0}]a_n + \tilde{k}[(\tilde{A}_{rz0}+1)(n+m) - 1]b_n = -\frac{P_n}{A}, \\ &-\tilde{k}[(\tilde{A}_{rz0}+1)(n+m) - \tilde{A}_{rz0}]a_n + [\tilde{A}_{zz0}(n+m)(n+m-1) - \tilde{k}^2]b_n = -\frac{Q_n}{A}, \end{aligned} \tag{4.69}$$

where

$$\begin{aligned} P_n = \beta^2\theta^2 a_{n-1} &- (1-A)[\tilde{k}^2 \tilde{A}_{rr0} - (n-N-1+m)(n-1+m)]a_{n-N-1} \\ &+ \tilde{k}(1-A)[N + (\tilde{A}_{rz0}+1)(n-N-1+m)]b_{n-N-1}, \end{aligned} \tag{4.70a}$$

$$\begin{aligned} Q_n = \beta^2\theta^2 b_{n-1} &- \tilde{k}(1-A)[N\tilde{A}_{rz0} + (\tilde{A}_{rz0}+1)(n-N-1+m)]a_{n-N-1} \\ &- (1-A)[\tilde{k}^2 - \tilde{A}_{zz0}(n-N-1+m)(n-1+m)]b_{n-N-1}. \end{aligned} \tag{4.70b}$$

Coefficients having negative indices should be considered as vanishing.

Having calculated the required number of series coefficients for each value m_j, one can consider the construction of the four fundamental solutions of the system of differential equations as complete. Convergence of the series slows down as parameter θ increases, and as the roots of equation $\tilde{G}(\xi) = 0$ approach unity. The required number of series terms is determined by test calculations.

In the case of $A = 1$, when relationship (4.43) reduces to a simple exponential relationship, the right-hand sides of system (4.70) are simplified. Analogously to representation (4.57), we have

$$\tilde{q} = \xi^m \sum_{n=0}^{\infty} a_n (\beta^2\theta^2)^n \xi^n , \tag{4.71}$$

$$w = \xi^m \sum_{n=0}^{\infty} b_n (\beta^2\theta^2)^n \xi^n . \tag{4.72}$$

In this case, we have in recurrent relationships (4.69)

$$P_n = a_{n-1} , \tag{4.73a}$$

$$Q_n = b_{n-1} . \tag{4.73b}$$

At $A = 1$, static solutions are expressed in the finite form

$$\tilde{q}_j = \xi^{m_j} , \tag{4.74a}$$

$$\tilde{w}_j = \frac{m_j(m_j - 1) - \tilde{k}^2 \tilde{A}_{rr0}}{\tilde{k}[1 - m_j(\tilde{A}_{rz0} + 1)]} \xi^{m_j} . \tag{4.74b}$$

For the case on an isotropic half-space, relationships (4.32) are employed. Values g_i, determined by (4.66), take the form

$$g_j = \tilde{k}^2 - (-1)^j \tilde{k}\, \mathrm{i} \sqrt{1 - 2\tau^2} . \tag{4.75}$$

Note that in the case of $\nu = 0\ (\tau^2 = 0.5)$, multiple roots are obtained; it is advisable to replace the value of Poisson's ratio $\nu = 0$ with a small non-zero value.

4.2.3 Vibrations of Half-Space under Action of Vertical Force Applied to Half-Space Surface

Following construction of the corresponding fundamental solutions, one can directly employ relationships (2.1). For the quantities c_{ij} determined from the fundamental solutions, we shall use presentations (4.34). In the dynamic problems, variable $\tilde{r}$ should be replaced with variable *a*, given in relationship (4.35). In the integrals having the form (2.1b), (2.1d), substitution (4.36) is carried out.

Further, a numerical study of properties of the half-space, with unbounded exponentially increasing stiffness, is performed for the case with the simple exponent ($A = 1$) for the isotropic half-space by using relationships (4.32). Fundamental solutions have the form (4.71), (4.72), whereas values $m = m_1,\ m_2$ are employed to construct a solution for the case of the load applied to the half-space surface. Note that for real values of $\tilde{k}$ and $\beta = 1$ (absence of internal friction), all quantities related to the second solution are complex conjugates with respect to the corresponding values for the first solution. Hence, values entering the integrals in expressions (2.1b), (2.1d) may be written as ($\tilde{k} > 0$, $\beta = 1$)

$$\frac{c_{22}\tilde{q}_1(\xi,\tilde{k}) - c_{21}\tilde{q}_2(\xi,\tilde{k})}{D} = \frac{\mathrm{Im}[c_{21}\tilde{q}_2(\xi,\tilde{k})]}{\mathrm{Im}[c_{12}c_{21}]}, \tag{4.76a}$$

$$\frac{c_{22}\tilde{w}_1(\xi,\tilde{k}) - c_{21}\tilde{w}_2(\xi,\tilde{k})}{D} = \frac{\mathrm{Im}[c_{21}\tilde{w}_2(\xi,\tilde{k})]}{\mathrm{Im}[c_{12}c_{21}]}. \tag{4.76b}$$

At small values of parameter θ, the given quantities have no singularities at $\tilde{k} > 0$, and the corresponding integrals may be calculated directly without using any special integration contour. Thus, in the absence of internal friction, the amplitudes of displacements appear to be real for small values of θ and, as a result, no energy is transferred to the half-space. The first pole of quantities (4.76) occurs in the vicinity of the point $\tilde{k} = 0$ at some value $\theta = \theta_0$. Taking the pole into account results in the appearance of an imaginary part in the representation for the amplitudes of vibrations; this implies that energy transfer to the half-space takes place. Frequency θ_0 serves as the so-called cutoff frequency. As parameter θ increases, additional poles appear. Values of θ at which new poles appear are related to the form of solution of the system of equations (4.50), (4.51) at $\tilde{k} = 0$, when the equations are separated

$$\xi\frac{\mathrm{d}^2\tilde{q}}{\mathrm{d}\xi^2} + \beta^2\theta^2\tilde{q} = 0, \tag{4.77a}$$

$$\tilde{A}_{zz0}\xi\frac{\mathrm{d}^2\tilde{w}}{\mathrm{d}\xi^2} + \beta^2\theta^2\tilde{w} = 0. \tag{4.77b}$$

Solutions of these equations are expressed in terms of the Bessel's functions [1]. In the case when no damping occurs ($\beta = 1$), we obtain the following pair of solutions of the considered system of differential equations

$$\tilde{q}_1 = \xi^{1/2}\,\mathrm{J}_1(2\theta\xi^{1/2}),\ \tilde{w}_1 = 0, \tag{4.78a}$$

$$\tilde{q}_2 = 0,\ \tilde{w}_2 = \xi^{1/2}\,\mathrm{J}_1\left(\frac{2\theta}{\sqrt{\tilde{A}_{zz0}}}\xi^{1/2}\right) = \xi^{1/2}\,\mathrm{J}_1(2\theta\tau\xi^{1/2}). \tag{4.78b}$$

According to relationships (4.34), coefficients c_{ij} take the form ($\tilde{k} = 0$)

$$c_{11} = 0,$$

$$c_{22} = 0,$$

$$c_{12} = -\widetilde{A}_{zz0}\frac{\mathrm{d}\widetilde{w}_j(1,\widetilde{k})}{\mathrm{d}\xi} = -\frac{\theta}{\tau}\mathrm{J}_0(2\theta\tau), \tag{4.79}$$

$$c_{21} = -\frac{\mathrm{d}\widetilde{q}_j(1,\widetilde{k})}{\mathrm{d}\xi} = -\theta\,\mathrm{J}_0(2\theta).$$

One can see that the quantity D entering the denominator of the integrands in relationships (2.1) becomes zero, simultaneously with the quantities $\mathrm{J}_0(2\theta)$ or $\mathrm{J}_0(2\theta\tau)$. Calculations show that for the corresponding values $\theta = \theta_0, \theta_1, \theta_2, \ldots$, at which these quantities vanish, a number of poles of the integrands entering the representation of the normalized amplitudes of vibrations increase; a new pole appears at $\widetilde{k} = 0$ and subsequently shifts to the right along the real axis of $\widetilde{k}$. The dimensionless cutoff frequency appears to be equal to half the value of the first zero of function $\mathrm{J}_0(x)$, i.e.

$$\theta_0 = 1.2028. \tag{4.80}$$

Note that at $\widetilde{k} = 0$, we have equations of 1-D motion of the rod under conditions of purely shear deformation (equation (4.77a)) and of longitudinal deformation when excluding transverse deformations (equation (4.77b)). The above values of θ correspond to the resonance frequencies of the rod. Regarding the half-space, to which the load harmonic in time is applied over a limited area of the half-space surface, the simple poles of the integrands in solution (2.1) do not lead to unbounded growth of the vibration amplitudes; they affect only the behavior of the amplitudes depending on the spatial variable *r*. However, the considered half-space with unbounded exponentially increasing stiffness has resonance frequencies due to the double zeros in denominator *D* that occur at some values of θ. These values are close to odd zeros of function $\mathrm{J}_0(2\tau\theta)$, e.g. for $\tau = 0.5$ ($\nu = 1/3$), the first three resonance values of θ equal 2.2840, 8.6316, 14.9181 (the first three of the previously mentioned odd zeros equal 2.4048, 8.6537, 14.9309). Values of $\widetilde{k}$ corresponding to the double roots of the denominator for given values of θ equal 0.399, 0.1882, 0.1581, respectively. For illustrative purposes, the graph for the quantity *D*, the denominator of the integrands in the representation of the normalized amplitudes of vibrations, is plotted at $\theta = 2.284$, $\nu = 1/3$, $\varepsilon = 0$ in Fig. 4.7. At θ-values close to the resonance values, the amplitudes may reach extremely high values if the calculation does not include dissipative properties of the half-space. Adopting $\varepsilon > 0$, integration may be done over the real axis of variable $\widetilde{k}$ at arbitrary values of θ. As follows from the previous discussion, the dynamic behavior of the half-space, with exponentially increasing shear modulus with depth, is analogous to the behavior of a layer of finite thickness, resting on an absolutely stiff foundation; in the last

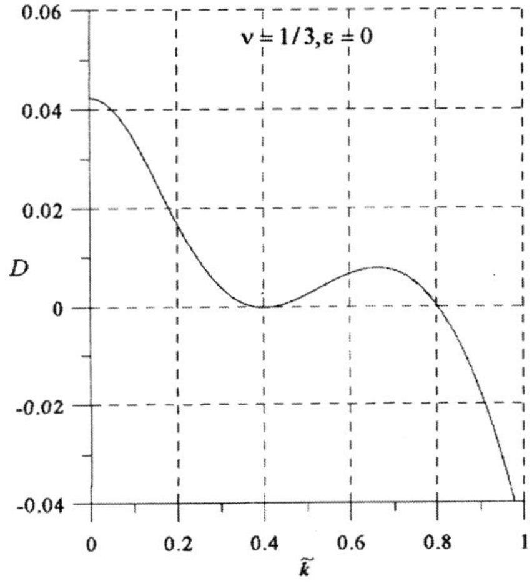

Fig. 4.7. Possibility of occurrence of double zero for denominator D in the integrand of form (2.1d) for amplitudes of vertical vibrations; in the corresponding point, double pole occurs in the integrand

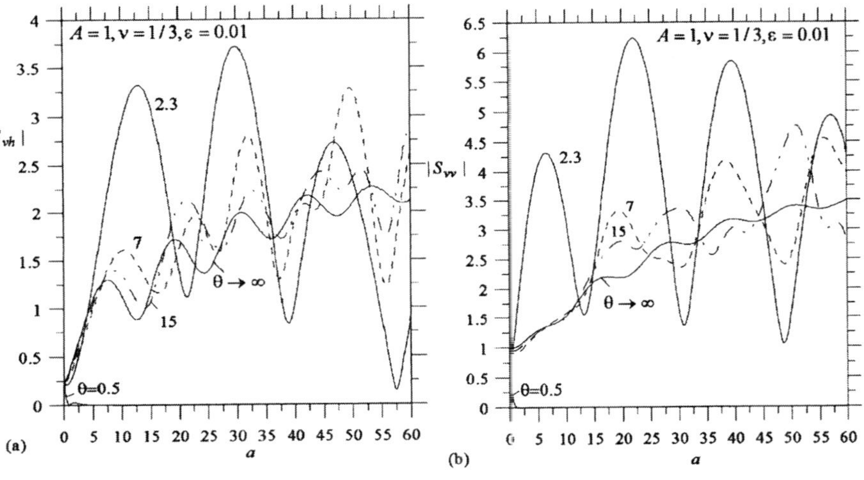

Fig. 4.8a,b. Absolute values of normalized horizontal (a) and vertical (b) amplitudes of vibrations of half-space surface points under action of vertical force

case, the cutoff frequency and resonance frequencies coincide with the natural frequencies of the vertical rod with a fixed bottom cross-section [17, 125].

Calculations indicate that at $\varepsilon = 0$, the poles of the integrands developing at $\theta > \theta_0$ are located in the interval $0 < \widetilde{k} < 1.2\theta$. When constructing the solution of the problem by numerical integration (at $\varepsilon > 0$), this interval is divided into a large number of smaller elements. Some considerations regarding integration have been given in section 2.1.3.

In Fig. 4.8, the results of calculations of the absolute values of S_{vh} and S_{vv} on the surface of an isotropic half-space are shown for $A=1, \nu=1/3, \varepsilon=0.01$; here, $\tilde{r}$ is replaced with parameter a having the form (4.35). The case of the homogeneous half-space corresponds to unbounded growth of parameter θ. The value $\theta=2.3$, being close to the first resonance value, leads to a significant increase of the normalized amplitudes of vibrations in comparison with the case of other values of θ. At $\theta = 0.5$, the amplitudes' behavior is only slightly different from the case of static action of the load.

4.2.4 Vibrations of Half-Space under Action of Horizontal Force Applied to Half-Space Surface

The general form of solution of the problem given in expressions (2.53)–(2.55) may be employed. One should substitute fundamental solutions constructed for the considered heterogeneous anisotropic half-space into these expressions. Expressions (4.34) for the coefficients c_{ij} and expression (4.41) for the quantity F entering functions $\tilde{H}_r$, $\tilde{H}_\vartheta$ must be accounted for.

The results of calculations of the absolute values for the normalized amplitudes of vibrations $S_{h\vartheta}$ and S_{hr} on the half-space surface for the case of an isotropic half-space and $A=1, \nu=1/3, \varepsilon=0.01$ are shown in Fig. 4.9. As a increases, the growth intensity of the normalized amplitudes $|S_{hr}|$ is lower than that of the amplitudes $|S_{h\vartheta}|$.

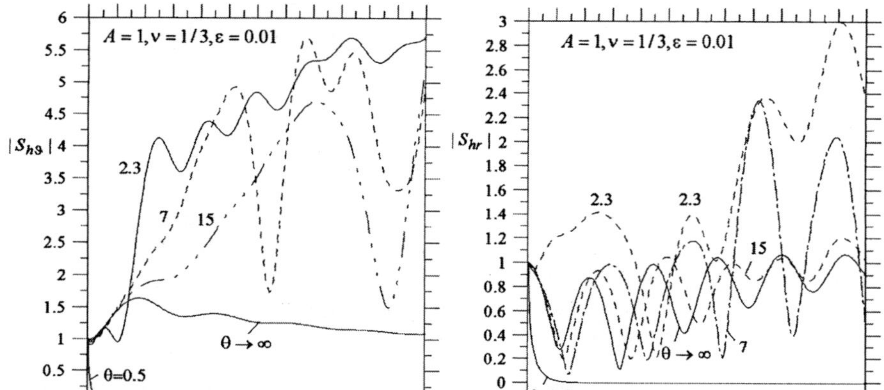

Fig. 4.9a,b. Absolute values of normalized tangential (a) and radial (b) amplitudes of vibrations of half-space surface points under action of horizontal force

Chapter 5. Application of Numerical–Analytical Methods to Static and Dynamic Problems for Heterogeneous Half-Space

5.1 Introduction

Static and dynamic problems for heterogeneous half-spaces may be solved by using the integral representations given in Chapter 1, where the integrands are expressed by fundamental solutions of equations (1.29)–(1.31), or equations (1.29), (1.126)–(1.129). In the case of a force acting within the half-space, besides two solutions of the homogeneous equations satisfying the condition of absence of sources of disturbance at infinity (as $z \to \infty$), an additional solution accounting for the specified force is introduced (a function having index a entering equations of the form (1.109), (1.120), (1.121), (1.157), (1.164), (1.165)). This additional solution has been assumed to vanish for the points of the half-space located below the force; a condition of crossing the horizontal plane in which the force is applied enables us to formulate initial conditions for the additional solution and to construct it in the part of the half-space above the point of application of the force. The construction of analytical representations for the solutions of specified systems of equations may be done only for some particular cases of heterogeneity for isotropic half-spaces. A number of such cases are considered in previous chapters. Here, we point to yet another case: an isotropic half-space with shear modulus increasing with depth by the square law; at Poisson's ratio $\nu = 0.25$, the equation of motion may be separated out and solved by employing modified Bessel's functions. This possibility has been stated for the plane problem [46] and for the axially symmetrical problem [47]. In [31, 33], the results reported by Gupta [46] were used to study some problems dealing with the dynamics of soil foundations.

A rather attractive approach is to construct solutions for the considered differential equations by using numerical methods. Such an approach was employed previously in a number of reports, e.g. [2, 114, 116]. In [2], half-space vibrations in time and space are treated by using a finite-difference scheme in the direction of the Z-axis with subsequent integration with respect to the parameter of Hankel's or Fourier's transformation. The well-known technique of replacement of a half-space with a set of homogeneous layers may be referred to as a numerical method. Parallel with the application of exact solutions of the equations of motion or equilibrium for each layer, the finite-element approach with approximation of displacements within the layers (for each value of the Hankel's or Fourier's transformation parameter) by simple functions of the z-coordinate (e.g. by a linear function) has been widely used [58, 59, 60, 66, 72, 73, 85, 102, 108, 122–124]. As a rule, when using the finite-element method, stiffness matrix formulations are employed. The reported results obtained on the basis of the approximate representation of solutions for separate layers prove the efficiency of this technique for a number of sample calculations. At the same time, special care is needed, apparently, when solving forced vibration problems. In this case, the behavior of Hankel's or Fourier's transformations of the solutions should be accounted for over a wide range of variation of the transformation parameter k ("horizontal" wave number). In the framework of the finite-element approach, this behavior differs from that of the exact solution. Consider, for example, the action of a vertical concentrated force applied to a half-space. In the half-space points located outside horizontal plane, to which the point of application of the force belongs, the exact result gives an exponentially decreasing solution (the multiplier to Bessel's function entering the corresponding integral is implied) in the space of wave numbers k (or the transformation parameter), as these numbers increase; when the observation point is located in the horizontal plane where the force is applied, the solution tends to a constant value (for horizontal displacements in the case of an incompressible medium, this constant equals zero). In the finite-element approach, we obtain a different behaviour: decreasing of the order k^{-1} for vertical displacements and tending to a constant for horizontal displacements as $k \to \infty$, regardless of the location of the observation point with respect to the plane in which the force is applied [58]. As noted in [58], in the case of action of the concentrated force, the logarithmic behavior of the singularity takes place at small distances ρ between the observation point and the vertical, on which the point of application of the force is located (i.e. even when the distance between these points is large). In the exact solution of the corresponding problem, singularity has the form r^{-1}, where r is a distance between the observation point and the point of application of the force. Differences between the stresses corresponding to the exact and approximate solutions may appear to be more significant. In addition, note that according to the solution presented in [58] (relationship (46b)), the transversal amplitudes U_θ, corresponding to the action of the horizontal force, should decrease as $\rho^{-1/2}$ with increasing ρ regardless of the structure of the layered foundation. According to the exact solution for a homogeneous half-space subjected to a horizontal force applied to its surface, the amplitudes of the surface points U_θ decrease as ρ^{-1} (for the homogeneous half-space), or even faster in the case when the stiffness decreases with depth (see the results of calculations for the normalized amplitudes of vibrations shown in Fig. 4.5; in order to obtain the complete amplitudes U_θ, one should multiply these results by the static displacements decreasing as $1/r$). Thus, in certain cases, the finite-element treatment may lead to results that differ significantly from those corresponding to exact solutions.

In the present chapter, the numerical–analytical approach is applied to construct solutions of dynamic and static problems for heterogeneous half-spaces having properties arbitrary varying with depth. Parallel to the Runge–Kutta method of 4th order, we employ the previously mentioned technique based on modeling the half-space as a set of thin homogeneous layers, constructing exact solutions of the corresponding equations for the layers. The action of vertical and horizontal forces applied to the half-space surface, or below it, is considered. The given continuously heterogeneous half-space is replaced with a sufficiently thick heterogeneous layer possessing properties of the original half-space. The layer rests on a homogeneous half-space having properties corresponding to those of the layer at its lower boundary. The discontinuity in properties may be taken into account. The choice of thickness of the heterogeneous layer depends on the frequency of vibrations (parameter θ in equation (1.32a)) and on the location of the point of application of the force. Complications arising due to high values of wave numbers are solved in the following way. Clearly, at high values of parameter k, the corresponding solutions should decrease quickly (by the exponential law) and tend to zero, as the distance from the horizontal plane of the action of the force increases. This fact follows from the form of differential equations (1.29)–(1.31), which may be treated as equations of longitudinal or shear vibrations of rods in an elastic medium, having a stiffness coefficient proportional to k^2 (in the case of equations (1.30), (1.31), certain "connections" exist between the rods; the characteristics of these connections are also dependent on parameter k). In the case of piecewise constant half-space properties, an exponential decrease of solutions with increasing distance to the plane of application of the force (at high values of k) follows from known transformations of the equations of motion for a layered half-space (e.g. [70]). These considerations show that the calculation domain where the solution is being constructed may be narrowed, as k increases, by moving the lower homogeneous half-space closer to the horizontal plane where the force is applied. If the force is applied within the half-space, then at sufficiently high values of k, it is advisable to move the free surface of the half-space closer to the plane, at which the point of application of the force is located. Thus, the initial locations of the lower and upper boundaries of the heterogeneous layer as well as its thickness are kept fixed only in the area $0 \le \mathrm{Re}(\widetilde{k}) \le \widetilde{k}_0$ (dimensionless parameter $\widetilde{k}$ is determined following (1.27)), where $\widetilde{k}_0$ is a quantity proportional to the frequency of vibrations (parameter θ in (1.32a)). This is the domain of propagating waves where parameter $\widetilde{k}$ is smaller than the corresponding frequency parameter (e.g. given in relationship (1.32a)) or having the same order of magnitude. For example, in the case employing the technique of piecewise homogeneous layers, the solutions include exponents with purely imaginary powers (in the absence of internal friction). For $\widetilde{k} > \widetilde{k}_0$, as noted previously, the boundaries of the heterogeneous layer contract to the plane where the force is applied. If the corresponding dimensionless distances (defined as the ratios of the dimensional distances to the reference length z_r) to the boundaries of the layer are limited, say, to the value $50/\sqrt{\widetilde{k}^2 - \widetilde{k}_0^2}$, then a decrease of solutions is provided at $\widetilde{k} > \widetilde{k}_0$ up to values having an order of magnitude $\exp(-50)$, on moving closer to the boundaries along a vertical. Clearly, consideration of parts of the given half-space located outside these boundaries is meaningless. If the point at which the amplitudes of vibrations or other values are to be determined is located outside these boundaries, then the sought values for this point are assumed to vanish (for the given value of $\widetilde{k}$). The technique of domain contraction is not only beneficial from a computational point of view (numerical solutions of the differential equations studied are more efficient), but also preferable as a tool that provides a higher accuracy. In addition, in the cases of forces applied within the half-space, in order to enhance the stability of numerical calculations at high values of $\widetilde{k}$, orthogonalization of the fundamental solutions of the considered systems of differential equations and of the particular solution, which occurs due to the action of the disturbing force, is applied [42]. When using the method of thin layers having constant properties, an alternative approach is proposed: the fundamental solutions are employed for each layer in order to remove exponentially increasing terms entering the particular solution. These additional tools for enhancing computational stability are needed only in the case with the acting forces applied within the half-space, when calculating the values for the points located above the point at which the force is applied. This subject is considered below in more detail.

5.2 Heterogeneous Half-Space Subjected to Vertical Force Applied to Half-Space Surface

In this section, we consider a half-space with shear modulus varying with coordinate z of the considered point by the power law, as an example for the application of the proposed techniques. Let the origin of coordinates be located at distance z_0 from the half-space surface with the Z-axis directed towards the half-space. By adopting $z_r = z_0$ (see relationships (1.27)), we make the half-space surface correspond to the value of dimensionless coordinate $\widetilde{z} = 1$. Consider the following relationship for the normalized shear modulus $\widetilde{G}$ given in (1.27):

$$\widetilde{G}(\widetilde{z}) = \widetilde{z}^{\alpha}, \tag{5.1}$$

where $\alpha \geq 0$. At $\alpha = 0$, the half-space becomes homogeneous, whereas the value $\alpha = 1$ corresponds to the linearly heterogeneous half-space described in Chap. 3. The results presented in that chapter provide a good validation of the accuracy of the applied numerical techniques. Note that the most significant cases for soil mechanics are when power α does not exceed unity, in accordance with well-known experimental results indicating that the soil stiffness is proportional to the power function of the average normal stress with the power less than unity [25, 35, 48].

5.2.1 Application of Thin Layer Technique for Numerical Solution of Differential Equations

In order to employ the general form of the solution given for the case of a vertical concentrated force in relationships (1.57), (2.1), it is sufficient to determine two linearly independent solutions of differential equations (1.30), (1.31), or (1.126)–(1.129); quantities c_{ij} determined from relationships (1.54a), (1.54b), (1.146), (1.147) are expressed by these solutions taken for the half-space surface. Further, we shall employ differential equations (1.136)–(1.139) or, in the isotropic case, equations (1.140)–(1.143).

The given heterogeneous half-space is being replaced with the system including a layer of thickness H_0 with characteristics of the given half-space and a homogeneous half-space having the properties corresponding to the plane $z = z_0 + H_0$ or $\widetilde{z} = 1 + \widetilde{H}_0$, where

$$\widetilde{H}_0 = \frac{H_0}{z_r} = \frac{H_0}{z_0} . \tag{5.2}$$

The scheme of the half-space with division of the layer of thickness H_0 into a set of thin layers, for which all parameters are assumed to be constant within the layer bounds and varying in going from one layer to another, is shown in Fig. 5.1. Here, a piecewise function is presented, which serves to approximate the continuously varying function representing the normalized shear modulus $\widetilde{G}$.

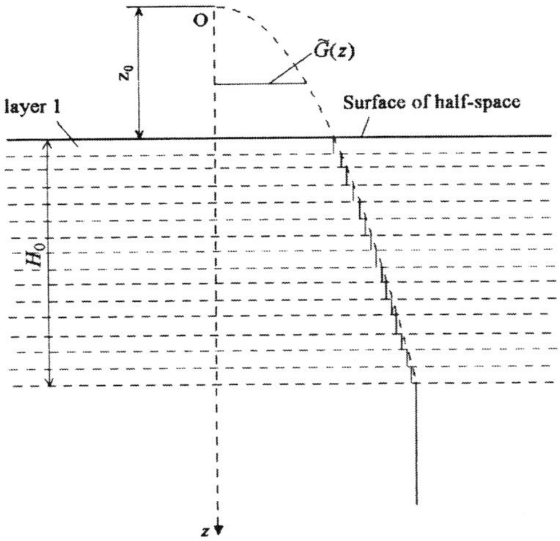

Fig. 5.1. Approximation of continuously heterogeneous half-space by set of layers

Further analysis is carried out for the isotropic half-space; no principal complications arise when treating the anisotropic case. Consider differential equations (1.140)–(1.143). When treating separate homogeneous layers, we shall use their parameters for the purpose of normalizing quantities entering relationships (1.27) and for determining values $\widetilde{\chi}$ and $\widetilde{e}$ in relationships (1.135). Parameter θ defined by (1.32a) is calculated for each layer by using its density, Poisson's ratio, shear modulus $G_{rz0} = G_0$, and reference length z_r (the last value is common for all layers). Here, the index of layer numbering is dropped. We have $\widetilde{G} = 1$, $\widetilde{\rho} = 1$; damping parameter β can vary in going from one layer to another. Equations (1.140)–(1.143) take the form

$$\frac{\mathrm{d}\widetilde{\chi}}{\mathrm{d}\widetilde{z}} = \widetilde{k}\widetilde{e} - \beta^2\theta^2\widetilde{q}\,, \tag{5.3}$$

$$\frac{\mathrm{d}\widetilde{e}}{\mathrm{d}\widetilde{z}} = \widetilde{k}\widetilde{\chi} - \beta^2\theta^2\widetilde{w}\,, \tag{5.4}$$

$$\frac{\mathrm{d}\widetilde{q}}{\mathrm{d}\widetilde{z}} = \widetilde{\chi} - \widetilde{k}\widetilde{w}\,, \tag{5.5}$$

$$\frac{\mathrm{d}\widetilde{w}}{\mathrm{d}\widetilde{z}} = \tau^2\widetilde{e} - \widetilde{k}\widetilde{q}\,. \tag{5.6}$$

One can easily check that the fundamental solutions of this system of differential equations have the form

$$\begin{bmatrix} \widetilde{\chi}_{\mathrm{I}} \\ \widetilde{e}_{\mathrm{I}} \\ \widetilde{q}_{\mathrm{I}} \\ \widetilde{w}_{\mathrm{I}} \end{bmatrix} = \begin{bmatrix} \beta^2\theta^2 \\ 0 \\ \alpha_s \\ \widetilde{k} \end{bmatrix} \exp(-\alpha_s\widetilde{z})\,, \tag{5.7a}$$

$$\begin{bmatrix} \widetilde{\chi}_{\mathrm{II}} \\ \widetilde{e}_{\mathrm{II}} \\ \widetilde{q}_{\mathrm{II}} \\ \widetilde{w}_{\mathrm{II}} \end{bmatrix} = \begin{bmatrix} 0 \\ \beta^2\theta^2 \\ \widetilde{k} \\ \alpha_p \end{bmatrix} \exp(-\alpha_p\widetilde{z})\,, \tag{5.7b}$$

$$\begin{bmatrix} \widetilde{\chi}_{\mathrm{III}} \\ \widetilde{e}_{\mathrm{III}} \\ \widetilde{q}_{\mathrm{III}} \\ \widetilde{w}_{\mathrm{III}} \end{bmatrix} = \begin{bmatrix} \beta^2\theta^2 \\ 0 \\ -\alpha_s \\ \widetilde{k} \end{bmatrix} \exp(\alpha_s\widetilde{z})\,, \tag{5.7c}$$

$$\begin{bmatrix} \tilde{\chi}_{\mathrm{IV}} \\ \tilde{e}_{\mathrm{IV}} \\ \tilde{q}_{\mathrm{IV}} \\ \tilde{w}_{\mathrm{IV}} \end{bmatrix} = \begin{bmatrix} 0 \\ \beta^2\theta^2 \\ \tilde{k} \\ -\alpha_p \end{bmatrix} \exp(\alpha_p \tilde{z}), \tag{5.7d}$$

where

$$\alpha_s = (\tilde{k}^2 - \beta^2\theta^2)^{1/2}, \tag{5.8a}$$

$$\alpha_p = (\tilde{k}^2 - \tau^2\beta^2\theta^2)^{1/2}. \tag{5.8b}$$

It is advisable to take the quantity θ for the upper layer as parameter $\tilde{\theta}$ which represents the properties of the entire half-space. According to (1.32a) ($z_r = z_0$),

$$\tilde{\theta} = \omega z_0 \sqrt{\frac{\rho(z_0)}{G_0}}. \tag{5.9}$$

Here, G_0 represents shear modulus (without accounting for damping) on the surface of the isotropic half-space (for the anisotropic half-space, value G_{rz0} is used). For the case corresponding to the scheme shown in Fig. 5.1, the value of parameter θ for the j-th layer, having an upper bound at Z_j, equals

$$\theta = \theta_j = \tilde{\theta} \sqrt{\frac{\rho(Z_j)}{\tilde{G}(Z_j)\rho(z_0)}}. \tag{5.10}$$

For constructing two solutions entering the integral representations for the amplitudes of vibrations (1.57), we propose the following simple procedure. In the lower homogeneous half-space, which extends unboundedly downward from the plane $z = z_0 + H_0$, these solutions are taken in the form (5.7a) (for the first solution) and (5.7b) (for the second solution). This choice provides decreasing solutions when $z \to \infty$ (at high values of $\tilde{k}$) or results in oscillations having the form of waves going to infinity (at small values of $\tilde{k}$). For the upper boundary of the lower homogeneous half-space, it is convenient to omit the exponential multipliers in solutions (5.7a) and (5.7b), placing an origin of coordinates at this boundary. Next, one should determine these solutions in each layer by continuing them up to the half-space surface, taking into account the conditions of transition through the common boundaries of two adjacent layers. In proceeding from one layer to another, quantities $\tilde{q}$ and $\tilde{w}$ are continuous, while values of $\tilde{\chi}$ and $\tilde{e}$ at the points located infinitesimally close to the boundary, both above and below it, should satisfy the condition of stress continuity. From relationships (1.144), (1.145),

$$G_j[\tilde{\chi}^- - 2\tilde{k}\tilde{w}] = G_{j+1}[\tilde{\chi}^+ - 2\tilde{k}\tilde{w}], \tag{5.11a}$$

$$G_j[\widetilde{e}^- - 2\widetilde{k}\widetilde{q}] = G_{j+1}[\widetilde{e}^+ - 2\widetilde{k}\widetilde{q}] \ . \tag{5.11b}$$

Here, quantities $\widetilde{\chi}^-$, $\widetilde{e}^-$ correspond to the lower boundary of *j*-th, and $\widetilde{\chi}^+$, $\widetilde{e}^+$ to the upper boundary of $(j+1)$-th layer; $\widetilde{q}$ and $\widetilde{w}$ correspond to the common boundary of the considered layers. In given relationships, the shear moduli include the influence of dissipative properties of the material analogously to relationship (1.28). As follows from relationships (5.11),

$$\widetilde{\chi}^- = \frac{G_{j+1}}{G_j}\widetilde{\chi}^+ + 2\widetilde{k}\widetilde{w}\left(1 - \frac{G_{j+1}}{G_j}\right), \tag{5.12a}$$

$$\widetilde{e}^- = \frac{G_{j+1}}{G_j}\widetilde{e}^+ + 2\widetilde{k}\widetilde{q}\left(1 - \frac{G_{j+1}}{G_j}\right). \tag{5.12b}$$

Thus, we have obtained relationships that enable a transition from layer number $j+1$ to layer number j. Further, we construct the solution in layer j, assuming that values of $\widetilde{\chi}$, $\widetilde{e}$, $\widetilde{q}$, $\widetilde{w}$ at its lower boundary are known. Representing the solution as a linear combination of four solutions (5.7) with coefficients B_i $(i = 1,...,4)$, respectively, we shall determine these coefficients by using known values of $\widetilde{\chi}$, $\widetilde{e}$, $\widetilde{q}$, $\widetilde{w}$ at the lower boundary of *j*-th layer. It is convenient to place an origin of coordinates at this boundary. Solving the corresponding system of equations results in the following values of coefficients B_i:

$$B_1 = \left(\widetilde{q} - \frac{\widetilde{k}\widetilde{e}}{\beta^2\theta^2}\right)\frac{1}{2\alpha_s} + \frac{\widetilde{\chi}}{2\beta^2\theta^2}, \tag{5.13a}$$

$$B_3 = -\left(\widetilde{q} - \frac{\widetilde{k}\widetilde{e}}{\beta^2\theta^2}\right)\frac{1}{2\alpha_s} + \frac{\widetilde{\chi}}{2\beta^2\theta^2}, \tag{5.13b}$$

$$B_2 = \left(\widetilde{w} - \frac{\widetilde{k}\widetilde{\chi}}{\beta^2\theta^2}\right)\frac{1}{2\alpha_p} + \frac{\widetilde{e}}{2\beta^2\theta^2}, \tag{5.13c}$$

$$B_4 = -\left(\widetilde{w} - \frac{\widetilde{k}\widetilde{\chi}}{\beta^2\theta^2}\right)\frac{1}{2\alpha_p} + \frac{\widetilde{e}}{2\beta^2\theta^2} \ . \tag{5.13d}$$

Here, index j is dropped when denoting layer parameters and coefficients B_i, whereas quantities $\widetilde{\chi}$, $\widetilde{e}$, $\widetilde{q}$, $\widetilde{w}$ refer to the lower boundary of the layer. Having determined coefficients B_i, we obtain the solution for layer number *j*, and, using it at the upper layer boundary, we can proceed to the next layer, having number $j-1$, by employing relationships (5.12), and then to the next layer, etc., eventually reaching the half-space surface. Thus, simple recurrent relationships provide the two sought solutions – the first one, proceeding from (5.7a), and the second one, proceeding from (5.7b). All operations performed for constructing these two solutions are identical. It is worth noting that, in order to construct two global fundamental solutions entering the integral representations of the vibrations amplitudes, four local fundamental solutions (5.7) are used within each layer; to distinguish global solutions from local ones, indices in the latter are shown as Roman numerals.

When applying the algorithm described above, we face complications at high values of integration parameter $\tilde{k}$. First, both solutions increase when moving upward (indeed, the linear combination of solutions (5.7) with coefficients B_i employed for each layer contains two terms having increasing exponential multipliers). Second, as $\tilde{k}$ increases, the initial solutions (5.7a), (5.7b) (for the lower homogeneous half-space) become progressively closer, which is especially noticeable at low frequencies (at small values of parameter $\tilde{\theta}$). Note that the use of extremely high values of parameter $\tilde{k}$ when calculating the integrals may be required at $\tilde{z} = 1$ (i.e. for the half-space surface), and for very small values of dimensionless distance $\tilde{r}$ (see the discussion concerning calculations of the integrals given in section 2.1.3). As noted previously, narrowing the calculation domain (making the heterogeneous part of the half-space thinner), as $\tilde{k}$ increases, serves as an efficient tool that enables us to overcome the first complication; correspondingly, the number of layers into which the heterogeneous part is divided decreases. Consequently, far fewer computations are required in order to provide the same accuracy. The form of solutions for layers (5.7) indicates that the fast variation of solutions (along vertical lines) occurs when the quantity $\mathrm{Re}(\tilde{k})$ exceeds the value of parameter θ for the considered layer. The value $\tilde{k}_0$, starting from which (i.e. at $\mathrm{Re}(\tilde{k}) > \tilde{k}_0$) an initially set thickness H_0 may be reduced, was taken as

$$\tilde{k}_0 = 1.2\tilde{\theta} + 0.1\,, \tag{5.14}$$

whereas the current value H of the heterogeneous layer thickness may be expressed as follows:

$$\tilde{H} = \begin{cases} \tilde{H}_0 & (\mathrm{Re}(\tilde{k}) \le \tilde{k}_0) \\ \min\left[\tilde{H}_0, \dfrac{50}{\sqrt{\tilde{k}^2 - \tilde{k}_0^2}}\right] & (\mathrm{Re}(\tilde{k}) > \tilde{k}_0), \end{cases} \tag{5.15}$$

where dimensionless quantities $\tilde{H} = H / z_0$ and $\tilde{H}_0$ are employed. Let N_0 denote an initial number of layers corresponding to the interval $0 < \mathrm{Re}(\tilde{k}) \le \tilde{k}_0$. As $\tilde{H}$ decreases, the number of layers is reduced (lower layers are removed), whereas the upper boundary of the lower homogeneous half-space, on which the heterogeneous layer rests, rises. The current number of layers is taken as

$$N = \max\left[\operatorname{round}\left(N_0 \frac{\widetilde{H}}{\widetilde{H}_0}\right), 1\right]. \tag{5.16}$$

Here, the value of $\widetilde{H}$ is corrected (if it exceeds the dimensionless thickness of the upper layer in the system of layers): when N is known, $\widetilde{H}$ is set equal to the sum of dimensionless thickness values of layers (according to N). When $\widetilde{H}$ in (5.15) becomes smaller than the dimensionless thickness of the upper layer (at this moment, following (5.16), we obtain $N = 1$), the thickness correction is not performed; further reduction of the layer thickness continues according to (5.15).

Next, we shall consider determination of the quantity $\widetilde{H}_0$. Comparison between the results of calculations for the case with $\alpha = 1$ in relationship (5.1), corresponding to exact solutions of the system of differential equations (Chap. 3) and to numerical solutions, indicates that the following expression for quantity $\widetilde{H}_0$ provides sufficiently high accuracy (4–5 correct digits):

$$\widetilde{H}_0 = \begin{cases} 100 + 2000(0.5 - \widetilde{\theta}) & (\widetilde{\theta} \le 0.5) \\ \dfrac{50}{\widetilde{\theta}} & (\widetilde{\theta} > 0.5). \end{cases} \tag{5.17}$$

A decrease of quantity $\widetilde{H}_0$ as parameter $\widetilde{\theta}$ increases is explained as follows. When the frequency of vibrations is kept constant, an increase of parameter $\widetilde{\theta}$ is equivalent to growth of quantity z_0, so that the considered half-space becomes closer to the homogeneous half-space, and, hence, transition to the lower heterogeneous half-space, on which the heterogeneous layer of thickness H_0 rests, becomes possible at smaller values of $\widetilde{H}_0$.

In the case with $\alpha < 1$, when deviation from homogeneity is less significant than in the case with $\alpha = 1$, quantity $\widetilde{H}_0$ decreases as compared with the value from (5.17), while in the case with $\alpha > 1$ this quantity increases. For the given half-space with relationship (5.1), we introduce the corresponding linearly heterogeneous half-space ($\alpha = 1$) having the same value of surface derivative of shear modulus with respect to the z-coordinate as the given half-space (with the rest of parameters kept equal). This requirement results in

$$\frac{\alpha}{z_0} = \frac{1}{z_{0l}}, \tag{5.18}$$

where z_{0l} is the distance between the point at which shear modulus of the equivalent linearly heterogeneous half-space becomes zero and the half-space surface (Fig. 5.2). From (5.18), we obtain the value z_{0l} and the corresponding parameter $\widetilde{\theta}_l$:

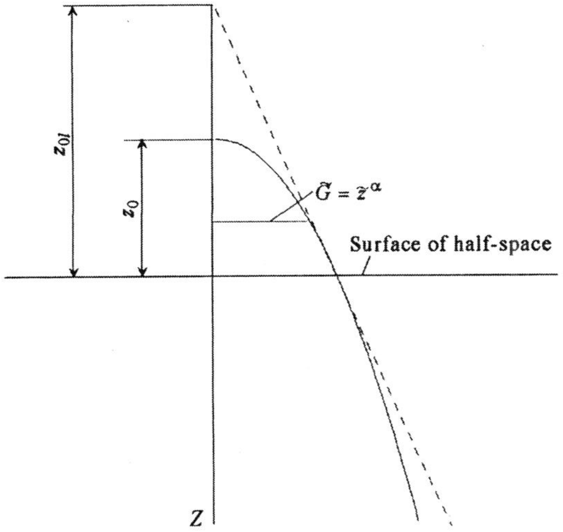

Fig. 5.2. Determination of linearly heterogeneous half-space corresponding to given half-space with stiffness varying with depth by power law

$$z_{0l} = \frac{z_0}{\alpha}, \tag{5.19}$$

$$\tilde{\theta}_l = \frac{\tilde{\theta}}{\alpha}. \tag{5.20}$$

For the half-space family with relationship (5.1), we shall apply (5.17) where $\tilde{\theta}$ is replaced with $\tilde{\theta}_l$ for the determination of $\tilde{H}_0$:

$$\tilde{H}_0 = \begin{cases} 100 + 2000(0.5 - \tilde{\theta}_l) & (\tilde{\theta}_l \le 0.5) \\ \dfrac{50}{\tilde{\theta}_l} & (\tilde{\theta}_l > 0.5). \end{cases} \tag{5.21}$$

As noted previously, loss of accuracy when performing calculations following the proposed algorithm may occur because the initial solutions (5.7a) and (5.7b) become increasingly close as $\tilde{k}$ increases. This is especially noticeable at low frequencies (small values of parameter $\tilde{\theta}$). One can switch to the static solutions in the layers for values of $\tilde{k}$ significantly exceeding $\tilde{\theta}$ (such a technique is employed in [4, 70]). However, there is a simpler alternative: starting with sufficiently high values of $\tilde{k}$, one can fix the lower boundary of the heterogeneous layer by using the following initial conditions for the components of the first and the second solutions on the lower boundary of this layer:

$$\begin{bmatrix} \widetilde{\chi}_{\rm I} \\ \widetilde{e}_{\rm I} \\ \widetilde{q}_{\rm I} \\ \widetilde{w}_{\rm I} \end{bmatrix} = \begin{bmatrix} 1 \\ 0 \\ 0 \\ 0 \end{bmatrix}, \quad \begin{bmatrix} \widetilde{\chi}_{\rm II} \\ \widetilde{e}_{\rm II} \\ \widetilde{q}_{\rm II} \\ \widetilde{w}_{\rm II} \end{bmatrix} = \begin{bmatrix} 0 \\ 1 \\ 0 \\ 0 \end{bmatrix}. \tag{5.22}$$

These values are used for $\widetilde{k} \geq 10\widetilde{\theta}$.

The considered piecewise constant approximation applied to solve the vibration problems of a continuously heterogeneous half-space is rather rough; therefore, in order to provide high accuracy, a large number of layers N_0 are required. Calculations indicate that at large N_0, the error varies as C/N_0 where C is a constant. This enables us to obtain a good approximation of the exact solution by employing Richardson's extrapolation: if T_1 and T_2 represent some results of solving the problem for N_0 and $N_0\,\gamma$ respectively, then the refined result T equals

$$T = \frac{\gamma T_2 - T_1}{\gamma - 1}, \tag{5.23}$$

which follows from the relationships

$$T = T_1 + \frac{C}{N_0}, \tag{5.24a}$$

$$T = T_2 + \frac{C}{\gamma N_0}. \tag{5.24b}$$

Consider some examples of calculations. For integration, we apply the integration contour shown in Fig. 2.3 and the considerations regarding the calculation of integrals given in section 2.1.3. If the material of the half-space possesses internal friction ($\varepsilon > 0$), one can integrate over the real axis [4, 70]. It is reasonable to perform substitution of the integration variable, analogously to (3.32),

$$\widetilde{k} = s\widetilde{\theta} \tag{5.25}$$

and employ variable a by (3.33)

$$a = \widetilde{\theta}\widetilde{r} = \omega r\sqrt{\frac{\rho}{G_0}}. \tag{5.26}$$

Note that the function which serves as the multiplier of Bessel's functions $\mathrm{J}_0(sa) = \mathrm{J}_0(\widetilde{k}\widetilde{r})$ or $\mathrm{J}_1(sa) = \mathrm{J}_1(\widetilde{k}\widetilde{r})$ in the integral representation for the amplitudes of vibrations contains neither variable a nor $\widetilde{r}$. It is advisable to construct the corresponding interpolation function of variables s or $\widetilde{k}$ for the mentioned multiplier on the main interval of variation of these variables, and then

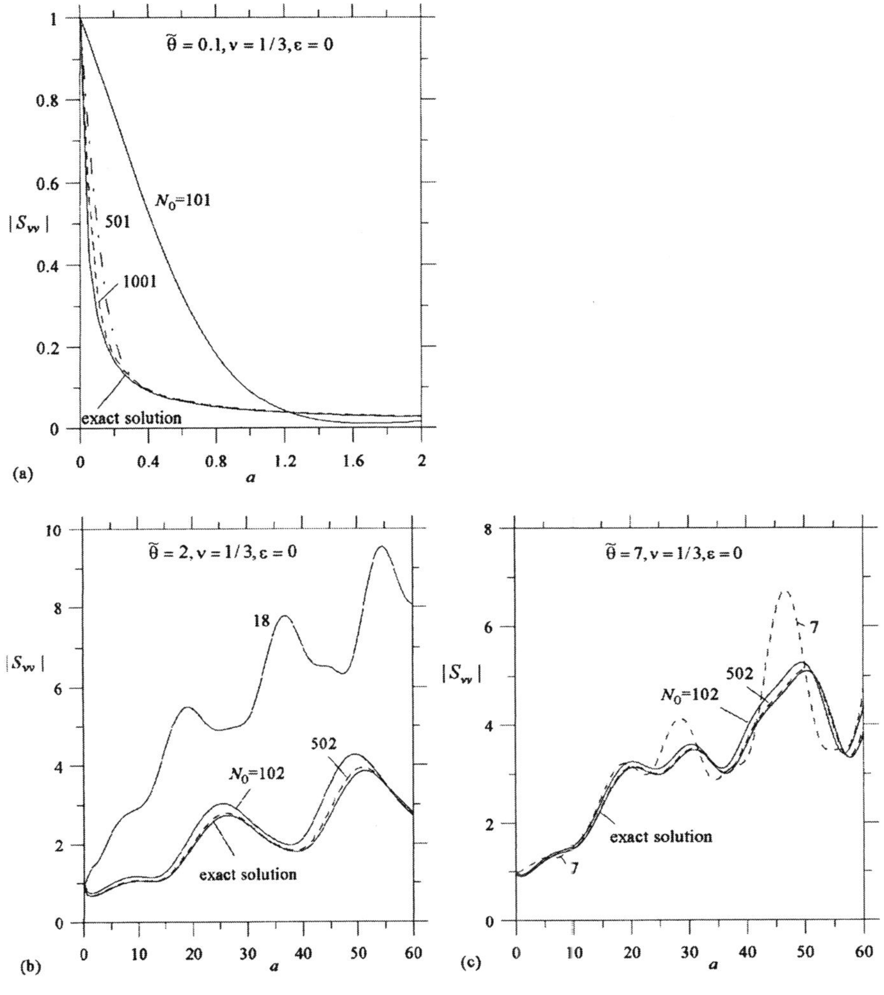

Fig. 5.3a,b,c. Influence of layers number N_0 on absolute values of normalized vertical amplitudes of vibrations for linearly heterogeneous half-space at $\widetilde{\theta}$ =0.1 (a), 2 (b), 7 (c)

to employ this function when performing calculations for different values of a or $\widetilde{r}$; this approach is especially efficient if the calculations are performed for a large number of values of a or $\widetilde{r}$ with fixed $\widetilde{z}$.

Let $\alpha = 1$, $\nu = 1/3$, $\varepsilon = 0$. Consider three values of $\widetilde{\theta}$, namely $\widetilde{\theta}$ = 0.1, 2, 7. Following relationship (5.17), at $\widetilde{\theta}$ = 0.1 we have $\widetilde{H}_0 = 900$. Absolute values of the normalized amplitudes of vertical vibrations for the points on the half-space

Table 5.1. $|S_{vv}|$ for a number of values of N_0 and a, with $\widetilde{\theta} = 0.1$, $\widetilde{H}_0 = 900$

	N_0 = 500	1000	2001	4001	Extr.	Exact	Runge–Kutta method
$a = 0.05$	0.7278	0.5840	0.4729	0.4382	0.4034	0.41469	0.41469
0.1	0.4964	0.3371	0.2957	0.2857	0.2757	0.27504	0.27504
0.2	0.2218	0.1772	0.1725	0.1688	0.1651	0.16526	0.16527
0.6	0.0677	0.0673	0.0665	0.0660	0.0656	0.06556	0.06557
1.0	0.0439	0.0434	0.0430	0.0428	0.0427	0.04264	0.04264
1.5	0.0318	0.0314	0.0312	0.0311	0.0310	0.03095	0.03094
2	0.0258	0.0255	0.0253	0.0252	0.0251	0.02512	0.02511

surface calculated by using (2.1d) where ν' is replaced with ν are shown in Fig. 5.3a. Further increase of parameter a does not lead to a significant change of vibration amplitudes that remain small (having an order of magnitude 0.01–0.02); the accuracy of the approximation of the continuously heterogeneous half-space by the layered half-space increases. At small values of parameter a, the results of calculations approach exact values only when the number of layers is extremely large. In Table 5.1, $|S_{vv}|$ is given for a number of values of N_0 and a. The 6th column in Table 5.1 corresponds to Richardson's extrapolation with N_0= 2001 and N_0= 4001; at small a, deviation from the exact solution is still noticeable. As parameters a or $\widetilde{r}$ decrease, the result of integration is increasingly influenced by the integrand behavior at high values of the integration variables $\widetilde{k}$ or s. In the case of a linearly heterogeneous half-space subjected to the static action of a concentrated force, asymptotic representations of the normalized displacements at small distances $\widetilde{r}$ have the form (3.71). Here, the logarithmic terms occur due to the specific behavior of the integrands at high values of $\widetilde{k}$. When applying the piecewise constant approximation, the behavior of the integrands is different; therefore, in order to provide sufficient accuracy, one should employ extremely thin layers (high values of N_0) in the case of small $\widetilde{r}$. Analogous considerations hold in the case of dynamic problems, especially for small values of parameter $\widetilde{\theta}$. The last column in Table 5.1 contains the results obtained by using the Runge-Kutta method of 4th order applied to the solution of differential equations (1.140)–(1.143). This method is discussed in the next subsection.

Significantly faster convergence to the exact results takes place for $\widetilde{\theta} = 2, 7$ with the corresponding values of $\widetilde{H}_0$ being equal 25 and 7.14286, respectively. The corresponding results are shown in Tables 5.2 and 5.3; in comparison with the case of $\widetilde{\theta} = 0.1$, the number of layers N_0 may be decreased (since the thickness of the heterogeneous layer is smaller). Richardson's extrapolation is performed by using (5.23) with the data given in the 3rd and 4th columns of Tables 5.2 and 5.3.

When studying vibrations of the layered half-space that serves to approximate a continuously heterogeneous half-space, the possibility exists of an abrupt

Table 5.2. $|S_{vv}|$ for a number of values of N_0 and a, with $\tilde{\theta}=2$, $\tilde{H}_0=25$

	$N_0=102$	502	1002	Extr.	Exact	Runge–Kutta method
$a=0.05$	0.9815	0.9407	0.9280	0.9153	0.91691	0.91692
0.1	0.9612	0.8932	0.8823	0.8713	0.87263	0.87263
0.2	0.9189	0.8334	0.8248	0.8162	0.81607	0.81608
0.6	0.7993	0.7349	0.7277	0.7204	0.72056	0.72058
1.0	0.7635	0.7043	0.6975	0.6906	0.69075	0.69076
1.5	0.7604	0.6994	0.6924	0.6854	0.68564	0.68565
2	0.7769	0.7125	0.7052	0.6979	0.69813	0.69814

Table 5.3. $|S_{vv}|$ for a number of values of N_0 and a, with $\tilde{\theta}=7$, $\tilde{H}_0=50/7$

	$N_0=102$	502	1003	Extr.	Exact	Runge–Kutta method
$a=0.05$	0.9956	0.9828	0.9789	0.9750	0.97540	0.97540
0.1	0.9905	0.9689	0.9653	0.9617	0.96203	0.96203
0.2	0.9795	0.9513	0.9481	0.9450	0.94487	0.94487
0.6	0.9509	0.9256	0.9225	0.9193	0.91934	0.91934
1.0	0.9524	0.9271	0.9240	0.9208	0.92080	0.92080
1.5	0.9746	0.9477	0.9444	0.9411	0.94105	0.94106
2	1.0099	0.9811	0.9776	0.9740	0.97399	0.97400

increase in amplitudes of vibrations in certain areas on the half-space surface for a certain number of layers N_0. This phenomenon (corresponding to $\tilde{\theta}=0.1$) is shown in Fig. 5.4. The value $N_0=31$ is a kind of resonance value; at adjacent values of N_0, an abrupt decrease (compared to $N_0=31$) of maximal amplitudes of vibrations occurs.

The cause of this phenomenon is the proximity of phases of the waves reaching the half-space surface, following reflections from the layer boundaries. In the considered example, the closest proximity of phases takes place at $N_0=31$, where parallel to the areas of high amplitudes of vibrations, areas with lowered amplitudes are observed. Such behavior corresponds to that of the integrand entering the integral which represents the amplitudes of vibrations. Indeed, in the example, the integrand has two poles on the real axis s (see (5.25)), namely $s_1=0.355$ and $s_2=0.639$, resulting in the occurrence of the corresponding Rayleigh waves. The contribution of the latter to the amplitudes of vibrations predominates at high values of parameter a. Summation of the vibration amplitudes corresponding to Rayleigh waves leads to the appearance of areas of amplification and weakening of vibrations shown in Fig. 5.4a for $N_0=31$. In addition, Fig. 5.4a contains the curve related to the corresponding homogeneous

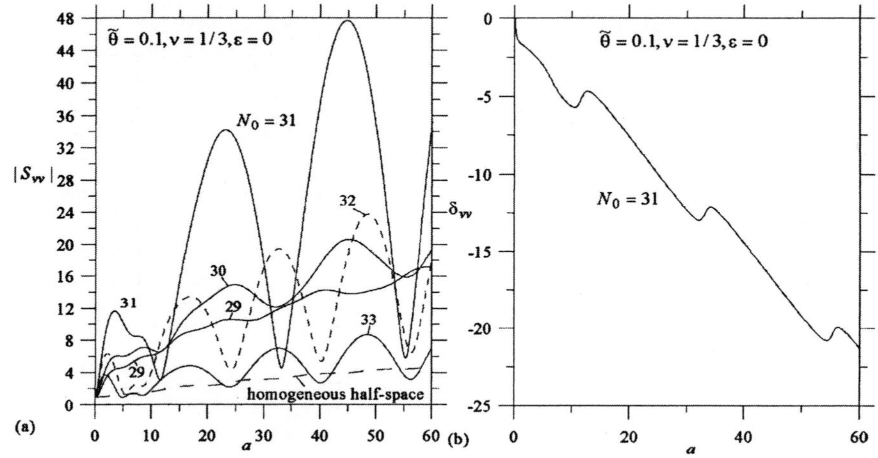

Fig. 5.4a,b. Abrupt increase of vertical amplitudes of vibrations (a) and "discontinuous" phase behavior of vertical amplitudes of vibrations (b) at N_0=31 (resonance number of layers)

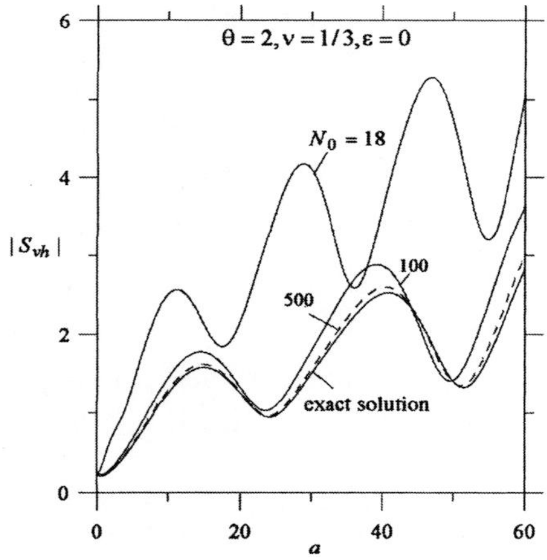

Fig. 5.5. Absolute values of normalized horizontal amplitudes of vibrations under action of vertical force at $\tilde{\theta} = 2$

half-space. One can see that stepwise increasing of the half-space stiffness may result in multiple increasing of the vibration amplitudes. Interestingly, in the graph for phase δ_{vv} of the complex amplitude of vibrations (Fig. 5.4b), the areas of small amplitudes undergo an abrupt phase change, whereas the phase curve deviates significantly from a straight line. This phenomenon is analogous to that discussed in the study of vibrations of the linearly heterogeneous half-space (Fig. 3.4).

At high values of $\widetilde{\theta}$, the "layer resonance" is less noticeable than in the case of $\widetilde{\theta} = 0.1$. In the case of $\widetilde{\theta} = 2$ ($\widetilde{H}_0 = 25$), the maximal amplitudes of vibrations occur at $N_0 = 18$. These maximal values are more than twice as great as the amplitudes corresponding to high values of N_0 (Fig. 5.3b). At $\widetilde{\theta} = 7$ ($\widetilde{H}_0 = 50/7$), a weakly expressed resonance value of N_0 equals 7.

Consider horizontal vibrations that occur due to the action of a vertical force applied to the half-space surface. Graphs for the normalized horizontal amplitudes of vibrations of the surface points of the heterogeneous half-space at $\alpha = 1$, $\nu = 1/3$, $\varepsilon = 0$, $\theta = 2$ are shown in Fig. 5.5. As in the case of the vertical vibrations, a value $N_0 = 18$ leads to a significant increase of vibration amplitudes for the surface points of the layered half-space.

5.2.2 Application of Runge-Kutta Method for Numerical Solution of Differential Equations

The two linearly independent solutions of the system of differential equations (1.30), (1.31) or (1.126)–(1.129) needed for construction of the solution according to relationships (1.57), (2.1) may be obtained by using an appropriate numerical method, e.g. the Runge–Kutta method. We shall use equations (1.126)–(1.129), or equations (1.130)–(1.133) for the isotropic case. Considerations regarding the scheme of calculation in the form of a heterogeneous layer resting on a homogeneous half-space, given in the previous subsection, remain valid. The thickness of this layer is calculated by (5.15), (5.17), (5.21) with $\widetilde{\theta}$ determined by (5.9) (in the case of transverse anisotropy, G_0 must be replaced with G_{rz0} in (5.9)). Relationship (5.17) corresponds to the linearly heterogeneous half-space ($\alpha = 1$), whereas for another values of α formula (5.21) is employed. For the lower homogeneous half-space, the solutions are taken (in the case of isotropy) according to (5.7a) and (5.7b) for the first and second fundamental solutions, respectively. The value of θ entering the solution for the lower homogeneous half-space is determined from (5.10) where the following substitution is performed:

$$Z_j = (1 + \widetilde{H}) z_0 \,. \tag{5.27}$$

Starting from the values given in (5.7a) and (5.7b) (with exponents dropped), the Cauchy problem is solved in order to construct two fundamental solutions entering expressions of the form (1.57), (2.1). As in the previous subsection, we fix the lower boundary of the heterogeneous layer at sufficiently high values of the integration variable $\widetilde{k}$ (see (5.22)). In vibration problems for the half-space subjected to a force applied to its surface, the technique employed provides a simple method of constructing solutions for the considered differential equations without any additional tools.

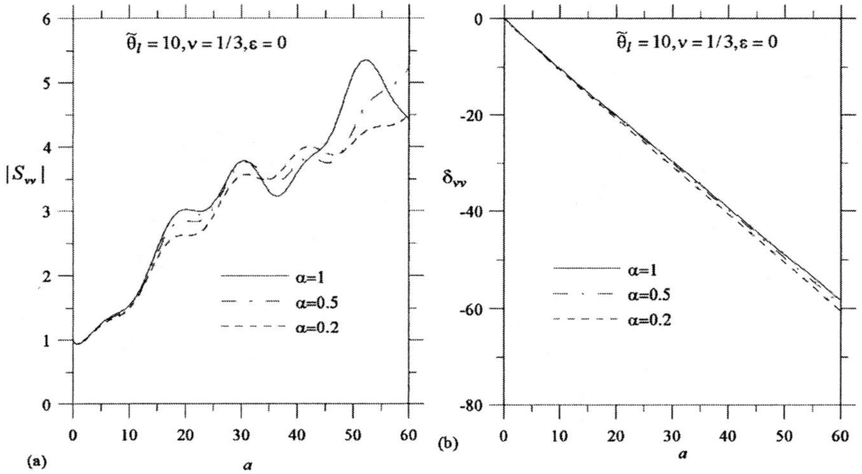

Fig. 5.6a,b. Absolute values of normalized vertical amplitudes of vibrations (a) and their phase (b) for three different values of parameter α at $\widetilde{\theta}_l = 10$

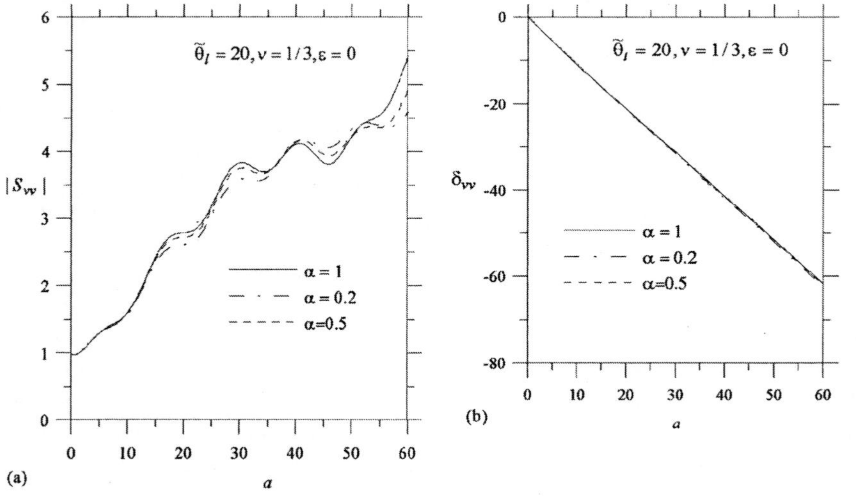

Fig. 5.7a,b. Absolute values of normalized vertical amplitudes of vibrations (a) and their phase (b) for three different values of parameter α at $\widetilde{\theta}_l = 20$

For the examples considered in the previous subsection, the results of calculations, obtained by using the Runge–Kutta method of 4th order, for a linearly heterogeneous half-space are presented in Tables 5.1–5.3. This method provides the same accuracy as the technique based on confluent hypergeometric functions developed in Chap. 3 for the linearly heterogeneous half-space. When integrating, the number of intervals was progressively doubled until the difference (for different number of intervals) between values of the sought integrand, containing the fundamental solutions of the system of differential equations, becomes sufficiently small.

Furthermore, we study the vibrations of an isotropic heterogeneous half-space where the shear modulus varies by the power law (5.1) with various values of parameter α. The developed numerical technique is applied to calculate the integrands entering the integrals that represent the amplitudes of vibrations. In the previous subsection, we introduced a linearly heterogeneous half-space having the same surface shear modulus and its derivative with respect to z as the considered half-space (see relationships (5.19), (5.20)). At high values of parameter $\widetilde{\theta}_l$, when the vibration amplitudes on the half-space surface become increasingly dependent on the half-space properties in the vicinity of its surface, the results that correspond to equal values of $\widetilde{\theta}_l$ (at different α) should be close. This fact is illustrated in the graphs plotted for two values of $\widetilde{\theta}_l$ shown in Figs. 5.6 and 5.7; one can see that the graphs expressing the phase of the complex amplitude of vertical vibrations are especially close to each other.

5.2.3 Parameter Determination for Isotropic Half-Space with Shear Modulus Varying by Power Law (Action of Vertical Force)

In Chap. 3, when dealing with vibrations of the linearly heterogeneous isotropic half-space, we consider determination of the half-space parameters when the characteristics of the surface waves that occur due to the action of vertical or horizontal forces are known from experiments. This technique is limited since linear variation of the shear modulus with depth is assumed. In the case when the heterogeneity characteristics differ from linear ones, the approach given in section 3.4 would yield approximate results valid in a specific range of vibration frequencies; the sought parameters (z_0 and G_0) may take different values for different intervals of variation of the frequency. Below, we present a technique of selecting the half-space parameters for the case of power-law heterogeneity, which, in addition, enables us to find the power index α in (5.1). We shall employ the parameter $\widetilde{\theta}_l$ discussed above. By performing the calculations for a large number of values of this parameter, one can study phases of the complex amplitudes of vibrations and, as described in section 3.4, determine the relationship between the dimensionless wave number, k_v, and parameter $\widetilde{\theta}_l$. It is advisable to determine k_v by using the second or third wave, as explained in section 3.4. In Fig. 5.9, graphs are plotted for the quantity k_v, determined for the third wave at $\nu = 1/3, \varepsilon = 0$, as a function of variable λ

$$\lambda = \frac{\widetilde{\theta}_l}{1+\widetilde{\theta}_l}\,. \tag{5.28}$$

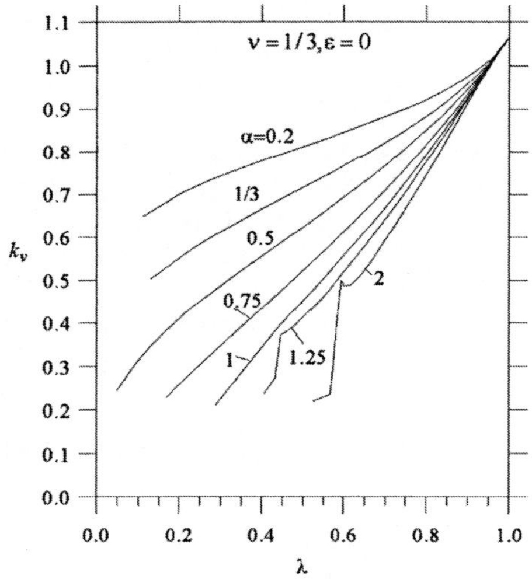

Fig. 5.8. Dimensionless wave numbers k_ν for various values of parameter α

When λ approaches unity, the graphs corresponding to different values of α become closer; values of k_ν tend to the value for the corresponding homogeneous half-space, which equals 1.066 for the third wave. This result is close to the value 1.0724 corresponding to a Rayleigh wave at $\nu = 1/3$. For other values of ν, graphs for k_ν have a similar form. In the cases with $\alpha = 1.25$ and $\alpha = 2$, the behavior of the dimensionless wave numbers becomes discontinuous at some values of $\widetilde{\theta}_l$, analogously to the discontinuities for the linearly heterogeneous half-space (Fig. 3.9). As the degree of heterogeneity increases, values of $\widetilde{\theta}_l$ corresponding to discontinuities increase: for $\alpha = 1.25$ we obtain $\widetilde{\theta}_l \approx 0.8$ whereas for $\alpha = 2$ we have $\widetilde{\theta}_l \approx 1.4$. For these values of α, one can establish a relationship between the vibration frequencies (or wavelengths) and the velocity of surface wave propagation in the range of sufficiently high frequencies (or short waves), for which values of $\widetilde{\theta}_l$ exceed the above given values that result in discontinuities. At small values of α, a regular behavior of wave numbers takes place also for low vibration frequencies. For example, at $\alpha = 0.5$, no abrupt changes of k_ν occur in the range $\widetilde{\theta}_l > 0.05$.

As in the case of a linearly heterogeneous half-space, we introduce the shear modulus of a homogeneous half-space, G_h, corresponding to velocities of the surface waves, which are determined by analyzing the forced vibrations of the heterogeneous half-space. We apply formulas (3.44)–(3.46), where z_0 and θ are replaced with z_{0l} and $\widetilde{\theta}_l$, respectively. Analogously to the case with the linearly

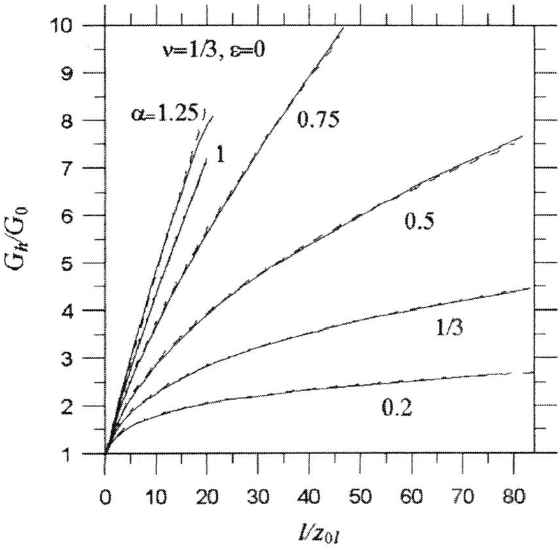

Fig. 5.9. Normalized shear modulus of homogeneous half-space corresponding to velocities of surface waves under action of vertical force

heterogeneous half-space (Fig. 3.10), we construct graphs for the relationship between the ratio G_h / G_0 and dimensionless wavelength l / z_{0l} for a number of values of α. These graphs are presented in Fig. 5.9 for the third wave at $\nu = 1/3, \varepsilon = 0$. Function (3.47), where $\tilde{l}$ is replaced with l / z_{0l}, still provides a very high accuracy. The results of approximation are represented in Fig. 5.9 by dashed lines. The values of parameters γ and ζ are given in Table 5.4.

Table 5.4. Values of parameters for approximate relationship (3.47) in the case of vertical force at various values of α ; $\tilde{l} = l / z_{0l}$

α	ζ	γ	$\gamma\zeta$
0.2	0.194	2.01	0.390
0.3333333	0.322	1.24	0.399
0.5	0.477	0.843	0.402
0.75	0.679	0.596	0.405
1	0.82	0.51	0.418
1.25	0.91	0.465	0.423

A good approximation of a relationship between power ζ and α can be presented as follows:

$$\zeta = \alpha - 0.177\alpha^{2.9} . \tag{5.29a}$$

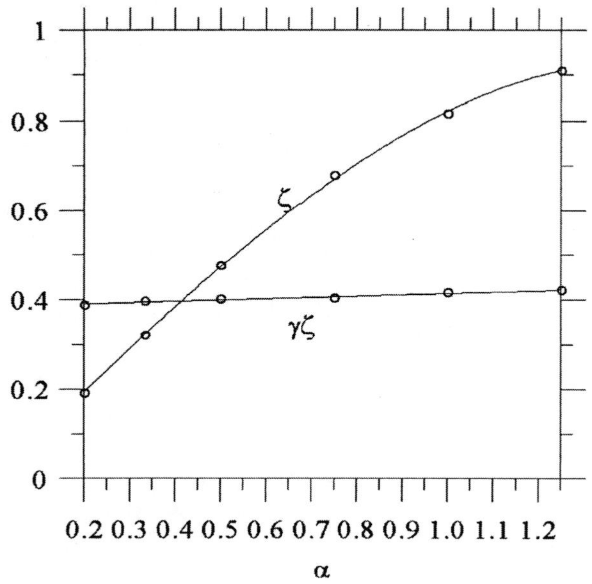

Fig. 5.10. Parameters of relationship (3.47) as function of parameter α in case of vertical force

A product of parameters $\gamma\zeta$ is expressed by the following linear function:

$$\gamma\zeta = 0.0299\alpha + 0.386\,. \tag{5.29b}$$

Relationships (5.29) are presented in Fig. 5.10 where circles represent the calculated values given in Table 5.4.

In order to find the equivalent depth H_e considered in section 3.4.1 for the linearly heterogeneous half-space, one should equate ratio G_h / G_0 to the quantity $\widetilde{G}$ by (5.1):

$$\frac{G_h}{G_0} = \left(1 + \frac{H_e}{z_0}\right)^{\alpha} = \left(1 + \frac{H_e}{\alpha z_{0l}}\right)^{\alpha}. \tag{5.30a}$$

This yields the following expression for the ratio of equivalent depth H_e to the wavelength l :

$$\frac{H_e}{l} = \alpha \frac{(G_h / G_0)^{1/\alpha} - 1}{(l / z_{0l})}\,. \tag{5.30b}$$

Ratio H_e / l is determined from the relationship between G_h / G_0 and the dimensionless wavelength l / z_{0l}. When using representation (3.47) (at $\widetilde{l} = l / z_{0l}$)

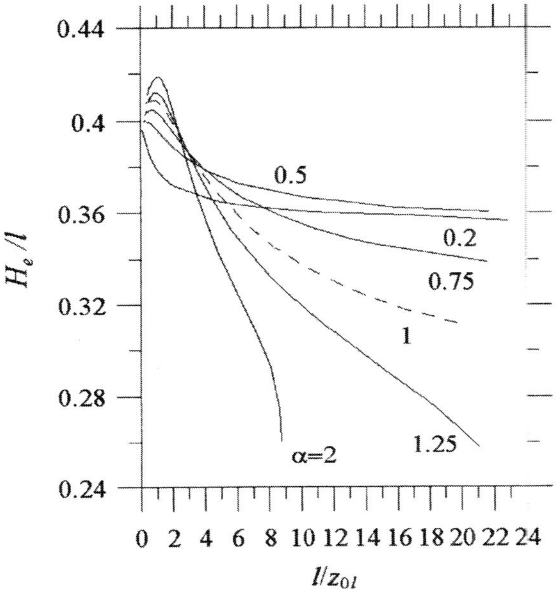

Fig. 5.11. Relationship between equivalent depth and surface wavelength in case of vertical force

we obtain

$$\frac{H_e}{l} = \alpha \frac{[1+\gamma(l/z_{0l})]^{\zeta/\alpha} - 1}{(l/z_{0l})} \,. \tag{5.31}$$

In this formula, one can express parameters γ and ζ in terms of α, according to (5.29). When argument l/z_{0l} tends to zero (high-frequency asymptotics), the limiting value H_e/l equals $\gamma\zeta$ (according to (5.31)), i.e. it is expressed by the quantity (5.29b) (being close to 0.4) which is weakly dependent on parameter α. In Fig. 5.11, quantity H_e/l is plotted versus l/z_{0l} by using values of G_h/G_0 calculated for the third wave. As parameter α decreases, i.e. as the considered half-space approaches the homogeneous one, the ratio H_e/l varies to a lesser extent with wavelength. At $\alpha = 2$, the term "equivalent depth" becomes meaningless at low frequencies of vibrations (at high values of l/z_{0l}), according to the discontinuous behavior of wave numbers described above.

With known relationship (3.47) (at $\tilde{l} = l/z_{0l}$) and expressions (5.29), one can determine the parameters of the half-space with power-law variation of shear modulus with depth, if the results of measurements of surface wave propagation velocities are available for various frequencies (various wavelengths). Suppose that the experimental results are processed in such a way that a relationship between the velocity of wave propagation along the foundation surface and the wavelength, $C = C(l)$, is derived. The value of α is obtained by a trial-and-error procedure. For a value of α, we find parameters γ and ζ by (5.29), z_0 by using formula (3.50b) with two selected values of wavelength and corresponding phase velocities, and further G_0, as shown in section 3.4.1. Following this procedure, one can derive a theoretical expression for function $C = C(l)$ by using formula (3.45). The selected value of α may be corrected, if necessary, in order to match better the experimental curve $C = C(l)$.

Since the value of C is proportional to $\sqrt{G_h}$, one should process experimental data for the considered case of the heterogeneous half-space by using relationship $C = C(l)$ in the following form (according to relationship (3.47)):

$$C = A(1 + Bl)^m , \tag{5.32}$$

where parameters A, B and m are determined from the available experimental data by employing a least-squares method. Decreasing A with corresponding increasing B results in a simple power law of the form (3.52). The following procedure is advisable: for each given value m, values A and B are easily found by the least-squares method (after raising to power $1/m$, this reduces to the selection of a linear function); next, we determine value m, at which the sum of the squares of deviations of the results obtained by (5.32) from the corresponding experimental data is minimized. If one determines the relationship of form (5.32), the procedure for selecting the half-space parameters is simplified. Indeed, we have $\zeta = 2m$ (since ratio G_h / G_0 is proportional to C^2) and value α may be found directly from equation (5.29a). Having determined parameters z_{0l} and G_0, we obtain relationship $C = C(l)$ coinciding with (5.32).

Next, we consider experimental data taken from [119]. Applying relationship (5.32) yields: A=113.5, B = 0.3, m = 0.34, whereas in formula (5.32), wavelength l is taken in meters, and velocity C in m/s. Equation (5.32) gives a better approximation than the simple power-law relationship of the form (3.52), according to which velocity C tends to zero as $l \to 0$. Note that the result of the simple power-law approximation done for the experimental data taken from [119] is somewhat different from (3.52):

$$C = 119.7 l^{0.158} . \tag{5.33}$$

A probable cause of deviation is the loss of accuracy that occurred when the graph from [119] was scanned. Let us apply the above technique of selection of the half-space parameters by using relationship (5.32) with the determined values of A, B, m; in formula (3.50b), we take l_1 = 5 m and l_2 = 2 m with the corresponding values of velocities C_1 and C_2. The results are as follows: $\alpha = 0.76$, $z_{0l} = 2.00$ m ($z_0 = z_{0l}\alpha = 1.523$ m), $G_0 = 26.35$ MN/m^2. Value $\zeta = 0.68$ and the found value of

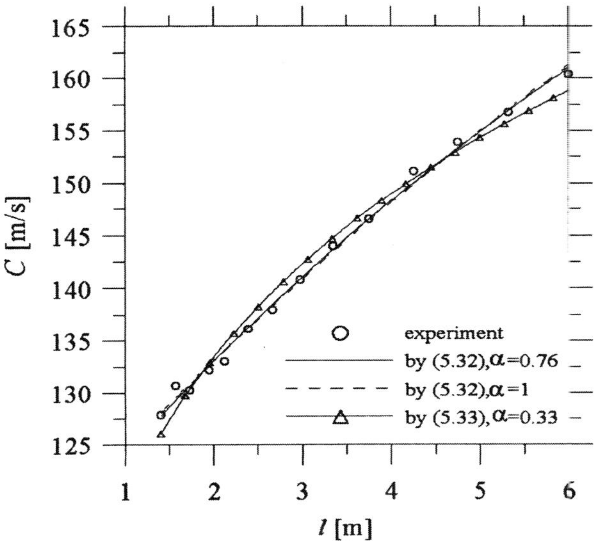

Fig. 5.12. Experimental and theoretical relationships between surface wave velocity and wavelength in case of vertical force

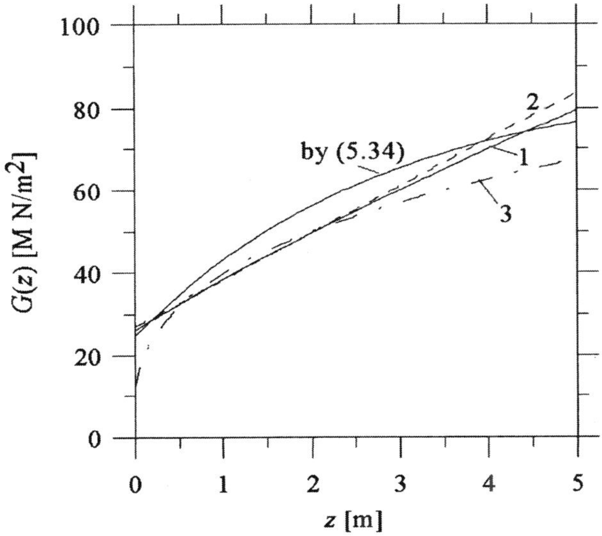

Fig. 5.13. Relationship between shear modulus and half-space depth for different methods of experimental data approximation

α satisfy equation (5.29a). When selecting $\alpha = 1$ in the case of relationship (5.32), the following values of parameters are obtained: $z_{0l} = z_0 = 2.39$ m, $G_0 = 27.13\,\text{MN/m}^2$. Experimental data (circles) and curves corresponding to the obtained parameters for $\alpha = 0.76$ (solid line) and for $\alpha = 1$ (dashed line) are shown in Fig. 5.12. The last two curves are practically identical. The selection of the law for the stiffness variation with depth is free, if the purpose is to provide a satisfactory approximation of experimental data in some specific range of the vibration frequencies. However, when determining an adequate value of parameter α by using the procedure described above, it becomes possible to extrapolate the results outside the bounds of the considered interval of frequencies. Further, we consider the approximation of the experimental data by using a simple power-law relationship, which, in this case, leads to relationship (5.33). The closest proximity to this relationship is provided by the following parameter values: $\alpha = 0.33$, $z_{0l} = 0.0818$ m, $G_0 = 12.04\,\text{MN/m}^2$. These values differ widely from those obtained on the basis of relationship (5.32). Thus, the form of functions used in approximating experimental relationship $C = C(l)$ has a strong influence on the sought half-space parameters. Compare the variation of shear modulus with depth for the above cases of selection of the half-space parameters (Fig. 5.13): 1) $\alpha = 0.76$, $z_0 = 1.523$ m, $G_0 = 26.35\,\text{MN/m}^2$; 2) $\alpha = 1$, $z_0 = 2.39$ m, $G_0 = 27.13\,\text{MN/m}^2$; 3) $\alpha = 0.33$, $z_0 = 0.027$m, $G_0 = 12.04\,\text{MN/m}^2$. The 4th curve in Fig. 5.14 corresponds to the following equation:

$$G = 62.5(1.4 - e^{-0.35H}), \tag{5.34}$$

derived in [119]. Curves 1–3 are rather close to each other for depths which serve as equivalent depths for the wavelengths used in the experiment. When adopting approximation $H \approx 0.39l$ (see Fig. 5.12), the experimental interval of wavelength $1.4\text{ m} < l < 6\text{ m}$ corresponds to the depth interval $0.55\text{ m} < H < 2.34\text{ m}$. Differences between the curves become significant in the vicinity of the half-space surface and at large depth. With the above considerations, the following set of parameter values is best justified: $\alpha = 0.76$, $z_0 = 1.523$m, $G_0 = 26.35\,\text{MN/m}^2$.

5.3 Heterogeneous Half-Space Subjected to Horizontal Force Applied to Half-Space Surface

A general form of solution for the case of a horizontal concentrated force applied to the half-space surface is given in relationships (1.86)–(1.88), (2.53)–(2.55). Besides the two fundamental solutions constructed when studying half-space vibrations under the action of vertical force, we shall employ the solution of equation (1.29), which satisfies the condition of absence of sources at infinity.

5.3.1 Application of Thin Layer Technique for Numerical Solution of Differential Equations

When applying the method of thin homogeneous layers, analogously to equations (5.3)–(5.6), we have

$$\frac{d^2 \tilde{p}}{d\tilde{z}^2} + (\beta^2\theta^2 - \tilde{k}^2\tilde{G}_{r\vartheta})\tilde{p} = 0, \tag{5.35}$$

where the characteristics of the considered layer should be used to determine parameter θ by (1.32a). We take $\tilde{G}=1$, $\tilde{\rho}=1$ (as in the case of equations (5.3)–(5.6)); for the isotropic case considered below, $\tilde{G}_{r\vartheta}=1$. The fundamental solutions of equation (5.35) and their derivatives have the form

$$\tilde{p}_{\mathrm{I}} = \exp(-\alpha_s\tilde{z}), \tag{5.36a}$$

$$\frac{d\tilde{p}_{\mathrm{I}}}{d\tilde{z}} = -\alpha_s \exp(-\alpha_s\tilde{z}), \tag{5.36b}$$

$$\tilde{p}_{\mathrm{II}} = \exp(\alpha_s\tilde{z}), \tag{5.37a}$$

$$\frac{d\tilde{p}_{\mathrm{II}}}{d\tilde{z}} = \alpha_s \exp(\alpha_s\tilde{z}), \tag{5.37b}$$

where α_s is calculated by (5.8a). Formulas (5.9) and (5.10) remain valid. In order to construct solution $\tilde{p}_1$ entering the integral representations of the vibration amplitudes, one can act analogously to the procedure given in section 5.2.1. Let the solution in the lower homogeneous half-space, extending unboundedly downward from the plane $z = z_0 + H$ (Fig. 5.1), take the form (5.36), thus providing a decrease of the solution as $z \to \infty$ (at high values of $\tilde{k}$) or resulting in vibrations in the form of waves going to infinity (at small values of $\tilde{k}$). At the upper boundary of the lower homogeneous half-space, it is convenient to drop the exponential multiplier in solutions (5.36), thus placing the origin of coordinates at the given boundary. Recall that for the lower homogeneous half-space, parameter θ is calculated by using formula (5.10) with substitution (5.27). Next, one should find a solution for each layer taking into account the conditions of transition at the common boundaries of two adjacent layers. For this transition, quantity $\tilde{p}$ is continuous, whereas its derivatives, taken at the points located above and below the boundary and infinitesimally close to it, are connected by the condition of stress continuity. Employing relationship (1.42a) yields

$$G_j \frac{d\tilde{p}^-}{d\tilde{z}} = G_{j+1} \frac{d\tilde{p}^+}{d\tilde{z}}, \tag{5.38}$$

Here, superscript "–" corresponds to the lower boundary of the j-th layer, and

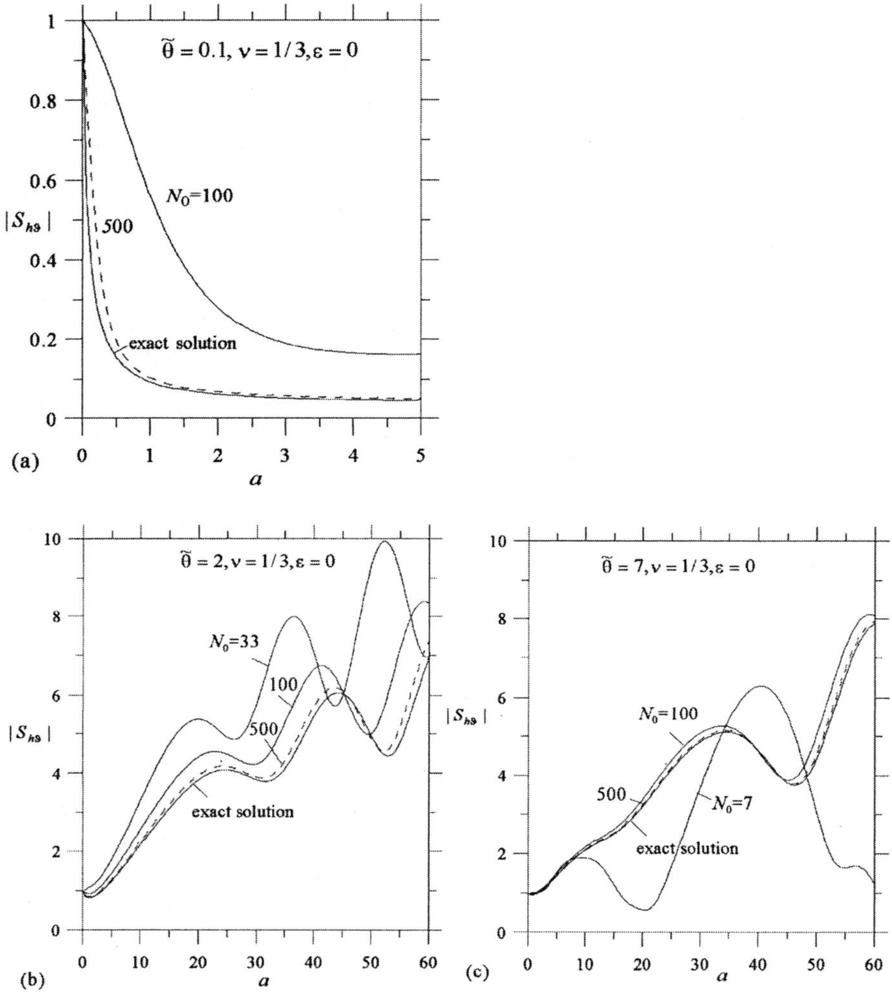

Fig. 5.14a,b,c. Influence of number of layers N_0 on absolute values of normalized tangential amplitudes of vibrations caused by horizontal force for linearly heterogeneous half-space at $\tilde{\theta} = 0.1$ (a), 2 (b), 7 (c)

superscript "+" to the upper boundary of the (j+1)-th layer. Shear modules entering relationship (5.38) include the influence of dissipative properties of the material analogously to relationship (1.28). Thus, with known values of function $\tilde{p}$ at the upper boundary of the (j+1)-th layer, one can determine the value of this function and its derivative at the lower boundary of the j-th layer. The solution for the j-th layer is constructed by using the known condition at its lower boundary in the form of a linear combination of functions (5.36) and (5.37). Denoting coefficients of the linear combination as S_i $(i = 1,2)$, respectively, we determine them by using the known condition at the lower boundary of the j-th layer (the origin of coordinates is placed at this boundary),

$$S_1 = \frac{1}{2\alpha_s}\left[\alpha_s \tilde{p} - \frac{d\tilde{p}}{d\tilde{z}}\right], \tag{5.39}$$

$$S_2 = \frac{1}{2\alpha_s}\left[\alpha_s \tilde{p} + \frac{d\tilde{p}}{d\tilde{z}}\right]. \tag{5.40}$$

Here, index j in the notation for the layer parameters is omitted; quantity $\tilde{p}$ and its derivative are related to the lower boundary of the layer, as noted previously. Having found coefficients S_i, we obtain the solution for layer j; determination of this solution at the upper boundary of the layer enables us to go to the next layer, of number j-1, etc., eventually reaching the half-space surface. Thus, simple recurrent relationships provide the construction of the sought solution over the entire half-space.

The technique of decreasing the depth of the heterogeneous layer, given in section 5.2.1 and related expressions (5.14)–(5.20) remain valid for the construction of solution $\tilde{p}$; considerations regarding the organization of computations also hold.

Consider examples of the calculation of amplitudes on the half-space surface. Let $\alpha = 1$, so that an exact solution of the problem is available. Analogously to Fig. 5.3, in Fig. 5.14 absolute values of the normalized amplitudes of vibrations $S_{h\vartheta}$ are shown for a number of values of parameter $\tilde{\theta}$ for different numbers of

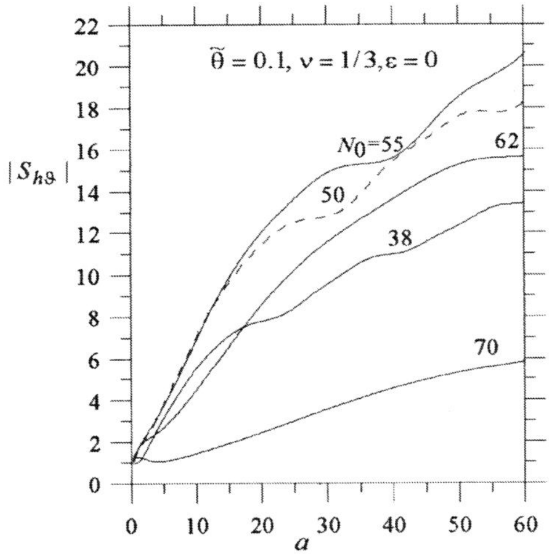

Fig. 5.15. Increase of tangential amplitudes of vibrations at $N_0 = 55$ ("resonance" number of layers)

layers. The corresponding values $\widetilde{H}_0$ were determined by formula (5.17). As in the case of action of a vertical force, a very significant increase in the vibration amplitudes is possible for some values of number of layers. At $\widetilde{\theta} = 0.1$, $N_0 = 51$, the amplitudes are many times greater than the static value and values for the corresponding continuously heterogeneous half-space (Fig. 5.15).

5.3.2 Application of Runge-Kutta Method for Numerical Solution of Differential Equations

In addition to the discussion given in section 5.2.2 regarding the construction of two fundamental solutions of system of equations (1.126)–(1.129) or equations (1.130)–(1.133) (for the isotropic case) by using the Runge–Kutta method of 4th order, we should determine function $\widetilde{p}_1$ that satisfies equation (1.29). In the lower homogeneous half-space, this function is taken in the same form (5.36) as in the case of constructing $\widetilde{p}_1$ in subsection 5.3.1. With known values of this function and its derivative at $\widetilde{z} = 1 + \widetilde{H}$, where $\widetilde{H}$ is determined by (5.15), integration of equation (1.29) is performed. As a result, all functions entering the integral representations of solutions of the considered problem, given in relationships of the type (1.86)–(1.88), (2.53)–(2.55), become known. Consider examples of the calculation of the vibration amplitudes for points at the half-space surface. Here, we emphasize especially quantities $\hat{u}_\vartheta$ that are influenced by Love-type waves. When studying vertical vibrations of the half-space with stiffness varying by

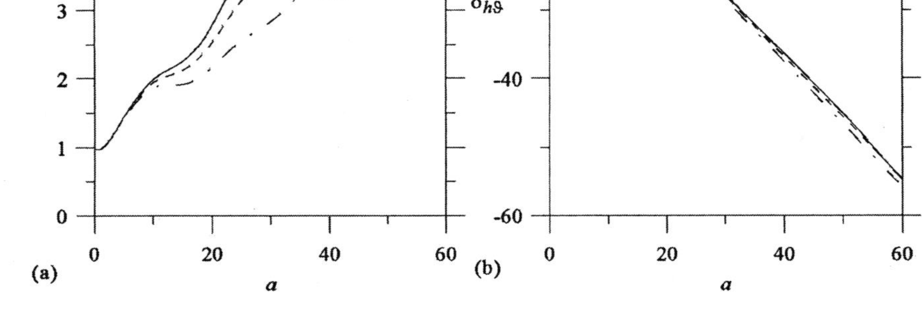

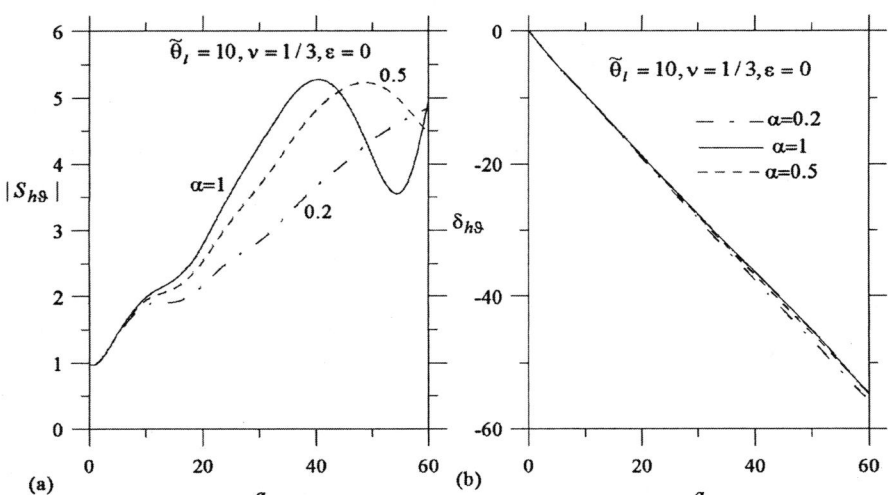

Fig. 5.16a,b. Absolute values of normalized tangential amplitudes of vibrations (a) and their phase (b) for three different values of parameter α at $\widetilde{\theta}_l = 10$

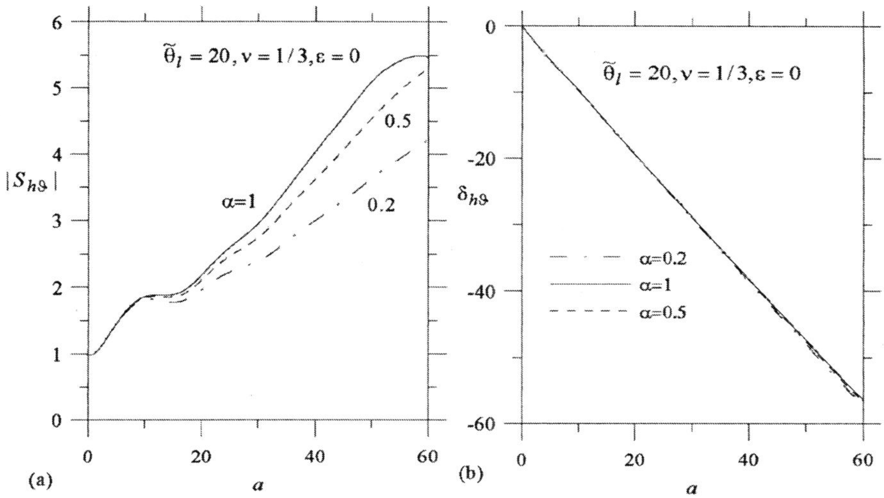

Fig. 5.17a,b. Absolute values of normalized tangential amplitudes of vibrations (a) and their phase (b) for three different values of parameter α at $\tilde{\theta}_l = 20$

power law with depth, we introduced a linearly heterogeneous half-space whose properties approach those of the given half-space at high frequencies (see relationships (5.19), (5.20)). In Figs. 5.16 and 5.17, graphs are plotted for the normalized amplitudes and phases of vibrations of the points on the half-space surface for three values of parameter α at $\tilde{\theta}_l = 10$ and $\tilde{\theta}_l = 20$. At sufficiently high values of $\tilde{\theta}_l$ and intermediate α, the results of calculations corresponding to different values of α, especially for phases $\delta_{h\vartheta}$, are close. Note that as parameter a increases, the value of $|S_{h\vartheta}|$ for the homogeneous half-space tends to 1.5 (see Fig. 3.7). This result differs substantially from the results obtained for the heterogeneous half-space, even in the case of high values of $\tilde{\theta}_l$.

5.3.3 Parameter Determination for Isotropic Half-Space with Shear Modulus Varying by Power Law (Action of Horizontal Force)

Below, the technique given in subsection 5.2.3 is applied to the amplitudes of vibrations $\hat{u}_\vartheta$ that occur due to the action of a horizontal force. As previously, it is convenient to employ parameters $\tilde{\theta}_l$ and z_{0l} related to the "equivalent" linearly heterogeneous half-space. First, for a number of values of parameter α, one should perform numerous calculations for various values of $\tilde{\theta}_l$, and determine dimensionless (i.e. related to parameter a) wave numbers k_ϑ considered previously for the linearly heterogeneous half-space in section 3.4.2. In order to

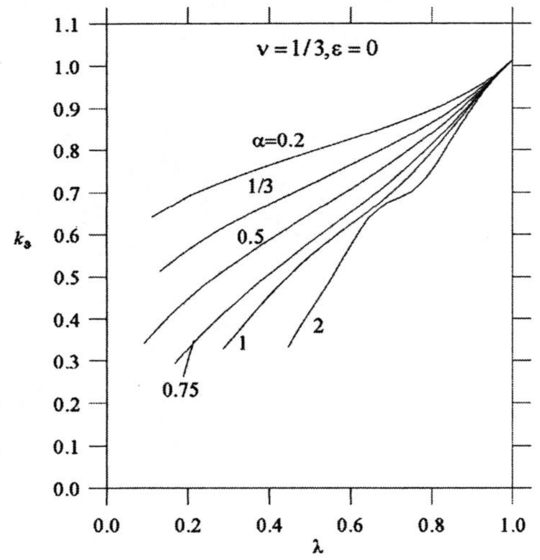

Fig. 5.18. Dimensionless wave numbers k_9 for various values of parameter α

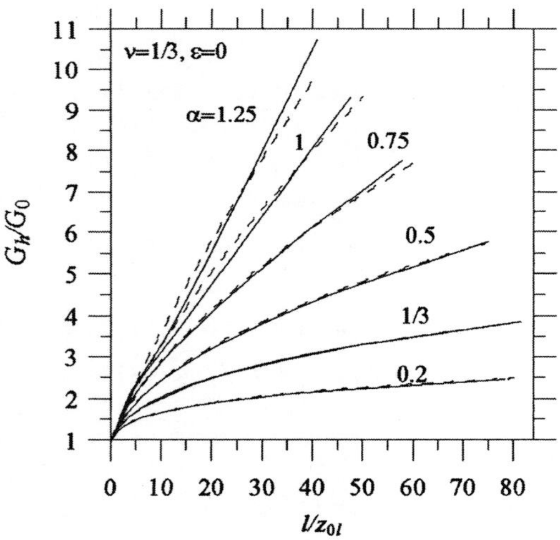

Fig. 5.19. Normalized shear modulus of homogeneous half-space corresponding to velocities of surface waves under action of horizontal force

determine values k_9, we shall use the third wave. As in the case of the linearly heterogeneous half-space, these values are weakly dependent on Poisson's ratio, due to the close proximity of the considered waves to pure shear waves. In Fig. 5.18, quantities k_9 are plotted versus parameter λ, determined by (5.28).

Analogously to formula (3.44), the expression for phase velocity may be written as follows:

$$C = \frac{1}{k_\vartheta}\sqrt{\frac{G_0}{\rho}} . \tag{5.41}$$

The expression for the shear modulus G_h of the homogeneous half-space, which corresponds to velocity C, has the same structure (3.45):

$$G_h = k_{\vartheta h}^2 C^2 \rho = \frac{k_{\vartheta h}^2}{k_\vartheta^2} G_0 . \tag{5.42}$$

Quantity $k_{h\vartheta}$ is related to the homogeneous half-space (see section 3.4.2); its value is close to unity (for the third wave with $\nu = 1/3$ we obtain $k_{h\vartheta} = 1.0123$). By using the relationship between k_ϑ and θ_l, we obtain a relationship between the wavelength and θ_l employing expression of the form (3.46):

$$\frac{l}{z_{0l}} = \frac{2\pi}{z_{0l} k_\vartheta(\theta_l)\omega\sqrt{\rho / G_0}} = \frac{2\pi}{k_\vartheta(\theta_l)\theta_l} . \tag{5.43}$$

Next, we construct graphs for the ratio G_h / G_0 versus the dimensionless wavelength l / z_{0l} for a number of values of α. These graphs are shown in Fig. 5.19 for the third wave with $\nu = 1/3, \varepsilon = 0$. Function (3.47) with $\tilde{l} = l / z_{0l}$ provides an acceptable accuracy over a wide interval of wavelengths. The dashed lines in Fig. 5.19 represent the results of approximation. The values of parameters γ and ζ are presented in Table 5.5

Table 5.5. Values of parameters for approximate relationship (3.47) in the case of horizontal force at various values of α ; $\tilde{l} = l / z_{0l}$

α	ζ	γ	$\gamma\zeta$
0.2	0.193	1.4	0.27
0.3333333	0.313	0.9	0.282
0.5	0.45	0.64	0.288
0.75	0.595	0.5	0.297
1	0.72	0.425	0.306
1.25	0.81	0.385	0.312

Analogously to relationships (5.29),

$$\zeta = \alpha - 0.275\alpha^{2.2} , \tag{5.44a}$$

$$\gamma\zeta = 0.038\alpha + 0.267 . \tag{5.44b}$$

These relationships are presented in Fig. 5.20 where the circles represent the calculated values given in Table 5.5.

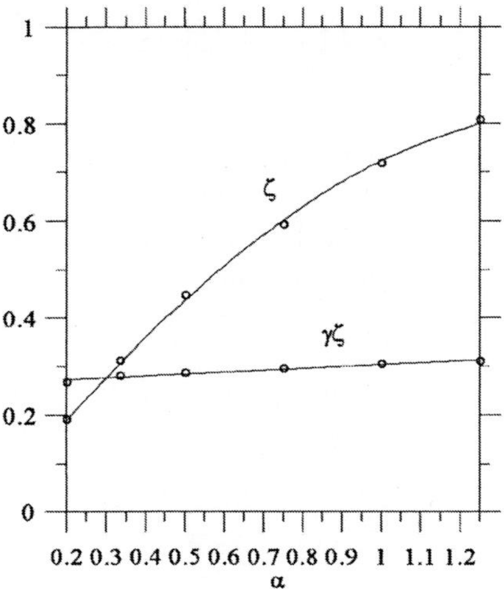

Fig. 5.20. Parameters of relationship (3.47) as function of parameter α in case of action of horizontal force

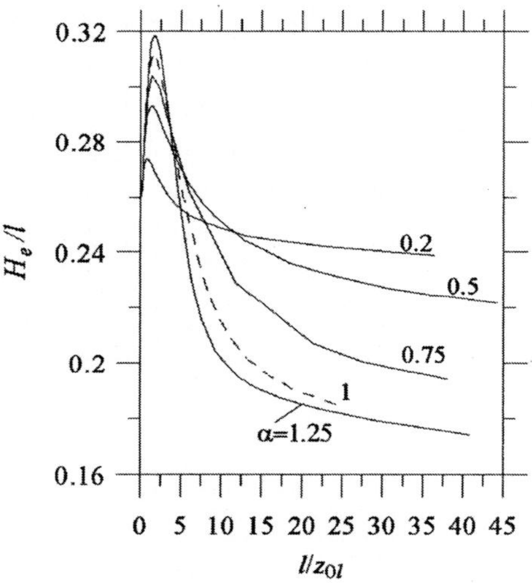

Fig. 5.21. Relationship between equivalent depth and surface wavelength in case of action of horizontal force

Consider equivalent depth H_e for the case of horizontal vibrations caused by a horizontal force. Formulas (5.30), (5.31) remain unchanged. The results of calculations by formula (5.30b) are shown in Fig. 5.21. The value of H_e/l is now much smaller than in the case of vertical vibrations; the interval of the non-monotonous behavior of H_e / l at small wavelengths is expressed more distinctly.

By employing relationships (3.47) (at $\tilde{l} = l / z_{0l}$) and relationships (5.44), one can determine the parameters of the half-space with shear modulus varying by power law with depth, if the results of measuring the velocities of propagation of the surface waves are available for various frequencies (various wavelengths). In the experiments, a source of horizontal vibrations is used. The amplitudes of vibrations are considered in the direction of the horizontal disturbing force at the points on the half-space surface lying on the straight line normal to this force and passing through the center of application of the load. For these amplitudes, the phase variation with distance from the source corresponds to the phase variation of the quantity $S_{h\vartheta}$ considered above Parameter determination for the half-space is performed analogously to the case of vertical vibrations considered in 5.2.3.

5.4 Vibration Problems for Heterogeneous Half-Space Subjected to Force Applied within Half-Space

The distinction between the problems considered in this section and those presented in section 5.2 and 5.3 lies in the fact that the solution now includes an additional term (particular solution) accounting for the force applied within the half-space. This particular solution vanishes below the point where the force is applied. In the upper part of the half-space it is determined from the same homogeneous differential equations that were used in section 5.2 and 5.3 to solve vibration problems for the half-space, subjected to a force applied to the half-space surface.

5.4.1 Action of Vertical Force. Application of Thin Layer Technique for Numerical Solution of Differential Equations

In the case of action of the concentrated force, the solution given by relationships (1.157) may be rewritten as

$$u_r = \frac{P_0(1-\nu')}{2G_{rz0}\pi r} S_{vh} , \tag{5.45a}$$

$$S_{vh} = \frac{\tilde{r}\beta^2}{(1-\nu')} \int_0^\infty \tilde{k}\, \mathrm{J}_1(\tilde{k}\tilde{r})[A_1^*(\tilde{k})\tilde{q}_1(\tilde{z},\tilde{k}) + A_2^*(\tilde{k})\tilde{q}_2(\tilde{z},\tilde{k}) + \tilde{q}_a^*(\tilde{z},\tilde{k})]\mathrm{d}\tilde{k} , \tag{5.45b}$$

$$u_z = \frac{P_0(1-\nu')}{2G_{rz0}\pi r} S_{vv} , \tag{5.46a}$$

$$S_{vv} = \frac{\tilde{r}\beta^2}{(1-\nu')}\int_0^{\infty} \tilde{k}\, J_0(\tilde{k}\tilde{r})[A_1^*(\tilde{k})\tilde{w}_1(\tilde{z},\tilde{k}) + A_2^*(k)\tilde{w}_2(\tilde{z},\tilde{k}) + \tilde{w}_a^*(\tilde{z},\tilde{k})]\mathrm{d}\tilde{k}\,, \tag{5.46b}$$

where the multipliers of the normalized vibration amplitudes S_{vh} and S_{vv} equal the static vertical displacements that occur due to the action of the vertical surface force on the surface of the homogeneous isotropic half-space with shear modulus G_{rz0} and Poisson's ratio ν'. When studying an isotropic heterogeneous half-space (in the following, this case is considered in detail), we use notations G_0 and ν to replace G_{rz0} and ν', respectively. According to the considerations given at the end of section 1.6.1, coefficients $A_1^*(\tilde{k})$, $A_2^*(\tilde{k})$ satisfy the following system of equations:

$$\begin{aligned} c_{11}A_1^* + c_{12}A_2^* &= d_1, \\ c_{21}A_1^* + c_{22}A_2^* &= d_2, \end{aligned} \tag{5.47}$$

where the coefficients c_{ij} are determined by relationships (1.146), (1.147) and the right-hand sides are

$$d_1 = -[\tilde{e}_a^*(\tilde{z}_0,\tilde{k}) - (\tilde{A}_{zz} - \tilde{A}_{rz})\tilde{k}\tilde{q}_a^*(\tilde{z}_0,\tilde{k})], \tag{5.48a}$$

$$d_2 = -[\tilde{\chi}_a^*(\tilde{z}_0,\tilde{k}) - 2\tilde{k}\tilde{w}_a^*(\tilde{z}_0,\tilde{k})], \tag{5.48b}$$

where for the isotropic half-space $\tilde{A}_{zz} - \tilde{A}_{rz} = 2$. As noted at the end of section 1.6.1, the particular solutions denoted by the asterisk are determined by (1.153) where b_e is replaced with unity.

The horizontal plane that contains the point of application of the force is determined by the coordinate $z = z_1$ or depth $h_0 = z_1 - z_0$; the corresponding dimensionless quantities are denoted as $\tilde{z}_1 = z_1 / z_r$ and $\tilde{h}_0 = h_0 / z_r$ whereas, as in section 5.2, we take $z_r = z_0$.

The given half-space is replaced with the half-space that includes the heterogeneous (upper) and homogeneous (lower) parts. Now, the heterogeneous part extends upward from the horizontal plane that contains the point of application of the force for a distance of h_0 (the depth corresponding to the point where the force is applied), and downward from this plane for the distance H_0 considered in section 5.2. As noted previously, as the integration variable $\tilde{k}$ increases, contraction of boundaries of the heterogeneous part to the horizontal plane of application of the force is performed. In addition to relationships (5.14), (5.15), we employ a relationship for the current dimensionless depth $\tilde{h}$ of the part of half-space located above the point at which the force is applied. Value $\tilde{h}$ is determined by $\widetilde{h}_0$ in the same manner as $\widetilde{H}$ is determined by $\widetilde{H}_0$ according to (5.15),

$$\widetilde{h} = \begin{cases} \widetilde{h}_0 & (\mathrm{Re}(\widetilde{k}) \le \widetilde{k}_0) \\ \min\left[\widetilde{h}_0, \dfrac{50}{\sqrt{\widetilde{k}^2 - \widetilde{k}_0^2}}\right] & (\mathrm{Re}(\widetilde{k}) > \widetilde{k}_0). \end{cases} \tag{5.49}$$

Coordinate z of the upper boundary of the half-space becomes equal to $z_1 - h$ instead of z_0; clearly, z_0 still represents the characteristic length. We carry out the division into layers in such a way that the horizontal plane containing the point at which the force is applied would serve as a boundary between two adjacent layers. Let N_1 be the number of the upper of these layers. Analogously to equation (5.16), as values of $\widetilde{H}$ and $\widetilde{h}$ decrease, the number of layers above and below the point at which the force is applied decreases (with the appropriate correction of $\widetilde{H}$ and $\widetilde{h}$ in order to keep the layer dimensions constant) until a single layer remains above and below this point. The thickness of these layers tends to zero as variable $\widetilde{k}$ increases. Regarding the estimation of $\widetilde{H}_0$, we introduce the following corrections into formula (5.21): instead of parameters $\widetilde{\theta}$ and $\widetilde{\theta}_l$ corresponding to the half-space surface, we shall use similar quantities $\widetilde{\theta}_h$ and $\widetilde{\theta}_{hl}$ for the horizontal plane with coordinate $z = z_1$,

$$\widetilde{\theta}_h = \widetilde{\theta}\sqrt{\frac{\rho(z_1)}{\widetilde{G}(\widetilde{z}_1)\rho(z_0)}}, \tag{5.50}$$

$$\widetilde{\theta}_{hl} = \frac{\widetilde{\theta}_h}{\alpha}. \tag{5.51}$$

To estimate the value of $\widetilde{H}_0$, quantity $\widetilde{\theta}_l$ in relationship (5.21) is replaced with $\widetilde{\theta}_{hl}$.

According to section 1.6.1, in addition to the two solutions studied in the previous section, we construct solution $\widetilde{\chi}_a^*(\widetilde{z},\widetilde{k})$, $\widetilde{q}_a^*(\widetilde{z},\widetilde{k})$, $\widetilde{e}_a^*(\widetilde{z},\widetilde{k})$, $\widetilde{w}_a^*(\widetilde{z},\widetilde{k})$ that equals zero when $z > z_1$. For the part $z \le z_1$ we use the initial condition: at the lower boundary of the layer number N_1, component $\widetilde{e}_a^*$ equals unity when the remainder of components vanishes.

Further construction of the particular solution is, in principle, similar to the procedure given in section 5.2.1 for the first and second solutions: with the known solution at the lower boundary of layer N_1, we determine coefficients B_i ($i = 1,\ldots,4$) by using formulas (5.13), find the solution for the upper boundary of the layer by using four fundamental solutions (5.7), and then employ relationships (5.12) when treating the transition between the layers.

Value θ entering the solution for the lower homogeneous half-space is determined by (5.10) with the substitution

$$Z_j = (1 + \widetilde{h}_0 + \widetilde{H}) z_0 \,. \tag{5.52}$$

Starting with values given in (5.7a) and (5.7b) (with exponents dropped), Cauchy problem is solved for the construction of two corresponding fundamental solutions; the particular solution is constructed in the part of half-space $z < z_1$. As in the previous subsection, we fix the lower boundary of the heterogeneous layer for sufficiently high values of the integration variable $\widetilde{k}$ (see (5.22)).

In the problems for a half-space subjected to the action of a force applied within the half-space, the employed technique provides a solution of the considered differential equations without any special tools, only if the point at which the solution is sought is located below the point of application of the force. The structure of the solution given in relationships (5.45), (5.46) indicates that at $z < z_1$, the result (extremely small at high values of $\widetilde{k}$) includes large terms (summands denoted by index a increase in an exponential manner, as moving upward from the plane $z = z_1$). Because of this, loss of accuracy still occurs even when the size of the part of the heterogeneous half-space located above the plane $z = z_1$ is reduced, according to formula (5.49).

Below, we present a technique that enables us to exclude the loss of accuracy. For each layer j, we consider simultaneously the two solutions with indices 1 and 2, and in the case of $j \le N_1$, the solution with index a (particular solution). These solutions enter integrals (5.45b), (5.46b). The origin of coordinates is located at the lower boundary of the layer. Assume that all three solutions for the lower boundary of the layer are known and the corresponding coefficients (5.13) are calculated for each of the solutions. Recall that when crossing the boundary between layers of number j+1 and number j, in addition to the condition of continuity of quantities q and w, relationships (5.11) are employed. Next, we find a linear combination of the first and second solutions in such a way that after subtracting this linear combination from the solution with index a, increasing (as moving upward) exponential terms would vanish. Finally, only that part of the particular solution which contains decreasing exponents is employed in the calculations. This transformation leads to the transformation of coefficients A_1^*, A_2^* (see (5.45), (5.46)) that become different in going from one layer to another. Here, we introduce the second index (1, 2 and a) in coefficients $B_i \; (i = 1,\ldots,4)$ in (5.13) for the first, second and the particular solutions, respectively. Coefficients D_{1j} and D_{2j} in the linear combination of the first and second fundamental solutions, the increasing part of which equals the increasing part of the particular solution, are determined from the system of equations

$$D_{1j}B_1^{(1)} + D_{2j}B_1^{(2)} = B_1^{(a)}, \tag{5.53a}$$

$$D_{1j}B_2^{(1)} + D_{2j}B_2^{(2)} = B_2^{(a)}. \tag{5.53b}$$

After subtraction of the linear combination of the first and second solutions with coefficients D_{1j} and D_{2j} from the particular solution (with index a), the fundamental solutions (5.7a) and (5.7b) (with indices I and II) which are responsible for an increase in the solution when moving upward, are removed from the particular solution.

In addition, the first and second solutions are transformed in order to prevent the possibility that they would increase enormously when moving upward. The following scaling multiplier is used for both solutions:

$$M_j = \exp(-\alpha_p \tilde{h}_j), \tag{5.54}$$

where α_p is determined by (5.8b), by using the parameters of the considered j-th layer, and $\tilde{h}_j$ is the dimensionless layer thickness. Since the origin of coordinates is placed at the lower boundary of the layer (when constructing the solution within the layer), as a result of such multiplication the exponential multiplier at the upper boundary of the layer in component (5.7b) entering the solution with multiplier B_2 becomes unity. In component (5.7a), its value tends to unity at high values of $\tilde{k}$, whereas in the remainder of the components, this multiplier becomes extremely small at the upper boundary of the layer at high values of $\tilde{k}$. Note that when moving upward, we consider solutions at the layer boundaries only.

Thus, the particular solution in the j-th layer becomes

$$\begin{bmatrix} \tilde{\chi}_a^0 \\ \tilde{e}_a^0 \\ \tilde{q}_a^0 \\ \tilde{w}_a^0 \end{bmatrix} = \begin{bmatrix} \tilde{\chi}_a^* \\ \tilde{e}_a^* \\ \tilde{q}_a^* \\ \tilde{w}_a^* \end{bmatrix} - D_{1j} \begin{bmatrix} \tilde{\chi}_1 \\ \tilde{e}_1 \\ \tilde{q}_1 \\ \tilde{w}_1 \end{bmatrix} - D_{2j} \begin{bmatrix} \tilde{\chi}_2 \\ \tilde{e}_2 \\ \tilde{q}_2 \\ \tilde{w}_2 \end{bmatrix}, \tag{5.55}$$

while the complete solution is expressed in the following form:

$$\begin{bmatrix} \tilde{\chi} \\ \tilde{e} \\ \tilde{q} \\ \tilde{w} \end{bmatrix} = \begin{bmatrix} \tilde{\chi}_a^0 \\ \tilde{e}_a^0 \\ \tilde{q}_a^0 \\ \tilde{w}_a^0 \end{bmatrix} + A_{1j}^* \begin{bmatrix} \tilde{\chi}_1^0 \\ \tilde{e}_1^0 \\ \tilde{q}_1^0 \\ \tilde{w}_1^0 \end{bmatrix} + A_{2j}^* \begin{bmatrix} \tilde{\chi}_2^0 \\ \tilde{e}_2^0 \\ \tilde{q}_2^0 \\ \tilde{w}_2^0 \end{bmatrix}, \tag{5.56}$$

Here, index 0 means that the solutions are transformed as described above. Prior to the calculation of solutions at the upper boundary of the current layer, they are transformed at the lower boundary of this layer. After accounting for the transition conditions (5.11), the procedure is repeated for the next layer. The solutions obtained for the upper boundary of the upper layer are used in the calculation of the coefficients and the right sides of the system of equations (5.47), and in the determination of coefficients A_1^*, A_2^* for the upper layer. As a result of transformation (5.55) and the introduced scaling, coefficients A_1^*, A_2^* in (5.45) and (5.46) vary in going from one layer to another; this variation is denoted in relationship (5.56) by the index indicating the layer number. The relationship between the coefficients for the j-th and (j+1)-th layers is found from comparison between representations (5.56) written for the j and j+1 layers (i.e. before and after the transformations),

$$A_{1,j+1}^* = M_j A_{1,j}^* - D_{1j}\,, \tag{5.57a}$$

$$A_{2,j+1}^* = M_j A_{2,j}^* - D_{2j}\,. \tag{5.57b}$$

Having determined coefficients A_{1j}^*, A_{2j}^* for the upper layer ($j = 1$) by using the system of equations (5.47), we can find these coefficients for all layers by employing recurrent relationships (5.57) (when moving downward); here, values M_j, D_{1j}, D_{2j} determined for each layer when moving upward are used. Recall that for the part of the half-space $z > z_1$ ($j > N_1$), the particular solution equals zero. Therefore, coefficients D_{1j}, D_{2j} in transformations (5.57) vanish. Application of relationships (5.57) for the number j that equals the number of the lower layer in the heterogeneous part of the half-space yields coefficients that enable us to obtain the solution in the lower homogeneous part of the half-space by using the first and second solutions (5.7) (the origin of coordinates is located at the upper boundary of the homogeneous part of the half-space).

Note that, when constructing the solution for each layer by using relationship (5.56), at high values of $\tilde{k}$, the transformed particular solution decreases in each layer in an exponential manner when moving upward. The transformed fundamental solutions decrease in each layer, starting with intermediate values at the upper boundary, when moving downward. Consequently, coefficients A_{1j}^*, A_{2j}^* at $j = 1$ determined from the system of equations (5.47) appear to be very small. These coefficients are also small for the other j except $j = N_1 + 1$ (see (5.57)) when coefficients $D_{1,j-1}, D_{2,j-1}$ are not small. According to relationship (5.56), the solution differs noticeably from zero only in those parts of the half-space being adjacent closely from above and below to the plane $z = z_1$ where the force is applied. These considerations prove once again how reasonable the applied technique is of contracting the calculation area of the heterogeneous half-space, as parameter $\tilde{k}$ increases.

Next, we present the results of calculations performed for the linearly heterogeneous half-space ($\alpha = 1$), showing the influence of the number of layers on the values of vertical vibration amplitudes. We consider the amplitudes of vibrations in the horizontal plane, in which the vertical force is applied, at

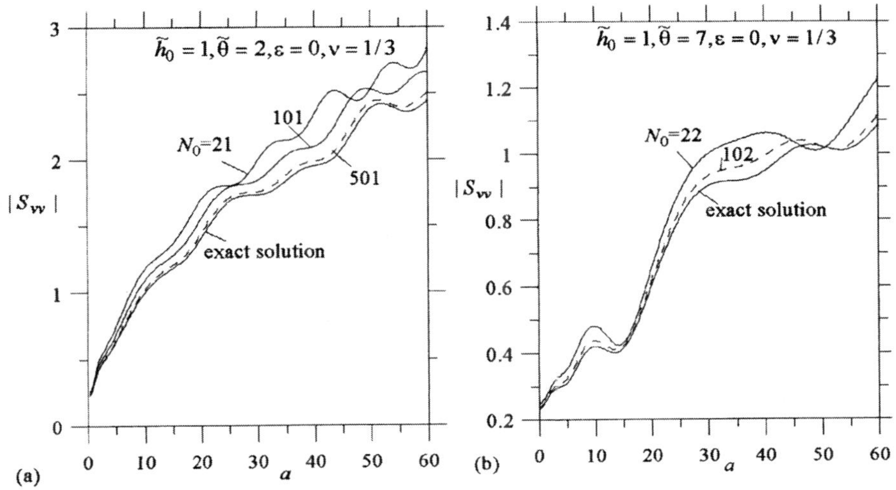

Fig. 5.22a,b. Influence of number of layers N_0 on absolute values of normalized vertical amplitudes of vibrations for linearly heterogeneous half-space when force application point is buried ($\widetilde{\theta} = 2$ (a), and $\widetilde{\theta} = 7$ (b)); $\widetilde{z} = 1 + \widetilde{h}_0$

$\widetilde{h}_0 = 1$, $\nu = 1/3$, $\varepsilon = 0$. The results of calculations for $\widetilde{\theta} = 2$ and some initial values of number of layer, namely $N_0 = 21$, 101, 501, are presented in Fig. 5.22a. A part of these layers is located above the point at which the force is applied; on the initial interval of variation of parameter $\widetilde{k}$, this part contains 1, 3, 14 layers, respectively. Value $\widetilde{H}_0$ calculated by using (5.21), (5.51) equals 35.355. In Fig. 5.22b, graphs are plotted for $\widetilde{\theta} = 7$, for $N_0 = 22$, 102. Above the horizontal plane of application of the force, the initial number of layers equals 2 and 10, respectively. In this case, we have a smaller value $\widetilde{H}_0 = 10.102$, resulting in a faster convergence to the exact solution.

5.4.2 Action of Vertical Force. Application of Runge–Kutta Method for Numerical Solution of Differential Equations

Consider the application of the Runge–Kutta method to construct solutions entering integrals (5.45b), (5.46b). In the simplest option of applying this method, it is sufficient to use the previously employed technique of contracting (as the integration variable $\widetilde{k}$ increases) the boundaries of the calculation area in the heterogeneous half-space to the horizontal plane, in which the force is applied, and to fix the lower boundary of the heterogeneous part of the half-space at high values of $\widetilde{k}$. Here, formulas (5.14), (5.15), (5.49), (5.22) are kept in mind. The recommended technique for determining quantity $\widetilde{H}_0$ (see the previous subsection) still holds. In the framework of this approach, we do not encounter a loss of accuracy in calculations, if the point at which the solution is sought is located below the point of application of the force. As a universal technique which allows us to overcome the difficulties that arise at high values of parameter $\widetilde{k}$, Godunov's method [42] can be recommended. In the following, we consider application of this method to the given half-space vibration problem while use of the structure of solution (5.45), (5.46) is kept unchanged. The variation interval of variable $\widetilde{z}$, from the lower boundary of the calculation area of the heterogeneous half-space ($\widetilde{z} = 1 + \widetilde{h}_0 + \widetilde{H}$) to its surface ($\widetilde{z} = 1 + \widetilde{h}_0 - \widetilde{h}$), is divided into a number of intervals ("layers"). Within the bounds of each "layer", three considered solutions (with indices 1, 2, a in integrals (5.45b), (5.46b)) are constructed by using the Runge–Kutta method (or any other numerical method) with initial data obtained at the previous stage. When reaching the upper boundary of the current "layer", orthogonalization and normalization of the first two solutions are performed, while the projection of the particular solution (with index a) to the linear subspace, formed by the first two solutions, is subtracted from this particular solution. These transformations prevent an increase in solutions and the related loss of accuracy. The values of transformed solutions are taken as initial data for integration in the next "layer" located above, and the procedure is repeated until the half-space surface is reached. Note that the transformations employed in the previous subsection with piecewise constant approximation differ from the method presented here in that the complete elimination of increasing components is done in the particular solution, while, in the framework of Godunov's method, the first two solutions are used to provide a maximal possible decrease of the particular solution for a specific point. According to Godunov's method, the transformed solutions denoted by superscript 0 take the form (the "layer" number is omitted),

$$\begin{bmatrix} \widetilde{\chi}_1^0 \\ \widetilde{e}_1^0 \\ \widetilde{q}_1^0 \\ \widetilde{w}_1^0 \end{bmatrix} = \frac{1}{\omega_{11}} \begin{bmatrix} \widetilde{\chi}_1 \\ \widetilde{e}_1 \\ \widetilde{q}_1 \\ \widetilde{w}_1 \end{bmatrix}, \tag{5.58a}$$

$$\begin{bmatrix} \widetilde{\chi}_2^0 \\ \widetilde{e}_2^0 \\ \widetilde{q}_2^0 \\ \widetilde{w}_2^0 \end{bmatrix} = \frac{1}{\omega_{22}} \left(\begin{bmatrix} \widetilde{\chi}_2 \\ \widetilde{e}_2 \\ \widetilde{q}_2 \\ \widetilde{w}_2 \end{bmatrix} - \omega_{21} \begin{bmatrix} \widetilde{\chi}_1^0 \\ \widetilde{e}_1^0 \\ \widetilde{q}_1^0 \\ \widetilde{w}_1^0 \end{bmatrix} \right), \tag{5.58b}$$

$$\begin{bmatrix} \tilde{\chi}_a^0 \\ \tilde{e}_a^0 \\ \tilde{q}_a^0 \\ \tilde{w}_a^0 \end{bmatrix} = \begin{bmatrix} \tilde{\chi}_a^* \\ \tilde{e}_a^* \\ \tilde{q}_a^* \\ \tilde{w}_a^* \end{bmatrix} - \omega_{a1} \begin{bmatrix} \tilde{\chi}_1^0 \\ \tilde{e}_1^0 \\ \tilde{q}_1^0 \\ \tilde{w}_1^0 \end{bmatrix} - \omega_{a2} \begin{bmatrix} \tilde{\chi}_2^0 \\ \tilde{e}_2^0 \\ \tilde{q}_2^0 \\ \tilde{w}_2^0 \end{bmatrix}. \tag{5.58c}$$

Here, ω_{21} represents the projection of the second solution (prior to its transformation) onto the normalized first solution, and ω_{a1}, ω_{a2} are projections of the particular solution onto the orthonormalized first and second solutions, respectively; ω_{11} and ω_{22} are the corresponding norms (the square root of the sum of squared absolute values of the components of the vector solution). Note that these projections are determined by the scalar products, whereas the components of the second multiplicand in the scalar product are taken to be complex conjugate with respect to their initial complex values. The structure of transformation (5.58c) is similar to transformation (5.55). In each "layer", the solution may be presented in the form (5.56) by using the transformed solutions. When "layers" are numbered downward (as in the previous subsection), the following relationship between coefficients A_1^*, A_2^* corresponding to the upper boundaries of adjacent layers holds, instead of relationships (5.57):

$$A_{2,j+1}^* = \frac{A_{2,j}^* - \omega_{a2}}{\omega_{22}}, \tag{5.59a}$$

$$A_{1,j+1}^* = \frac{A_{1,j}^* - \omega_{21} A_{2,j+1}^* - \omega_{a1}}{\omega_{11}}. \tag{5.59b}$$

Here, coefficients $\omega_{11}, \omega_{22}, \omega_{21}, \omega_{a1}, \omega_{a2}$ refer to the j-th "layer". Starting with the upper "layer" ($j = 1$) where coefficients A_1^*, A_2^* are determined from the boundary condition at the half-space surface after the last transformation of the solutions, we find these coefficients for the rest of the "layers" (for their upper boundaries) by using recurrent relationships (5.59). Thus, the solution may be found for an arbitrary point in the half-space. It is advisable to perform the division into "layers" in such a way that the point where the solution is sought should be located at the "layer" boundary. Then, the sought solution is found directly by using the corresponding coefficients A_1^*, A_2^* and the transformed solutions for this boundary. The technique of contracting the calculation area to the horizontal plane in which the force is applied, and fixing the lower boundary of the heterogeneous part of the half-space at high values of $\tilde{k}$ are still advisable. Recall that the solution with index a is accounted for, starting with $\tilde{z} = \tilde{z}_1$, where: $\tilde{\chi}_a^* = 0$, $\tilde{e}_a^* = 1$, $\tilde{q}_a^* = 0$, $\tilde{w}_a^* = 0$. Verification of the considered numerical method, including the above recommendations, is performed for the linearly heterogeneous half-space; the results of calculations match (up to 4–5 digits) the corresponding results obtained by using the solutions presented in section 3.5.

5.4.3 Action of Horizontal Force. Application of Thin Layer Technique for Numerical Solution of Differential Equations

For the case of a concentrated force, the vibration amplitudes $\hat{u}_r, \hat{u}_\vartheta, \hat{u}_z$ presented in relationships (1.164) may be rewritten as follows:

$$\hat{u}_r = \frac{Q_0}{2G_{rz0}\pi r} S_{hr}, \tag{5.60a}$$

$$S_{hr} = \beta^2 \tilde{r} \int_0^\infty \left\{ \left[\tilde{q}(\tilde{z},\tilde{k}) - \frac{\tilde{p}(\tilde{z},\tilde{k})}{\tilde{G}(\tilde{z}_1)} \right] \frac{\mathrm{J}_1(\tilde{k}\tilde{r})}{\tilde{k}\tilde{r}} - \tilde{q}(\tilde{z},\tilde{k}) \mathrm{J}_0(\tilde{k}\tilde{r}) \right\} \tilde{k}\,\mathrm{d}\tilde{k}, \tag{5.60b}$$

$$\hat{u}_\vartheta = -\frac{Q_0(1-\nu')}{2G_{rz0}\pi r} S_{h\vartheta}, \tag{5.61a}$$

$$S_{h\vartheta} = -\frac{\beta^2 \tilde{r}}{1-\nu'} \int_0^\infty \left\{ \left[\tilde{q}(\tilde{z},\tilde{k}) - \frac{\tilde{p}(\tilde{z},\tilde{k})}{\tilde{G}(\tilde{z}_1)} \right] \frac{\mathrm{J}_1(\tilde{k}\tilde{r})}{\tilde{k}\tilde{r}} + \frac{\tilde{p}(\tilde{z},\tilde{k})}{\tilde{G}(\tilde{z}_1)} \mathrm{J}_0(\tilde{k}\tilde{r}) \right\} \tilde{k}\,\mathrm{d}\tilde{k}, \tag{5.61b}$$

$$\hat{u}_z = -\frac{Q_0(1-\nu')}{2G_{rz0}\pi r} S_{hz}, \tag{5.62a}$$

$$S_{hz} = -\frac{\beta^2 \tilde{r}}{1-\nu'} \int_0^\infty \tilde{k}\tilde{w}(z,k) \mathrm{J}_1(\tilde{k}\tilde{r})\,\mathrm{d}\tilde{k}, \tag{5.62b}$$

where nomenclature (1.165) is used and static displacements on the surface of a homogeneous half-space, generated by the force applied to the half-space surface, are set apart. Functions $\tilde{q}$ and $\tilde{w}$ are constructed analogously to the case of a vertical force considered in previous subsections. The only change is introduced in the initial condition for the particular solution $\tilde{\chi}_a^*(\tilde{z},\tilde{k})$, $\tilde{e}_a^*(\tilde{z},\tilde{k})$, $\tilde{q}_a^*(\tilde{z},\tilde{k})$, $\tilde{w}_a^*(\tilde{z},\tilde{k})$: now, at $\tilde{z} = \tilde{z}_1$ (at the lower boundary of layer N_1), component $\tilde{\chi}_a^*$ equals -1, while the rest of the components vanish. Application of the transformation technique for the particular solution, and scaling of the first and second solutions, are suitable for this case also.

In addition, one should construct parts of the solution which contain function $\tilde{p}$ having the form (1.165c); according to (1.119), quantity $C^*(\tilde{k})$ is

$$C^*(\tilde{k}) = -\frac{\mathrm{d}\,p_a^*(\tilde{z}_0,\tilde{k})}{\mathrm{d}\tilde{z}} \bigg/ \frac{\mathrm{d}\,p_1(\tilde{z}_0,\tilde{k})}{\mathrm{d}\tilde{z}}. \tag{5.63}$$

The determination of function $\widetilde{p}_1(\widetilde{z},\widetilde{k})$ is considered in section 5.3.1; one should take into account that parameter θ for the lower homogeneous half-space (for which the solution is constructed in the form (5.36)) is determined by (5.10) with substitution (5.52); the origin of coordinates should be placed at the upper boundary of the homogeneous half-space. Next, we go to higher layers, according to the presentation given in section 5.3.1. Function $\widetilde{p}_a^*(\widetilde{z},\widetilde{k})$ vanishes at the lower boundary of the layer adjacent from above to the horizontal plane, in which the force is applied, whereas its derivative equals here -1. The determination of function $\widetilde{p}_a^*(\widetilde{z},\widetilde{k})$ in the upper part of the half-space is analogous to that given in section 5.3.1, with respect to the determination of function $\widetilde{p}_1(\widetilde{z},\widetilde{k})$. The considerations regarding contraction of the calculation area of the half-space to the plane $z = z_1$, in which the force is applied, as parameter $\widetilde{k}$ increases, remain valid.

When considering the action of a vertical force applied within the half-space, we discussed a possible loss of accuracy due to an increase in the particular solution (denoted by subscript a), as moving upward from the plane $z = z_1$. This possibility still holds with respect to the part of the solution which contains function $\widetilde{p}$. Transformations of solutions similar to those done for the solution with components $\widetilde{\chi},\widetilde{e},\widetilde{q},\widetilde{w}$ are useful. Consider solutions $\widetilde{p}_a^*$ and $\widetilde{p}_1$ and their derivatives at the lower boundary of the j-th layer (following the transition of the form (5.38)). The coefficients are calculated by (5.39), (5.40) for each solution; further additional indices a and 1 for coefficients S_i ($i = 1,2$) will be introduced. Solution $\widetilde{p}_a^*$ and its derivative are transformed as follows:

$$\begin{bmatrix} \widetilde{p}_a^0 \\ \dfrac{\mathrm{d}\,\widetilde{p}_a^0}{\mathrm{d}\,\widetilde{z}} \end{bmatrix} = \begin{bmatrix} \widetilde{p}_a^* \\ \dfrac{\mathrm{d}\,\widetilde{p}_a^*}{\mathrm{d}\,\widetilde{z}} \end{bmatrix} - D_{pj} \begin{bmatrix} \widetilde{p}_1 \\ \dfrac{\mathrm{d}\,\widetilde{p}_1}{\mathrm{d}\,\widetilde{z}} \end{bmatrix}, \tag{5.64}$$

where multiplier D_{pj} is selected in such a way that terms containing increasing exponents (i.e. the part containing (5.36)) vanish in the particular solution,

$$D_{pj} S_1^{(1)} = S_a^{(1)}. \tag{5.65}$$

Further, solution $\widetilde{p}_1$ and its derivative are multiplied by a factor K_j that is similar to (5.54),

$$\begin{bmatrix} \widetilde{p}_1^0 \\ \dfrac{\mathrm{d}\,\widetilde{p}_1^0}{\mathrm{d}\,\widetilde{z}} \end{bmatrix} = K_j \begin{bmatrix} \widetilde{p} \\ \dfrac{\mathrm{d}\,\widetilde{p}}{\mathrm{d}\,\widetilde{z}} \end{bmatrix}, \tag{5.66}$$

where

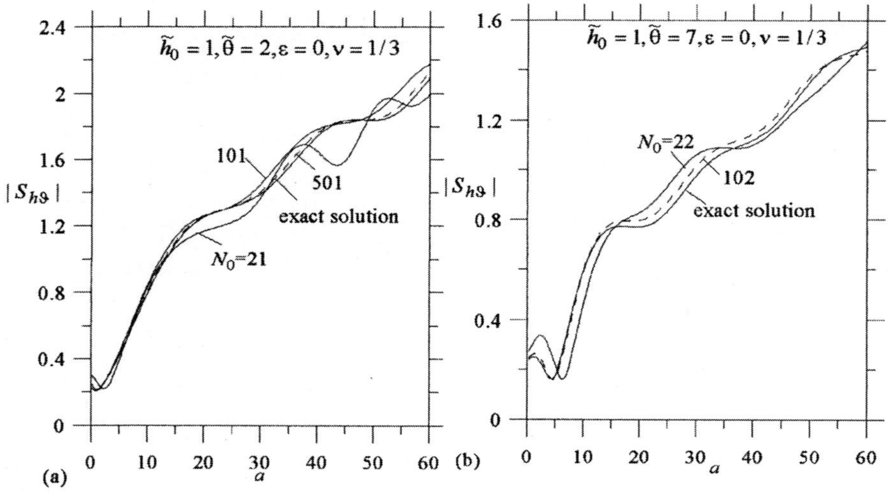

Fig. 5.23a,b. Influence of number of layers N_0 on absolute values of normalized tangential amplitudes of vibrations for linearly heterogeneous half-space when force application point is deepened at $\widetilde{\theta} = 2$ (a), and $\widetilde{\theta} = 7$ (b) for $\widetilde{z} = 1 + \widetilde{h}_0$

$$K_j = \exp(-\alpha_s \widetilde{h}_j)\,. \tag{5.67}$$

Here, value α_s is determined by (5.8a) by using the characteristics of the j-th layer. The solution for each layer may be presented analogously to (5.56) by employing transformed solutions:

$$\begin{bmatrix} \widetilde{p} \\ \dfrac{\mathrm{d}\widetilde{p}}{\mathrm{d}\widetilde{z}} \end{bmatrix} = C_j^* \begin{bmatrix} \widetilde{p}_1^0 \\ \dfrac{\mathrm{d}\widetilde{p}_1^0}{\mathrm{d}\widetilde{z}} \end{bmatrix} + \begin{bmatrix} \widetilde{p}_a^0 \\ \dfrac{\mathrm{d}\widetilde{p}_a^0}{\mathrm{d}\widetilde{z}} \end{bmatrix}, \tag{5.68}$$

where C_j^* varies in going from one layer to another. Coefficient C_j^* for the upper layer is calculated by using transformed solutions by formula (5.63). The relationship between these coefficients for two adjacent layers is found by comparing representations (5.68) before and after the transformations. As a result, the relationship analogous to (5.57) is obtained,

$$C_{j+1}^* = K_j C_j - D_{pj}\,. \tag{5.69}$$

Employing (5.69) yields the value of C_j^* for all layers, which enables the determination of function $\widetilde{p}$ for each layer by using equation (5.68). Note that for the lower homogeneous part of the half-space, the coefficient is determined by (5.69) for value j equal to the number of the lower layer in the heterogeneous part of the half-space. In the homogeneous part, we use solution (5.36), placing the origin of coordinates at the upper boundary of this homogeneous part of the half-space.

Next, we consider sample calculations done for the linearly heterogeneous half-space ($\alpha = 1$), following the above technique. The given examples represent the convergence of the solution to that corresponding to a continuously heterogeneous half-space, as layer number N_0 increases. The amplitudes of vibrations $S_{h\vartheta}$ are calculated for the points in the horizontal plane, in which the force is applied, at $\alpha = 1,\ \nu = 1/3,\ \varepsilon = 0,\ \widetilde{h}_0 = 1$. The calculations were performed for two values of parameter $\widetilde{\theta}$, namely 2 (Fig. 5.23a) and 7 (Fig. 5.23b).

5.4.4 Action of Horizontal Force. Application of Runge–Kutta Method for Numerical Solution of Differential Equations

In this subsection, we use the solution in the form (5.60)–(5.62) again. Instead of using the piecewise constant approximation of the coefficients for the corresponding differential equations and constructing exact solutions of the simplified equations within each layer, we employ the Runge–Kutta method of 4th order to solve the equations numerically. Part of the solution containing components $\widetilde{\chi}(\widetilde{z},\widetilde{k}),\ \widetilde{q}(\widetilde{z},\widetilde{k}),\ \widetilde{e}(\widetilde{z},\widetilde{k}),\ \widetilde{w}(\widetilde{z},\widetilde{k})$ is determined analogously to section 5.4.2; the only difference is in the initial conditions for the part of solution $\widetilde{\chi}_a^*(\widetilde{z},\widetilde{k}),\ \widetilde{q}_a^*(\widetilde{z},\widetilde{k}),\ \widetilde{e}_a^*(\widetilde{z},\widetilde{k}),\ \widetilde{w}_a^*(\widetilde{z},\widetilde{k})$: at $\widetilde{z} = \widetilde{z}_1$ (at the lower boundary of layer N_1 adjacent from above to the horizontal plane in which the force is applied), component $\widetilde{\chi}_a^*$ equals -1, whereas the rest of the components vanish.

Function $\widetilde{p}$ entering the complete solution is determined by using the numerical solution of differential equation (1.29). For an isotropic half-space ($\widetilde{G}_{r\vartheta} = \widetilde{G}$), in the case of relationship (5.1), this equation becomes (at $z_r = z_0$)

$$\frac{d^2 \widetilde{p}}{d\widetilde{z}^2} + \frac{\alpha}{\widetilde{z}}\frac{d\widetilde{p}}{d\widetilde{z}} + \left(\frac{\widetilde{\rho}\beta^2\theta^2}{\widetilde{z}^{\alpha}} - \widetilde{k}^2\right)\widetilde{p} = 0 . \tag{5.70}$$

As in the previous subsection, in the lower homogeneous half-space, function $\widetilde{p}_1$ is taken in the form (5.36) with the corresponding determination of parameter θ. This yields initial data for numerical integration over the upper heterogeneous part of the half-space ($1 + \widetilde{h}_0 - \widetilde{h} < \widetilde{z} < 1 + \widetilde{h}_0 + \widetilde{H}$). Function $\widetilde{p}_a^*$ is determined by numerical integration over the domain $1 + \widetilde{h}_0 - \widetilde{h} < \widetilde{z} < 1 + \widetilde{h}_0$ starting with initial values at $\widetilde{z} = 1 + \widetilde{h}_0$,

$$\widetilde{p}_a^* = 0,\quad \frac{d\widetilde{p}_a^*}{d\widetilde{z}} = -1 . \tag{5.71}$$

Employing Godunov's method [42] may still be relevant for improving the numerical solution. In the present case, after numerical integration in each "layer", solution $\widetilde{p}_1$ at the upper boundary of a "layer" is normalized, whereas projection of solution $\widetilde{p}_a^*$ onto the normalized solution $\widetilde{p}_1$ is subtracted from solution $\widetilde{p}_a^*$. This procedure gives the following analog to transformations (5.58):

$$\begin{bmatrix} \widetilde{p}_1^0 \\ \dfrac{d\,\widetilde{p}_1^0}{d\,\widetilde{z}} \end{bmatrix} = \frac{1}{\omega_{11}} \begin{bmatrix} \widetilde{p}_1 \\ \dfrac{d\,\widetilde{p}_1}{d\,\widetilde{z}} \end{bmatrix}, \tag{5.72a}$$

$$\begin{bmatrix} \widetilde{p}_a^0 \\ \dfrac{d\,\widetilde{p}_a^0}{d\,\widetilde{z}} \end{bmatrix} = \begin{bmatrix} \widetilde{p}_a^* \\ \dfrac{d\,\widetilde{p}_a^*}{d\,\widetilde{z}} \end{bmatrix} - \omega_{a1} \begin{bmatrix} \widetilde{p}_1^0 \\ \dfrac{d\,\widetilde{p}_1^0}{d\,\widetilde{z}} \end{bmatrix}. \tag{5.72b}$$

Here, the meaning of notations ω_{11}, ω_{a1} is the same as in formulas (5.58); the vector solution has two components: the function and its derivative. The complete solution for a "layer" may be written in a form similar to (1.165c) by employing transformed solutions and coefficient C_j^*, which varies in going from one "layer" to another,

$$\widetilde{p}(\widetilde{z},\widetilde{k}) = C_j^*(\widetilde{k})\widetilde{p}_1^0(\widetilde{z},\widetilde{k}) + \widetilde{p}_a^0(\widetilde{z},\widetilde{k}). \tag{5.73}$$

The solution for the upper boundary of the j-th "layer" should be determined by using the solutions transformed at this boundary; coefficient C_j^* corresponds to the same boundary. Comparing representations (5.73), written for two adjacent "layers", yields the following recurrent formula for coefficients C_j^*:

$$C_{j+1}^* = \frac{C_j^* - \omega_{a1}}{\omega_{11}}. \tag{5.74}$$

Having found coefficient C_1^* by using formula (5.63) and transformed solutions in the upper point, we then calculate the coefficients for the underlying "layers" by relationship (5.74). The solution is determined from equation (5.73).

5.5 Static solutions

In order to construct static solutions by using numerical methods, one can employ the approaches developed in previous sections. When applying the Runge–Kutta method, the technique and formulas (in which we take $\widetilde{\theta} = 0$) may be used in their original form, if the lower boundary of the heterogeneous part of the half-space is fixed for all $\widetilde{k}$. Thus, one applies conditions (5.22) and the analogous condition for function $\tilde{p}_1$ as initial conditions at the lower boundary,

$$\begin{bmatrix} \tilde{p}_1 \\ \dfrac{d\tilde{p}_1}{d\tilde{z}} \end{bmatrix} = \begin{bmatrix} 0 \\ 1 \end{bmatrix} . \tag{5.75}$$

In the case when the half-space is approximated by a set of homogeneous layers, one should introduce more significant corrections, primarily, since the fundamental solutions (5.7) are not valid at $\tilde{\theta} = 0$. Indeed, in this case, the first two solutions coincide, whereas the third and the fourth solutions are equal in magnitude but opposite in sign. To obtain the static solutions, the system of equations (5.3)–(5.6) should be solved from the beginning with $\tilde{\theta} = 0$, which results in

$$\begin{bmatrix} \tilde{\chi}_I \\ \tilde{e}_I \\ \tilde{q}_I \\ \tilde{w}_I \end{bmatrix} = \begin{bmatrix} 0 \\ 0 \\ \tilde{k} \\ \tilde{k} \end{bmatrix} \exp(-\tilde{k}\tilde{z}) , \tag{5.76a}$$

$$\begin{bmatrix} \tilde{\chi}_{II} \\ \tilde{e}_{II} \\ \tilde{q}_{II} \\ \tilde{w}_{II} \end{bmatrix} = \begin{bmatrix} -4\tilde{k} \\ 4\tilde{k} \\ 1+\tau^2 - 2\tilde{k}\tilde{z}(1-\tau^2) \\ -1-\tau^2 - 2\tilde{k}\tilde{z}(1-\tau^2) \end{bmatrix} \exp(-\tilde{k}\tilde{z}) , \tag{5.76b}$$

$$\begin{bmatrix} \tilde{\chi}_{III} \\ \tilde{e}_{III} \\ \tilde{q}_{III} \\ \tilde{w}_{III} \end{bmatrix} = \begin{bmatrix} 0 \\ 0 \\ -\tilde{k} \\ \tilde{k} \end{bmatrix} \exp(\tilde{k}\tilde{z}) , \tag{5.76c}$$

$$\begin{bmatrix} \tilde{\chi}_{IV} \\ \tilde{e}_{IV} \\ \tilde{q}_{IV} \\ \tilde{w}_{IV} \end{bmatrix} = \begin{bmatrix} 4\tilde{k} \\ 4\tilde{k} \\ 1+\tau^2 + 2\tilde{k}\tilde{z}(1-\tau^2) \\ 1+\tau^2 - 2\tilde{k}\tilde{z}(1-\tau^2) \end{bmatrix} \exp(\tilde{k}\tilde{z}) . \tag{5.76d}$$

Here, the first two solutions are taken from the corresponding solutions (5.7) at $\theta = 0$; the last two solutions are obtained from system (5.3)–5.6) at $\theta = 0$, when the first two equations are independent of the last two. Additional corrections are introduced into formulas (5.13) for coefficients B_i. The sum of the products of these coefficients by solutions (5.76) at $\tilde{z} = 0$ (the origin of coordinates is placed at the lower boundary of the considered layer) is set equal to the vector with components $\tilde{\chi}, \tilde{e}, \tilde{q}, \tilde{w}$. The system of equations obtained yields the static equivalent of relationships (5.13):

$$B_1 = \frac{4\tilde{k}(\tilde{q}+\tilde{w})-(1+\tau^2)(\tilde{\chi}+\tilde{e})}{8\tilde{k}^2}, \tag{5.77a}$$

$$B_2 = \frac{\tilde{e}-\tilde{\chi}}{8\tilde{k}}, \tag{5.77b}$$

$$B_3 = -\frac{4\tilde{k}(\tilde{q}-\tilde{w})+(1+\tau^2)(\tilde{\chi}-\tilde{e})}{8\tilde{k}^2}, \tag{5.77c}$$

$$B_4 = \frac{\tilde{e}+\tilde{\chi}}{8\tilde{k}}. \tag{5.77d}$$

The rest of the relationships, obtained in the previous sections and related to the approximation of the half-space by a set of homogeneous layers (including relationships containing function $\tilde{p}$), may be applied directly for the static case with the substitution $\theta = 0$.

Consider the results of calculations corresponding to the power law (5.1) for the case of an isotropic half-space at $\nu = 1/3$. In Fig. 5.24, the normalized displacements of the half-space surface, caused by a concentrated force applied to this surface, are shown for a number of values of parameter α in relationship (5.1). The value ξ introduced in (3.61) is used as an argument. Analysis of the results obtained at high values of $\tilde{r}$ indicates that the following asymptotic representations of the normalized static displacements hold at $\tilde{r} \to \infty$:

$$S_{vv} \approx \frac{C_{vv}(\nu,\alpha)}{\tilde{r}^{\alpha}}, \tag{5.78a}$$

$$S_{vh} \approx \frac{C_{vh}(\nu,\alpha)}{\tilde{r}^{\alpha}}, \tag{5.78b}$$

$$S_{hr} \approx \frac{C_{hr}(\nu,\alpha)}{\tilde{r}^{\alpha}}, \tag{5.78c}$$

$$S_{h\vartheta} \approx \frac{C_{h\vartheta}(\nu,\alpha)}{\tilde{r}^{\alpha}}. \tag{5.78d}$$

Equations (3.62b), (3.65) obtained for the case with $\alpha = 1$ (linearly heterogeneous half-space) may be considered as a particular case of these relationships. In Fig. 5.25, the coefficients entering the numerators in the right-hand sides of relationships (5.78) at $\nu = 1/3$ are represented as functions of parameter α (the dashed curves for coefficients $C_{vv}(\nu,\alpha)$ and $C_{vh}(\nu,\alpha)$ correspond to the exact values of these coefficients, which are discussed below). Note that at $\nu = 1/3$,

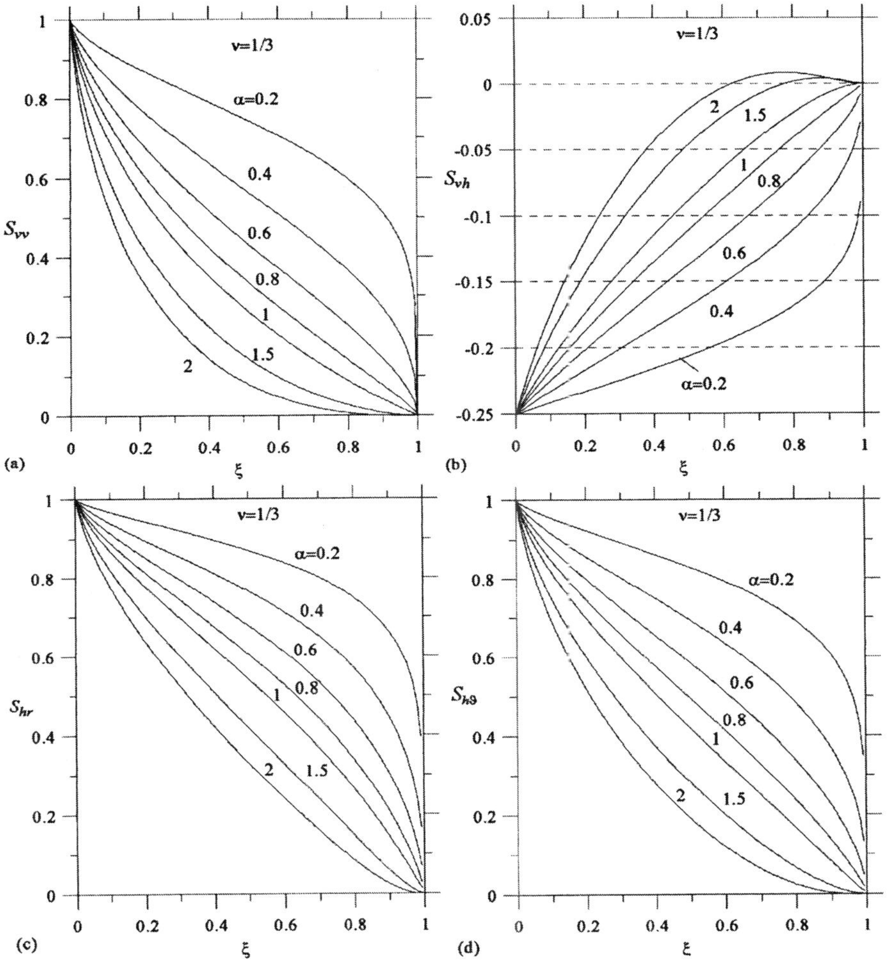

Fig. 5.24a,b,c,d. Static normalized displacements of half-space surface caused by concentrated force applied to the surface

$\alpha = 1$, according to the numerical method employed here, the values of these coefficients equal 0.3745, –0.0048, 1.49, 0.749 respectively for C_{vv}, C_{vh}, C_{hr}, $C_{h\vartheta}$, whereas the corresponding exact solutions for the linearly heterogeneous half-space yield 0.375, 0, 1.5, 0.75 (as shown in section 3.5.1, the first term in the asymptotic representation for S_{vh} at $\nu = 1/3$ has the form $-0.75/\tilde{r}^2$). When parameter α tends to zero, we have the case of an isotropic half-space: the coefficients C_{vv}, C_{hr}, $C_{h\vartheta}$ tend to unity whereas coefficient C_{vh} tends to $-0.5(1-2\nu)/(1-\nu)$, according to the employed normalization of displacements

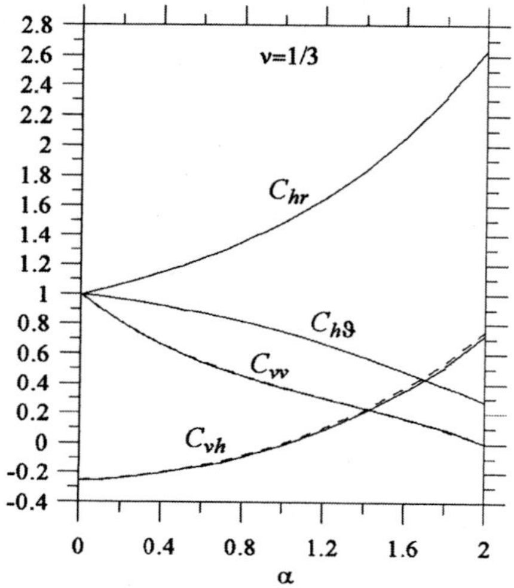

Fig. 5.25. Coefficients entering relationships (5.78) determining asymptotic behavior of displacements in the case of large distances between half-space surface points and force application point

u_r in relationship (5.45a). By using relationships (5.78), one can study the half-space properties in the limiting case, when the surface shear modulus G_0 tends to zero. Let the following quantity be kept constant:

$$m_\alpha = \frac{G_0}{z_0^\alpha}. \tag{5.79}$$

According to relationship (5.1), the shear modulus may be written as

$$G(z) = m_\alpha z^\alpha . \tag{5.80}$$

As seen from the last equation, the dimensions of coefficient m_α are force/(length)$^{2+\alpha}$. In formulas for the displacements of the half-space points, we substitute $G_0 = m_\alpha z_0^\alpha$ and allow z_0 to approach zero, thus moving the origin of coordinates toward the half-space surface. For the case of the uniform loading of a circular area of radius R on the surface of a linearly heterogeneous half-space, the corresponding analysis is performed in section 3.5.1. This analysis indicated that at $\nu < 0.5$, displacements of the points of application of the load increase unboundedly at constant m_α, as z_0 approaches zero. Now, we consider values of α that differ from unity. Taking into account the general expressions of the form (2.1), (2.53)–(2.55) and relationships (5.78) for the normalized displacements, we rewrite expressions for the displacements caused by a unit concentrated force at $z_0 = 0$,

$$w_{vv} = \frac{(1-\nu)C_{vv}(\nu,\alpha)}{2\pi m_\alpha r^{\alpha+1}}, \tag{5.81a}$$

$$w_{vh} = \frac{(1-\nu)C_{vh}(\nu,\alpha)}{2\pi m_\alpha r^{\alpha+1}}, \tag{5.81b}$$

$$w_{hr} = \frac{C_{hr}(\nu,\alpha)}{2\pi m_\alpha r^{\alpha+1}}, \tag{5.81c}$$

$$w_{h\vartheta} = -\hat{u}_\vartheta = \frac{(1-\nu)C_{h\vartheta}(\nu,\alpha)}{2\pi m_\alpha r^{\alpha+1}}. \tag{5.81d}$$

The solution of the static problem dealing with the action of a vertical load applied to the half-space surface, whereas the shear modulus varies according to (5.80), is given in [99, 100]. For this particular case, when the simplifying relationship $\alpha\nu = 1-2\nu$ is adopted, the solution is as reported in [51, 55]. Indeed, the form of the relationships for Green's functions w_{vv} and w_{vh} given in [99, 100] corresponds to relationships (5.81a), (5.81b). Employing the results obtained in these works, one can find the exact values of coefficients $C_{vv}(\nu,\alpha)$ and $C_{vh}(\nu,\alpha)$:

$$C_{vv}(\nu,\alpha) = \frac{\pi C q \sin(\pi q/2)\Gamma(0.5+0.5\alpha)}{2(1+\alpha)\sqrt{\pi}\Gamma(1+0.5\alpha)}, \tag{5.82}$$

$$C_{vh}(\nu,\alpha) = \frac{\pi C \cos(\pi q/2)\Gamma(1+0.5\alpha)}{\alpha\sqrt{\pi}\Gamma(0.5+0.5\alpha)}, \tag{5.83}$$

where

$$q = \left[(1+\alpha)\left(1-\frac{\alpha\nu}{1-\nu}\right)\right]^{1/2}, \tag{5.84}$$

$$C = \frac{2^{1+\alpha}\Gamma[1.5+0.5(\alpha+q)]\Gamma[1.5+0.5(\alpha-q)]}{\pi\Gamma(2+\alpha)}. \tag{5.85}$$

The dashed curves shown in Fig. 5.25 correspond to relationships (5.82) and (5.83). A discrepancy between the exact and approximate results, noticeable for coefficient C_{vh}, occurs since the numerical method used for the calculation of the coefficients requires use of extremely high values of parameter $\tilde{r}$ (see relationships (5.78)) whereas the normalized displacements are very small (especially for a high value of power index α). When calculating small values, it is difficult to provide a high relative precision. Note that the model of a foundation corresponding to relationships (5.81) has a physical meaning for $\alpha < 1$, when the singularities are integrable at the zero point; for these values of α, the accuracy is acceptable for practical applications.

Next, we consider the formulas obtained in section 1.7 for the case of loading rectangular and circular areas with various loads. According to these formulas, in the case of the surface load, at $\alpha \geq 1$, displacements become unbounded within the loaded area. When $0 \leq \alpha < 1$, Green's functions (5.81) represent a simple static model of a deformable foundation, having properties that are intermediate between those of the elastic homogeneous half-space and those of Winkler's foundation ($\alpha \to 1$). Consider, for example, vertical displacements caused by constant vertical load p_0 applied to the circular area of radius R on the half-space surface. According to formulas (1.213b), (5.81a),

$$\delta_z(r) = \frac{p_0(1-\nu)RC_{\nu\nu}(\nu,\alpha)}{\pi G_R}\Delta_z(\eta), \tag{5.86a}$$

$$\Delta_z(\eta) = \int_{|\eta-1|}^{\eta+1} \varphi\lambda^{-\alpha}\, d\lambda + \begin{cases} \pi\dfrac{(1-\eta)^{1-\alpha}}{1-\alpha} & (r < R) \\ 0 & (r \geq R). \end{cases} \tag{5.86b}$$

In (5.86a), the shear modulus at depth R is introduced:

$$G_R = m_\alpha R^\alpha. \tag{5.87}$$

In the case of $\alpha = 0,\ G_R = G_0$, we obtain the result corresponding to the homogeneous half-space; as $\alpha \to 1$, displacements within the area in which the load is applied increase unboundedly; for $\eta > 1$, these displacements remain bounded. In the center of the circular area in which the load is applied, displacements are expressed by a simple formula:

$$\Delta_z(0) = \frac{\pi}{1-\alpha}. \tag{5.88}$$

In Fig. 5.26, the behavior of quantity $\Delta_z(\eta)/\Delta_z(0)$ is shown for several values of α. The properties of the foundation are determined by parameters α and m_α, which enter the expression for shear modulus (5.80); taking $m_\alpha = A/(1-\alpha)$ (where A is a constant) and allowing α to approach unity, we obtain the relationship for Winkler's foundation (by using formulas (5.86), (5.88)):

$$\delta_z(r) = \begin{cases} \dfrac{p_0}{k_z} & (r < R) \\ 0 & (r \geq R), \end{cases} \tag{5.89}$$

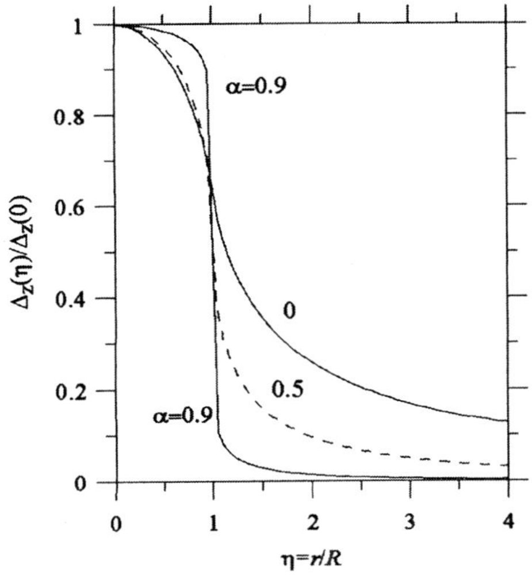

Fig. 5.26. Influence of parameter α on static displacements of half-space surface under action of vertical load distributed uniformly over circular area

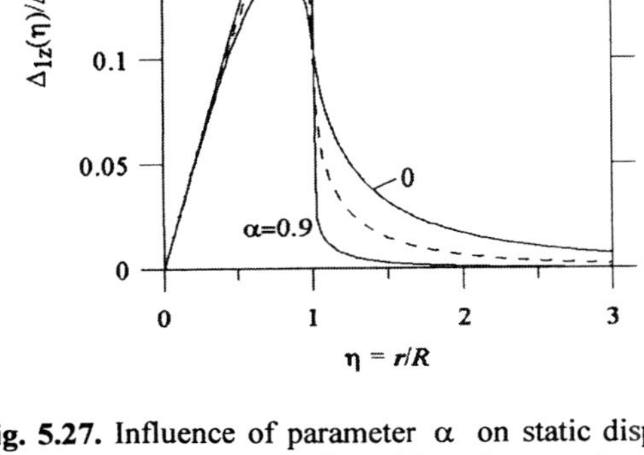

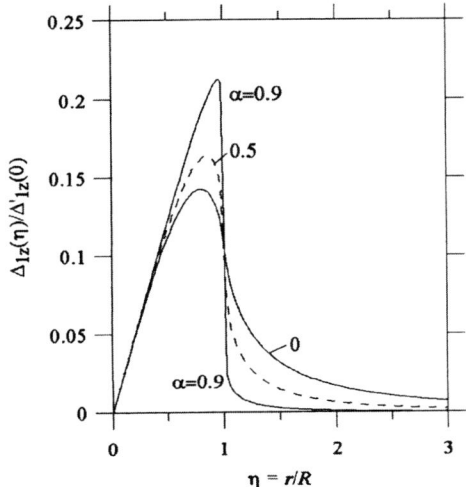

Fig. 5.27. Influence of parameter α on static displacements of half-space surface under action of vertical load distributed linearly over circular area

where

$$k_z = \frac{A}{(1-\nu)C_{\nu\nu}(\nu,1)} \; . \tag{5.90}$$

In the second example, we consider the loading studied in section 1.7.6 (Fig. 1.12). Vertical displacements are expressed by formula (1.227), which in the case of Green's function from relationship (5.81a) takes the form

$$\delta_{1z}(r) = \frac{p_0(1-\nu)R^2 C_{vv}(\nu,\alpha)}{\pi G_R}\Delta_{1z}(\eta), \tag{5.91a}$$

$$\Delta_{1z}(\eta) = \int_{|\eta-1|}^{\eta+1} \lambda^{-\alpha}[\eta\varphi - \lambda\sin(\varphi)]d\lambda + \begin{cases} \pi\eta\dfrac{(1-\eta)^{1-\alpha}}{1-\alpha} & (r < R) \\ 0 & (r \geq R). \end{cases} \tag{5.91b}$$

The derivative of quantity $\Delta_{1z}(\eta)$ with respect to η at $\eta = 0$, which determines the slope (to the X-axis) of the deformed half-space surface at the point of origin of the coordinates, is

$$\Delta'_{1z}(0) = 2\pi\frac{1+\alpha}{1-\alpha}. \tag{5.92}$$

Note that the contribution of the first term on the right side of equation (5.91b) to this result equals $-\pi/2$. In Fig. 5.27, the graphs represent the quantity $\Delta_{1z}(\eta)/\Delta'_{1z}(0)$ for a number of values of α. Again, we see how displacements become similar to the applied load as parameter α approaches unity, i.e. the foundation becomes Winkler's one.

Analogous results occur for other types of load applied to the surface of the heterogeneous half-space. In the case of a horizontal load distributed uniformly over a circular area, employing formulas (1.216), (1.217) and functions (5.81c), (5.81d) gives, analogously to (2.191), (2.192),

$$\delta_{1x}(r,R) = \frac{q_0 R}{2\pi G_R}\Delta_{1x}(\eta), \tag{5.93a}$$

$$\begin{aligned}\Delta_{1x}(\eta) = &\int_{|\eta-1|}^{\eta+1}[C_{hr}(\nu,\alpha)(\varphi + 0.5\sin 2\varphi) + C_{h\vartheta}(\nu,\alpha)(1-\nu)(\varphi - 0.5\sin 2\varphi)]\lambda^{-\alpha}\,d\lambda \\ &+ \begin{cases} \pi\dfrac{(1-\eta)^{(1-\alpha)}}{1-\alpha}[C_{hr}(\nu,\alpha) + (1-\nu)C_{h\vartheta}(\nu,\alpha)] & (r < R) \\ 0 & (r \geq R), \end{cases}\end{aligned} \tag{5.93b}$$

$$\delta_{2x}(r,R) = \frac{q_0 R}{2\pi G_R}\Delta_{2x}(\eta), \tag{5.94a}$$

$$\Delta_{2x}(\eta) = \int_{|\eta-1|}^{\eta+1}[C_{hr}(\nu,\alpha)(\varphi - 0.5\sin 2\varphi) + C_{h\vartheta}(\nu,\alpha)(1-\nu)(\varphi + 0.5\sin 2\varphi)]\lambda^{-\alpha}\,d\lambda$$

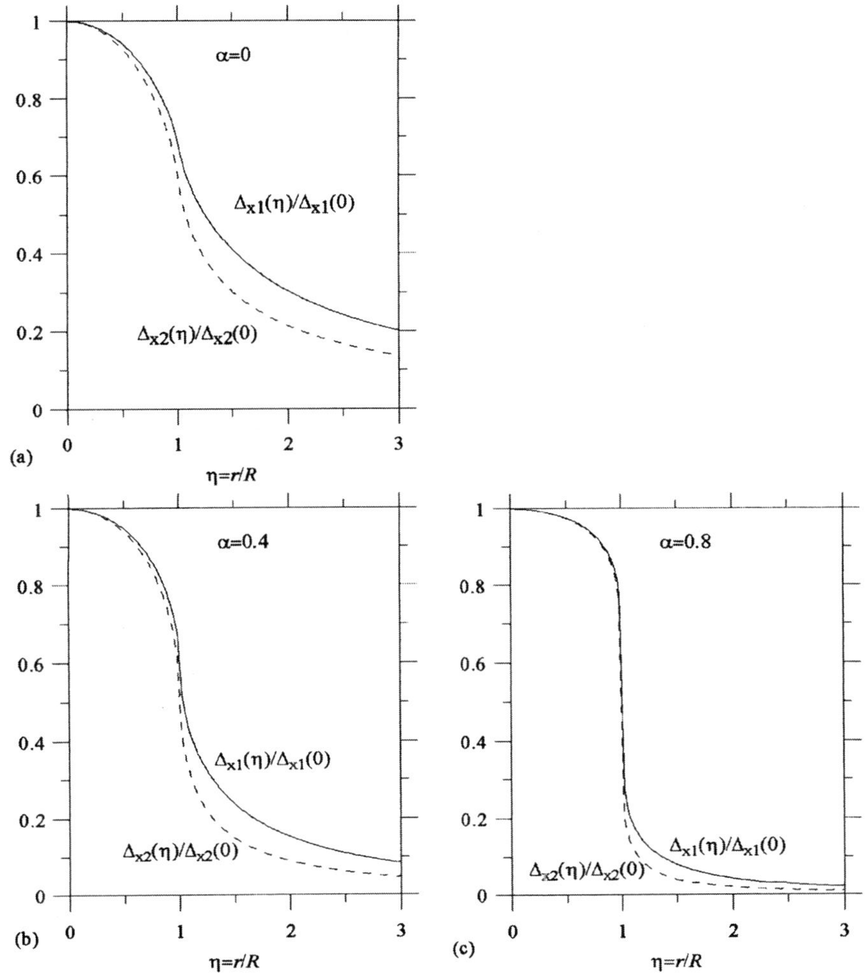

Fig. 5.28a,b,c. Static horizontal displacements of half-space surface under action of horizontal load distributed uniformly over circular area

$$+\begin{cases}\pi\dfrac{(1-\eta)^{(1-\alpha)}}{1-\alpha}[C_{hr}(\nu,\alpha)+C_{h\vartheta}(\nu,\alpha)(1-\nu)] & (r<R)\\ 0 & (r\geq R).\end{cases} \tag{5.94b}$$

In the center of the circle, normalized displacements take the form

$$\Delta_{1x}(0)=\Delta_{2x}(0)=\frac{\pi}{1-\alpha}[C_{hr}(\nu,\alpha)+C_{h\vartheta}(\nu,\alpha)(1-\nu)]\ . \tag{5.95}$$

Figure 5.28 represents the behavior of quantities Δ_{1x} and Δ_{2x} referred to the initial value (5.95) at $\nu = 1/3$ for several values of α.

For the load shown in Fig. 1.13, the tangential displacements δ_{1y} are considered in section 1.7.7. Taking into account expressions (5.81c), (5.81d) for Green's functions, we express this displacement in the form analogous to (2.235):

$$\delta_{1y}(r,R) = \frac{q_0 R^2}{2G_R \pi} \Delta_{1y}(\eta), \tag{5.96a}$$

$$\Delta_{1y}(\eta) = \int\limits_{|\eta-1|}^{\eta+1} \left\{ C_{h\vartheta}(1-\nu)\left[\eta\varphi + \frac{\eta\sin(2\varphi)}{2} - 2\lambda\sin(\varphi)\right] + C_{hr}\eta\left[\varphi - \frac{\sin(2\varphi)}{2}\right]\right\} \lambda^{-\alpha}\,d\lambda$$

$$+\begin{cases} \pi\eta[C_{h\vartheta}(1-\nu') + C_{hr}]\dfrac{(1-\eta)^{1-\alpha}}{1-\alpha} & (\eta \le 1) \\ 0 & (\eta > 1). \end{cases} \tag{5.96b}$$

Next, we present expressions corresponding to the self-balanced tangential load applied to the circle of radius *R*. In the following, these expressions will be employed to determine horizontal stiffness of the disk which is in contact with the half-space. Using relationships (1.231), (1.232), (1.235), we obtain, analogously to (2.194), (2.195),

$$\delta_{1x}(r,R) = \frac{q_0 R^3}{2G_R \pi} \Delta_{1x}^{(2)}(\eta), \tag{5.97a}$$

$$\Delta_{1x}^{(2)}(\eta) = \int\limits_{|\eta-1|}^{\eta+1} \Big\{ C_{hr}\left[(\lambda^2 + \eta^2)(\varphi + 0.5\sin(2\varphi)) - 4\lambda\eta\sin(\varphi)\right]$$

$$+ C_{h\vartheta}(1-\nu)(\eta^2 - \lambda^2)(\varphi - 0.5\sin(2\varphi))\Big\} \lambda^{-\alpha}\,d\lambda$$

$$+\begin{cases} \pi(1-\eta)^{1-\alpha}\left\{ C_{hr}\left[\dfrac{(1-\eta)^2}{3-\alpha} + \dfrac{\eta^2}{1-\alpha}\right] + C_{h\vartheta}(1-\nu)\left[\dfrac{\eta^2}{1-\alpha} - \dfrac{(1-\eta)^2}{3-\alpha}\right]\right\} & (\eta \le 1) \\ 0 & (\eta > 1), \end{cases} \tag{5.97b}$$

$$\delta_{2x}(r,R) = \frac{q_0 R^3}{2G_R \pi} \Delta_{2x}^{(2)}(\eta), \tag{5.98a}$$

$$\Delta_{2x}^{(2)}(\eta) = -\int_{|\eta-1|}^{\eta+1}\Big\{C_{h\vartheta}(1-\nu)[(\lambda^2+\eta^2)(\varphi+0.5\sin(2\varphi))-4\lambda\eta\sin(\varphi)]$$
$$+C_{hr}(\eta^2-\lambda^2)[\varphi-0.5\sin(2\varphi)]\Big\}\lambda^{-\alpha}\,\mathrm{d}\lambda$$
$$-\begin{cases}\pi(1-\eta)^{1-\alpha}\left\{C_{h\vartheta}(1-\nu)\left[\dfrac{(1-\eta)^2}{3-\alpha}+\dfrac{\eta^2}{1-\alpha}\right]+C_{hr}\left[\dfrac{\eta^2}{1-\alpha}-\dfrac{(1-\eta)^2}{3-\alpha}\right]\right\} & (\eta\le 1)\\ 0 & (\eta>1).\end{cases}$$
(5.98b)

Plane problems

Next, we consider some plane problems for the model of a deformable foundation determined by relationships (5.81). For the loads distributed along an infinite line on the half-space surface (Fig. 1.4), corresponding results are obtained by integrating Green's functions (5.81). For the vertical load shown in Fig. 1.4a, we obtain for a point on the half-space surface:

$$u_z(x) = 2p_0\int_0^{\infty} w_{vv}(\sqrt{x^2+y^2})\mathrm{d}y = \frac{p_0(1-\nu)C_{vv}(\nu,\alpha)}{\pi m_\alpha}\int_0^{\infty}(x^2+y^2)^{-0.5(1+\alpha)}\mathrm{d}y$$
$$=\frac{p_0(1-\nu)C_{vv}(\nu,\alpha)}{2\sqrt{\pi}m_\alpha x^\alpha}\frac{\Gamma(0.5\alpha)}{\Gamma(0.5+0.5\alpha)}, \qquad (5.99a)$$

$$u_x(x) = 2p_0\int_0^{\infty} w_{vh}(\sqrt{x^2+y^2})\frac{x}{\sqrt{x^2+y^2}}\mathrm{d}y = \frac{p_0(1-\nu)C_{vh}(\nu,\alpha)x}{\pi m_\alpha}$$
$$\times\int_0^{\infty}(x^2+y^2)^{-1-0.5\alpha}\mathrm{d}y = \frac{p_0(1-\nu)C_{vh}(\nu,\alpha)}{\alpha\sqrt{\pi}m_\alpha x^\alpha}\frac{\Gamma(0.5+0.5\alpha)}{\Gamma(0.5\alpha)}. \qquad (5.99b)$$

For the horizontal load shown in Fig. 1.4b,

$$u_y(x) = 2q_0\int_0^{\infty}[w_{h\vartheta}(\sqrt{x^2+y^2})\cos^2(\gamma)+w_{hr}(\sqrt{x^2+y^2})\sin^2(\gamma)]\mathrm{d}y$$
$$=\frac{q_0}{\pi m_\alpha}\int_0^{\infty}[(1-\nu)C_{h\vartheta}x^2(x^2+y^2)^{-1.5-0.5\alpha}-C_{hr}y^2(x^2+y^2)^{-1.5-0.5\alpha}]\mathrm{d}y$$
$$=\frac{q_0}{2\sqrt{\pi}m_\alpha}\frac{\Gamma(0.5\alpha)}{\Gamma(1.5+0.5\alpha)x^\alpha}[\alpha(1-\nu)C_{h\vartheta}+0.5C_{hr}]. \qquad (5.100)$$

For the horizontal load shown in Fig. 1.4c,

$$u_x(x) = 2q_0 \int_0^{\infty} [w_{h\vartheta}(\sqrt{x^2+y^2})\sin^2(\gamma) + w_{hr}(\sqrt{x^2+y^2})\cos^2(\gamma)]\mathrm{d}y$$

$$= \frac{q_0}{\pi m_\alpha} \int_0^{\infty} [(1-\nu)C_{h\vartheta}y^2(x^2+y^2)^{-1.5-0.5\alpha} + C_{hr}x^2(x^2+y^2)^{-1.5-0.5\alpha}]\mathrm{d}y$$

$$= \frac{q_0}{2\sqrt{\pi}m_\alpha} \frac{\Gamma(0.5\alpha)}{\Gamma(1.5+0.5\alpha)x^\alpha}[0.5(1-\nu)C_{h\vartheta} + \alpha C_{hr}]. \tag{5.101}$$

Displacements u_z caused by the considered load may be found by using the principle of reciprocity if displacements u_x by (5.99b) are known; for equal p_0 and q_0 we have $u_z = -u_x$.

The Green's functions for the plane problem, represented for the surface displacements by relationships (5.99)–(5.101) at unit values of loads p_0 and q_0, enable us to calculate displacements for various types of loads, which are constant in the direction of variation of the y-coordinate. Examples of the corresponding formulas are given in section 1.7.9, where the employed Green's functions $w_{vz}, w_{vx}, w_{hx}, w_{hz} = -w_{vx}, w_{hy}$ are obviously related to displacements (5.99)–(5.101). Note that as $\alpha \to 0$, the displacements increase unboundedly; while at $\alpha \geq 1$, loss of integrability of Green's functions occurs as $x \to 0$. Feasible results correspond to the case $0 < \alpha < 1$; this is the same interval as for the spatial case, excluding the point $\alpha = 0$. As an example, we consider the vertical displacements of the half-space surface caused by a vertical load p applied to an infinite strip of width $2R$ on the half-space surface. Employing formulas (1.237) and (5.99a) (with $p_0 = 1$) determining $w_{vz}(x,0)$ gives

$$\delta_z(r) = \frac{p\ (1-\nu)RC_{vv}(\nu,\alpha)}{2\sqrt{\pi}G_R} \frac{\Gamma(0.5\alpha)}{\Gamma(0.5+0.5\alpha)}\Delta_z(\eta), \tag{5.102a}$$

$$\Delta_z(\eta) = \frac{(\eta+1)^{1-\alpha} - |\eta-1|^{1-\alpha}}{1-\alpha} + \begin{cases} 2\dfrac{(1-\eta)^{1-\alpha}}{1-\alpha} & (0 < r < R) \\ 0 & (r \geq R). \end{cases} \tag{5.102b}$$

The asymptotic representation of quantity $\Delta_z(\eta)$ at $\eta \to \infty$ takes the form

$$\Delta_z(\eta) \approx \frac{2}{\eta^\alpha} \tag{5.103}$$

corresponding to relationship (5.99a) for the strip contracted to the line. At the central point, value Δ_z is

$$\Delta_z(0) = \frac{2}{1-\alpha}. \tag{5.104}$$

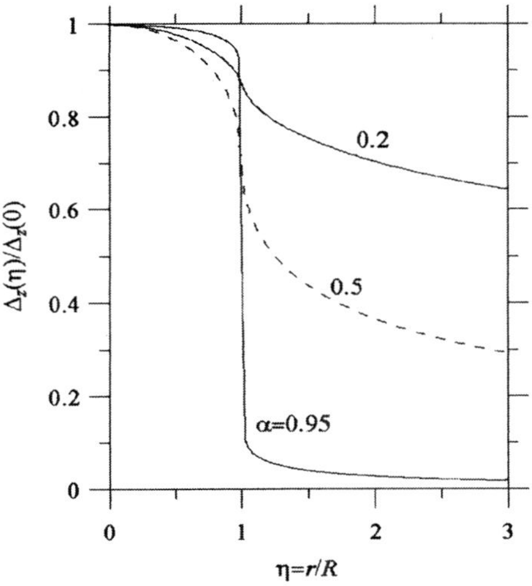

Fig. 5.29. Influence of parameter α on static displacements of half-space surface under action of vertical load distributed uniformly over infinite strip

In Fig. 5.29, quantity $\Delta_z(\eta)/\Delta_z(0)$ is shown for a number of values of parameter α. When α approaches unity, one can see that the behavior of the displacements corresponds to Winkler's foundation. In comparison with a 3-D case, the displacements approach zero in a slower fashion as the distance between the considered points and the area of application of the load increases, especially at small values of α.

5.6 Contact Problems for Circular Disk Resting on Half-Space with Stiffness Increasing with Depth by Power Law

5.6.1 Static Stiffnesses for Circular Disk

To solve the contact problems for a circular disk resting on a half-space surface, we use the technique given in section 2.7.1. In order to construct solutions for various problems, it is sufficient to have four Green's functions or the corresponding normalized displacements $S_{vv}, S_{vh}, S_{hr}, S_{h\vartheta}$ used in section 2.7.1, for expressing displacements caused by the loading of the circular area on the half-space surface by various loads. Besides normalized displacements, the formulas contain the shear modulus on the half-space surface and Poisson's ratio.

First, we study contact problems for the half-space model, which is a particular case of the power-law model with $z_0 = 0$. This model, considered in the previous section, is of interest at values of α that satisfy the condition $0 \le \alpha < 1$.

Vertical Stiffness in the Case $z_0 = 0$

For the case of relaxed contact conditions, i.e. neglecting tangential stresses in the area of contact between the disk and the half-space, the solution of the present problem has been constructed by using either dual integral equations [55, 56, 63, 76] or Fredholm's equations of the 1st kind [98]. Below, the solution of this problem is constructed by employing the method of ring-shaped elements, with the purpose to provide an additional verification for this method. It is further used in the cases for which no alternative solutions are reported in the literature. The system of equations for determining the pressures acting on the ring-shaped elements has the form (2.190), with the corresponding changes in expressions for the system coefficients and in the meaning of variables. These changes occur due to the differences between representations of the vertical displacements in (2.177) and in (5.86). Dimensionless pressures $\widetilde{p}_j$ determined previously by (2.182) now have the following form:

$$\widetilde{p}_j = \frac{p_j R(1-\nu) C_{\nu\nu}(\nu,\alpha)}{\pi G_R} . \tag{5.105}$$

Matrix elements K_{ij} are calculated by comparing expressions (2.177) and (5.86). Instead of relationship (2.184), we obtain

$$K_{ij} = \overline{R}_j^{1-\alpha} \Delta_z(\overline{r}_i / \overline{R}_j) - \overline{R}_{j-1}^{1-\alpha} \Delta_z(\overline{r}_i / \overline{R}_{j-1}) . \tag{5.106}$$

Having determined values $\widetilde{p}_j$ from system (2.190), we can find the vertical stiffness analogously to expression (2.188),

$$K_z = \pi \sum_{j=1}^{N} p_j (R_j^2 - R_{j-1}^2) = \frac{\pi^2 R G_R}{C_{\nu\nu}(\nu,\alpha)(1-\nu)} \sum_{j=1}^{N} \widetilde{p}_j (\overline{R}_j^2 - \overline{R}_{j-1}^2) . \tag{5.107}$$

When studying the stiffnesses of the disk resting on a heterogeneous half-space (section 3.5.2), we introduced the notion of stiffness of the disk resting on the "equivalent" homogeneous half-space having shear modulus G_μ which equals the shear modulus of the given heterogeneous half-space at depth μR. For the half-space model considered,

$$G_\mu = m_\alpha (\mu R)^\alpha = G_R \mu^\alpha . \tag{5.108}$$

As for the case of the linearly heterogeneous half-space, it is reasonable to take the value of coefficient μ as equal to unity for determination of the vertical

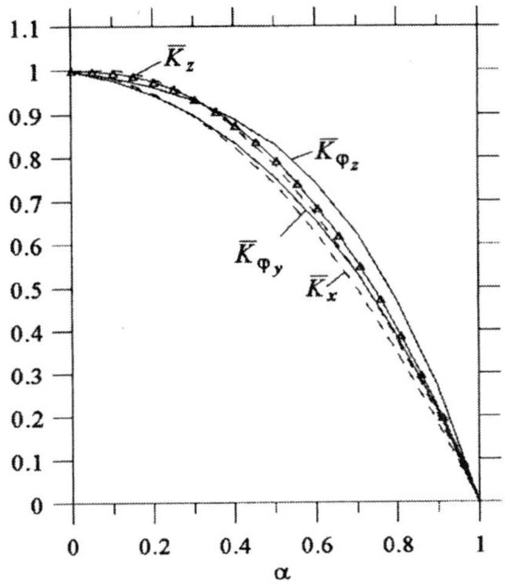

Fig. 5.30. Normalized static stiffnesses of disk on half-space with shear modulus increasing with depth by power law (5.80) starting with zero value on half-space surface; $\nu = 1/3$

stiffness, 0.35 for determination of the horizontal and rocking stiffness, and 0.1 for determination of the torsional stiffness of the disk resting on a half-space.

The normalized stiffness of the disk is calculated analogously to (3.82a), referring K_z to the corresponding stiffness k_z^{hom} for the "equivalent" homogeneous half-space,

$$\overline{K}_z = \frac{K_z}{k_z^{\text{hom}}}, \tag{5.109}$$

$$k_z^{\text{hom}} = \frac{4G_\mu R}{1-\nu} \ . \tag{5.110}$$

Taking into account (5.107) gives

$$\overline{K}_z = \frac{\pi^2}{4\mu^\alpha C_{\nu\nu}(\nu,\alpha)} \sum_{j=1}^{N} \widetilde{p}_j (\overline{R}_j^2 - \overline{R}_{j-1}^2) \ . \tag{5.111}$$

Note that quantities $\widetilde{p}_j$ do not depend on Poisson's ratio. In Fig. 5.30, the quantity $\overline{K}_z$ is presented as a function of parameter α for $\mu = 1$, $\nu = 1/3$. As parameter α approaches unity, the stiffness tends to zero, according to the Green's function properties considered in the previous section. Next, we compare the obtained solution with the known solution [98] constructed for an elliptic stamp. For the particular case of a circular stamp,

$$K_z = \frac{4RG_R}{1-\nu}\frac{\pi}{C_{\nu\nu}(\nu,\alpha)(1+\alpha)\Gamma(0.5+0.5\alpha)\Gamma(0.5-0.5\alpha)}. \tag{5.112}$$

Hence,

$$\overline{K}_z = \frac{\pi}{\mu^\alpha C_{\nu\nu}(\nu,\alpha)(1+\alpha)\Gamma(0.5+0.5\alpha)\Gamma(0.5-0.5\alpha)}. \tag{5.113}$$

Table 5.6 shows the values of quantity $\overline{K}_z$ calculated by the exact formula (5.113) with value $C_{\nu\nu}(\nu,\alpha)$ by (5.82) and values of $\overline{K}_z$ found by using formula (5.111), whereas coefficient $C_{\nu\nu}(\nu,\alpha)$ is taken from the corresponding numerical solution given in the previous section (($\mu=1$, $\nu=1/3$).

Table 5.6. Comparison between approximate (by (5.111)) and exact (by (5.113)) solutions

α	0	0.2	0.4	0.6	0.8	1
$\overline{K}_z$ by (5.111)	0.9994	0.9817	0.8735	0.6723	0.3783	0
$\overline{K}_z$ by (5.113)	1	0.9822	0.8734	0.6708	0.3776	0

The numerical error expressed in Table 5.6 is caused by two factors: the error that occurs when calculating coefficient $C_{\nu\nu}(\nu,\alpha)$, and the error that occurs as a result of the numerical solution of the contact problem.

Rocking Stiffness in the Case $z_0 = 0$

Here, we employ the technique for determination of the stiffness presented in section 2.7.1. Neglecting tangential stresses in the contact area leads to an equation analogous to equation (2.226),

$$\sum_{j=1}^{N} C_{ij}\tilde{p}_{0j} = \overline{r}_i \quad (i=1,\ldots,N), \tag{5.114}$$

where

$$\tilde{p}_{0j} = \frac{p_{0j}(1-\nu)RC_{\nu\nu}(\nu,\alpha)}{G_R\pi}, \tag{5.115}$$

$$C_{ij} = \overline{R}_j^{2-\alpha}\Delta_{1z}(\overline{r}_i/\overline{R}_j) - \overline{R}_{j-1}^{2-\alpha}\Delta_{1z}(\overline{r}_i/\overline{R}_{j-1}). \tag{5.116}$$

Expressions (5.115) and (5.116) correspond to expressions (2.223c) and (2.210). The rocking stiffness of the disk may be written in a form analogous to (2.220), (2.224),

$$K_{\varphi_y} = \frac{\pi}{4}\sum_{j=1}^{N} p_{0j}(R_j^4 - R_{j-1}^4) = \frac{\pi^2 G_R R^3}{4(1-\nu)C_{\nu\nu}(\nu,\alpha)}\sum_{j=1}^{N}\tilde{p}_{0j}(\overline{R}_j^4 - \overline{R}_{j-1}^4). \tag{5.117}$$

Analogously to (3.94), the normalized stiffness takes the form

$$\overline{K}_{\varphi_y} = \frac{K_{\varphi_y}}{k_{\varphi_y}^{\text{hom}}}, \tag{5.118}$$

where

$$k_{\varphi_y}^{\text{hom}} = \frac{8G_\mu R^3}{3(1-\nu)}. \tag{5.119}$$

Taking into account (5.117) and (5.108) yields

$$\overline{K}_{\varphi_y} = \frac{3\pi^2}{32\mu^\alpha C_{\nu\nu}(\nu,\alpha)} \sum_{j=1}^{N} \widetilde{p}_{0j}(\overline{R}_j^4 - \overline{R}_{j-1}^4) . \tag{5.120}$$

Quantity $\overline{K}_{\varphi_y}$ is shown in Fig. 5.30 at $\mu = 0.35, \nu = 1/3$. Let us compare the obtained results with the exact results reported in [98], according to which the rocking stiffness may be written as

$$K_{\varphi_y} = \frac{8R^3 G_R}{3(1-\nu)} \frac{3\pi}{2C_{\nu\nu}(\nu,\alpha)(1+\alpha)(3+\alpha)\Gamma(1.5+0.5\alpha)\Gamma(0.5-0.5\alpha)} . \tag{5.121}$$

Hence,

$$\overline{K}_{\varphi_y} = \frac{3\pi}{2\mu^\alpha C_{\nu\nu}(\nu,\alpha)(1+\alpha)(3+\alpha)\Gamma(1.5+0.5\alpha)\Gamma(0.5-0.5\alpha)} . \tag{5.122}$$

A comparison between values of $\overline{K}_{\varphi_y}$ by (5.120) (with $C_{\nu\nu}(\nu,\alpha)$ found by using the corresponding numerical solution) and by (5.122) (with $C_{\nu\nu}(\nu,\alpha)$ calculated by the exact formula (5.82)) is shown in Table 5.7 ($\mu = 0.35, \nu = 1/3$).

Table 5.7. Comparison between approximate (by (5.120)) and exact (by (5.122)) solutions

α	0	0.2	0.4	0.6	0.8	1
$\overline{K}_{\varphi_y}$ by (5.120)	0.9984	0.9453	0.8372	0.6570	0.3841	0
$\overline{K}_{\varphi_y}$ by (5.122)	1	0.9466	0.8376	0.6559	0.3836	0

Horizontal Stiffness in the Case $z_0 = 0$

Under relaxed contact conditions, when tangential stresses are only considered in the contact area, we apply equations of the form (2.212) that contain uniformly distributed $\widetilde{q}_j$ and self-balanced $\widetilde{q}_{0j}$ loads,

$$\sum_{j=1}^{N} C_{ij}^{(1)} \widetilde{q}_j + \sum_{j=1}^{N} C_{ij}^{(2)} \widetilde{q}_{0j} = 1,$$

$$\sum_{j=1}^{N} C_{ij}^{(4)} \widetilde{q}_j + \sum_{j=1}^{N} C_{ij}^{(5)} \widetilde{q}_{0j} = 1, \tag{5.123}$$

where the normalized loads and coefficients of the system of equations differ from those entering expressions (2.201a), (2.201b), (2.202), (2.203), (2.205), (2.206) according to the form of expressions for displacements (5.93), (5.94), (5.97), (5.98):

$$\widetilde{q}_j = \frac{q_j R}{2G_R \pi}, \tag{5.124}$$

$$\widetilde{q}_{0j} = \frac{q_{0j} R^3}{2G_R \pi}, \tag{5.125}$$

$$C_{ij}^{(1)} = \overline{R}_j^{1-\alpha} \Delta_{1x}(\overline{r}_i / \overline{R}_j) - \overline{R}_{j-1}^{1-\alpha} \Delta_{1x}(\overline{r}_i / \overline{R}_{j-1}), \tag{5.126}$$

$$C_{ij}^{(2)} = \overline{R}_j^{3-\alpha} \Delta_{1x}^{(2)}(\overline{r}_i / \overline{R}_j) - \overline{R}_{j-1}^{3-\alpha} \Delta_{1x}^{(2)}(\overline{r}_i / \overline{R}_{j-1}), \tag{5.127}$$

$$C_{ij}^{(4)} = \overline{R}_j^{1-\alpha} \Delta_{2x}(\overline{r}_i / \overline{R}_j) - \overline{R}_{j-1}^{1-\alpha} \Delta_{2x}(\overline{r}_i / \overline{R}_{j-1}), \tag{5.128}$$

$$C_{ij}^{(5)} = \overline{R}_j^{3-\alpha} \Delta_{2x}^{(2)}(\overline{r}_i / \overline{R}_j) - \overline{R}_{j-1}^{3-\alpha} \Delta_{2x}^{(2)}(\overline{r}_i / \overline{R}_{j-1}). \tag{5.129}$$

In order to find the horizontal stiffness for the disk, stresses q_j are summed analogously to relationship (2.188),

$$K_x = \pi \sum_{j=1}^{N} q_j (R_j^2 - R_{j-1}^2) = 2\pi^2 G_R R \sum_{j=1}^{N} \widetilde{q}_j (\overline{R}_j^2 - \overline{R}_{j-1}^2) . \tag{5.130}$$

The corresponding normalized stiffness $\overline{K}_x$ is determined as the ratio of quantity K_x to the stiffness of the "equivalent" homogeneous half-space calculated by (3.87b) and (5.108):

$$\overline{K}_x = \frac{\pi^2 (2-\nu)}{4\mu^\alpha} \sum_{j=1}^{N} \widetilde{q}_j (\overline{R}_j^2 - \overline{R}_{j-1}^2) . \tag{5.131}$$

In Fig. 5.30, the normalized stiffness $\overline{K}_x$ is shown at $\mu = 0.35, \nu = 1/3$.

Torsional Stiffness in the Case $z_0 = 0$

Employing the method of ring-shaped elements, we obtain equations of the form (2.236)

$$\sum_{j=1}^{N} C_{ij}\widetilde{q}_{0j} = \overline{r}_i, \tag{5.132}$$

where normalized loads $\widetilde{q}_{0j}$ and coefficients C_{ij} may be written analogously to expressions (2.237), (2.238), whereas expression (5.96) for displacements $\delta_{1y}(r, R)$ is taken into account,

$$\widetilde{q}_{0j} = \frac{q_{0j}R}{2G_R\pi}, \tag{5.133}$$

$$C_{ij} = \overline{R}_j^{2-\alpha}\Delta_{1y}(\overline{r}_i / \overline{R}_j) - \overline{R}_{j-1}^{2-\alpha}\Delta_{1y}(\overline{r}_i / \overline{R}_{j-1}). \tag{5.134}$$

The torsional stiffness is found by using the equation analogous to (3.96):

$$K_{\varphi_z} = \pi^2 G_R R^3 \sum_{j=1}^{N} \widetilde{q}_{0j}(\overline{R}_j^4 - \overline{R}_{j-1}^4). \tag{5.135}$$

Hence, the following expression holds for the normalized stiffness (see (3.97), (3.98)):

$$\overline{K}_{\varphi_z} = \frac{3\pi^2}{16\mu^\alpha} \sum_{j=1}^{N} \widetilde{q}_{0j}(\overline{R}_j^4 - \overline{R}_{j-1}^4). \tag{5.136}$$

Although Green's functions entering the expressions for displacements Δ_{1y} and these expressions themselves contain Poisson's ratio, the final results appear to be independent of Poisson's ratio. For $\mu = 0.1$, normalized stiffness $\overline{K}_{\varphi_z}$ is shown in Fig. 5.30 as a function of parameter α.

Owing to the applied normalization, the obtained values of the normalized stiffness are in close proximity to each other; for practical purposes, one can use the following relationship for all four values:

$$\overline{K} = 1 - \alpha^{2.3}. \tag{5.137}$$

The corresponding curve in Fig. 5.30 is denoted by triangles.

Static Stiffnesses for the Circular Disk at $z_0 \neq 0$

In the case $z_0 \neq 0$, Green's functions have a complicated structure. Therefore, in order to determine the coefficients in the corresponding equations, we employ the results of calculating values $S_{vv}(\widetilde{r})$, $S_{vh}(\widetilde{r})$, $S_{hr}(\widetilde{r})$, $S_{h\vartheta}(\widetilde{r})$ with a sufficiently small step with respect to variable $\widetilde{r} = r / z_0$. This is necessary to enable interpolation when calculating the corresponding integrals. The displacements for various types of loads, which are used to determine the coefficients in the equations for contact problems, are given in section 2.7.1. In the considered isotropic case, we have to replace in these formulas G_{rz0} with G_0 and ν' with ν.

The results of calculations are presented for $\nu = 1/3$ and $\alpha = 0.5$. Values $S_{vv}(\tilde{r}), S_{vh}(\tilde{r}), S_{hr}(\tilde{r}), S_{h\vartheta}(\tilde{r})$ were calculated with a step of 0.004 on the interval $0 < \tilde{r} < 1$ where their abrupt changes take place, and with a step of 0.05 on the interval $1 < \tilde{r} < 48$. In order to determine the displacements by using the formulas given in section 2.7.1, in which integration is required, local parabolic interpolation was used for quantities $S_{vv}(\tilde{r}), S_{vh}(\tilde{r}), S_{hr}(\tilde{r}), S_{h\vartheta}(\tilde{r})$. Recall that in the formulas for displacements constructed by using the principle of superposition, argument r in Green's functions is replaced with λR, where R is the radius of the area of application of the load; arguments R_j and R_{j-1} in the matrix elements of the systems of equations (e.g. of the form (2.181)) show that in the integrals determining the displacements, argument $\tilde{r}$ entering the normalized amplitudes is replaced with $\lambda\tilde{R}R_j$ and $\lambda\tilde{R}R_{j-1}$, respectively (see (2.180)). Here, $\tilde{R} = R/z_0$ denotes the dimensionless radius of the disk. Furthermore, we calculate the stiffnesses of the disk corresponding to various types of its motion, as described in section 3.5.2; the same values of coefficient μ are used to determine the normalized stiffnesses. When performing normalization, value G_μ, representing the shear modulus at depth μR, is found by the formula

$$G_\mu = G_0(1 + \mu\tilde{R})^\alpha \tag{5.138}$$

rather than by using (3.82c). As a result, formulas for the normalized stiffness are modified: in expressions (3.85), (3.89), (3.91), (3.94), (3.98), quantity $(1 + \mu\tilde{R})$ is replaced with $(1 + \mu\tilde{R})^\alpha$. In Fig. 5.31, the relationship between the normalized stiffnesses and variable ξ_R, introduced in (3.84), is represented for relaxed contact conditions. Value $\xi_R = 0$ corresponds to the homogeneous half-space, for which values of the normalized stiffnesses should equal unity. Owing to the error introduced by applying the method of ring-shaped elements, the values obtained deviate slightly from unity (the magnitudes of deviations are indicated in section 3.5.2). As ξ_R approaches unity, this half-space approaches the half-space with zero shear modulus on its surface, which has been studied previously. For such a half-space, the values of the normalized stiffnesses at $\nu = 1/3, \alpha = 0.5$ equal 0.7849, 0.7409, 0.7578, 0.8318 for $\overline{K}_z, \overline{K}_x, \overline{K}_{\varphi_y}, \overline{K}_{\varphi_z}$, respectively. The ordinates of the points on the graphs in Fig. 5.31 approach these values as $\xi_R \to 1$. Recall that for linearly heterogeneous half-space, the corresponding limiting values equal zero for $\nu < 0.5$ (at $\nu = 0.5$, the limit stiffnesses $\overline{K}_z$ and $\overline{K}_{\varphi_y}$ differ from zero).

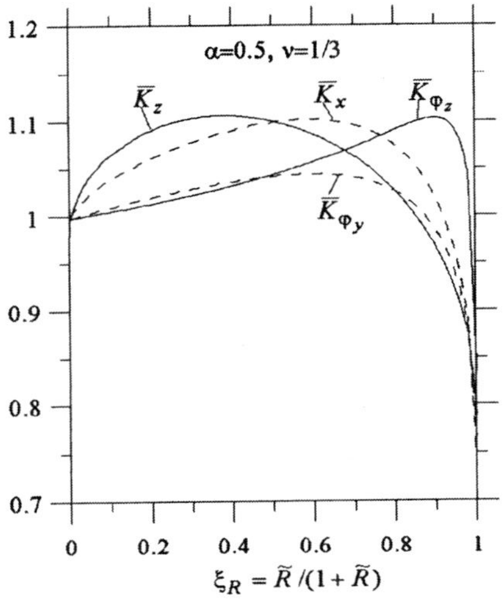

Fig. 5.31. Normalized static stiffnesses of disk on half-space with power-law heterogeneity having non-zero shear modulus on half-space surface

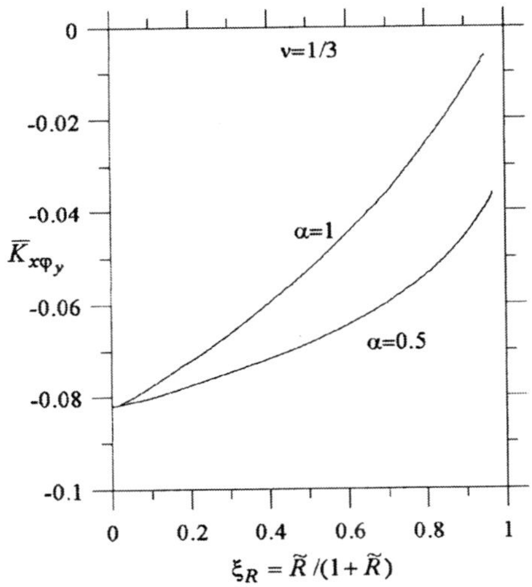

Fig. 5.32. Normalized horizontal rocking stiffness at two values of parameter α

Taking into account the welded contact results in an insignificant change of the values shown in Fig. 5.31. However, the welded contact may appear to be important if the horizontal rocking stiffness $K_{x\varphi_y} = K_{\varphi_y x}$, expressed in (3.90), is a matter of interest. In Fig. 5.32, the normalized stiffness $\overline{K}_{x\varphi_y}$ given in relationship (3.91) is shown with the above-mentioned modification by accounting for expression (5.138). In addition, this figure contains the graph corresponding to the linearly heterogeneous half-space ($\alpha = 1$).

5.6.2 Dynamic Stiffnesses for Circular Disk

For the case of the linearly heterogeneous half-space, the dynamic stiffnesses are studied in section 3.6. The technique presented in this section is also applicable for the case of a half-space with stiffness varying by power law with depth. Argument a_μ introduced in relationship (3.101) has the following form, corresponding to expression (5.138) for quantity G_μ (representing shear modulus at depth $\mu\tilde{R}$):

$$a_\mu = \omega R\sqrt{\frac{\rho}{G_\mu}} = \omega R\sqrt{\frac{\rho}{G_0(1+\mu\tilde{R})^\alpha}} \,. \tag{5.139}$$

The value of coefficient μ is taken to be the same as when solving the corresponding static problems, namely unity for determination of the vertical stiffness, 0.35 for determination of the horizontal and rocking stiffness, and 0.1 for determination of the torsional stiffness of the disk on a half-space. The selection of quantity a_μ for the argument results in the close proximity of the normalized dynamic stiffnesses corresponding to different values of $\tilde{R}$ on a wide interval of variation of parameter a_μ . Below, the results of calculations are given for the case $\alpha = 0.5$, $\nu = 1/3$, $\varepsilon = 0$. Preliminarily, the normalized amplitudes of vibrations for

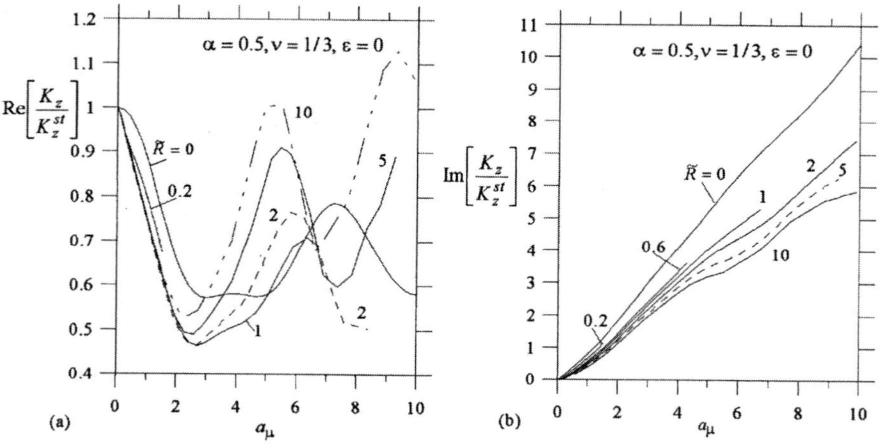

Fig. 5.33a,b. Real (a) and imaginary (b) parts of normalized vertical dynamic stiffness of disk on half-space under relaxed contact conditions

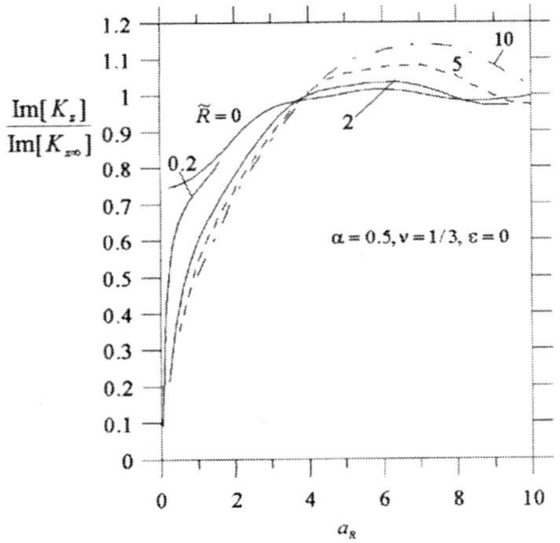

Fig. 5.34. Ratio of imaginary part of vertical stiffness to its limiting value for $\omega\to\infty$

the concentrated force were calculated for a large number of values θ, analogously to the case of the linearly heterogeneous half-space. In Fig. 5.33, the results are presented for the case of relaxed contact, for the normalized dynamic vertical stiffness, which is determined as the ratio of quantity K_z found by relationship (3.100) to the corresponding static value K_z^{st}; static stiffnesses were studied in section 5.6.1 (see the graphs in Fig. 5.31). The nodes of the broken lines shown in Fig. 5.33 correspond to discrete values of parameter θ (more than 30 values were used). Note the similarity between these curves and the corresponding curves presented in Fig. 3.23, which are related to the linearly heterogeneous half-space. The proximity of the curves shown in Fig. 5.33 to each other (especially noticeable at small values of the argument) is due to the selection of parameter a_μ for the argument. Since the half-space considered in the present subsection has a lower degree of heterogeneity than the linearly heterogeneous half-space, deviation of the results of calculations from the corresponding results for the homogeneous half-space is less than in the case of the linearly heterogeneous half-space. The latter correspond to $\tilde{R}=0$. The ratios of the imaginary part of the dynamic stiffness to the limiting value when $\omega\to\infty$, determined by formula (3.103), are presented in Fig. 5.34. These graphs are similar to those shown in Fig. 3.25; however, now, the ordinates approach the limiting value (being equal to unity) at smaller values of parameter a_R.

In Fig. 5.35, values of the normalized horizontal stiffness are shown for the case of relaxed contact conditions, i.e. calculated by using equations (2.212), with $\mu=0.35$. The stiffness was determined by (3.106) and then divided by the corresponding static value considered in the previous subsection. As before, the curves obtained are similar to those corresponding to the linearly heterogeneous half-space (shown in Fig. 3.30) though the latter correspond to welded contact

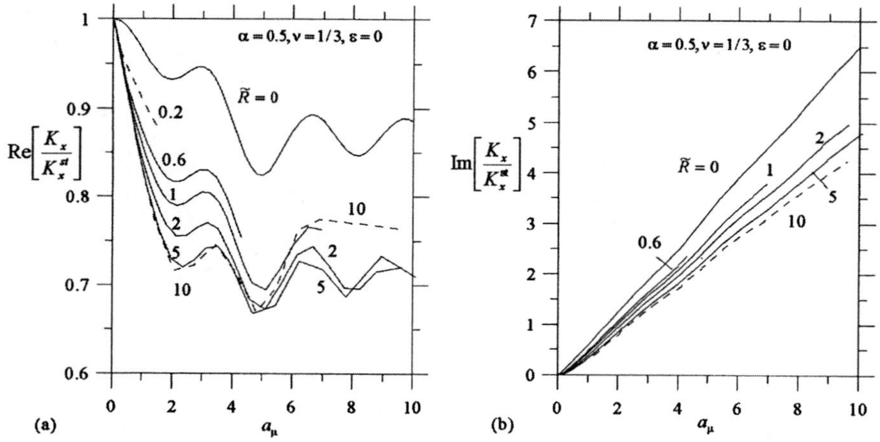

Fig. 5.35a,b. Real (a) and imaginary (b) parts of normalized horizontal dynamic stiffness of disk on half-space under relaxed contact conditions

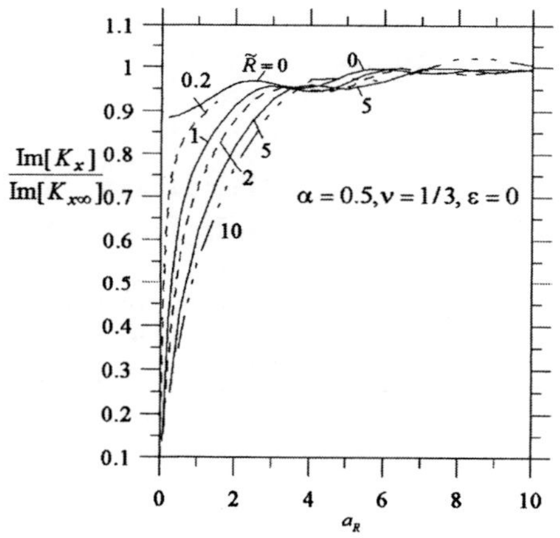

Fig. 5.36. Ratio of imaginary part of horizontal stiffness to its limiting value for $\omega \to \infty$

conditions, in contrast to the curves presented in Fig. 5.35. In Fig. 5.36, the normalized imaginary part of the dynamic stiffness is shown approaching its limiting value determined by formula (3.108). As can be seen, in the considered case, it approaches the limiting value (which equals unity) faster than in the linearly heterogeneous half-space (see Fig. 3.31).

Consider rocking (with respect to the *Y*-axis) vibrations of the disk. The normalized stiffness is presented in Fig. 5.37. Note that the curves corresponding to different values of parameter $\widetilde{R}$ are in close proximity to each other, especially for the real part of the stiffness. In engineering practice, these results enable us to

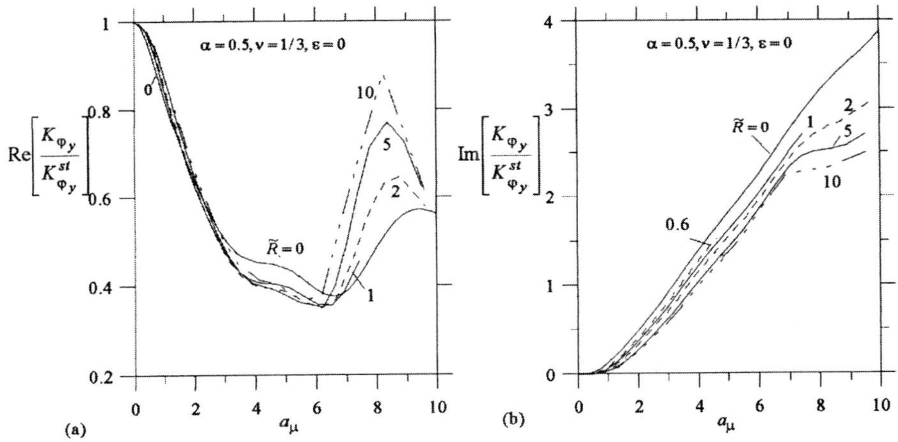

Fig. 5.37a,b. Real (a) and imaginary (b) parts of normalized rocking dynamic stiffness of disk on half-space under relaxed contact conditions

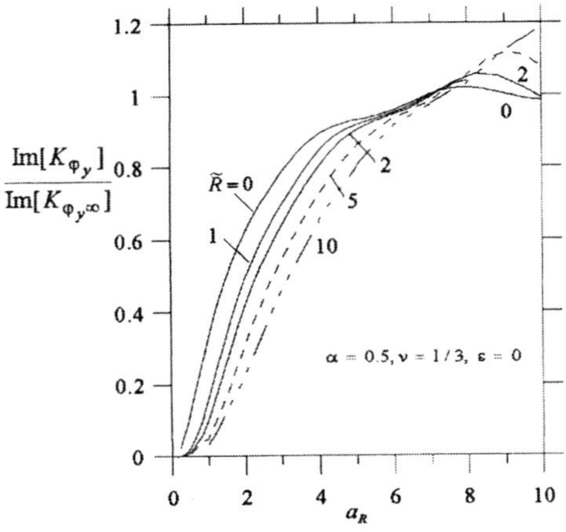

Fig. 5.38. Ratio of imaginary part of rocking stiffness to its limiting value for $\omega \to \infty$

go to the "equivalent" homogeneous half-space with shear modulus G_μ by (5.138) at $\mu = 0.35$, not only in static contact problems, but also in the corresponding dynamic problems at low and medium frequencies. The behavior of the normalized imaginary part of the rocking stiffness, with respect to its limiting values by (3.111), as the frequency of vibrations goes to infinity, is shown in Fig. 5.38. A comparison of these results with the corresponding results obtained for the linearly heterogeneous half-space shown in Fig. 3.35, indicates that lowering the degree of heterogeneity leads to a reduced influence of parameter $\widetilde{R}$ and to a faster approach to the limiting value.

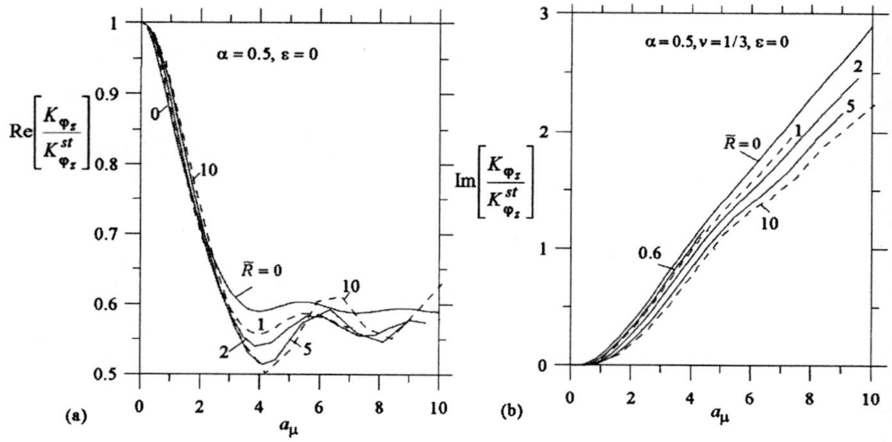

Fig. 5.39a,b. Real (a) and imaginary (b) parts of normalized torsional dynamic stiffness of disk on half-space

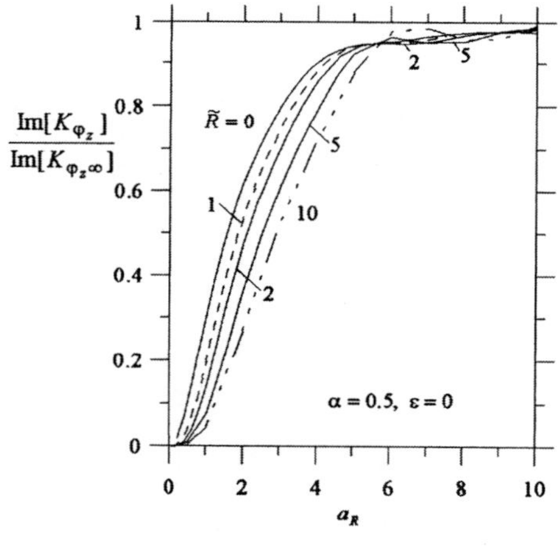

Fig. 5.40. Ratio of imaginary part of torsional stiffness to its limiting value for $\omega \to \infty$

The results of calculations for the case of torsional vibrations with $\mu = 0.1$ are given in Figs. 5.39 and 5.40. Here, the plotted curves are rather close to the curves corresponding to the rocking vibrations of the disk, and to the corresponding curves for the linearly heterogeneous half-space shown in Figs. 3.39 and 3.40. When $a_\mu \to \infty$, the limits for the real values of the normalized stiffness decrease (starting with a value of 0.589 at $\widetilde{R} = 0$ given in relationship (2.258)), as parameter $\widetilde{R}$ increases. In addition, oscillations occur in the vicinity of the limiting value. In Fig. 5.40, the normalized imaginary part of the dynamic torsional stiffness is shown approaching the corresponding limiting value. It is

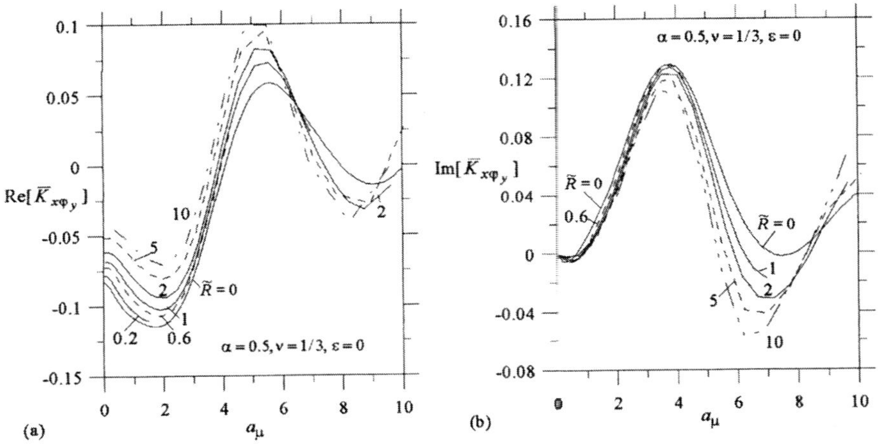

Fig. 5.41a,b. Real (a) and imaginary (b) parts of normalized horizontal rocking dynamic stiffness of disk on half-space

useful to compare these results with those obtained for the linearly heterogeneous half-space shown in Fig. 3.40 and with the results for rocking vibrations (Fig. 5.38).

In conclusion, consider the horizontal rocking stiffness $K_{x\varphi_y}$ (lateral dynamic stiffness), representing a moment which should be applied to the disk in order to provide the translational horizontal vibrations. The expression for $K_{x\varphi_y}$ given in (3.90) still holds; the previously employed normalization (dividing $K_{x\varphi_y}$ by the value Rk_x^{hom}) remains valid. Here, k_x^{hom} corresponds to expression (3.87b):

$$k_x^{\text{hom}} = \frac{8G_\mu R}{2-\nu} = \frac{8G_0 R(1+\mu\widetilde{R})^\alpha}{2-\nu} \,. \tag{5.140}$$

In order to determine the horizontal rocking stiffness, the complete system of equations of the form (2.200) is employed. In Fig. 5.41, the quantity $\bar{K}_{x\varphi_y}$ is shown for a number of values of parameter $\widetilde{R}$ and $\nu=1/3$, $\alpha=0.5$, $\mu=0.35$. For low and intermediate frequencies ($a_\mu<3$), taking into account the heterogeneity leads to a decreasing magnitude of $\bar{K}_{x\varphi_y}$, as for the case of the linearly heterogeneous half-space at $\nu=1/3$ (see Fig. 3.32). Owing to the normalization used and the selection of parameter a_μ for the argument, the curves plotted for different values of parameter $\widetilde{R}$ are rather close to each other.

References

1. Abramowitz, M. and Stegun, I. A.: Handbook of Mathematical Functions, New York: Dover 1965.
2. Alekseyev, A. S. and Mikhaylenko, B. G.: Solution of Lamb's problem for a vertically inhomogeneous elastic halfspace. Izv. Akad. Nauk. Physics of the solid Earth 12 (1976/1977) 748–755.
3. Alverson, R. C., Gair, F. C., and Hook, J. F.: Uncoupled equations of motion in inhomogeneous elastic media. Bull. Seism. Soc. Am. 53 (1963) 1023–1030.
4. Apsel, R. J. and Luco, J. E.: On the Green's functions for a layered half–space. Part II. Bull. Seism. Soc. Am. 73 (1983) 931–51.
5. Awojobi, A. O.: Approximate solution of high–frequency–factor vibrations of rigid bodies on elastic media. J. Appl. Mech. ASME 38 (1971) 111–117.
6. Awojobi, A. O.: Vertical vibration of a rigid circular foundation on Gibson soil. Geotechnique 22 (1972) 333–343.
7. Awojobi, A. O.: Vibration of rigid bodies on non–homogeneous semi infinite elastic media. Q. J. Mech. Appl. Math. 26 (1973) 483–498.
8. Awojobi, A. O. and Gibson, R.E.: Plane strain and axially symmetric problems of a linearly nonhomogeneous elastic half–space. Q. J. Mech. Appl. Math. 26 (1973) 285–302.
9. Bahar, L. Y.: Transfer matrix approach to layered systems. J. Eng. Mech. Div. ASCE 98 (1972) 1159–1172.
10. Barden, L.: Stresses and displacements in a cross–anisotropic soil. Geotechnique 13 (1963) 198–210.
11. Biot, M. A.: Continuum dynamics of elastic plates and multilayered solids under initial stress. J. Math. Mech. 12 (1963) 793–810.
12. Biot, M. A.: Mechanics of incremental deformations, New York: John Wiley & Sons 1965.
13. Biot, M. A.: Fundamentals of generalized rigidity matrices for multi–layered media. Bull. Seism. Soc. Am. 73 (1983) 749–763.
14. Brown, P. T. and Gibson, R. E.: Surface settlement of a deep elastic stratum whose modulus increases linearly with depth, Can. Geotech. J. 9 (1972) 467–476.
15. Brown, P. T. and Gibson, R. E.: Rectangular loads on inhomogeneous elastic soil. J. Soil Mech. Found. Div. ASCE 10 (1973) 917–920.
16. Brown, P. T.: Influence of soil inhomogeneity on raft behaviour. Soil Found. 14 (1974) 61–70.
17. Bycroft, G. N.: Forced vibrations of a rigid circular plate on a semi–infinite elastic space and on an elastic stratum. Phil. Trans. Roy. Soc. London 248 A (1956) 327–368.
18. Bycroft, G. N.: Soil–structure interaction at higher frequency factors. Earthquake Eng. Struct. Dyn. 5 (1977) 235–248.
19. Carrier, G. F.: The propagation of waves in orthotropic media. Q. Appl. Math. 4 (1946) 160–165.
20. Carrier, W. D. and Christian, J. T.: Rigid circular plate resting on a non–homogeneous elastic half–space. Geotechnique 23 (1973) 67–84.

21. Chapel, F. and Tsakalidis, C.: Computation of the Green's functions of elastodynamics for a layered half–space through a Hankel transformation – application to foundation vibration and seismology. Proc. Fifth Int. Conf. Numer. Meth. Geomech., Nagoya 3 (1985) 1311–1318.
22. Chuaprasert, M. F. and Kassir, M. K.: Torsion of nonhomogeneous solid. J. Eng. Mech. Div. ASCE 99 (1973) 703–714.
23. Chuaprasert, M. F. and Kassir, M. K.: Displacements and stresses in non–homogeneous solid. J. Eng. Mech. Div. ASCE 100 (1974) 861–872.
24. Dempsey, J. P. and Li, H.: Rectangular footing on nonhomogeneous elastic half space. In Foundation Engineering: Current principles and practice, ASCE 2 (1989) 1212–1225.
25. Dobry, R., Ladd, R. S.,Yokel, F. Y., Chang, R. M. and Powell, D.: Prediction of pore water pressure buildup and liquefaction of sands during earthquakes by the cyclic strain method. NMS Building Science Series 138 (1982).
26. Dobry, R. and Gazetas, G.: Dynamic stiffness and damping of foundations by simple methods. In Vibration Problems in Geotechnical Engineering, ed. G. Gazetas and E. T. Selig, ASCE (1985) 77–107.
27. Dunkin, J. W.: Computation of modal solutions in layered elastic media at high frequencies. Bull. Seism. Soc. Am. 55 (1965) 335–348.
28. Eringen, A. C. and Suhubi, S.: Elastodynamics, Vol. 2, New York: Academic Press 1975.
29. Ewing, W. M., Jardetzky, W. S. and Press, F.: Elastic Waves in Layered Media, New York: McGraw–Hill 1957.
30. Fröhlich, O. K.: Druckverteilung in Baugrunde, Berlin: Springer–Verlag OHG 1934.
31. Gazetas, G.: Static and dynamic displacements of foundations on heterogeneous multilayered soils. Geotechnique 30 (1980) 159–177.
32. Gazetas, G.: Stresses and displacements in cross–anisotropic soils. J. Geotech. Eng. Div. ASCE 108 (1982) 532–553.
33. Gazetas, G.: Vibrational characteristics of soil deposits with variable wave velosity. Int. J. Numer. Anal. Methods Geomech. 6 (1982) 1–20.
34. Gazetas, G.: Simple physical methods for foundation impedances. Dynamic behaviour of foundations and buried structures. (Developments in soil mechanics and foundation engineering; 3) (1987) 45–93.
35. Gazetas, G.: Foundation vibration. In Foundation Engineering Handbook, ed. Hsai–Yang Fang (1991) 553–593.
36. Gerrard, C. M. and Harrison, W. J.: Circular loads applied to a cross–anisotropic half–space. In H. G. Poulos and E. H. Davis: Elastic solutions for soil and rock mechanics, New York: John Wiley & Sons 1974.
37. Gerrard, C. M.: Stresses and displacements in layered cross–anisotropic elastic systems. Proc. Fifth Australia–New Zealand Conf. Soil Mech. Found. Eng. (1976) 187–197.
38. Ghosh, M. L.: Reflection and diffraction of SH waves due to a line source in a heterogeneous medium. Appl. Sci. Res. 23 (1971) 373–392.
39. Gibson, R. E.: Some results concerning displacements and stresses in a non–homogeneous elastic half–space. Geotechnique 17 (1967) 58–67.
40. Gibson, R. E.: The analytical method in soil mechanics. Geotechnique 24. (1974) 115–140.
41. Gilbert, F. and Backus, G. E.: Propagator matrices in elastic wave and vibration problems. Geophysics 31 (1966) 326–332.
42. Godunov, S. K.: On a numerical solution of boundary problems for systems of linear ordinary differential equations. Usp. Mat. Nauk 16 (1961) 171–174.

43. Gradsteyn, I. S. and Ryzhik, I. M.: Table of Integrals, Series & Products. New York & London: Academic Press 1965.
44. Gucunski, N. and Woods, R. D.: Inversion of Rayleigh wave dispersion curve for SASW test. Proc. First Int. Conf. Soil Dyn. Earthquake Eng., Karlsruhe (1991) 127–138.
45. Gucunski, N. and Woods, R. D.: Numerical simulation of the SASW test. Soil Dyn. Earthquake Eng. 11 (1992) 213–227.
46. Gupta, R. N.: Reflection of elastic waves from a linear transition layer. Bull. Seism. Soc. Am. 56 (1966) 511–526.
47. Gusina, B. B. and Pak, R. Y. S.: Vertical vibration of a circular footing on a linear–wave–velocity half–space. Geotechnique 48 (1998) 159–168.
48. Hardin, B. O. and Drnevich, V. P.: Shear modulus and damping in soils: Design equations and curves. J. Soil Mech. Found. Div. ASCE 98 (1972) 667–692.
49. Harkrider, D. G.: Surface waves in multilayered elastic media – I: Rayleigh and Love waves from buried sources in sources on a multilayered elastic half–space. Bull. Seism. Soc. Am. 54 (1964) 627–679.
50. Haskell, N. A.: The dispersion of surface waves on multilayered media. Bull. Seism. Soc. Am. 43 (1953) 17–34.
51. Holl, D. L.: Stress transmission in earths. Proc. High Res. Board 20 (1940) 709–721.
52. Hook, J. F.: Separation of the vector wave equation of elasticity for certain types of inhomogeneous, isotropic media. J. Acoust. Soc. Am. 33 (1961) 302–313.
53. Hull, S.W. and Kausel, E.: Dynamic loads in layered half–spaces. In Engineering Mechanics in Civil Engineering, ed. A. P. Boresi and K. P. Chong, ASCE, New York 1984.
54. Karlsson, T. and Hook, J. F.: Lamb's problem for an inhomogeneous media with constant veliocities of propagation. Bull. Seism. Soc. Am. 53 (1963) 1007–1022.
55. Kassir, M. K.: Boussinesq problems for nonhomogeneous solid. J. Eng. Mech. Div. 98 (1972) 457–470.
56. Kassir, M. K. and Chuaprasert, M. F.: A rigid punch in contact with a nonhomogeneous elastic solid. J. Appl. Mech. ASME 41 (1974) 1019–1024.
57. Kausel, E. and Rosset, J. M.: Stiffness matrices for layered soils. Bull. Seism. Soc. Am. 71 (1981) 1743–1761.
58. Kausel, E and Peek, R.: Dynamic loads in the interior of a layered stratum: An explicit solution. Bull. Seism. Soc. Am. 72 (1982) 1459–1481.
59. Kausel, E.: Wave propagation in anisotropic layered media. Int. J. Numer. Methods Eng. 23 (1986) 1567–1578.
60. Kausel, E. and Seale, S. H.: Dynamic and static impedances of cross–anisotropic halfspaces. Soil Dyn.Earthquake Eng. 9 (1990) 172–178.
61. Kirkner, D. J.: Vibration of a rigid disk on a transversely isotropic elastic half space. Int. J. Numer. Methods Geomech. 6 (1982) 293–306.
62. Knopoff, L.: A matrix method for elastic wave problems. Bull. Seism. Soc. Am. 54 (1964) 431–438.
63. Korenev, B. G.: Punch lying on an elastic half–space whose modulus of elasticity is a function of the depth. Dokl. Akad. Nauk SSSR 112 (957) 823–826.
64. Lekhnitskii, S. G.: Radial distribution of stresses in a wedge and in a half–plane with variable modulus of elasticity. Prikl. Mat. Mekh 26 (1962) 146–151.
65. Lekhnitskii, S. G.: Theory of Elasticity of an Anisotropic Elastic Body, San Francisco: Holden–Day 1963.
66. Leung, K. L., Vardoulakis, I. G., Beskos, D. E. and Tassoulas, J. L.: Vibration isolation by trenches in continuously nonhomogeneous soil by the BEM. Soil Dyn. Earthquake Eng. 10 (1991) 172–179.

67. Luco, J. E. and Westmann, R. A.: Dynamic response of circular footings. J. Eng. Mech. Div. ASCE 97 (1971) 1381–1395.
68. Luco, J. E.: Impedance functions for a rigid foundation on a layered medium. Nucl. Eng. Des. 31 (1974) 204–217.
69. Luco, J. E.: Vibration of a rigid disk on a layered viscoelastic medium. Nucl. Eng. Des. 36 (1976) 325–340.
70. Luco, J. E. and Apsel, R. J.: On the Green's functions for a layered half–space. Part I. Bull. Seism. Soc. Am. (1983) 909–929.
71. Lysmer, J.: Vertical Motion of Rigid footings. Ph.D. Diss. Univ. of Michigan 1965.
72. Lysmer, J.: Lumped mass method for Rayleigh waves. Bull. Seism. Soc. Am. 60 (1970) 89–104.
73. Lysmer, J. and Waas, G.: Shear waves in plane infinite structures. J. Eng. Mech. Div. ASCE 98 (1972) 85–105.
74. Michell, J. H.: The stress distribution in an aelotropic solid with an infinite plane boundary. Proc. London Math. Soc. 32 (1900) 247–258.
75. Mindlin, R. D.: Force at a point in the interior of a semi–infinite solid. J. Phys. 79 (1936) 195–202.
76. Mossakovskii, V. I.: Pressure of a circular punch on an elastic half–space whose modulus of elasticity is an exponential function of the depth. Prikl. Mat. Mekh. 22 (1958) 123–125.
77. Muravskii, G. B.: On a model of an elastic foundation (in Russian). Structural mechanics and analysis of structures (Stroitel'naia Mekhanika i Raschet Sooruzhenii) (1967), no. 6, 14–17.
78. Muravskii, G. and Operstein, V.: Time–harmonic vibration of an incompressible linearly non–homogeneous half–space. Earthquake Eng. Struct. Dyn. 25 (1996) 1195–1209.
79. Muravskii, G. B.: Green function for an incompressible linearly non–homogeneous half–space. Arch. of Appl. Mech. 67 (1996) 81–95.
80. Muravskii, G.: Time–harmonic problem for a non–homogeneous half–space with exponentially varying shear modulus. Int. J. Solids Struct. 34 (1997) 3119–3139.
81. Muravskii, G.: On time–harmonic problem for non–homogeneous elastic half–space with shear modulus limited at infinite depth. Eur. J. Mech. A/Solids 16 (1997) 227–294.
82. Muravskii, G.: Green functions for a compressible linearly inhomogeneous half–space. Arch. Appl. Mech. 67 (1997) 521–534.
83. Muravskii, G.: On time–harmonic vibration of an inhomogeneous elastic half–space: numeric–analytic solutions. Proc. Sixth Pan–Am. Congr. Appl. Mech. and Eighth Int. Conf. Dyn. Prob. Mech. 7 (1999) 809–812.
84. Nazarian, S.and Stokoe, K. H.: In situ shear wave velocities from spectral analysis of surface waves. Proc. Eighth World Conf. Earthquake Eng. 3 (1984) 31–38.
85. Oner, M.and Dong, S. B.: Analysis of in–plane waves in layered half–space by global–local finite element method. Soil Dyn. Earthquake Eng. 7 (1988) 2–8.
86. Pan, Y. C. and Chou, T. W.: Point force solution for an infinite transversally isotropic solid. J. Appl. Mech. 43 (1976) 608–612.
87. Pan, Y. C. and Chou, T. W.: Green's function solutions for semi–infinite transversally isotropic material. Int. J. Eng. Sci. 17 (1979) 545–551.
88. Pennington, D. S., Nash, D. F. T. and Lings, M. L.: Anisotropy of G_0 shear stiffness in Gault Clay. Geotechnique 47 (1997) 391–398.
89. Plevako, V. P.: On the theory of elasticity of non–homogeneous media. Prikl. Mat. Mekh. 35 (1971) 853–860.

90. Plevako, V. P.: On a possibility of using of harmonic functions for solving problems of theory of elasticity of non–homogeneous media. Prikl. Mat. Mekh. 36 (1972) 886–894.
91. Poulos, H. G. and Davis, E. H.: Elastic Solutions for Soil and Rock Mechanics, New York: John Wiley & Sons 1974.
92. Prange, B.and Huber, G.: Oberflächenwellenfelder zur Bestimmung der dynamischen Untergrundparameter. Symposium Meßtechnik im Erd– Grundbau, DGEG, München (1983) 63–69.
93. Protsenko, V. S.: Torsion of an elastic half–space whose modulus of elasticity varies by the power law. Prikl. Mekh. 3 (1967) 1.
94. Rao, C. R.: Rayleigh waves in a half–space with bounded variation in density and rigidity. Bull. Seism. Soc. Am. 64 (1974) 1263–1274.
95. Rao, C. R. A. and Goda, M. A. A.: Generalization of Lamb's problem to a class of inhomogeneous elastic half–spaces. Proc. Roy. Soc. London A359 (1978) 93–110.
96. Richart., F. E., Jr., Hall, J. R., Jr. and Woods, R. D.: Vibrations of soils and foundations, Englewood Cliffs, NJ: Prentice Hall 1970.
97. Rix, G. J. and Leipski, E. A.: Accuracy and resolution of surface wave inversion. Conf. Recent Adv. in Instrum., Data Acquisition and Testing in Soil Dyn. (1991) 17–32.
98. Rostovtsev, N. A.: On some solutions of an integral equation in the theory of linearly deformable base. Prikl. Mat. Mekh. 28 (1964) 119–127.
99. Rostovtsev, N. A.: On the theory of elasticity of non–homogeneous medium. Prikl. Mat. Mekh. 28 (1964) 601–611.
100. Rostovtsev, N. A. and Khranevskaia, I. E.: The solution of the Boussinesq problem for a half space whose modulus of elasticity is a power function of the depth. Prikl. Mat. Mekh. 35 (1971) 1053–1061.
101. Schleicher, F.: Zur Theorie des Baugrundes. Bauigenieur 48, 49 (1926).
102. Seale, S. H. and Kausel, E.: Point loads in cross–anisotropic, layered halfspaces. J. Eng. Mech. 115 (1989) 509–524.
103. Sezawa, K.: Futher studies on Rayleigh–waves having some azimuthal distribution. Bull. Earthquake Res. Inst. 6 (1929) 1–18.
104. Stallybrass, M. P.: On the Reissner–Sagoci problem at high frequencies. Int. J. Eng. Sci. 5 (1967) 689–703.
105. Stokoe, K. H. and Nazarian, S.: Use of Rayleigh waves in liquefaction studies. In Measurements and Use of Shear Wave Velocity, ed. R. D. Woods, ASCE, New York 1985.
106. Stoneley, R.: The transmission of Rayleigh waves in a heterogeneous medium. Geoph. Suppl., Mon. Not. Roy. Astron. Soc. 3 (1934) 222–232.
107. Sveklo, V. A.: Concentrated force in a transversely isotropic half–space and in a composite space. Prikl. Mat. Mekh. 33 (1969) 532–537.
108. Tajimi, H.: A contribution to theoretical prediction of dynamic stiffness of surface foundations. Proc. Seventh World Conf. Earthquake Eng., Istanbul 5 (1980) 105–112.
109. Ter–Mkrtichian, L. N.: Some problems in the theory of non–homogeneous elastic media. Prikl. Mat. Mekh. 25 (1961) 1667–1675.
110. Thomas, D. P.:Torsional oscillation of an elastic half–space. Q. J. Mech. Appl. Math. 21 (1968) 51–65.
111. Thomson, W. T.: Transmission of elastic waves through a stratified solid medium. J. Appl. Phys. 21 (1950) 89–93.
112. Vardoulakis, I.: Surface waves in a half–space of submerged sand. Earthquake Eng. Struct. Dyn. 9 (1981) 329–342.

113. Vardoulakis, I.: Torsional surface waves in inhomogeneous elastic media. Int. J. Numer. Anal. Methods Geomech. 8 (1984) 287–296.
114. Vardoulakis, I. and Vrettos, C.: Dispersion–law of Rayleigh–type waves in a compressible half–space. Int. J. Numer. Anal. Methods Geomech. 12 (1988) 639–655.
115. Veletsos, A. S. and Wei, Y. T.: Lateral and rocking vibration of footings. J. Soil Mech. Found. Div. ASCE 97 (1971) 1227–1248.
116. Vrettos, Ch.: Dispersive *SH*–surface waves in soil deposits of variable shear modulus. Soil Dyn. Earthquake Eng. 9 (1990) 255–264.
117. Vrettos, Ch.: In–plane vibration of soil deposits with variable shear modulus: I. Surface waves. Int. J. Numer. Anal. Methods Geomech. 14 (1990) 209–222.
118. Vrettos, Ch.: In–plane vibration of soil deposits with variable shear modulus: II. Line load. Int. J. Numer. Anal. Methods Geomech. 14 (1990) 649–662.
119. Vrettos, Ch. and Prange, B: Evaluation of in situ effective shear modulus from dispersion measurements. J. Geotech. Eng. ASCE 116 (1990) 1581–1585.
120. Vrettos, Ch.: Forced anti–plane vibrations at the surface of an inhomogeneous half–space. Soil Dyn. Earthquake Eng. 10 (1991) 230–235.
121. Vrettos, Ch.: Time–harmonic Boussinesq problem for a continuously non–homogeneous soil. Earthquake Eng. Struct. Dyn. 20 (1991) 961–977.
122. Waas, G.: Linear two–dimensional analysis of soil dynamics problems in semi–infinite layered media. Dissertation, Univ. of California, Berkeley, 1972.
123. Waas, G.: Dynamisch belastete Fundamente auf geschichtete Baugrund. VDI–Berichte no. 381 (1980) 185–191.
124. Waas, G., Riggs, H. R. and Werkle, H.: Displacements solutions for dynamic loads in transversely–isotropic stratified media. Earthquake Eng. Struct. Dyn. (1985) 329–342.
125. Warburton, G. B.: Forced vibration of a body upon an elastic stratum. J. Appl. Mech. ASME 24 (1957) 55–58.
126. Westergaard, H. M.: A problem of elasticity suggested by a problem in soil mechanics. In Contribution to Mechanics of Solids, Timoshenko 60th Anniversary Volume. New York: Macmillan 1938.
127. Wilson, J. T.: Surface waves in a heterogeneous medium. Bull. Seism. Soc. Am. 32 (1942) 297–304.
128. Wolf, J. P.: Dynamic Soil–Structure Interaction, Englewood Cliffs, NJ: Prentice Hall 1985.
129. Wolf, J. P.: Soil–Structure Interaction Analysis in Time Domain, Englewood Cliffs, NJ: Prentice Hall 1988.
130. Wolf, K.: Ausbreitung der Kraft in der Halbebene und im Halbraum bei anisotropen Material. Z. Angew. Math. Mech. 15 (1935) 249–254.
131. Xu, P. C. and Mal, A. K.: Calculation of the in–plane Green's functions for a layered viscoelastic solid. Bull. Seism. Soc. Am. 77 (1987) 1823–1837.

Subject Index

Foundations of Engineering Mechanics

Palmov	Vibrations of Elasto-Plastic Bodies (1998, ISBN 3-540-63724-9)
Babitsky	Theory of Vibro-Impact Systems and Applications (1998, ISBN 3-540-63723-0)
Skrzypek/ Ganczarski	Modeling of Material Damage and Failure of Structures Theory and Applications (1999, ISBN 3-540-63725-7)
Kovaleva	Optimal Control of Mechanical Oscillations (1999, ISBN 3-540-65442-9)
Kolovsky	Nonlinear Dynamics of Active and Passive Systems of Vibration Protection (1999, ISBN 3-540-65661-8)
Guz	Fundamentals of the Three-Dimensional Theory of Stability of Deformable Bodies (1999, ISBN 3-540-63721-4)
Alfutov	Stability of Elastic Structures (2000, ISBN 3-540-65700-2)
Morozov/ Petrov	Dynamics of Fracture (2000, ISBN 3-540-64274-9)
Astashev/ Babitsky/ Kolovsky	Dynamics and Control of Machines (2000, ISBN 3-540-63722-2)

Foundations of Engineering Mechanics

Series Editors: Vladimir I. Babitsky, Loughborough University
Jens Wittenburg, Karlsruhe University

Svetlitsky	Statics of Rods (2000, ISBN 3-540-67452-7)
Landa	Regular and Chaotic Oscillations (2001, ISBN 3-540-41001-5)
Muravskii	Mechanics of Non-Homogeneous and Anisotropic Foundations (2001, ISBN 3-540-41631-5)
Gorshkov/ Tarlakovsky	Transient Aerohydroelasticity of Spherical Bodies (2001, ISBN 3-540-42151-3)